Necessary Condition Analysis (NCA)

"Jan Dul has written a wonderful book that provides researchers with a rigorous and practical framework for explicitly theorizing the necessary conditions that must be present to make an outcome possible, and that their absence guarantees the absence of the outcome. By systematically integrating theory development, research design, and data analysis, the book shows how to identify "must-have" conditions. Importantly, this book equips researchers with concrete tools to generate more precise, impactful contributions that advance cumulative knowledge."

-**Herman Aguinis**, Avram Tucker Distinguished Scholar and Professor of Management, The George Washington University, USA

Necessary Condition Analysis (NCA) is a new research method for data analysis. Its unique feature is the use of 'necessary but not sufficient' logic to detect the critical 'must have' factors in datasets, rather than average effects. This book introduces this methodology at an advanced level, taking readers through the theoretical underpinning and its varied application in empirical research across the social sciences and beyond. Aimed at PhD students and researchers, data analysts, practitioners, and others, the book helps readers develop a deeper understanding of the NCA approach and apply it effectively and with high-quality standards. It brings together the fundamentals of NCA and integrates the most recent methodological advancements and tools. No prior knowledge of NCA is required but readers would benefit from familiarity with the basics of research methodology.

Key features:

- Addresses a new data analysis method.
- Provides a deeper understanding of the method.
- Includes hands on practical recommendations for application.

Jan Dul is the founder of NCA. He is professor of Technology and Human Factors at the Rotterdam School of Management, Erasmus University, Netherlands. His background spans the technical, medical, and social sciences. His more than 200 publications include articles in leading journals and books translated into multiple languages. He is a frequent speaker at academic and professional events worldwide and has shared his expertise with over 50 companies. He has received several awards and recognitions in business and management and in human factors and ergonomics.

Necessary Condition Analysis (NCA)

Principles and Application

Jan Dul

CRC Press
Taylor & Francis Group
Boca Raton London New York

CRC Press is an imprint of the
Taylor & Francis Group, an **informa** business
A CHAPMAN & HALL BOOK

Designed cover image: Jan Dul, Renata Bauer, Ger Haan

First edition published 2027
by CRC Press
2385 NW Executive Center Drive, Suite 320, Boca Raton FL 33431

and by CRC Press
4 Park Square, Milton Park, Abingdon, Oxon, OX14 4RN

CRC Press is an imprint of Taylor & Francis Group, LLC

ISBN: 978-1-0325-5447-1 (hbk)
ISBN: 978-1-032-55448-8 (pbk)
ISBN: 978-1-003-43076-6 (ebk)

DOI: 10.1201/9781003430766

Typeset in Latin Modern font
by KnowledgeWorks Global Ltd.

Publisher's note: This book has been prepared from camera-ready copy provided by the author.

Contents

List of figures

List of tables

Preface

"Is coffee a necessary condition for you?" Someone asked me this question after hearing that I had taken time away in Italy to focus on writing my book, and that I like Italy's coffee culture. "No, it is not; it only helps." Focus, time, perseverance, and several other things are true necessary conditions. Without any of them, the book would not exist.

And here it is! In the book I present the principles of Necessary Condition Analysis (NCA) and their application in empirical studies in research and practice. NCA is an approach that uses necessity causal logic to identify necessary conditions from data. The core idea is that a necessary condition must be present to make the outcome possible, and that its absence guarantees the absence of the outcome. Compensation is not possible.

My intention with this book is to help researchers, data analysts, practitioners, and others develop a deeper understanding of the NCA approach and apply it effectively and with high-quality standards. The book brings together the fundamentals of NCA and integrates the most recent methodological advancements and tools. It synthesizes ideas from my earlier publications that describe NCA at introductory levels (e.g., Dul, 2016b, 2020) as well as more advanced treatments (Dul, Van der Laan, et al., 2020; Dul, 2021, 2024a), while also introducing more background, new topics, and recent developments. From 2021 to 2026, a regularly updated online predecessor of this book was available (Dul, 2021), which is now archived and integrated into the present book.

The first chapter of this book provides a general introduction to the possibilities and current use of NCA. Next, Part I contains five chapters on the *principles* of NCA, covering NCA's causality, theory, mathematics, statistics, and its credibility for identifying necessity from data. Part II has six chapters about the *application* of these principles in empirical studies discussing NCA's hypothesis formulation, data collection, data analysis, reporting, and how to apply NCA in multimethod research and in practice. I finish the book with a summary and a personal reflection.

The book's content expresses my current understanding and interpretation of NCA. NCA develops over time, just like other methodological approaches do. As NCA is applied more widely, new questions, topics, insights, and ideas will emerge. An online version of the book with updates and additional materials is available via the book-page: https://jandul.github.io/NCA/.

I hope this book serves those who wish to gain a comprehensive understanding of the principles of NCA and their application, both in academic research and practical contexts.

Enjoy discovering and applying NCA!

Leiden, March 2026

Acknowledgments

I created NCA during a long process that spans more than a decade. NCA and the content of this book are not only the result of my own thinking and reading but also of many discussions with fellow researchers during in-person and online meetings, at conferences, workshops, seminars, webinars, and in email exchanges. These interactions have played an important role in shaping NCA.

I wish to express my gratitude to the universities and other research organizations that invited me to deliver NCA seminars and workshops, both in its early years and more recently. These engagements enabled valuable face-to-face discussions that contributed to the development of NCA. I would like to mention the following institutions in alphabetical order: AGH University of Krakow (Poland), Bocconi University (Italy), Chalmers University (Sweden), Deakin University (Australia), Eindhoven University of Technology (Netherlands), Erasmus University (Netherlands), Federal University of Pernambuco (Brazil), George Washington University (USA), Helmut Schmidt University (Germany), INSEAD Business School (France), Jagiellonian University (Poland), LUISS (Italy), Mines Paris Tech (France), Nottingham University (UK), Oslo University (Norway), Oxford University (UK), Penn State University (USA), Radboud University of Nijmegen (Netherlands), RIVM (Netherlands), Tilburg University (Netherlands), University of Amsterdam (Netherlands), TNO (Netherlands), University of Bayreuth (Germany), University of Beira Interior (Portugal), University of Bologna (Italy), University of Coimbra (Portugal), University of Côte d'Azur (France), University of Durham (UK), University of Groningen (Netherlands), University of Kassel (Germany), University of Lincoln (UK), University of Maastricht (Netherlands), University of Paris (France), University of Porto (Portugal), University of Sevilla (Spain), University of Trento (Italy), University of Valencia (Spain), and Wroclaw University of Economics and Business (Poland).

I would also like to express my special appreciation to the Rotterdam School of Management at Erasmus University (Netherlands). Since the incubation of NCA, the school has supported me in developing and disseminating NCA. In particular, I wish to thank Erik van Raaij and Finn Wynstra for their enduring support.

I extend my acknowledgments to the dedicated team of NCA ambassadors and supporters who have contributed in different roles to the dissemination and advancement of this approach, ranging from student assistants to co-authors and co-developers: Jorick Alberga, Florence Allard-Poesie, Tatiana Andreeva, Joost Baart, Gijs van Biezen, Ruben Blom, Stefan Breet, Jon Bokrantz, Govert Buijs, Silvia Dello Russo, Monique van Donzel, Ger Haan, Tony Hak†, Sven Hauff, Patrycja Klimas, Wilfred Knol, Roelof Kuik, Erwin van der Laan†, Wangoo Lee, Igor Marchetti, Magdalena Marczewska, Justine Massu, Indy Oostrom, Erik van Raaij, Henk van Rhee, Nicole Richter, Krista Schellevis, Chloé Schwitzgebel, Daniele Spinelli, Zsófia Tóth, and Caroline Witte. Their support greatly accelerated the adoption of NCA in research.

Throughout the development of NCA, I engaged in extensive and ongoing discussions with numerous scholars and practitioners regarding the positioning of NCA in relation to mainstream approaches. In particular, I learned a lot from my conversations with Gary Goertz, regarding the positioning of NCA vis-à-vis regression analysis and statistics; Barbara Vis and Claude Rubinson on using NCA within the context of QCA; Oskar Kosch on the relationship between NCA and machine learning; Roelof Kuik on the detailed mathematical and statistical foundations of NCA, and Ger Haan on strategies for making NCA appealing to practitioners.

Finally, I would like to thank Bob Bastian, Gijs van Biezen, Jon Bokrantz, Patrycja Klimas, Roelof Kuik, Igor Marchetti, and José Luis Roldán for commenting on a draft version of this book.

Thank you all. Without your help, NCA would not be what it is today.

1

Introduction

NCA is a methodological approach that uses a *necessity* causal perspective for identifying necessary conditions from data. Necessity causal logic in NCA implies that if a certain level of the condition is not present, a certain level of the outcome will not occur: If not X, then not Y. No other factor can compensate for the absence of the necessary condition (or a required level of it). The necessary condition *allows* the outcome to exist, but does not produce it.[1] A necessary condition operates independently of the rest of the causal structure and serves as a 'critical success factor' or 'must-have'. When absent, it creates a 'bottleneck' or 'constraint' that prevents the target outcome, which cannot be compensated by other factors. For example, a student must have a high school diploma for admission to a university. Without such a diploma, the student will not be admitted. Other factors like a student's high motivation cannot compensate for the lack of a diploma.

Necessity causal logic and NCA differ from *sufficiency* causal logic and related methods. In sufficiency causal logic the condition *produces* the outcome: if X, then Y. A single factor may be sufficient for the outcome, for example when students graduate, they receive a diploma. Often, not a single factor but a combination of factors (a 'configuration') may be sufficient for the outcome: if [combination of X's], then Y. A university diploma alone is not sufficient for admission to a PhD program; however, combining it with strong motivation, good grades, and a recommendation letter will secure it. Such a configurational sufficiency logic is used in methodologies and frameworks like Rothman's Sufficient-Component Cause model in medicine (SCC, Rothman, 1976), Wright's Necessary Element of a Sufficient Set in law (NESS, Wright, 1985), and Ragin's Qualitative Comparative Analysis in the social sciences (QCA, Ragin, 1987).

Necessity causal logic also differs from *probabilistic* causal sufficiency logic where the single factor is a 'contributing cause' that helps to produce the outcome and contributes to the outcome 'on average': 'if X, then probably Y.' Highly motivated people are more likely to be admitted to a PhD program than those with lower motivation, although there are highly motivated people who are not admitted and less motivated people who are admitted. Such a probabilistic causal sufficiency logic is applied in statistical methodologies for causal inference, such as in Pearl's Directed Acyclic Graphs (DAGs) and do-operator approach (Pearl, 2009) and Angrist et al.'s Instrumental Variable approach (IV, Angrist et al., 1996).

The different causal lenses (single necessity causes, configurational sufficiency causes, and single probabilistic sufficiency causes) offer fundamentally distinct perspectives on causality. These lenses reflect different theoretical expectations (hypotheses) about how outcomes arise. This means that phenomena and data can be approached with different causal lenses (Chapter 2). The choice of lens may be guided by the specific nature of the phenomenon,

[1]This book uses specific terminology, such as condition and outcome, that is explained in the glossary of Appendix A.

DOI: 10.1201/9781003430766-1

the study question, or the analyst's theoretical priorities[2]. For instance, if the question is whether a certain resource (e.g., access to capital) is required for a firm to enter a market, a necessity lens is appropriate. This lens helps identify bottleneck factors: conditions that must be in place for success. In contrast, if the aim is to identify combinations of national conditions that are sufficient for strong climate policy commitment (e.g., public support + high income, or climate risk + high fossil fuel dependence), a configurational sufficiency perspective, as used in QCA, may be appropriate. This approach assumes that different pathways can each lead to the same outcome. If the goal is to understand whether self-efficacy increases the likelihood of engaging in a behavior, a probabilistic sufficiency lens is fitting. This approach helps estimate how much more likely the outcome is when a cause is present.

No lens is inherently superior; each has unique strengths and limitations. Thus, there are no better or worse lenses; there are no more or less relevant lenses. These lenses are different and shed a different light on the same issues. What matters is that the chosen lens aligns with the goals of the study, the nature of the causal claim being made, and indeed the preference of the analyst. Once a causal lens is selected, it is essential to apply a methodological approach that matches the logic of that lens, ensuring *theory–method fit*, which is essential for the credibility and usefulness of the findings (Chapter 11).

This book's emphasis on causality has both theoretical and practical reasons. Theoretically, scientific theories aim to explain why outcomes occur, not merely to describe patterns of co-occurrence (Chapters 3 and 7). Practically, changing or influencing an outcome requires intervention on causes of the outcome, not on co-occurring factors (Chapter 12). The assumption that X is a necessary cause of Y implies the assumption that X temporally precedes Y (Chapter 2).

NCA is an approach that is specifically designed to analyze necessity. It has two intertwined parts: a 'soft' part and a 'hard' part. The soft part is NCA's soul: the causal perspective of necessity (NCA methodology). The hard part is NCA's data analysis method (NCA method) that fits this perspective. Therefore, the term 'NCA' refers to a combination of the NCA methodology with its causal necessity perspective, and the corresponding NCA method for analyzing data.[3] NCA's necessity causal lens extends beyond common binary conditional necessity logic. Specifically, NCA adds causal reasoning to conditional logic, incorporates a continuous view on necessity, and permits exceptions (the 'typicality' perspective, see Chapter 2).

The book is organized as follows. Part I discusses the methodological principles of NCA, explaining its causality (e.g., compared to other causal perspectives), theory (e.g., main elements of a necessity theory, different types of necessity theories), mathematics (e.g., mathematical estimation of ceiling line, NCA parameters such as effect size, and measures of fit) and statistics (e.g., NCA's null hypothesis tests, Monte Carlo simulations with NCA), and NCA's credibility to identify necessity (e.g., using statistical and empirical criteria).

Part II of the book outlines how to conduct NCA. Specifically, this part discusses NCA's hypothesis formulation, data collection, data analysis, reporting, how to apply NCA in a

[2] In this book the general term for a user of NCA or other methodological approaches is *analyst*. The analyst can be an academic scholar or a practitioner. The general term for an academic research activity or a project in practice is *study*.

[3] In previous publications I have inconsistently used the terms 'approach', 'methodology' and 'method' in the context of NCA. In this book, I follow the common distinction between *methodology*, referring to the philosophical and theoretical framework, and *method* referring to the specific techniques, tools, or procedures used to collect and analyze data. The combination of the two parts is called *approach* or *methodological approach*.

multimethod study, and how NCA can be used in practice. The book concludes with a summary and personal reflection on 10 years of NCA, and Part III provides additional materials.

With its innovative approach to causality and data analysis, NCA has rapidly entered the social, medical, and technical sciences.[4] The publication in 2016 of NCA's core paper (Dul, 2016b) marks the start of NCA, although proto-versions of NCA were introduced earlier (Dul et al., 2010; Dul & Hak, 2008).[5] Subsequently, several extensions were developed, including a statistical significance test (Dul, Van der Laan, et al., 2020), and specialized software (Appendix B). Various publications, such as a textbook (Dul, 2020) and an online book (Dul, 2021) with practical guidelines on applying the method were developed to facilitate the understanding and dissemination of NCA. Furthermore, publications introducing and summarizing NCA have emerged across various fields of business and management, including human resource management (Hauff et al., 2021), marketing and sales (Conde, 2025; Dul, Hauff, et al., 2021), tourism management (Tóth et al., 2019), entrepreneurship (Linder et al., 2023), supply chain management (Bokrantz & Dul, 2023), and international business (Richter & Hauff, 2022). Also outside the business and management field, the method was introduced in fields such as public health (Greco et al., 2022), clinical psychology (Marchetti et al., 2026), education (Tynan et al., 2020), and creativity (Dul, Karwowski, et al., 2020).

Editors of academic journals have recognized the value of NCA by publishing editorial comments on the approach (e.g., Dul, 2025; Fainshmidt et al., 2020; Hernaus & Černe, 2022; S. Robinson et al., 2022; Sarstedt & Liu, 2023). Review articles (e.g., Aguinis et al., 2020; Boon et al., 2019; Dabić et al., 2021; Del Sordo & Zattoni, 2025; Frazier et al., 2017; Rozenkowska, 2023) and a plethora of journal articles suggest using NCA in future studies, for example, in the fields of education (e.g., Credé & Tynan, 2021), human resource management (e.g., Hauff, 2021; Parker & Knight, 2024), information systems (e.g., P. N. Sharma et al., 2024), international business/management (e.g., Richter et al., 2022; Zahoor et al., 2023), marketing (e.g., Guenther et al., 2023; Jovanovic & Morschett, 2022), operations and supply chain management (e.g., Acquah, Quaicoe, et al., 2023), strategy (e.g., Bouncken, Fredrich, Ritala, et al., 2020; Deist et al., 2023; Klimas et al., 2023; Ogundipe et al., 2022), and tourism and hospitality (e.g., Becker et al., 2023). At the time of writing this book, hundreds of articles that apply NCA have been published already. The number of publications in 'Web of Science' ranked journals[6] that apply NCA has increased steadily. Appendix C shows these articles that have appeared through the end of 2025.

Published articles use NCA as the sole or primary method (Appendix C; Table C.1), or combine NCA in a multimethod study (Chapter 11), often with Qualitative Comparative Analysis (QCA, Table C.5) or with regression-based methods. Examples of regression-based methods include multiple linear regression (MLR, Table C.2) and structural equation modeling (SEM, Tables C.3 and C.4). NCA is also combined with a variety of other methods (Tables C.6 and C.7). Although the number of articles with NCA has grown rapidly, NCA is not always applied according to minimum standards (Chapter 10, Dul et al., 2023), and

[4]The NCA methodology and method can be used in any type of investigation that aims to describe causal relationships. Although usually quantitative data (numbers) are used in NCA, NCA can also be used with qualitative data (words, symbols, see Chapter 8) as long as scores (values, levels) for the condition and the outcome are available.

[5]The dissemination of NCA's core paper (Dul, 2016b) has been welcomed by Bergh et al. (2022): "The significant expansion of an unknown method along with its usability (including an illustration, user-friendly recommendations, and a software tool) makes the article a gold standard for a contribution via transfer" (p. 1840).

[6]Journals ranked in Clarivate's Journal Citation Reports.

several misconceptions have been published.[7] Therefore, **the main goal of this book is to promote proper understanding and high-quality application of NCA**.

Many articles that apply NCA conclude that necessity was identified, although the evidence is not always convincing. Other articles that apply NCA concluded that necessity was not found (e.g., Batey et al., 2021; Golini et al., 2016; Gu et al., 2022; L. Luo et al., 2022; B. Peng et al., 2022). Not finding a necessary condition might be a valuable result, because such a result shows that a supposed essential factor for an outcome does not need to be present and its absence can be compensated (e.g., Arenius et al., 2017). On the other hand, finding a large number of potential necessary conditions, for example in an exploration study (e.g., Gantert et al., 2022; Klimas et al., 2022; Stek & Schiele, 2021) does not mean that all are informative because several conditions may not have a theoretical meaning or may be trivial, which could have been easily identified before the analysis was done (Chapter 7).[8]

Conducting NCA consists of four stages:

1. Formulate the formal necessity hypothesis.
2. Collect data.
3. Analyze data.
4. Report results.

In the first stage (Chapter 7), necessity causal logic is employed to formulate a theoretical expectation about conditions that are necessary for the outcome[9].

The second stage (Chapter 8) consists of collecting new or existing data for testing the hypothesis, which includes the consideration of study design, sampling or case selection, and measurement. NCA does not have a 'measurement model', as SEM and QCA have (in QCA it is called 'calibration'). This means that any type of data can be used as input to NCA as long as the scores of X and Y are meaningful, valid, and reliable. NCA poses no new requirements on the data, although some aspects of data collection (case selection in qualitative studies; power analysis in quantitative studies, see, for example, Dul (2024b)) may differ. NCA can also be used with archival data to give a new perspective on previous findings (Dul et al., 2024).

In the third stage (Chapters 9, 11), NCA's data analysis is conducted, which differs entirely from common types of data analysis.

In the fourth stage (Chapter 10), the results of NCA are reported.

Currently, most applications of NCA are in academic research, where researchers use the NCA approach to develop and test theories. NCA is also entering practical settings where

[7] Most misconceptions relate to a lack of awareness about the principles of NCA. Example misconceptions are: "NCA should specify the full data generation process (DGP)" (given its purpose, NCA specifies only a part of the DGP), "NCA is just a statistical method" (NCA is a broad methodological approach that combines theory, mathematics (geometry) and uses statistical tools), "NCA's statistical test is a test of H1" (NCA's statistical tests are null hypothesis tests, testing H0), "NCA is the necessity analysis of QCA" (the necessity analysis of NCA and that of fuzzy-set QCA are fundamentally different), "Necessary conditions according to NCA are the same as INUS conditions" (INUS conditions are local necessary parts of a sufficient configuration that produces the outcome; NCA considers overall necessary conditions for the outcome), "NCA cannot test sufficiency" (NCA can test the absence of X being necessary for absence of Y, which is equivalent to X is sufficient for Y), etc. Several misconceptions are addressed in specific publications (Dul et al., 2019; Dul, Vis, et al., 2021; Dul, 2022).

[8] A positive or negative conclusion about necessity depends on the analyst's judgement of many elements, summarized in NCA's SCoRe checklist for conducting and reporting an NCA study (Chapter 10).

[9] In this book, it is assumed that hypothesis formulation is done before data collection. The reasons are discussed in Chapter 7.

NCA is applied by practitioners such as consultants and data analysts as an innovative approach for addressing challenges and exploring new opportunities. This is discussed in Chapter 12.

After discussing the fundamental backgrounds of the NCA methodology in Part I, the four stages of the NCA method are discussed in Part II.

Part I

Principles

This part of the book presents the theoretical background and assumptions of NCA. Understanding the principles of NCA is important for recognizing the method's possibilities and limitations. It is also essential for avoiding unrealistic expectations and misinterpretations.

The first chapter (Chapter 2) discusses NCA's causality perspective and how it is different from conventional perspectives. The next chapter (Chapter 3) discusses what a necessity theory is. The subsequent two chapters present the mathematical (Chapter 4) and statistical (Chapter 5) backgrounds of NCA. The final chapter (Chapter 6) evaluates NCA's credibility for identifying necessary conditions from data.

2

Causality

2.1 Summary of this chapter

NCA uses a causal perspective that is (partly) different from common views on causality. This chapter presents NCA's causal perspective by comparing it to other causal perspectives. The chapter begins by discussing what a cause is (Section 2.2), followed by a discussion of the commonly used sufficiency causality (Section 2.3). Next, necessity causality is discussed (Section 2.4). For both types of causality, three perspectives are distinguished: deterministic, probabilistic, and typicality (deterministic with exceptions). NCA uses the deterministic (or typicality) necessity causal perspective. In the next section (Section 2.5), NCA's use of conditional logic (assuming a causal direction) is compared with the conventional view on conditional logic (in which no causal direction is assumed). Subsequently, another unique feature of NCA is discussed: the concept of necessary condition in degree (NiD), which extends the conventional binary approach (absence or presence of X is necessary for absence or presence of Y) to a continuous approach: level x of X is necessary for level y of Y (Section 2.6). Finally, Section 2.7 discusses that, due to causal underdetermination, no causal perspective is superior to another.

2.2 What is a cause?

Causality is often loosely defined as the *influence* of a cause X on an effect Y. General expressions like X has an effect on Y, X affects Y, or X contributes to Y without specifying the type of causality are commonplace. Causality is a complex subject that is extensively debated in the philosophy of science literature and elsewhere. The Scottish philosopher David Hume (1711–1776) laid the foundation for modern discussions about causation. Hume defined a cause as follows:

> "... we may define a cause to be an object, follow'd by another, and where all the objects, similar to the first, are follow'd by objects, similar to the second: Or in other words, where, if the first object had not been, the second never had existed" (Hume, 1756, p. 121)

DOI: 10.1201/9781003430766-2

The first part of this definition refers to *sufficiency* causality (if X, then Y).[1] It includes a specification of the temporal order: Y follows X, and a generalization: the statement applies also to "similar objects." The second part of the definition refers to *necessity* causality (if not X, then not Y)[2] although without the generalization (for a further discussion see Dul, 2024a). The two types of causality are illustrated in Figure 2.1, showing observed XY patterns where X is the condition and Y is the outcome that can have two values (absent/present).[3] When X is a sufficient cause for Y (Figure 2.1-left), observations (points, cases) can exist in all corners except corner 4: If X is present, then Y is present. When X is a *necessary* cause for Y (Figure 2.1-right), observations can exist in all corners except in corner 1: If X is absent, then Y is absent. The corner where observations are not possible is called the *empty space* and the corner where observations are possible is called the *feasible area.*

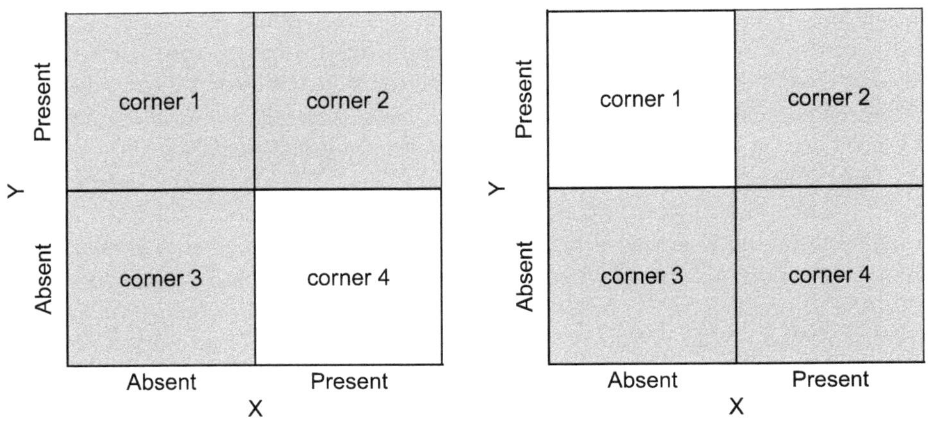

FIGURE 2.1 XY-tables illustrating sufficiency causality (Left) and necessity causality (Right) with binary concepts. Gray = observations (points, cases) are possible. White = no observations are possible.

2.3 Sufficiency causality

Most phenomena in the social, biomedical, and other sciences are studied from the perspective of sufficiency causality (if X, then Y). When people talk about a cause or causality, they (almost always) implicitly mean sufficiency causality.

Three fundamentally different perspectives on sufficiency causality can be distinguished: deterministic, probabilistic, and typicality (Dul, 2024a). Table 2.1 summarizes these perspectives and lists some main methods that (implicitly) use this perspective when inferring causality.

[1] If X, then Y means: if we see X, we also see Y, and if we add X, we get Y.

[2] If not X, then not Y means: if we do not see X, we also do not see Y, and if we remove X, we do not see Y anymore.

[3] Causality itself cannot be observed, see Section 2.7 and Chapter 7.

TABLE 2.1 Perspectives on sufficiency causality: if X, then ... Y.

Perspective	Logic	Main methods
Deterministic	if X, then *always* Y	QCA, SCC, NESS
Probabilistic	if X, then *probably* Y	LR, MLR, SEM
Typicality	if X, then *typically* Y	Methods in physics

Note:
QCA = Qualitative Comparative Analysis
SCC = Sufficient Component Cause
NESS = Necessary Element of a Sufficient Set
LR = Logistic (Logit) Regression
MLR = Multiple Linear Regression
SEM = Structural Equation Modeling

2.3.1 Deterministic sufficiency

Hume's sufficiency definition has been interpreted as being about deterministic causality: if X, then *always* Y (e.g., Baumgartner, 2009). Many phenomena in the physical world may be described by such a *deterministic sufficiency causal perspective.* For example, reaching the speed of sound produces a sonic boom. In the social sciences, single deterministic sufficiency causes are rare, as normally several factors together produce an outcome. Such a sufficiency perspective is used in 'parts-whole' methods and frameworks (Machamer et al., 2000; Varzi, 2019). The Sufficient-Component Cause model in medicine [SCC; Rothman (1976)], the Necessary Element of a Sufficient Set approach in law (NESS, Wright, 1985), and Qualitative Comparative Analysis in social sciences [QCA; Ragin (1987); Campbell & Fiss (2026)], for example, use such a sufficiency perspective. The combination of causal factors (called configurations in QCA) is considered deterministically sufficient for the outcome: If [combination of X's], then Y. Several combinations may be sufficient ('equifinality'), such that none of these configurations is necessary for the outcome. The causal factors that are parts of the configurations are called INUS conditions: Insufficient but Necessary parts of Unnecessary but Sufficient configurations (Mackie, 1965). INUS conditions should not be confused with necessary conditions; INUS conditions are 'locally' necessary for the configuration to produce the outcome, and usually not overall necessary for the outcome itself (Dul, Vis, et al., 2021). NCA considers only the overall necessary conditions for the outcome. Although the theoretical foundations of SCC, NESS, and QCA are based on a deterministic ontology, in practice the determinism may be imperfect.[4] Deterministic sufficiency means that corner 4 of Figure 2.1-left remains empty.

2.3.2 Probabilistic sufficiency

As single deterministic sufficiency causes are rare in the social sciences, a non-deterministic causal approach is often adopted: probabilistic sufficiency: 'if X, then *probably* Y', or 'if X, then *likely* Y.' Pearl states that:

[4]See, for example, QCA's use of the consistency score < 1 to allow imperfect sufficiency.

> "According to one of the basic tenets of probabilistic causality, a cause should raise the probability of the effect" (Pearl, 2009, p. 254)

The dominant framework of the *probabilistic sufficiency* causal perspective focuses on the average effects of 'parts' of the whole. The lower-right corner is not empty but is less populated than the other cells. It identifies the Average Causal Effect (ACE) or Average Treatment Effect (ATE) of a causal factor. The approach is currently mainstream in quantitative studies using statistical methods. Statistical models are based on the assumption of the probability distributions of the cause and the effect. The effect is produced by an unknown 'data generation process' that describes the underlying relationships between concepts and the stochastic processes involved in generating the observed data. When statistical models with regression-based methods are used to infer causality, the underlying causal perspective is probabilistic sufficiency. Examples of regression-based methods that are commonly used for inferring causality are multiple linear regression (MLR), logistic (logit) regression, and structural equation modeling (SEM). Probabilistic sufficiency may be inferred if the lower-right corner of XY-tables or XY-plots is less densely populated with observations than the other corners (Figures 2.1 and 2.3).

2.3.3 Typicality sufficiency

A recently developed non-deterministic and non-probabilistic perspective on causality is the 'typicality' perspective. The *typicality sufficiency causal perspective* was introduced in physics, including quantum mechanics and thermodynamics (e.g., Goldstein, 2012; Wilhelm, 2022). It challenges the deterministic ontology but does not adopt a probabilistic ontology. The typicality interpretation also gets attention in other scientific disciplines, including the social sciences (Wagner, 2020). This perspective can be described as deterministic with exceptions. It accepts the perspective if X, then Y but fundamentally allows exceptions: if X, then *typically* Y, or if X, then *almost always* Y. The typicality perspective does not describe exceptions in terms of likelihood of occurrence, but in terms of cardinality (Wilhelm, 2022): the number of elements that deviates from the majority of the elements in the set. Typicality sufficiency means that corner 4 of Figure 2.1-left remains empty, but a few observations may be exceptions.

2.4 Necessity causality

The three types of perspectives on causality can also be applied to necessity causality. This is shown in Table 2.2.

2.4.1 Deterministic necessity

The second sentence of Hume's definition of causality can be considered a deterministic view on necessity causality (Goertz & Mahoney, 2012). If X does not occur, Y will not happen. Whereas single sufficient causes are seldom on their own sufficient for an effect, single necessity causes alone can stop an effect from occurring if absent. For example, oxygen is necessary for human life but not sufficient on its own. There are multiple necessity causes for human life, and each will stop the outcome from occurring if it is absent. Deterministic necessity means that corner 1 of Figure 2.1-right remains empty.

2.4.2 Probabilistic necessity

The probabilistic necessity perspective can be phrased as: If not X, then probably not Y. Probabilistic necessity means that corner 1 of Figure 2.1-right may have observations but is less densely populated than the other corners.

2.4.3 Typicality necessity

The idea of typicality was developed for sufficiency causality but can also be applied to necessity causality (Dul, 2024a). It accepts a deterministic perspective with exceptions: if not X, then *typically* not Y, or if not X, then *almost always* not Y. Again, the typicality perspective does not describe exceptions in terms of probabilities but in terms of cardinality. Typicality necessity means that corner 1 of Figure 2.1-right remains empty, but a few observations may be exceptions. Exceptions are rare without quantifying what is "rare." If an exception is not an error, it could be a unique counterexample (e.g., an innovative case), worth studying in detail. However, the phenomenon of interest is still described with deterministic necessity, excluding the exception.[5] Note that NCA has a deterministic view on causality and that the ceiling line is a sharp border between an area with cases and an area without cases (Section 4.3.2). The presence of outliers (in the "empty" space) indicates that necessity does not exist, rather than that necessity exists with high probability.

With a deterministic view on necessity causality, a case with high Y and low X rejects necessity. The data point may be mistakenly in the empty area because of measurement error, sampling, or case selection error. Such errors may be corrected, or the case may be removed from further consideration if the error cannot be repaired. However, if the case is not an error it may be a special case: a rare exception in the otherwise empty space, representing another phenomenon not captured by necessity. Only the presence of *such* exception may be the reason for adopting the typicality perspective. NCA keeps its deterministic approach, but allows rare exceptions. Having many outlier cases in the "empty" space that cannot be explained by errors is a reason to reject necessity (or to redefine the formal hypothesis; see Chapter 7 if these cases have a common characteristic), but it is not a reason to accept necessity with a lenient view on (typicality) necessity.

TABLE 2.2 Perspectives on necessity causality: if not X, then ... not Y.

Perspective	Logic	Main methods
Deterministic	if not X, then *always* not Y	NCA, (QCA)
Probabilistic	if not X, then *probably* not Y	(PN)
Typicality	if not X, then *typically* not Y	NCA

Note:

NCA = Necessary Condition Analysis

QCA = Qualitative Comparative Analysis (only in kind; typicality implicit)

PN = Probability of Necessity (only binary variables; no generic necessity)

[5]Note the difference between Exception and Noise as discussed in Section 4.5.4.

Table 2.2 maps different methods on the three causal perspectives. It shows that NCA can be used for the deterministic and typicality perspectives. The table also includes QCA for identifying necessity using a deterministic or typicality perspective. Although QCA's main focus is on sufficiency, and QCA studies often ignore necessity, QCA can also identify necessity. The necessity analysis of QCA is less detailed than that of NCA, as it only identifies *necessity-in-kind* (NiK)[6] and often misses necessary conditions (Torres & Godinho, 2022). The differences between necessity approaches of NCA and QCA are explained in Dul (2016a) and Vis & Dul (2018) and summarized in Section 11.5. The table also mentions Pearl's Probability of Necessity (PN) approach for identifying probabilistic necessity using a counterfactual analysis. Pearl (2009) introduced the Probability of Necessity (PN) concept (where X and Y are binary) to address the question: "Was event X necessary for event Y to occur?" PN is the probability that Y would not have occurred if X had not occurred, given that both X and Y actually occurred.[7] This is a counterfactual probability conditioned on the actual world where X and Y happened. PN evaluates a hypothetical world where X did not happen. PN is used for identifying necessity causes at the individual level, for example in legal reasoning, where the question is not whether in general X is a probabilistic cause of Y, but rather whether X was the necessity cause of Y in a specific situation that has happened.

2.5 Causal versus conditional logic

NCA extends conditional logic with causality. In conditional logic, as used in philosophy and mathematics, relationships are described in terms of if-then statements. In the statement 'if A, then B', A is the antecedent and B is the consequent, and A and B usually have only two values: true or false. The truth or falsity of B depends on the truth or falsity of A. In conditional logic, 'if A, then B' means that A implies B, or if A occurs, then B occurs. This can be interpreted as A being a *sufficient condition* for B. For example, being born in Italy (A) implies having Italian citizenship (B). This can be written as:

$$A_s \Rightarrow B \tag{2.1}$$

Here the subscript 's' emphasizes the interpretation of A as a sufficient condition, and the double arrow $\Rightarrow$ means 'implies'.

In conditional logic, the statement that A is a *necessary condition* for B implies that B can only be true if A is true: if B, then A. For example, oxygen is necessary for human life because being alive implies that the person has access to oxygen. Therefore, truth of B (being alive) implies the truth of A (oxygen). This can be written as:

$$B \Rightarrow A_n \tag{2.2}$$

[6] Necessity-in-kind (NiK) refers to the qualitative statement that X is necessary for Y without specifying the level of X and Y other than presence/absence or low/high. This contrasts with NCA's necessity-in-degree (NiD) where specific levels of X and Y are specified.

[7] Formally, PN is defined as: $PN = P(Y_{X=0} = 0 \mid X = 1, Y = 1)$.

where the subscript 'n' emphasizes the necessary condition interpretation of A. This can be rewritten as:

$$A_n \Leftarrow B \tag{2.3}$$

where the double arrow $\Leftarrow$ means "is implied by."

Because B can only be true when A is true, the falsity of A implies the falsity of B. This can be written as:

$$\neg A_n \Rightarrow \neg B \tag{2.4}$$

or

$$\neg B \Leftarrow \neg A_n \tag{2.5}$$

where the symbol $\neg$ indicates negation, or 'no' or 'not'.

Equations (2.2)–(2.5) are four equivalent expressions for A_n being necessary for B from the perspective of conditional logic.

The phrase B implies A_n (2.2) can also be understood as B is *sufficient* for A_n. Therefore, the expression A_n is necessary for B can be written in four equivalent ways:

1. A_n is necessary for B (oxygen is necessary for being alive).
2. B is sufficient for A_n (being alive is sufficient for oxygen).
3. $\neg A_n$ is sufficient for $\neg B$ (no oxygen is sufficient for not being alive).
4. $\neg B$ is necessary for $\neg A_n$ (not being alive is necessary for no oxygen).

A major difference between necessity in *causal logic* as used in NCA, and necessity logic as used in *conditional logic* is that NCA assumes a causal direction between A_n and B whereas conditional logic does *not* assume such a causal direction. This means that NCA assumes a temporal order: first A_n, followed by B. Conditional logic is ignorant about whether there is first oxygen and then life or first life and then oxygen. Therefore, NCA's causal logic can be seen as an extension of conditional logic where a causal direction is assumed between A_n and B. This temporal order excludes the possibility of 'first B, then A_n'. The assumption of a temporal order may be justified when first the occurrence of A_n is observed and afterwards the occurrence of B, or that it is theoretically implausible that A_n occurs after B (see Section 7.5.3 for a further discussion on temporality). It means that when NCA formulates that A_n is a *necessary condition* for B, it is assumed that A_n is a *causal* necessary condition for B. Assuming also no reverse or reciprocal causality (Section 5.7), it implies that only two of the four conditional expressions are causal statements:

1. A_n is a necessity cause of B (oxygen is necessary for life).
3. $\neg A_n$ is a sufficient cause of $\neg B$ (no oxygen is sufficient for no life).

Therefore, in NCA the antecedent A_n is considered the *necessity cause* of the consequent B (the effect). Statements 2 and 4 still apply as logical statements, but not as causal statements, assuming no reverse or reciprocal causality.

To stress that NCA makes a causal assumption about the relationship between antecedent and consequent, the symbols A and B or P and Q that are common in conventional conditional logic, are replaced by the symbols X and Y to express the cause-effect relationship, where X is the cause and Y is the effect. Then the two equivalent causal expressions for a necessity cause are:

$$X \text{ is necessary for } Y \tag{2.6}$$

$$\neg X \text{ is sufficient for } \neg Y \tag{2.7}$$

In the remainder of this book, the expression 'X is a necessary condition for Y' implies that X is a necessity cause of Y and both expressions (2.6) and (2.7) apply. Dul (2016b) refers to expression (2.6) as the 'necessity formulation' of the necessary condition and expression (2.7) as the 'sufficiency formulation' of the necessary condition. In this book the statement X is necessary for Y is understood in a deterministic or typicality causal perspective: 'X is always necessary for Y' or 'X is almost always/typically necessary for Y'. Consequently, all or nearly all observations with the outcome Y do have the necessity cause X, and all or nearly all observations that do not have the necessity cause X do not have the outcome Y.[8]

2.6 From binary necessity to continuous necessity

Another extension that NCA introduces to necessity logic involves the levels of condition X and outcome Y. Until now, the book considered the levels of X and Y as binary. Only two values are possible: absent or present (like in Figure 2.1-right), 0 or 1, or false or true, etc. This binary approach is common in conditional logic as well as in the SCC, NESS, and QCA methods and frameworks.[9]

NCA extends the binary levels of condition and outcome to multiple possible levels by allowing X and Y to be discrete concepts (with a number of finite levels of more than 2) or continuous concepts (with an infinite number of levels) within minimum and maximum *bounds* of X and Y. This has several consequences.

First, the XY-table is extended to an XY-plot as shown in Figure 2.2. Instead of mapping observations on the cells of the XY-table, observations are now mapped on a 2D Euclidean coordinate system. The bounds of X and Y are again 0 and 1, but could be any other minimum and maximum values. The bounds on X and Y create a *bounding box* (Section 4.3). In a *dichotomous necessity relationship* where both the condition and the outcome are binary, the observations are mapped at corners of the XY-plot (Figure 2.2-left). When necessity applies, no observations are possible in the upper-left corner [0,1].

[8] Until recently, I was not aware of the existence of the typicality perspective on causality. What I previously called a 'probabilistic view' on necessity causality (Dul, 2020) refers to what I now (Dul, 2024a) call the 'typicality perspective' of necessity causality (based on cardinality). The current description 'probabilistic perspective' of causality refers to the use of probabilities to describe causality, but this is not part of NCA.

[9] Also in fuzzy set QCA (fsQCA) the necessary condition in the sufficiency solution is binary, see Section 11.5.

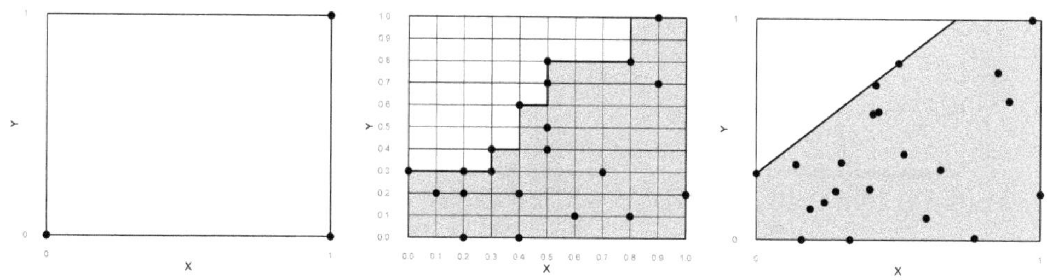

FIGURE 2.2 XY-plot with dichotomous (Left), discrete (Middle), and continuous (Right) values of condition X and outcome Y.

In a *discrete necessity relationship* observations can be on the intersections of a grid representing the possible discrete values of X and Y (Figure 2.2-middle), and in a *continuous necessity relationship* observations can exist anywhere in the unit box (Figure 2.2-right), but not in the upper-left corner. Combinations of dichotomous, discrete, and continuous values are also possible in NCA. This is shown in Figure 2.3.

Second, whereas in the dichotomous situation the bounding box [0,1] × [0,1] can be either full or empty, in the discrete and continuous situation it can be partly empty. The size of the empty space may vary, depending on the location of the observations. A border line (*ceiling line*) separates the empty space (*ceiling zone*) from the area where observations are possible (*feasible area*). Referring to the XY-plot in Figure 2.2-left where both the condition and the outcome are dichotomous, the ceiling line between empty and feasible area is predefined: the step line connecting the observations [0,0], [0,1], and [1,1] is always the same. The bounding box (unit box) is completely empty, such that points can only be on the ceiling line. No observations can be above or to the left of the ceiling line. In the discrete and continuous situations, the location of the ceiling line depends on the location of the observations in the unit box. This means that the location of the ceiling line and the size of the empty space vary. In any case, there can be no points above or to the left of the ceiling line. If necessity applies, points can only be on, below, or to the right of the ceiling: the feasible area. The ceiling line can be assumed (predefined) or estimated from data. In all situations, the assumption is that the ceiling line represents the sharp border between the empty space and the feasible area to represent deterministic necessity (see model fit measure *Sharpness* discussed in Section 4.5.6). When a relevant empty space exists that is caused by necessity, the necessary condition is formulated qualitatively as necessity-in-kind (NiK): X is necessary for Y.

Third, the discrete and continuous situations add the possibility of a quantitative formulation of a necessary condition statement: *level x of X* is necessary for *level y of Y*. This is called *necessity-in-degree* (NiD) and is shown for the continuous situation in Figure 2.4. Any observation C on the ceiling line defines a specific necessity-in-degree: level $X = x_c$ is necessary for level $Y = y_c$. Assuming an increasing ceiling line, the condition must have at least level x_c for outcome level y_c. If the condition is less than x_c, the outcome will be less than y_c. If $X = x_c$, it cannot be concluded that $Y = y_c$, but only that $Y \leq y_c$. This situation can be described by level x_c of X is *necessary but not sufficient* for level y_c of Y (Section 3.8 for a further discussion on the meaning of necessary but not sufficient).

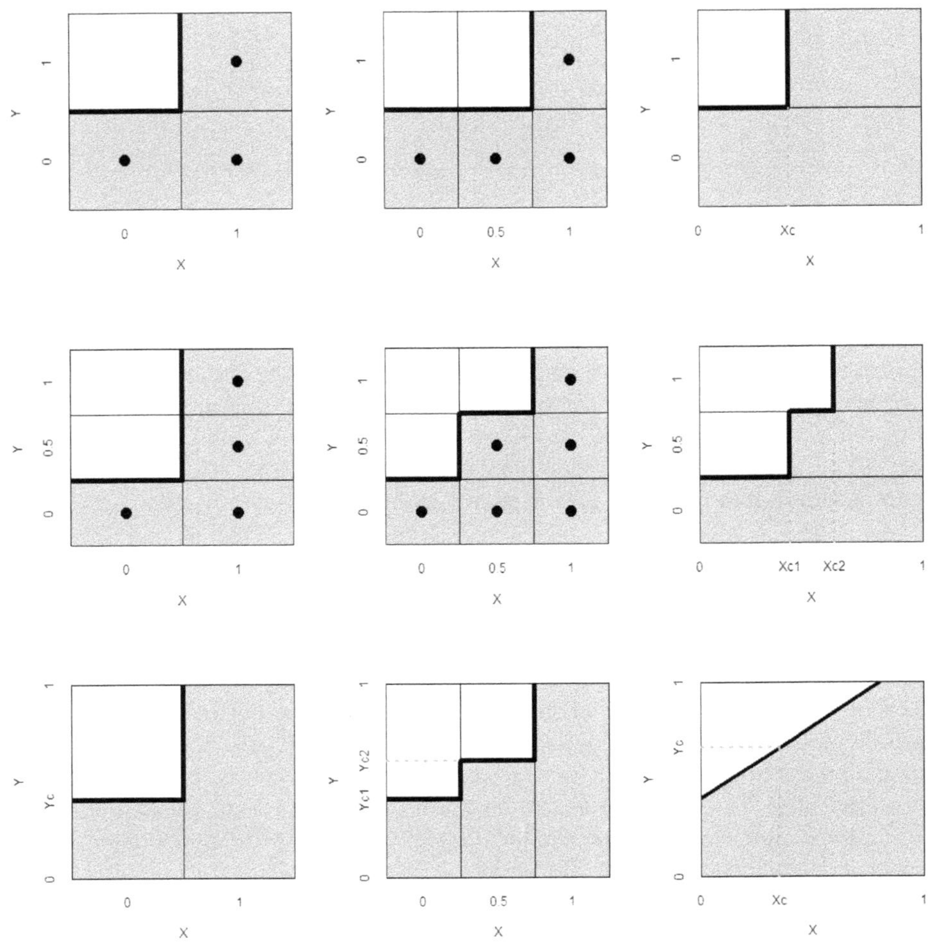

FIGURE 2.3 XY-plot with combinations of dichotomous, discrete, and continuous values of condition X and outcome Y. The upper-left corner is empty if the presence/high value of X is necessary for the presence/high value of Y.

2.7 Causal underdetermination

A causal relationship cannot be directly observed. What can be observed are only sequences and patterns. For example, when a glass falls from the table, we see the motion of the hand, the contact with the glass, the glass's acceleration, and its eventual impact on the ground. Nowhere in this stream of sensory data is there an observable entity called "causation." The causal link must instead be *inferred*. Was the push responsible for the fall, was the table moved at the same moment, or did the glass fall by coincidence?

Because causality cannot be observed, a human interpretation of observed data must be made. The same observational data can support multiple causal explanations. This is the phenomenon known as *causal underdetermination* (e.g., Stanford, 2023). An observed relationship between X and Y may reflect that X causes Y, Y causes X, an unobserved

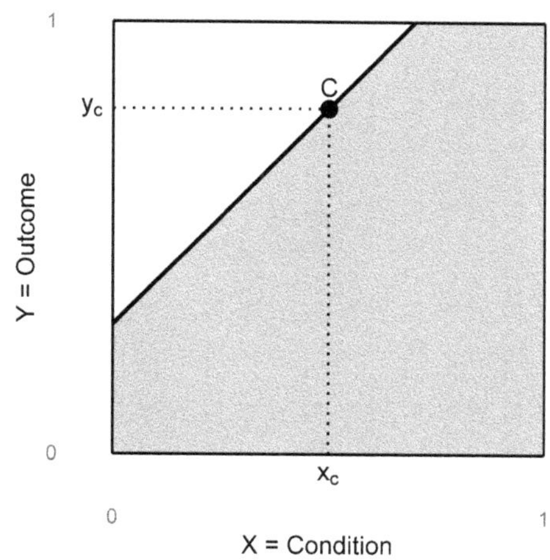

FIGURE 2.4 XY-plot illustrating necessity causality when X and Y are continuous (necessity-in-degree, NiD).

cause, or mere chance. Underdetermination implies that a single dataset can be explained by different causal models, whether framed in terms of necessity, sufficiency, probabilistic tendencies, or typicality.

The idea that the same evidence can be "seen" differently is illustrated by the classic ambiguous picture below. Some viewers immediately see a young woman, while others see an old woman, and it can be hard to switch between the two. Likewise, the same data can support different causal interpretations, and once someone becomes familiar with one perspective, it may be difficult to adopt another.

A related implication is the problem of non-identifiability (or observational equivalence): distinct models can generate the same observable data patterns and therefore fit the data equally well, even though they imply different underlying structures (e.g., Rothenberg, 1971). When this happens, the evidence cannot distinguish between these competing explanations. For example, when a correlation between X and Y is observed, this could be interpreted as a result of a probabilistic sufficiency relationship, or as a result of a necessity relationship (Appendix G). In principle, both interpretations are possible.

Given the inherent underdetermination of causal claims from observations, theory becomes indispensable. Theory provides the assumptions and background knowledge needed to justify a causal model that is consistent with the observations. Without such guiding principles, data remain compatible with infinitely many interpretations. Theory specifies, for example, which mechanisms are plausible or which temporal order is even possible. The central role of theory in data analysis for causal inference is illustrated by Nancy Cartwright's slogan:

"No causes in, no causes out" (Cartwright, 1989, pp. 39–90)

FIGURE 2.5 The same image can be seen in different ways. Likewise, the same data can be viewed through different causal lenses. From: My wife and my mother-in-law. They are both in this picture – find them. Prints & Photographs Online Catalog, Library of Congress. Retrieved 1 October 2025. https://www.loc.gov/pictures/item/2010652001/.

Some causal theories may be more plausible than others. For example, if we observe that plants exposed to more sunlight grow more, the explanation that sunlight causes plant growth is far more credible than the idea that plant growth causes sunlight. An observed relationship is considered spurious when an alternative explanation is clearly more credible because the theory is more plausible or because of better model fit.[10]

However, several causal theories may also be equally credible. This is the situation known as causal pluralism: different causal perspectives may each illuminate different aspects of the same phenomenon (Section 11.2). Causal pluralism can help to better understand the underlying causal mechanisms of the phenomenon of interest. If several causal theories are equally credible, it is also possible to select one theory for pragmatic reasons. Such reasons could be simplicity (Section 3.3), usefulness for prediction, computational elegance, compatibility with other accepted theories, interpretability, or practical actionability. A necessity theory could be a candidate.

Underdetermination and pluralism imply that no single causal perspective is inherently the "proper" one for every study question. In particular, commonly used probabilistic sufficiency perspectives and their associated regression tools are not inherently preferred. It is the analyst's freedom to select an appropriate perspective and tools, and the analyst's responsibility to justify the corresponding hypothesis and interpretation (Chapter 7).

[10] Section 6.4.2 discusses model fit in the context of NCA.

This book is about the use of the necessity causal perspective and tools when studying phenomena. It therefore follows that a necessity-based theoretical framework is essential for describing such phenomena coherently. What constitutes such a necessity theory is discussed in the next chapter.

3

Theory

3.1 Summary of this chapter

This chapter discusses causal theory in general and necessity theory in particular. A causal theory helps us understand phenomena, predict effects, and explain empirical findings. The chapter begins (Section 3.2) by discussing the four fundamental elements of a theory: (1) focal unit (e.g., person, organization, country), (2) concepts (varying characteristics of the focal unit), (3) propositions (causal relationships between the concepts), and (4) theoretical domain (generalization to where the theory holds). This is followed by a discussion of parsimonious necessity theories (Section 3.3) and types of necessity theories (Section 3.4), for example a pure necessity theory where all relationships between the concepts are described in terms of necessity, and an embedded necessity theory, where part of the theory has one or more necessity relationships. Next, the direction of a necessity proposition is discussed (Section 3.5), which depends on whether X and Y are present/have high value or are absent/have low value. The most common direction is that the presence or high value of X is necessary for the presence or high value of Y (high-high direction), which is the default when the direction is not specified. Other possible directions are high-low, low-high, and low-low. The following section (Section 3.6) discusses the expected data pattern when necessity holds. A consequence of a high-high necessary condition is that the upper-left corner in an XY-plot is expected to be empty. The next section (Section 3.7) discusses the possibility of having double necessity when multiple corners are expected to be empty. The chapter finishes with a discussion of the theoretical meaning of 'necessary but not sufficient' (Section 3.8).

3.2 Main elements of a theory

Because there are multiple ways to conceptualize theory, there is no consensus on what a theory is. This book adopts the 'four elements' approach to theory (Dul & Hak, 2008), which enables straightforward empirical testing. It uses a broad and inclusive definition of what constitutes a theory, ranging from established theories that are broadly accepted to theoretical assertions by an analyst made in a specific study. Not only academic theories but also 'theories-in-use' qualify as a theory. At a minimum, a theory must have four fundamental elements: focal unit, concepts, propositions, and theoretical domain. The *focal unit* is the object to which the theory applies: country, person, firm, project, etc. The *concepts* are the variant characteristics of the focal unit representing properties of a phenomenon that can increase or decrease. In NCA, a concept that is the cause is called the 'condition' and a concept that is the effect is called the 'outcome'. The *proposition* describes the relationship

DOI: 10.1201/9781003430766-3

between cause and effect, including the causal explanation. In NCA, the causal relationship is a necessity relationship: X is necessary for Y, meaning that X is (almost always) necessary for Y (Dul, 2024a). The words 'almost always' may be omitted from the necessity proposition, allowing both a deterministic and a typicality perspective. As theories in the social sciences are usually semantic (expressed in words), specific levels of the condition (cause) and the outcome (effect) are not specified. The *theoretical domain* is the set of cases of the focal unit where the theory is supposed to hold, and is defined by the boundary conditions of the theory. A *case* is one particular example of the focal unit: a specific country, a specific person, etc. For example, a theory that proposes a relationship between physical exercise and stress might be applicable to working adults, but not to children. In that case, the theoretical domain includes only working adults.

For example, the established Theory of Planned Behavior (TPB) explains how an individual's intention for a specific non-habitual behavior relates to the performance of that behavior (Ajzen, 1991). The *focal unit* is 'individual' or 'person', two of its *concepts* are 'Intention for the behavior' (the cause) and 'Performing the behavior' (the effect), the *proposition* describes the causal relationship between these concepts, and one possible *theoretical domain* is 'all people in the world'.

The relationship between the two TPB concepts is usually specified as a *probabilistic sufficiency relationship*, for example 'Intention to perform a behavior likely to result in Performing the behavior', or, as the founder of TPB puts it: "As a general rule, the stronger the intention to engage in a behavior, the more likely should be its performance" (Ajzen, 1991, p. 181). Recent interpretations of TPB suggest formulating the relationship between the two concepts in terms of *necessity* (Eccarius & Chen, 2024; e.g., Frommeyer et al., 2022; Rozenkowska, 2023): for example 'Intention to perform a behavior is necessary for Performing the behavior'. The key difference between a conventional (probabilistic sufficiency) theory and a necessity theory is the difference in specification of the causal relationship, in other words the *causal perspective* (Dul, 2024a) that is used to describe the relationship between the concepts (i.e., the proposition).

A theory is only complete if it is accompanied by a plausible substantive explanation of why the cause affects the outcome. For a necessity proposition, the theory should explain (1) why all cases from the theoretical domain with the outcome have the condition, (2) why without the condition, the outcome does not exist, (3) why the absence of the condition is not compensable by other factors (no possibility of substitution), and (4) that the condition precedes the outcome (fulfilling the temporality requirement of causality). Details about how to formulate a formal necessity hypothesis embedded in a necessity theory are explained in Chapter 7. When applying NCA, a necessity theory that causally explains the necessity relationship is essential.

3.3 Parsimonious necessity theories

Necessity theories are inherently parsimonious: single concepts make a strong, actionable prediction: the absence of a necessary condition perfectly predicts the absence of the outcome. Starting with William of Ockham's parsimony principle[1] (also known as Occam's Razor),

[1] An important element of parsimony is that theories should be simple: "... *frustra fit per plura, quod potest fieri per pauciora*" (it is pointless to do with more what can be done with fewer) (Ockham, 1323).

many famous scholars including Newton[2], Einstein[3], Popper[4], and Simon[5] emphasized the importance of keeping theories simple, such that they are understandable, practically useful, and falsifiable.

Because necessary conditions operate in isolation from the rest of the causal structure (see Section 4.6), even the simplest necessity theory with only one or a few necessary conditions can be actionable. Figure 3.1 shows a necessity theory with two concepts: one condition X and one outcome Y. To emphasize the necessity perspective used to describe the relationship, nc (necessary cause or necessary condition) is placed above the arrow.

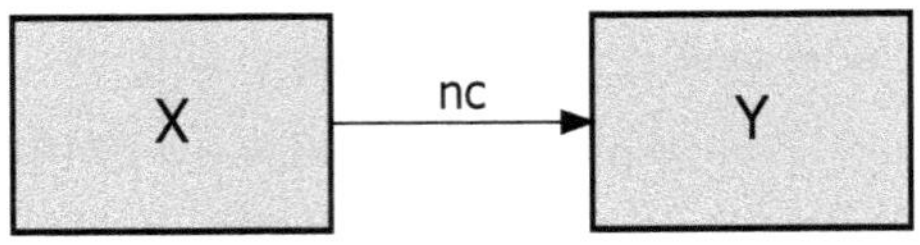

FIGURE 3.1 Conceptual model of a necessity theory with one condition and one outcome.

It is also possible that multiple concepts (e.g., X_1, X_2, X_3) are necessary for the same outcome (Y) as shown in Figure 3.2-left, or that one or more conditions are necessary for several outcomes (e.g., Y_1, Y_2, Y_3), which is shown in Figure 3.2-right. Note that conditions X_1, X_2, and X_3 are all individual necessary conditions. If any one condition is missing (below its required level) the outcome cannot occur (at its target level), while the presence of all conditions is normally not sufficient for the outcome because other contributing factors must be in place as well.

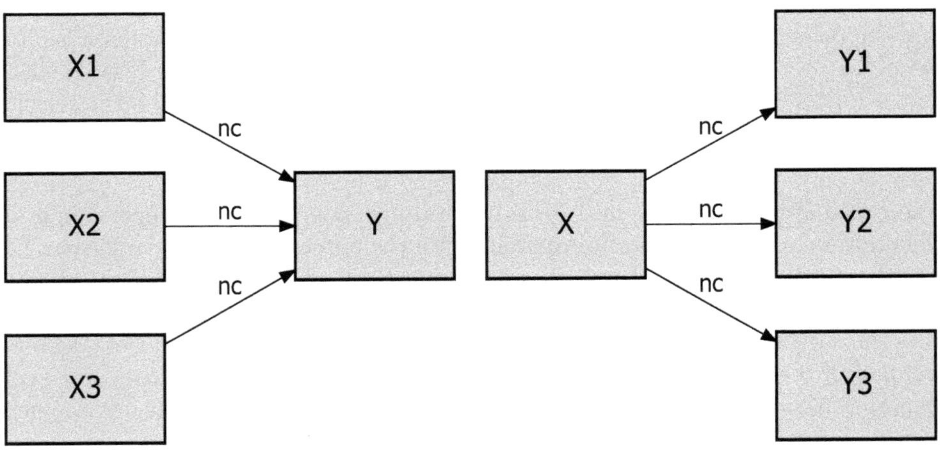

FIGURE 3.2 Conceptual models of a necessity theory with three conditions and one outcome (Left), and of a theory with one condition and three outcomes (Right).

---

[2] "*Causas rerum naturalium non plures admitti debere, quam quae & verae sint & earum phaenomenis explicandis sufficiant*" (the causes of natural things should not be multiplied beyond what is true and sufficient to explain their appearances) (Newton, 1687).

[3] "*It can scarcely be denied that the supreme goal of all theory is to make the irreducible basic elements as simple and as few as possible without having to surrender the adequate representation of a single datum of experience*" (Einstein, 1934).

[4] "*We do not seek highly probable theories, but explanations; and the best explanations are often simple explanations*" (K. Popper, 1959).

[5] "*Simplicity is not an optional accessory in human thought and action; it is a necessity*" (Simon, 1957).

Furthermore, several necessary conditions can combine into a necessity causal chain where X_1 is necessary for X_2, and X_2 is necessary for Y. This situation is shown in Figure 3.3 and is called 'chain necessity' of X_1 for Y. Chain necessity differs from 'direct necessity' as shown in Figure 3.1. In the dichotomous necessity-in-kind (NiK) situation, the necessity of X_1 for X_2 and of X_2 for Y implies the necessity of X_1 for Y. However, in the necessity-in-degree (NiD) situation it is possible that the chain necessity of X_1 for Y is absent despite the necessity between the elements of the chain (X_1 for X_2 and X_2 for Y), as discussed in Section 4.6.4.

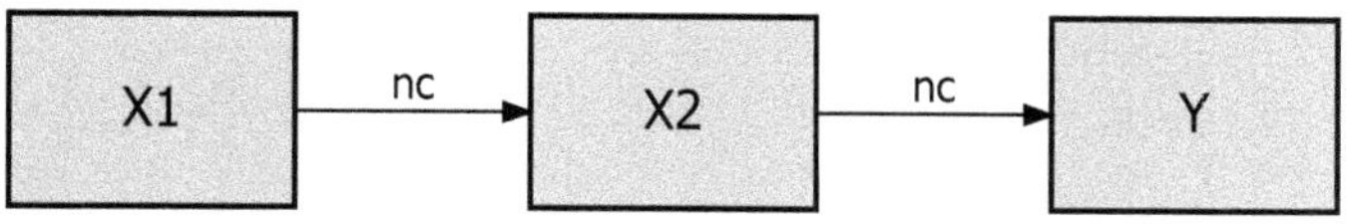

FIGURE 3.3 Conceptual model of a chain of necessary conditions. X_1 is necessary for X_2 and X_2 is necessary for Y.

Necessity models are inherently parsimonious. A single necessary condition can stop the outcome, independently of other conditions and other contributing factors. However, in sufficiency-based perspectives, a single factor is seldom enough to produce the outcome. For a probabilistic or configurational sufficiency model, a series of factors are included to model causal complexity. This applies to both probabilistic sufficiency (e.g., structural causal models) and configurational sufficiency (e.g., as in QCA). Parts of the relationships of sufficiency models could (also) be necessity relationships. For the necessity analysis these parts can be isolated (as in Figure 3.1) from the rest of the causal structure and independently be analyzed for necessity (Chapter 11).

3.4 Types of necessity theories

Necessity theories can be classified according to the extent to which necessity propositions are part of the theory. Bokrantz & Dul (2023) distinguish between *pure necessity theories*, which are theories that consist only of necessity relationships between the concepts and *embedded necessity theories*, which are theories that consist of a mix of necessity relationships and other relationships (e.g., probabilistic sufficiency or configurational sufficiency). In embedded necessity theories, some of the relationships may be necessity relationships (e.g., $X_1 \xrightarrow{nc} Y$), whereas others may be probabilistically sufficient relationships ($X_2 \xrightarrow{+} Y$). It is also possible that one condition has both causal roles in a model such that the presence of X is necessary for the presence of Y, but also that X likely increases Y ($X \xrightarrow{nc\ ,\ +} Y$).[6]

Figure 3.4 shows the four possible types of necessity theories by also considering if the necessity theory is established or emerging.

Established pure necessity theories, where all relationships in the theory are necessity relationships as in Figures 3.1, 3.2, and 3.3 in Section 3.2 are relatively scarce. One example is the early theory of Guilford (1967) that intelligence is necessary for creativity. Shortly

[6]See also Section 3.5 for various possible directions of necessity and probabilistic sufficiency relationships and their notation. It is possible that X is necessary for Y with or without being also probabilistically sufficient for Y. NCA only considers the necessity role of the concept. Assuming a double role of a concept is commonplace in theories that are analyzed both by NCA and a regression-based model (see Section 11.3).

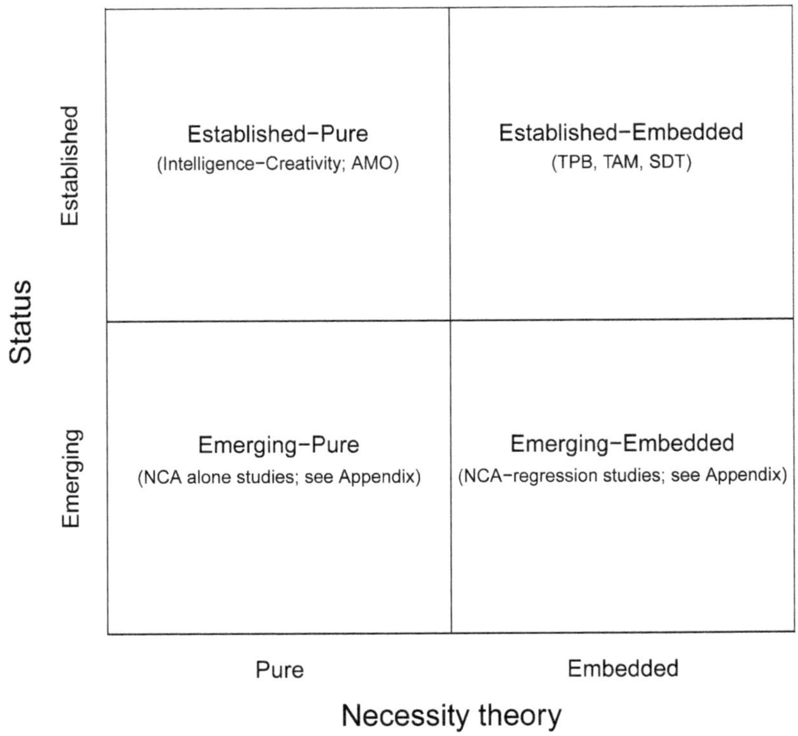

FIGURE 3.4 Typology of necessity theories with examples. AMO = Ability-Motivation-Opportunity. TPB = Theory of Planned Behavior. TAM = Technology Acceptance Model. SDT = Self-Determination Theory.

after NCA was introduced (Dul, 2016b), this theory was empirically tested with NCA and supported (Karwowski et al., 2016). Another established pure necessity theory is the original Ability, Motivation, Opportunity model (AMO) or Motivation, Opportunity, Ability (MOA) model of human behavior. Necessity lies at the core of this theory assuming that A, M, and O are single necessary conditions for behavior.[7]

Established embedded necessity theories are broadly accepted theories consisting of both necessity and probabilistic sufficiency relationships. Often, these theories are originally formulated as probabilistic sufficiency theories for all their relationships, and revisited to suggest a necessity perspective for one or more relationships. An example is Frommeyer et al. (2022)'s interpretation of one relationship of the Theory of Planned Behavior (TPB): Intention of the behavior is necessary for Performance of the behavior, while keeping the other relationships as probabilistic sufficiency.[8] Another example is the Technology Acceptance Model (TAM) that is usually interpreted and studied as a probabilistic sufficiency theory and that was recently revisited with a necessity causal perspective for all relationships while maintaining also the probabilistic sufficiency perspective (Erdmann & Toro-Dupouy, 2025; Hassan et al.,

[7] In this framework it is also assumed that the three necessary conditions are jointly sufficient. Referring to the origins of the AMO model, it has recently been analyzed as a pure necessity theory (Hauff et al., 2021; Tuuli & Rhee, 2021; Van Rhee & Dul, 2017) although AMO studies until then (re)formulated the original necessity relationships as probabilistic sufficiency relationships and used regression analyses for testing them (e.g., Siemsen et al., 2008).

[8] In another example, Eccarius & Chen (2024) added Trust to the TPB model as a necessary antecedent of the drivers of Intention.

2025; Kopplin, 2023b; Low & Ramayah, 2023; e.g., Richter, Schubring, et al., 2020; D. N. Su et al., 2023).[9] Also the Self-Determination Theory (SDT) was recently revisited from the perspective of necessity only (H. Ding & Kuvaas, 2023, 2025) or in combination with the probabilistic sufficiency perspective (Cassia & Magno, 2024). SDT suggests that humans have three basic psychological needs (motivation, competence, relatedness) that must be satisfied for optimal human development and well-being.

Many established theories exist (not shown in Figure 3.4) that *could be* revisited from a necessity perspective to develop pure or embedded necessity theories because their founders clearly employed necessity causal reasoning to formulate these theories. Examples include Porter's theory on competitive advantage of nations (Porter, 1985), Wernerfelt's resource-based view of the firm (Wernerfelt, 1984), Teece's theory on dynamic capabilities (Teece et al., 1997; Teece, 2007), Rogers' theory on person-centered psychotherapy (Rogers, 1957), Turner's framework for project management success (Turner, 2009), and Meehl's concept of 'specific etiology' in medicine (Meehl, 1962). Until recently, such theories were mostly tested using regression-based methods that implicitly assume probabilistic sufficiency, suggesting theory-method misfit (Table 11.1). With NCA available, testing necessity with a dedicated methodology becomes possible.

Since the introduction of NCA, the number of *emerging pure necessity theories* has risen. Appendix C.1 includes examples of NCA studies in which only necessity relationships are theoretically formulated and tested with NCA. After replication and further theorizing, these emerging necessity theories may become established necessity theories. An example of an *empirical* study with an emerging pure necessity theory is a study by B. Yan et al. (2023) that proposes and finds four necessary conditions for a country's high level of early COVID-19 mortality rate: high levels of a delayed first response, political decentralization, elderly populations, and urbanization. An example of a *theoretical* study with an emerging pure necessity theory is the study by Andrevski & Miller (2022) about defining the concept 'strategic forbearance'. When firms are attacked by competitors, forbearance is defined by three necessary conditions: the presence of awareness of the attack, the presence of capabilities to react, and the absence of a motivation to react.

Many studies formulate *emerging embedded necessity theories.* Such studies often start with a probabilistic sufficiency theory based on earlier empirical regression-based studies and then suggest that one or more of the relationships could (also) be formulated as necessity relationships. In an *empirical* study W. Lee & Jeong (2021) propose an emerging embedded necessity theory in which two dimensions of tourist happiness experience (Hedonic enjoyment and Personal expressiveness) are related to Satisfaction (customer's judgment about their fulfillment with a product or service) and Place attachment (bonding between individuals and places). This is shown in the conceptual model of Figure 3.5. By conducting both NCA and structural equation modeling (SEM), they conclude that Hedonic enjoyment (pleasure, comfort) has both a necessity relationship and a probabilistic sufficiency relationship with Satisfaction, but only a necessity relationship with Place attachment. Furthermore, they conclude that Personal expressiveness (a subjective experience of self-realization that one is acting in such a way that one is truly being oneself) has no probabilistic sufficiency relationship with Satisfaction, and both a necessity relationship and a probabilistic sufficiency relationship with Place attachment. Other examples of emerging embedded necessity theories may be found in empirical studies that apply NCA in combination with regression-based methods (Tables C.2, C.3, and C.4 in Appendix C). Liehr & Hauff (2022) propose in a *theoretical* study an emerging embedded necessity theory about leadership competencies for

[9] These are examples of 'causal pluralism' studies where both perspectives are studied at the same time, see Chapter 11.

employee innovative behavior. They review a large number of competencies that contribute to innovative behavior and conclude that three competencies (according to the literature) have a probabilistic sufficiency relationship with innovative behavior but are not necessary (rewards, sharing expertise, resource management) and four competencies are (also) necessary: providing support, communicating a vision, granting autonomy and discretion, and providing feedback.

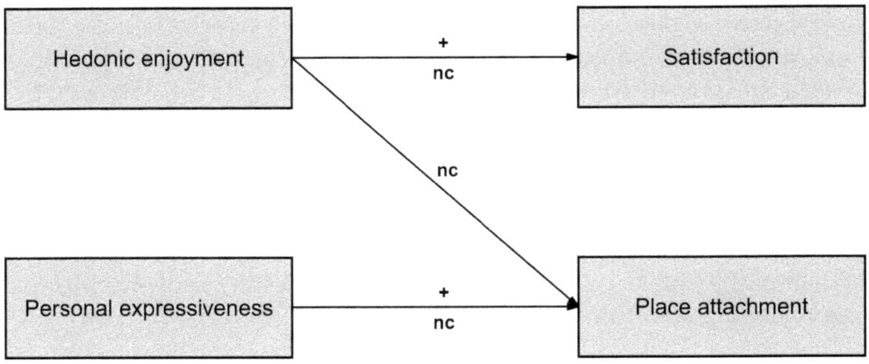

FIGURE 3.5 Example of an emerging embedded necessity theory for the effect of two dimensions of happiness experiences (Hedonic enjoyment and Personal expressiveness) on two outcomes (Satisfaction and Place attachment). The symbol + represents a positive probabilistic relationship and the symbol nc represents a necessity relationship with high-high direction (After W. Lee & Jeong, 2021).

3.5 Direction of a necessity relationship

In Figure 3.5 the direction of the probabilistic sufficiency relationship is given by a + symbol above the arrow. This convention indicates a positive effect: X increases Y or X has a positive effect on Y or more precisely, that a higher value of X increases the probability of a higher value of Y. Similarly, a – symbol indicates a negative effect: X decreases or has a negative effect on Y; a higher value of X decreases the probability of a higher value of Y.

Such single symbols do not convey meaning for a necessity relationship. In a necessity proposition the direction is not described by a verb as in X *increases* Y or X *has* a positive effect on Y. For a necessity proposition a noun or adverb (or a combination) is used to describe relationships between levels of X and Y. For example, in the dichotomous situation *absence* of X is necessary for *presence* of Y or in the continuous situation *low* level of X is necessary for *high* level of Y. For a necessity relationship four possible directions exist (Figure 3.6).

First, for a necessary condition with a *high-high* direction, the presence/high value of X is necessary for presence/high value of Y. This is symbolized as + nc + (Figure 3.6). For example, high level of Urbanization is necessary for a high level of Early COVID-19 mortality (B. Yan et al., 2023).

Second, for a necessary condition with a *low-high* direction, the absence/low value of X is necessary for the presence/high value of Y (- nc +). For example, low Intensity of production is necessary for high Environmental sustainability in agriculture (Lankoski & Lankoski, 2023).

Third, for a necessary condition with a *high-low* direction, the presence/high value of X is necessary for the absence/low value of Y (+ nc -). For example, the presence/high value of social support is necessary for the absence/low value of stress.

Fourth, for a necessary condition with a *low-low* direction, the absence/low value of X is necessary for the absence/low value of Y (- nc -). For example, the absence of rain is necessary for the absence of a wet surface.

The third and fourth formulations of a necessary condition are less common than the first and second. Often the absence/low value of the outcome is redefined as the presence/high level of the opposite of the outcome. For example, the presence/high value of social support is necessary for the presence/high value of relaxation or the absence of rain is necessary for the presence of a dry surface. Thus, the direction of the necessity relationship depends on the definition of the concepts that are part of the proposition.

In the absence of a specified direction, the implied direction is + nc +, as shown in Figure 3.5. Throughout this book, the default assumption is the + nc + direction, unless stated otherwise.

Figure 3.6 shows the four possible conceptual models depending on the direction of the necessity relationship.

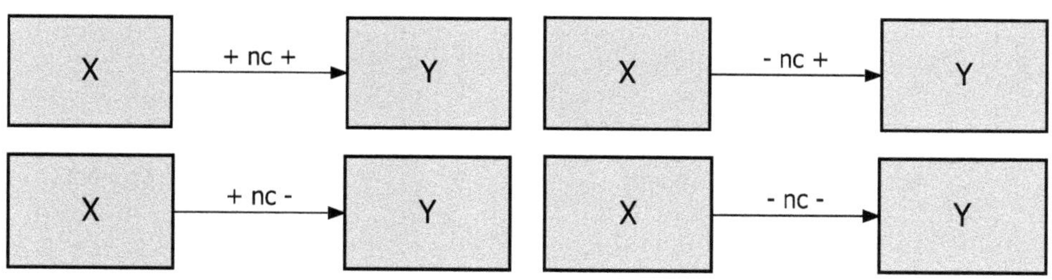

FIGURE 3.6 Conceptual models for necessary conditions. Top-Left: high-high = presence/high value of X is necessary for presence/high value of Y. Top-Right: low-high = absence/low value of X is necessary for presence/high value of Y. Bottom-Left: high-low = presence/high value of X is necessary for absence/low value of Y. Bottom-Right: low-low = absence/low value of X is necessary for absence/low value of Y.

3.6 Expected empty corner

Depending on the direction of necessity, different corners in the XY-table or XY-plot are expected to be empty. For the *dichotomous situation*, the XY-table of Figure 2.1-right illustrates necessity causality when the presence of X is necessary for the presence of Y. Corner 1 cannot have observations (points, cases) whereas the other corners may have observations. This is also shown in Figure 3.7-top-left. In Figure 3.7 rows represent the two values of the outcome Y (Absent/Present), and columns the two values of the condition X.

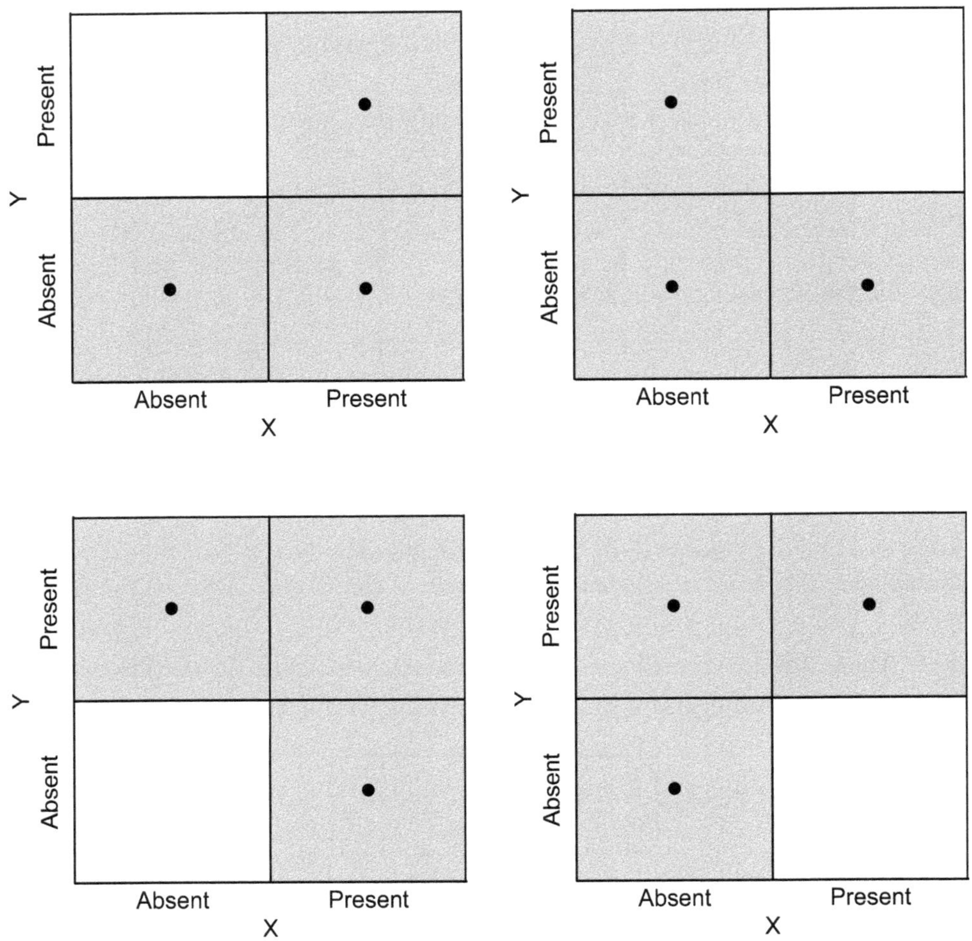

FIGURE 3.7 XY-tables showing empty corners without observations depending on the direction of the necessity relationship. Top-Left: high-high = presence of X is necessary for presence of Y. Top-Right: low-high = absence of X is necessary for presence of Y. Bottom-Left: high-low = presence of X is necessary for absence of Y. Bottom-Right: low-low = absence of X is necessary for absence of Y.

The condition X increases to the right, and the outcome Y increases upward. Each corner in the table represents a certain combination of X and Y. For example, the upper-left corner represents absence of X and presence of Y. A gray corner with a black dot indicates that the corner may contain observations. A white corner without dots indicates that no observations are possible.

When the presence of X is necessary for the presence of Y, the upper-left corner is empty. This corresponds to the necessity statement 'if not X, then not Y' (see the necessity statement of expression (2.6) in Chapter 2). The outcome is absent when the condition is absent. This means that the upper-left corner does not have observations, and that the lower-left corner has observations. Similarly, when the outcome is present, the condition must also be present: the upper-right corner has observations. The content of the lower-right corner is irrelevant for + nc + necessity: this corner may or may not have observations when the necessary condition applies. For + nc + necessity, the emptiness of the upper-left corner is essential.

For other directions of the necessity relationship, other corners in the XY-table remain empty when necessity applies. When − nc + necessity applies, the upper-right corner remains empty. Similarly, for + nc − necessity, the lower-left corner is empty, and for − nc − necessity, the lower-right corner stays unoccupied. Therefore, the empty corner depends on how the necessary condition is formulated in the theory (and how the direction of the axes and the concepts are defined). Each formulation of the necessity proposition results in a specific corner of the XY-table being empty, referred to as the *expected empty corner.*

For the continuous situation, Figure 3.8 shows the four directions and corresponding empty corners. Here an XY-plot instead of an XY-table is used as a visualization method for representing that a low/high value of the condition is necessary for a low/high value of the outcome. The condition X increases to the right, and the outcome Y increases upward. Each corner in the plot represents a certain combination of low/high X- and Y-values. For example, the upper-left corner represents low level of X and high level of Y. A gray area indicates that the area may contain observations.

3.7 Double necessity

It is possible that a single factor has two necessity relationships with the outcome at the same time, implying that more than one corner is expected to be empty. When the concepts are continuous, it is possible that two *adjacent* corners are empty. This allows theorizing that an *optimum* (not low, not high) level of X is necessary for a *high* level of Y, implying that the upper-left *and* the upper-right corners are expected to be empty (Figure 3.9-left). For example, in team performance, moderate levels of conflict (X) may be necessary for high team innovation performance (Y). Low conflict leads to group think, which may limit performance; high conflict leads to dysfunction, which also limits team innovation performance. Therefore, a moderate level of conflict enables teams to challenge each other productively, providing the possibility for high innovation performance.

Similarly, it may be theorized that an *optimum* (not low, not high) level of X is necessary for a *low* level of Y, implying that the lower-left and lower-right corners are expected to be empty (Figure 3.9-right). For example, in stress management, moderate levels of workload (X) may be necessary for low employee burnout (Y). Low workload can result in boredom and lack of purpose, which may limit the possibility of low burnout; high workload can cause stress and exhaustion, which also limits the possibility of low burnout. Therefore, a moderate workload is necessary for low burnout because both too low and too high workload create psychological strain that prevents burnout from remaining low.

It is also possible to theorize that a *high* level of X is necessary for an *extreme* (low or high) level of Y. Then the upper-left and the lower-left corners are expected to be empty (Figure 3.10-left). Similarly, it may be theorized that a *low* level of X is necessary for an *extreme* (low or high) level of Y. Then the upper-right and the lower-right corners are expected to be empty (Figure 3.10-right).

The situations shown in Figures 3.9 and 3.10 do not have dichotomous equivalents, as this would imply that the outcome or the condition is always absent

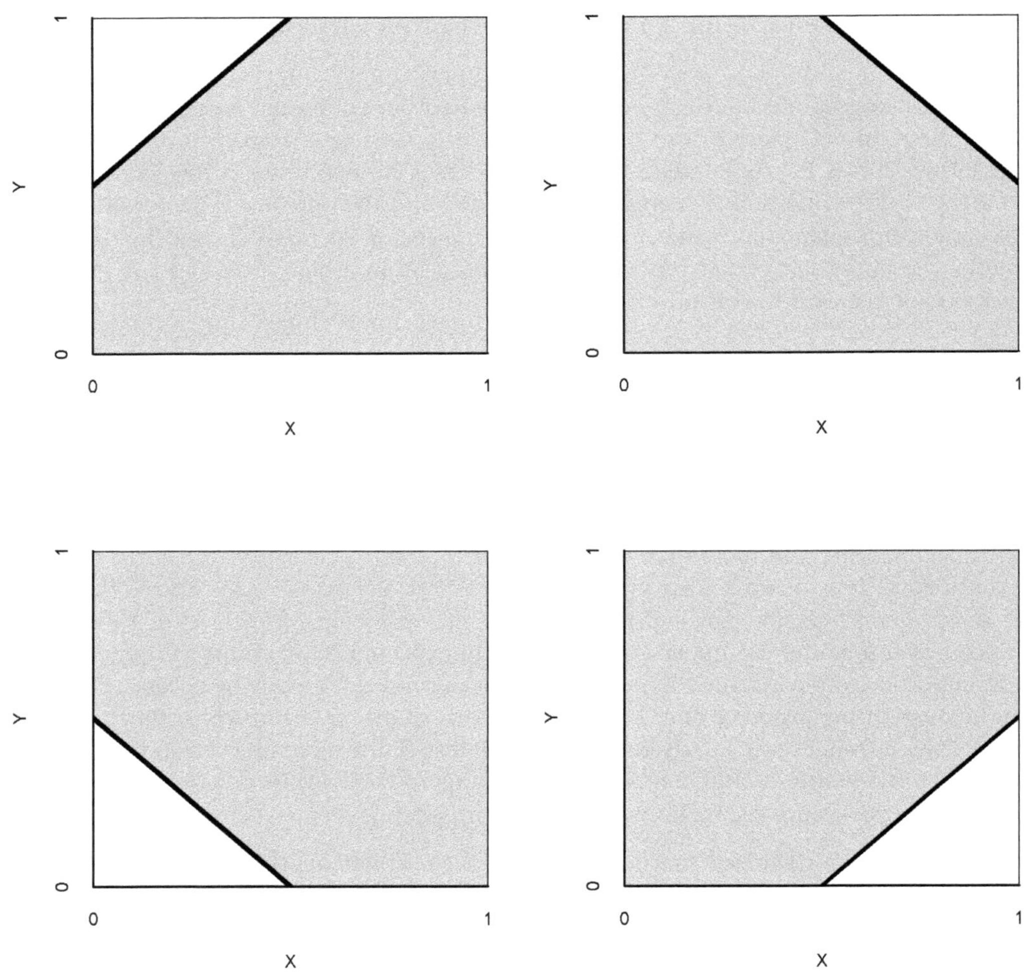

FIGURE 3.8 XY-plots showing the direction of a necessity relationship. Top-Left: a high value of X is necessary for a high value of Y (+ nc +, high-high). Top-Right: a low value of X is necessary for a high value of Y (- nc +, low-high). Bottom-Left: a high value of X is necessary for a low value of Y (+ nc -, high-low). Bottom-Right: a low value of X is necessary for a low value of Y (- nc -, low-low).

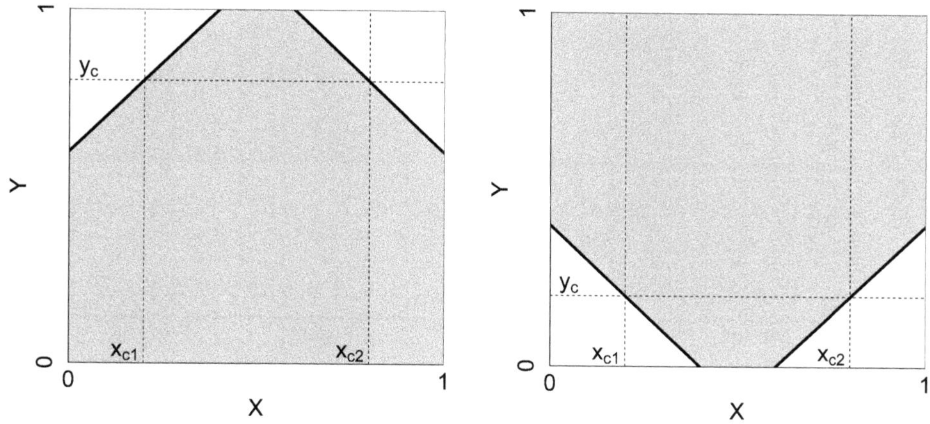

FIGURE 3.9 *XY*-plot showing an optimum level of the necessary condition. Left: An optimum level of $x_{c1} < x < x_{c2}$ is necessary for a high level of $y = y_c$. Right: An optimum level of $x_{c1} < x < x_{c2}$ is necessary for a low level of $y = y_c$.

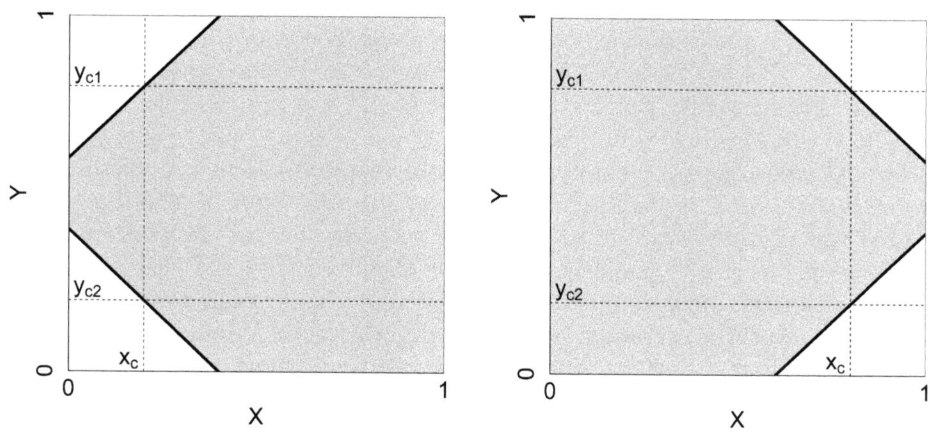

FIGURE 3.10 *XY*-plot showing necessity for an extreme level of the outcome. Left: A high level of $x > x_c$ is necessary for an extreme (low or high) level of $y = y_{c1}$ or $y = y_{c2}$. Right: A low level of $x < x_c$ is necessary for an extreme (low or high) level of $y = y_{c1}$ or $y = y_{c2}$.

(Figure 3.11-top-left) or always present (Figure 3.11-top-right), or that the condition is always present (Figure 3.11-bottom-left), or always absent (Figure 3.11-bottom-right).

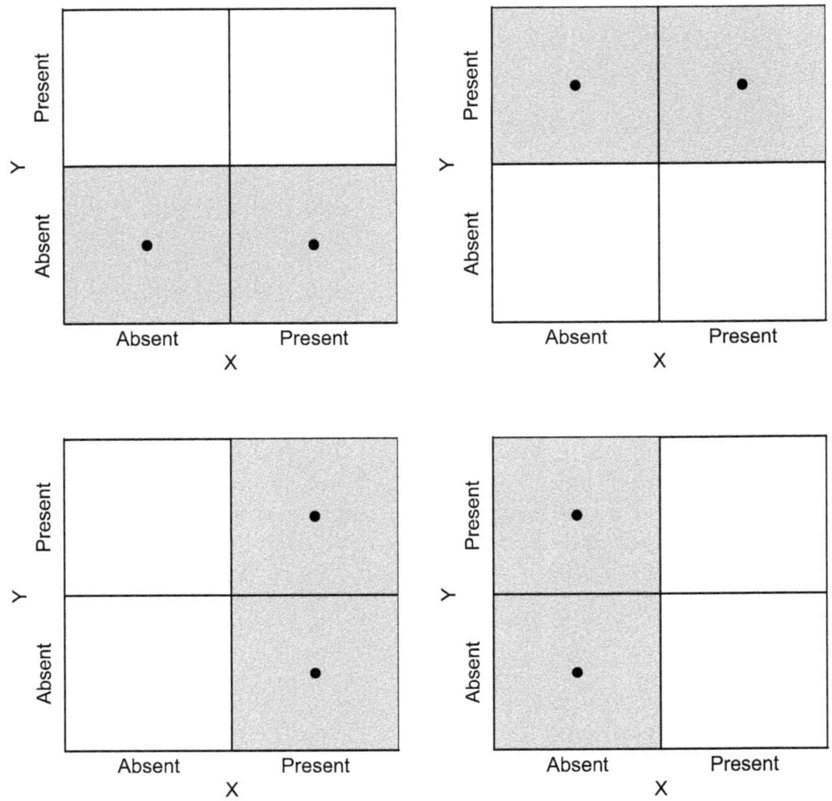

FIGURE 3.11 XY-tables showing two adjacent empty corners without observations. Top-Left: presence of Y is not possible. Top-Right: absence of Y is not possible. Bottom-Left: absence of X is not possible. Bottom-Right: presence of X is not possible.

When two *opposite* corners are empty, as shown in Figure 3.12-left, two claims apply: the *presence/high level* of X is necessary for the *presence/high level* of Y (empty upper-left corner) and that the *absence/low level* of X is necessary for the *absence/low level* of Y (empty lower-right corner). Similarly, it may be theorized that the *absence/low level* of X is necessary for the *presence/high level* of Y (empty upper-right corner) and that the *presence/high level* of X is necessary for the *absence/low level* of Y (empty lower-left corner). This is shown in Figure 3.12-right. This situation has dichotomous equivalents, as shown in Figure 3.13.

3.8 The meaning of 'necessary but not sufficient'

To understand what necessary *but not* sufficient means, this section first explains what is meant by necessary *and* sufficient. In the dichotomous situation, 'necessary *and* sufficient' means that the upper-left corner and the lower-right corner of the XY-plot are both empty.

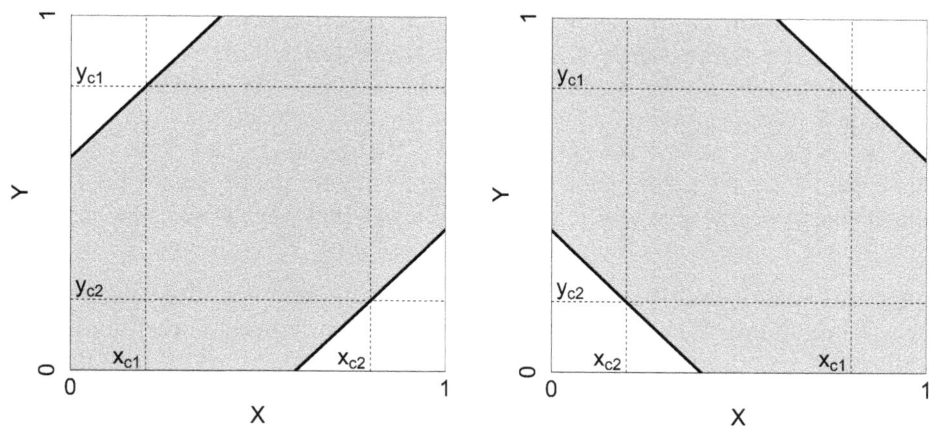

FIGURE 3.12 Double necessity. Left: A high level of $x > x_{c1}$ is necessary for a high level of $y = y_{c1}$, and a low level of $x < x_{c2}$ is necessary for a low level of $y = y_{c2}$. Right: A low level of $x < x_{c1}$ is necessary for a high level of $y = y_{c1}$, and a high level of $x > x_{c2}$ is necessary for a low level of $y = y_{c2}$.

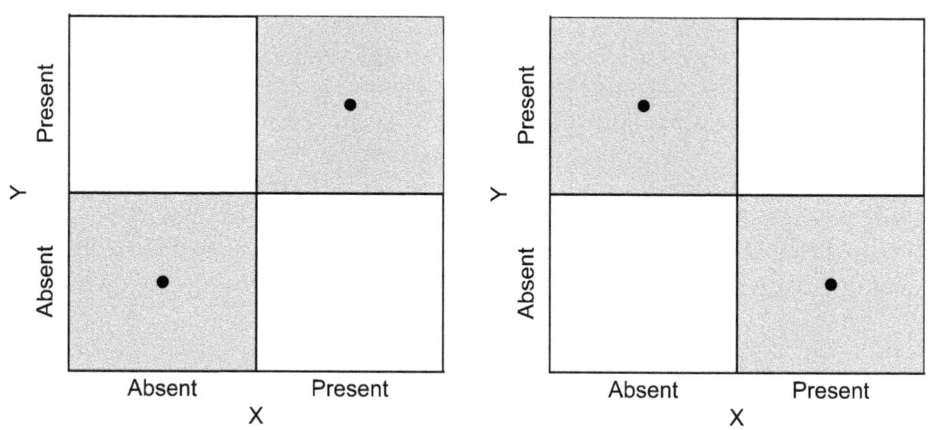

FIGURE 3.13 XY-tables showing two opposite empty corners without observations. Left: the presence of X is necessary for the presence of Y and the absence of X is necessary for the absence of Y. Right: the absence of X is necessary for the presence of Y and the presence of X is necessary for the absence of Y.

This is shown in Figure 3.13, which is interpreted as a *double necessity* relationship of X for Y: a necessity relationship where the presence of X is necessary for the presence of Y (upper-left corner empty) and a necessity relationship where the absence of X is necessary for the absence of Y (lower-right corner empty). By logic, in the dichotomous situation the second relationship can be reformulated as a *sufficiency* relationship where the presence of X is sufficient for the presence of Y, as shown in Figure 2.1-left. The first relationship (necessity) and the second relationship (sufficiency) now have the same X-value (presence) and the same Y-value (presence). This allows a single statement that the presence of X is necessary *and* sufficient for the presence of Y. If the sufficiency relationship does not hold, the formulation of the combined relationship would be: the presence of X is necessary *but not* sufficient for the presence of Y. This means that the lower-right corner is not empty.

In discrete and continuous situations, necessity *and* sufficiency are more complex. Like in the dichotomous situation, the upper-left corner and the lower-right corner are both empty. This is shown in Figure 3.12-left, which is interpreted as a double necessity relationship of X for Y: a necessity relationship where a high level of X is necessary for a high level of Y (upper-left corner empty) and a necessity relationship where a low level of X is necessary for a low level of Y (lower-right corner empty). By logic, the second relationship can again be reformulated as a sufficiency relationship: a not-low level of X is sufficient for a not-low level of Y. However, a 'not-low' level is usually not the same as a 'high' level. For example, in Figure 3.12-left, level $X = x_{c1}$ is necessary for level $Y = y_{c1}$, whereas level $X = x_{c2}$ is sufficient for level $Y = y_{c2}$. There are no common X-levels and common Y-levels that would allow a single statement that level $X = x_c$ is necessary *and* sufficient for level $Y = y_c$, unless the upper ceiling line and the lower ceiling line (floor line) overlap or intersect at the point (1,1). If the lines overlap there are only observations on the ceiling line and the necessity *and* sufficiency statement holds for all observations. If the lines intersect at point [1,1], only at that point the necessity *and* sufficiency statement holds. Both situations are, however, exceptional.

In all other situations the statement that level $X = x_c$ is necessary *but not* sufficient for level $Y = y_c$ has a broader meaning than that the lower-right corner is not empty. If the lower-right corner can have observations, observations are normally possible in the entire area below the ceiling line, not just in the lower-right corner. This allows a single statement that $X = x_c$ is necessary *but not* sufficient for $Y = y_c$. When $X = x_c$, no observations are above point C on the ceiling line, whereas observations below point C_1 are possible. In other words, for a given level $X = x_c$, it is not possible to have values of $Y > y_c$ (necessity applies) but it is possible to have values $Y \leq y_c$ ("non-sufficiency" applies). Consequently, where in the dichotomous situation the statement X is necessary *but not* sufficient for Y means that observations are possible in the lower-right corner, in the discrete and continuous situations it means that there are observations below the ceiling line. NCA focuses on the identification of necessity (rather than on the identification of non-sufficiency) in any corner of the XY-table or the XY-plot. If the focus is on the lower-right corner, the emptiness of this corner is interpreted as indicating that the absence or a low level of X is necessary for the absence or a low level of Y.

4

Mathematics

4.1 Summary of this chapter

This chapter describes the mathematics of a necessity relationship between a pair of variables: X is necessary for Y. The chapter focuses on necessity-in-degree (NiD) that describes the general situation of two continuous variables with an empty space in the upper-left corner of the XY-plot. Using a conventional Euclidean coordinate system where the X-axis runs horizontally with increasing values from left to right, and the Y-axis runs vertically with increasing values from bottom to top, this indicates that a high level of X is necessary for a high level of Y. The mathematical approach described here can be extended to situations with dichotomous or discrete variables, or to analyzing other corners, though this is not addressed in this chapter. The chapter first presents the branches of mathematics applied in NCA (Section 4.2). In the subsequent section NCA's mathematical model specification is presented (Section 4.3). It includes the specification of the bounding box, the ceiling line, and the model's degrees of freedom. This is followed by a discussion on how to estimate a necessity model by estimating parameters related to the necessity effect size and model fit. Next, the situation of multiple necessary conditions is discussed, explaining why NCA considers multiple X_jY-planes (multiple projections), rather than one multidimensional space, and explaining how a chain of necessity can be analyzed (Section 4.6).

4.2 Branches of mathematics

Rather than relying on statistics and probability theory, NCA employs two branches of mathematics: geometry and algebra. Geometry describes relationships between objects and spaces including points, lines, angles, and planes. NCA focuses on points in the two-dimensional space (XY-plane), describes relationships between points (e.g., ceiling line) and considers the size of an area (e.g., effect size). Algebra uses symbols, letters, and numbers and describes their relationships and operations including variables, constants, functions, equations, and inequalities. NCA uses functions to describe the ceiling line with constants and variables, uses equations to calculate effect size, and uses inequalities to specify (in)feasible areas. When describing NCA, geometry and algebra are intertwined. NCA is not a statistical data analysis method per se, although it uses tools from statistics (Chapter 5).

DOI: 10.1201/9781003430766-4

4.3 Model specification

In NCA, variables X and Y can be dichotomous (two possible levels), discrete (more than two possible levels) or continuous variables (infinite number of possible levels). With numeric variables (including ratio- and interval-scale variables), effect size and other necessity parameters can be calculated. With categorical variables, these calculations are not possible unless equal distances between ordered categories are assumed (e.g., "low", "medium", "high"). NCA can handle both non-random and random variables. When NCA uses random variables for X and Y, NCA only assumes that the univariate distributions are bounded, but no assumption is needed about the shape of the distribution.[1]

NCA's model specification consists of specifying the bounding box and the ceiling line.

4.3.1 Bounding box

NCA assumes that the condition and the outcome are bounded. There are two reasons for this. First, variables represent properties of phenomena. Many properties are finite, represented by minimum and maximum values of the variables, rather than being infinite (e.g., height of people). Second, the assumption that X and Y are bounded allows the definition and calculation of the necessity effect size (Section 4.4.1) and other NCA parameters (Sections 4.4.2 and 4.5). Assuming bounds on (X, Y) means that there are four constants that define the ranges of the condition X and the outcome Y:

$$x_{\text{min}} \leq x \leq x_{\text{max}} \quad \text{and} \quad y_{\text{min}} \leq y \leq y_{\text{max}} \tag{4.1}$$

These bounds define the bounding box in an XY-plot. The *bounding box* is defined by the minimum and maximum values of X and Y, that is $[x_{\text{min}}, x_{\text{max}}] \times [y_{\text{min}}, y_{\text{max}}]$. The feasible and empty areas, as well as the ceiling line and its extensions, are defined relative to this bounding box.

The *scope* (S) is the area of the bounding box and is mathematically expressed as:

$$S = (x_{\text{max}} - x_{\text{min}}) \cdot (y_{\text{max}} - y_{\text{min}}) \tag{4.2}$$

The bounding box can be constructed in two ways: theoretically and empirically. The *theoretical scope* is defined by theoretical bounds of X and Y when the theoretical bounds are known or presumed.[2] The *empirical scope* is defined by the observed minimum and maximum values of X and Y. This option is often selected when theoretical bounds are unknown.

[1]Also, NCA's null hypothesis tests for estimation of the p-value do not make assumptions about distributions (non-parametric tests, see Section 5.3). However, when statistical simulations are done with NCA such as in Monte Carlo simulations for conducting NCA's power analysis, the analyst must assume variable distributions including the shape. Any bounded distribution could be selected, for example uniform or truncated normal (Chapters 5 and 6).

[2]A theoretical scope may apply, for example, when points are obtained from scales with meaningful minimum and maximum values, such as percentage scales, or Likert scales.

Without loss of generality (for the purpose of NCA), the variables can be linearly transformed to the unit box $[0, 1] \times [0, 1]$ to obtain a scope value of 1 (Section 8.5.2).[3] Then a point (x, y) gets transformed by min-max normalization into (x', y') as follows:

$$x' = \frac{x - x_{\text{min}}}{x_{\text{max}} - x_{\text{min}}} \quad \text{and} \quad y' = \frac{y - y_{\text{min}}}{y_{\text{max}} - y_{\text{min}}} \tag{4.3}$$

4.3.2 Ceiling line

When necessity exists, the bounding box has an area where points are possible (*feasible area*) and an area where points are impossible (*empty space* or *ceiling zone*). The ceiling line[4] is a line within the bounding box that separates these two areas. Specifying the ceiling line involves choosing its *location* (in which corner) and its *form*.

The *location* depends on the direction of the hypothesis (Section 3.5), which defines the *expected empty corner*. For a high-high hypothesis (presence/high value of X is necessary for presence/high value of Y), the expected empty corner is the upper-left corner of the XY-plot (corner 1).[5]

To specify the *form* of the ceiling line, several options exist. Figure 4.1 shows three classes of non-decreasing ceiling lines in a bounding box. The first class is the Ceiling Envelopment Free Disposal Hull (CE-FDH), which is a stepwise linear ceiling line consisting of several horizontal and vertical line segments (Figure 4.1-top-left), the second class is Ceiling Envelopment Variable Returns to Scale (CE-VRS), which is a concave piecewise linear ceiling consisting of several oblique line segments (Figure 4.1-top-right), and the final one is a linear ceiling line consisting of one segment (Figure 4.1-bottom). The ceiling line represents an upper bound on the outcome in the feasible area (the gray area). The ceiling line is considered a part of the feasible area.

A ceiling line is assumed to be *non-decreasing* for two reasons.[6] First, it allows the expression that *at least* a certain level of X is necessary for a certain level of Y. Second, a non-decreasing ceiling line has the *'no-reversal'* (no change of direction) property. This means that if a point (x, y) is in the ceiling zone, then any point (x', y') to the left or above it (a point with $x' \leq x$ and $y' \geq y$) and within the bounding box is also in the ceiling zone. Similarly, if a point (x, y) is in the feasible area, then any point (x', y') in the bounding box with $x' \geq x$ and $y' \leq y$ is also in the feasible area. This property is used for the mathematical expression of the ceiling line (see below) and for two measures of imperfect necessity, namely purity and solidity (Sections 4.5.4 and 4.5.5, respectively).

4.3.2.1 Stepwise linear ceiling line

NCA uses the stepwise linear ceiling line (CE-FDH) as the base ceiling model. It refers to the Free Disposal Hull (FDH) technique (Deprins et al., 1984) that originates in

---

[3]A proof of the acceptability of an affine transformation including linear transformation is presented in Appendix D.

[4]Although in mathematics the term line is usually reserved for a straight line and curve for a non-straight one, line is used here in the broader, graphical sense to refer to both. Specific shapes of ceiling lines are distinguished with adjectives, such as linear, stepwise linear, or concave piecewise linear.

[5]For a low-high hypothesis (the absence/low value of X is necessary for the presence/high value of Y) the expected empty corner is the upper-right corner (corner 2), etc.

[6]When a ceiling line has increasing and decreasing parts as for a parabolic ceiling, multiple analyses can be done for each increasing and decreasing part separately. For the increasing part, the empty area is in the upper-left corner, whereas it is in the upper-right corner for the decreasing part. With a parabola-shaped ceiling, an optimum (rather than just low or high) level of X is necessary for a high level of Y (Section 3.7).

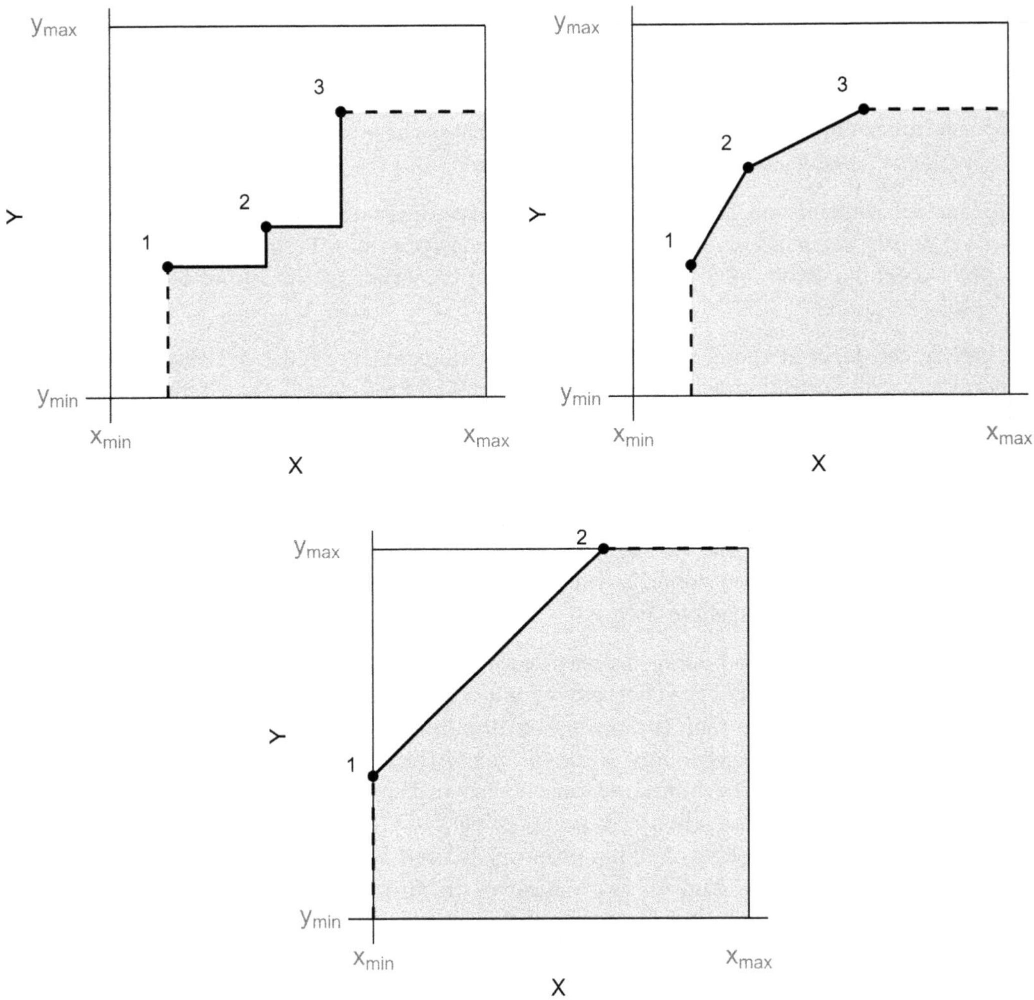

FIGURE 4.1 Three classes of ceiling lines within a bounding box for high-high necessity. Top-Left: stepwise linear. Top-Right: concave piecewise linear. Bottom: linear. White: ceiling zone. Gray: feasible area. Numbered points are peers (dominant point with no cases above and to the left of it) connected by solid line segments constructing the ceiling line. The dashed lines are extensions such that the ceiling line envelopes the feasible area.

production economics. This application of FDH identifies the efficiency of decision-making units (DMUs) with a certain production input (X) and output (Y). The technique assumes the free disposability property meaning that if a DMU can produce a certain output, it can also reduce the amount of output without using additional input, and use more input without increasing output. FDH identifies the most efficient DMUs (points with maximum output for a given input and minimum input for a given output). The outer boundary of the hull ('envelope' or 'frontier') connects the most efficient points (called 'peers'). Points below the frontier are considered inefficient.

NCA uses the FDH technique in 2D. In NCA's context, the hull is called the feasible area, and the disposability property is called the no-reversal property. The feasible area is the smallest area defined by *peers*: a strictly increasing set of points, $(x_i, y_i)_{i=1,\dots,P}$ for which

$$x_i < x_{i+1} \quad \text{and} \quad y_i < y_{i+1} \quad \text{for} \quad i = 1, \dots, P-1 \tag{4.4}$$

Where P is the number of peers. Peers are points that dominate other points. When the upper-left corner is the expected empty corner (as in Figure 4.1), no points exist to the left and above the peers.

A class of ceiling lines is defined by the way that peers are defined and connected. The CE-FDH-class of ceiling lines consists of all models with a feasible area generated by a set of P peers connected by horizontal and vertical line segments (Figure 4.1-top-left).

Let the peers be $(x_i, y_i)_{i=1,\dots,P}$. The feasible space generated by these peers is the hypograph[7] of the ceiling function $f(x)$, defined for $x \in [x_1, x_{\max}]$ by

$$f_{\text{CE-FDH}}(x) = \begin{cases} y_i & \text{if } x_i \le x < x_{i+1} \quad (i = 1, \dots, P-1), \\ y_P & \text{if } x_P \le x \le x_{\max} \end{cases} \tag{4.5}$$

and $f_{\text{CE-FDH}}(x)$ is not defined for $x < x_1$.

The generalized inverse is:

$$\begin{aligned} f^{-}_{\text{CE-FDH}}(y) &= \inf\{x \mid f_{\text{CE-FDH}}(x) \ge y\} \\ &= \begin{cases} x_1 & \text{if } y \le y_1, \\ x_{i+1} & \text{if } y_i < y \le y_{i+1} \quad (i = 1, \dots, P-1) \end{cases} \end{aligned} \tag{4.6}$$

and $f^{-}_{\text{CE-FDH}}(y)$ is undefined (infeasible) if $y > y_P$.

The associated feasible space is generated by all points on the line segments from (x_i, y_i) to (x_{i+1}, y_{i+1}) for $i = 1, \dots, P-1$.

The ceiling line is given by the sequence of line segments from $(x_1, y_{\min})$ to (x_1, y_1), to (x_2, y_1), to (x_2, y_2), and so on, ending at (x_P, y_P) and $(x_{\max}, y_P)$. Possibly, the first segment from $(x_1, y_{\min})$ to (x_1, y_1) and/or the last segment from (x_P, y_P) to $(x_{\max}, y_P)$ has length equal to zero (starts and ends at the bounding box). When the first peer starts at $x = x_{min}$ and the last peer ends at $y = y_{max}$ the bounding box is called a *tight* bounding box.

[7] A hypograph is the set of all points that lie on or below the graph of a function.

4.3.2.2 Concave piecewise linear (CE-VRS) ceiling line

The CE-VRS-class of ceiling line consists of all models with a set of P peers such that the sequence of peers forms a concave sequence, i.e., the slopes between consecutive peers strictly decrease:

$$\frac{y_{i+1}-y_i}{x_{i+1}-x_i} > \frac{y_{i+2}-y_{i+1}}{x_{i+2}-x_{i+1}} \quad \text{for} \quad i = 1, \ldots, P-2 \tag{4.7}$$

The ceiling function is given by linear interpolation between consecutive peers:

$$f_{\mathrm{CE-VRS}}(x) = y_{i-1} + \frac{y_i - y_{i-1}}{x_i - x_{i-1}}(x - x_{i-1}) \quad \text{if} \quad x_{i-1} \leq x \leq x_i \quad (i = 2, \ldots, P) \tag{4.8}$$

and $f_{\mathrm{CE-VRS}}(x) = y_P$ if $x \geq x_P$, while $f_{\mathrm{CE-VRS}}(x)$ is not defined for $x < x_1$.

The generalized inverse is:

$$\begin{aligned} f^{-}_{\mathrm{CE-VRS}}(y) &= \inf\{x \mid f_{\mathrm{CE-VRS}}(x) \geq y\} \\ &= \begin{cases} x_1 & \text{if } y \leq y_1, \\ x_i + \dfrac{x_{i+1}-x_i}{y_{i+1}-y_i}(y - y_i) & \text{if } y_i < y \leq y_{i+1} \\ & \quad (i = 1, \ldots, P-1) \end{cases} \end{aligned} \tag{4.9}$$

and $f^{-}_{\mathrm{CE-VRS}}(y)$ is undefined (infeasible) if $y > y_P$.

The associated feasible space is generated by all points on the line segments from (x_i, y_i) to (x_{i+1}, y_{i+1}) for $i = 1, \ldots, P-1$.

The ceiling line follows the linear segments from $(x_1, y_{\min})$ to (x_1, y_1), then to (x_2, y_2), and so on, ending at (x_P, y_P) and $(x_{\max}, y_P)$. Possibly, the first and/or last segment has length equal to zero.

4.3.2.3 Linear ceiling line

The class of linear ceiling lines consists of models with a ceiling that is a (positive-length) segment of a line. The ceiling function is

$$f_{\mathrm{LIN}}(x) = \max\{y_{\min}, \min\{y_{\max}, a + bx\}\} \tag{4.10}$$

with $b > 0$ to ensure the line is strictly increasing and with $(a+bx_{\min}, a+bx_{\max}] \cap [y_{\min}, y_{\max}]$ to ensure the line intersects the bounding box with a segment of positive length. This is equivalent to

$$a + bx_{\max} > y_{\min} \quad \text{and} \quad a + bx_{\min} < y_{\max} \tag{4.11}$$

which in turn is equivalent to

$$y_{\min} - bx_{\max} < a < y_{\max} - bx_{\min}. \tag{4.12}$$

The line $y = a + bx$ intersects the boundary of the bounding box in two points (x_1, y_1) and (x_2, y_2) with

$$(x_1, y_1) = (\max\{x_{\text{min}}, (y_{\text{min}} - a)/b\},\ \max\{y_{\text{min}}, a + bx_{\text{min}}\}) \tag{4.13}$$

and

$$(x_2, y_2) = (\min\{x_{\text{max}}, (y_{\text{max}} - a)/b\},\ \min\{y_{\text{max}}, a + bx_{\text{max}}\}). \tag{4.14}$$

The set of two points (x_1, y_1) and (x_2, y_2) forms a set of peers. In terms of these peers, the ceiling can be written as

$$f_{\text{LIN}}(x) = \begin{cases} y_1 + \dfrac{y_2 - y_1}{x_2 - x_1}(x - x_1) = a + bx & \text{if } x_1 \leq x \leq x_2, \\ y_2 & \text{if } x \geq x_2 \end{cases} \tag{4.15}$$

with $f_{\text{LIN}}(x)$ not defined for $x < x_1$.

The generalized inverse is:

$$\begin{aligned} f_{\text{LIN}}^{-}(y) &= \inf\{x \mid f_{\text{LIN}}(x) \geq y\} \\ &= \begin{cases} x_1 & \text{if } y \leq y_1, \\ x_1 + \dfrac{x_2 - x_1}{y_2 - y_1}(y - y_1) = \dfrac{y - a}{b} & \text{if } y_1 < y \leq y_2 \end{cases} \end{aligned} \tag{4.16}$$

and $f_{\text{LIN}}^{-}(y)$ is undefined (infeasible) if $y > y_2$.

The ceiling line consists of the line segments from (x_1, y_{min}) to (x_1, y_1), then to (x_2, y_2), and finally to (x_{max}, y_2). Possibly, the first and/or last segment has length equal to zero.

4.3.3 Model degrees of freedom

A necessity model specifies the bounding box with the ceiling line that is inside it. The ceiling line is assumed not to be outside the bounding box. Model degrees of freedom refers to the minimum number of parameters that are needed to describe the necessity model. The bounding box has 4 degrees of freedom. It can be described by, for example, the coordinates of the lower-left and the upper-right corner points: $x_{\text{min}}, y_{\text{min}}, x_{\text{max}}, y_{\text{max}}$.

Since a sequence of peers in the two classes of segmented ceiling lines (CE-FDH and CE-VRS) is defined in terms of strict inequalities, each peer can change position slightly without violating the conditions. Hence, the degrees of freedom of each of the CE-FDH and CE-VRS classes is $2P$.

Since the class of linear ceiling lines only imposes restrictions on a and b through strict inequalities, the degrees of freedom equals 2.

Therefore the general expression for the degrees of freedom of a necessity model is:

$$df_{\text{model}} = \begin{cases} 6, & \text{linear ceiling line.} \\ 2P+4, & \text{segmented ceiling line.} \end{cases} \tag{4.17}$$

where P is the number of peers of the segmented ceiling lines (CE-FDH and CE-VRS).

4.4 Mathematical estimation

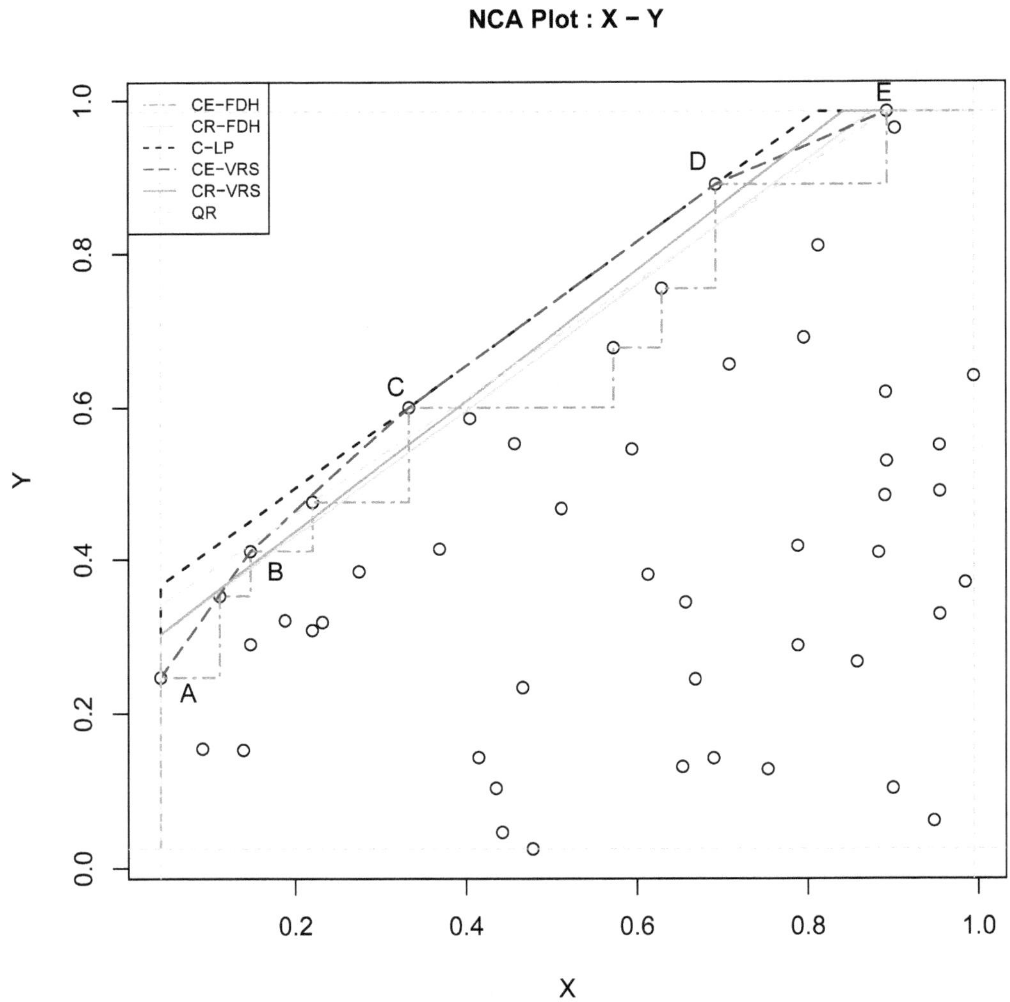

FIGURE 4.2 Mathematical estimation techniques using different ceiling lines. CE-FDH: stepwise linear ceiling line. CR-FDH: linear ceiling line (shown as the lower solid line). C-LP: linear ceiling line (shown as dashed line). CE-VRS: concave piecewise linear ceiling line (shown as segments passing through A, B, C, D, and E). CR-VRS: linear ceiling line (shown as the upper solid line). QR (Quantile Regression): linear ceiling line (shown as the dashed-dotted line in the middle). Population ceiling line: Y = 0.3 + X. Population effect size = 0.25. 50 points randomly selected from the population.

The mathematical estimation of the necessity model consists of estimating the parameters of the bounding box and the ceiling line from data. This can be done with the NCA software (Appendix B).

The estimation of the bounding box requires that the analyst selects the type of bounding box (defined by the theoretical or empirical scope). For the theoretical scope, the analyst specifies x_{min}, y_{min}, x_{max} and y_{max}. For the empirical scope the software estimates these values by taking the observed extremes.

The estimation of the ceiling line requires that the analyst selects the form of the ceiling line. If a segmented ceiling line is selected (CE-FDH or CE-VRS), the software provides the estimated x- and y-coordinates of the peers. To select a linear ceiling line there are several options. The Ceiling Regression Free Disposal Hull ceiling line (CR-FDH) is a trend line through the CE-FDH peers using least squares optimization. The Ceiling Regression Variable Returns to Scale (CR-VRS) ceiling line is the trend line through the CE-VRS peers using least squares optimization. The Ceiling - Linear Programming (C-LP) ceiling line is the line that is tight to the CE-FDH peers such that the sum of heights of the CE-FDH peers is minimized ("least sum"). The different lines are shown in Figure 4.2.[8]

After selecting the form of the line, the software provides the values of the parameters to describe the ceiling line (e.g., intercept and slope for the linear ceiling line). Criteria for selecting the type of bounding box and form of the ceiling line are discussed in Section 9.3.

NCA's mathematical estimations with the above ceiling techniques are based on the location of the peers. CE-FDH and CE-VRS use peers directly, while the linear ceiling lines CR-FDH, CR-VRS and C-LP ceiling line use them indirectly. The use of only points near the border is motivated by several fundamental considerations.[9] First, NCA's primary object of interest is the *empty space* above the ceiling: the region where observations are impossible. This follows necessity logic: if not X, then not Y. The ceiling zone is the complement of the feasible area in the bounding box. NCA quantifies the size of this zone and relates it to the size of the bounding box, which requires knowing the maximum feasible area but does not require modeling how points are distributed within that feasible area. Consequently, only points close to the ceiling are informative for locating the boundary. Points far below it do not contribute to identifying the boundary.

Second, conceptually, in NCA's deterministic logic, the ceiling is an abrupt transition from a region that can contain points to a region that cannot, more like a cliff edge than a gradual slope. The characteristic feature of a ceiling is thus a *sharp boundary* between feasibility and infeasibility. Methods that estimate the ceiling indirectly via a gradual (non-abrupt) distributional model of all data risk misrepresenting this discontinuity. A gradual model can blur the boundary and thereby estimate a "soft" transition that is not characteristic of a true ceiling.

Third, ceiling estimates should be *insensitive to data far below the ceiling*. For example, increasing the density of points in the lower-right corner should not have an effect on how the ceiling is constructed. Such points would pull the estimated ceiling estimates downward. This inflates the estimated empty space and thus overstates the true effect size. For this reason, ceiling estimation techniques that rely on fitting the full distribution under the

---

[8] For better visibility, the plot can be produced by the `NCA` software in `R` as follows: `library(NCA); set.seed(123); data <- nca_random (50, 0.3, 1); model <- nca_analysis(data, "X", "Y", ceilings = c("ce_fdh", "cr_fdh", "c_lp", "ce_vrs", "cr_vrs", "QR")); nca_output(model, summaries = FALSE, pdf = TRUE)`.

[9] It is also possible as an option in the `NCA` software that the analyst selects a custom line using a theoretical intercept and the slope.

ceiling are generally risky and should be applied with caution.[10] This is illustrated with the distribution-based Quantile Regression (QR) ceiling line. In Figure 4.2, this line lies among the other ceilings (that all only use points near the ceiling) and approximates the true ceiling line well. However, after altering the distribution of points by adding observations in the lower-right corner, the QR-based ceiling becomes biased, whereas ceilings estimated from on upper points remain stable (4.3). Broader evidence from Monte Carlo simulations likewise suggests that a distribution-based method like QR can exhibit bias, even at larger sample sizes, warranting prudence in their use (Section 5.5).

4.4.1 Effect size

When a high value of X is necessary for a high value of Y, the ceiling line determines the maximum possible y-value ($y = y_c$) for a given x-value, while the bounding box determines the absolute maximum possible y-value ($y = y_{\max}$). The difference between the two can be considered the *constraint* that X puts on Y. The increase of the ceiling line expresses the change in constraint under varying values of X. The maximum constraint occurs for $x = x_{\min}$. The effect builds up with increasing x at a rate determined by the increase in the ceiling line. The total effect is then the area between the ceiling line and the maximum possible $y = y_{\max}$.

Referring to Figure 4.1, the general expression for the necessity effect size is as follows. Let the bounding box be $B = [x_{\min}, x_{\max}] \times [y_{\min}, y_{\max}]$ with $x_{\max} > x_{\min}$ and $y_{\max} > y_{\min}$ and let $c(x)$ be the ceiling value of defined on $[x_1, x_{\max}]$ (with x_1 the leftmost x where the ceiling is defined).

The *scope* S is the area of the bounding box:

$$S = (x_{\max} - x_{\min})(y_{\max} - y_{\min}) \tag{4.18}$$

The feasible area is the area under the ceiling line $f(x)$ (only for $x \in [x_1, x_{\max}]$):

$$F = \int_{x_1}^{x_{\max}} (f(x) - y_{\min})\, dx \tag{4.19}$$

The ceiling zone C is the complement of the feasible area within the bounding box:

$$C = S - F \tag{4.20}$$

Equivalently, C can be written as the sum of the left strip (where the ceiling is undefined) and the area above the ceiling for $x \in [x_1, x_{\max}]$:

$$C = (x_1 - x_{\min})(y_{\max} - y_{\min}) + \int_{x_1}^{x_{\max}} (y_{\max} - f(x))\, dx \tag{4.21}$$

Consequently the effect size d is:

[10] Early NCA work proposed quantile regression and stochastic frontier analysis as statistical ceiling estimators (Dul, 2016b; Goertz et al., 2013). Since these approaches lower the ceiling line when more cases appear below the ceiling they are no longer recommended. Stochastic frontier analysis has not been adopted in practice of NCA and is deprecated in the NCA software. Quantile regression remains available, but is rarely used in NCA (for an exception see, P.-K. A. Chen, 2026).

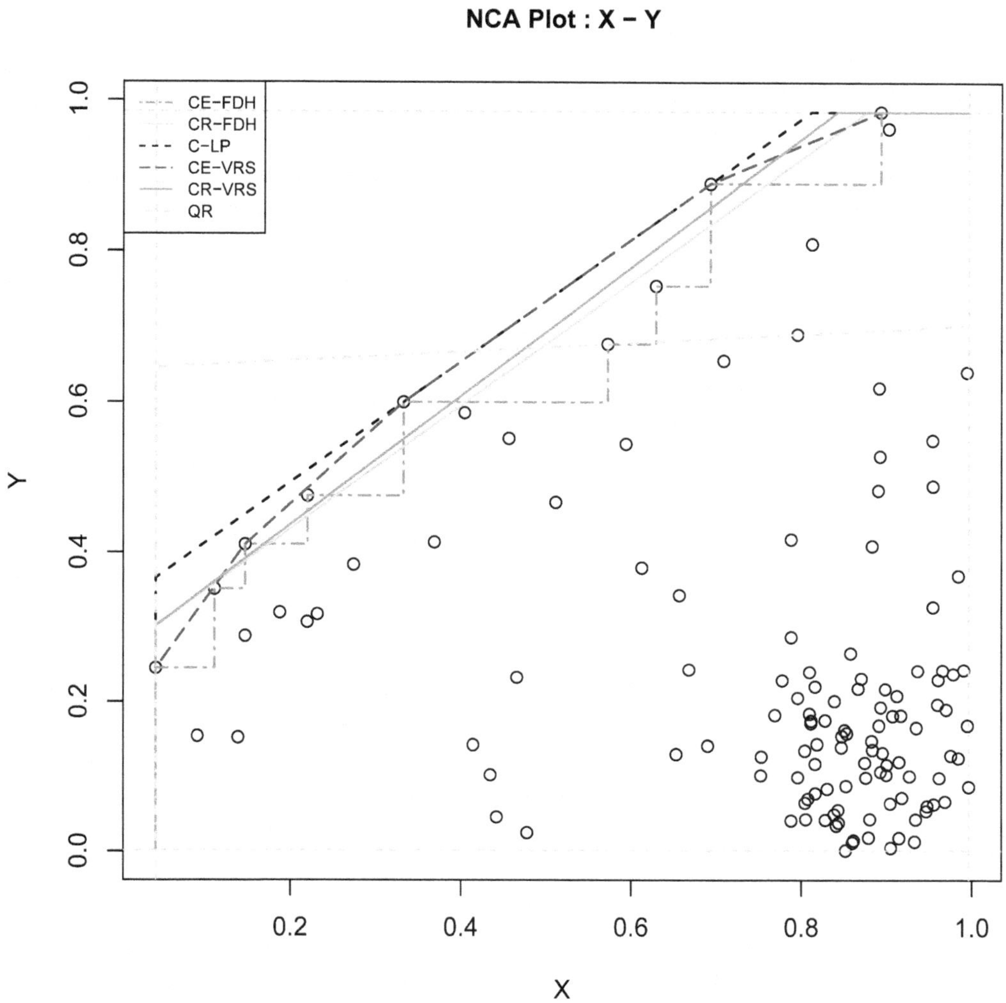

FIGURE 4.3 Comparison of an estimation techniques based on the distribution of all data below the ceiling: QR = Quantile Regression (shown as the near-horizontal dash-dotted line) and based on upper peers (the other lines). Population ceiling line = $Y = 0.3+X$. Population effect size is 0.25. First 50 points randomly selected from the population, next 50 points added in the lower-right corner.

$$d = \frac{C}{S} = 1 - \frac{F}{S} \tag{4.22}$$

The effect size for the different ceiling classes can be derived from the ceiling line equations.

4.4.1.1 Class of stepwise linear ceiling lines

For the CE-FDH class of ceiling line, see Equation (4.5), the feasible area is:

$$F_{\text{CE-FDH}} = \sum_{i=1}^{P-1} (x_{i+1} - x_i)(y_i - y_{\min}) + (x_{\max} - x_P)(y_P - y_{\min}) \tag{4.23}$$

Therefore, the effect size is:

$$\begin{aligned} d_{\text{CE-FDH}} &= 1 - \frac{F_{\text{CE-FDH}}}{S} \\ &= 1 - \frac{\sum_{i=1}^{P-1}(x_{i+1}-x_i)(y_i-y_{\min}) + (x_{\max}-x_P)(y_P-y_{\min})}{(x_{\max}-x_{\min})(y_{\max}-y_{\min})} \end{aligned} \tag{4.24}$$

4.4.1.2 Class of concave piecewise linear ceiling lines

For the CE-VRS class of ceiling line, see Equation (4.8), the feasible area is:

$$F_{\text{CE-VRS}} = \sum_{i=1}^{P-1}(x_{i+1}-x_i)\left(\frac{y_i+y_{i+1}}{2} - y_{\min}\right) + (x_{\max}-x_P)(y_P-y_{\min}) \tag{4.25}$$

Therefore, the effect size is:

$$\begin{aligned} d_{\text{CE-VRS}} &= 1 - \frac{F_{\text{CE-VRS}}}{S} \\ &= 1 - \frac{\sum_{i=1}^{P-1}(x_{i+1}-x_i)\left(\frac{y_i+y_{i+1}}{2} - y_{\min}\right) + (x_{\max}-x_P)(y_P-y_{\min})}{(x_{\max}-x_{\min})(y_{\max}-y_{\min})} \end{aligned} \tag{4.26}$$

4.4.1.3 Class of linear ceiling lines

For the class of linear ceiling line, see Equation (4.15), let the ceiling line intersect the bounding box at two points (x_1, y_1) and (x_2, y_2) with $x_{\min} \leq x_1 < x_2 \leq x_{\max}$. The feasible area under the ceiling within the bounding box is:

$$F_{\text{LIN}} = (x_2 - x_1)\left(\frac{y_1+y_2}{2} - y_{\min}\right) + (x_{\max}-x_2)(y_{\max}-y_{\min}) \tag{4.27}$$

Therefore, the effect size is:

$$\begin{aligned} d_{\text{LIN}} &= 1 - \frac{F_{\text{LIN}}}{S} \\ &= 1 - \frac{(x_2-x_1)\left(\frac{y_1+y_2}{2} - y_{\min}\right) + (x_{\max}-x_2)(y_{\max}-y_{\min})}{(x_{\max}-x_{\min})(y_{\max}-y_{\min})} \end{aligned} \tag{4.28}$$

4.4.2 Necessity inefficiency

As illustrated in Figure 4.4, for the standard case of a linear ceiling line above the diagonal, X may not constrain Y over the entire ranges of X and Y within the bounding box. The ceiling line intersects with the bounding box: $[x_{\min}, y_{c\min}]$ and $[x_{c\max}, y_{\max}]$. The vertical extension is the distance $y_1 - y_{\min}$, and the horizontal extension $x_{\max} - x_2$.

When (x_c, y_c) is a point on the ceiling line and $x < x_c$, then x is a bottleneck for $y = y_c$. Only an increased value of x such that $x \geq x_c$ enables a value $y = y_c$. For reaching the value of $y = y_{\max}$ it is necessary to have a value of $x \geq x_2$, where x_2 is the value of x of the point where the ceiling line crosses the $y = y_{\max}$ line. Thus, for enabling the maximum possible value $y_{\max}$, x should be at least $x = x_2$. Increasing x beyond $x = x_2$ for enabling

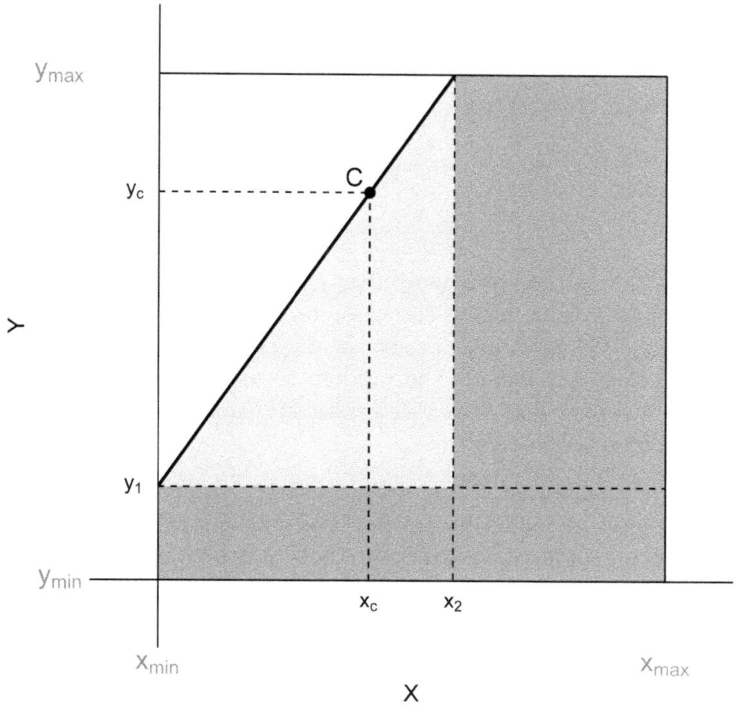

FIGURE 4.4 Necessity inefficiency refers to the part of the feasible area where Y is not constrained by X in the area $[x_{min}, x_{max}] \times [y_{min}, y_1]$ (outcome inefficiency), and to the part of the feasible area where X does not constrain Y in the area $[x_2, x_{max}] \times [y_{min}, y_{max}]$ (condition inefficiency).

$y = y_{\max}$ is not needed as y is no longer constrained. This is called 'inefficiency' regarding enabling the maximum outcome.

Condition inefficiency specifies the extent to which x does not constrain y for levels of $x_2 \leq x \leq x_{\max}$. Only for $x_{\min} \leq x \leq x_2$, x constrains y (ceiling line exists for $x_{\min} \leq x \leq x_2$). Condition inefficiency is expressed as a percentage:

$$\text{condition inefficiency} = \frac{x_{\max} - x_2}{x_{\max} - x_{\min}} * 100\% \tag{4.29}$$

where $x_{\max}$ and $x_{\min}$ are the minimum and maximum values of x (bounds) and x_2 is the value of x of the point where the ceiling line crosses the $y = y_{\max}$ line.

Similarly, *outcome inefficiency* specifies the extent to which Y is not constrained by X for lower levels of $y_{\min} \leq y \leq y_1$, where y_1 is the value of y of the point where the ceiling line crosses the $x = x_{\min}$ line. Only for $y_1 \leq y \leq y_{\max}$, y is constrained by X (ceiling line exists for $y_1 \leq y \leq y_{\max}$). Outcome inefficiency is expressed as a percentage:

$$\text{outcome inefficiency} = \frac{y_1 - y_{\min}}{y_{\max} - y_{\min}} * 100\% \tag{4.30}$$

where $y_{\min}$ and $y_{\max}$ are the minimum and maximum values of Y (bounds), and y_1 is the value of y of the point where the ceiling line crosses the $x = x_{\min}$ line. Outcome inefficiency is

0% if X constrains Y for all values of Y; outcome inefficiency is 100% if Y is not constrained for any value of X.

4.5 Model fit

NCA's *mathematical* model specification differs from a *statistical* model specification. The latter includes a random term called the error term or disturbance (ε) representing the variation in the data. NCA does not specify the distribution of data under the ceiling and therefore the model does not include the random term. Thus, NCA's model consists of a bounding box with a ceiling line and is deterministic (possibly with noise and exceptions, see Section 4.5.4, not probabilistic).

Since NCA cannot rely on statistical (distribution-based) approaches for evaluating model fit, NCA-specific metrics of model fit are developed to quantify how well an NCA model captures the underlying necessity pattern in the data, and how closely the model's predicted values match observed values. In this section eight metrics of model fit are presented: complexity, fit, ceiling accuracy, exceptions, noise, support, spread, and sharpness.

4.5.1 Complexity

The *complexity* metric (cp) can be used to balance parsimony against accuracy. A complex model fits data well (accuracy) but needs more parameters, whereas the model with low complexity is attractive for generalization and practical usefulness, but may lack accuracy.

The proposed complexity measure relates to the model degrees of freedom (Section 4.3.3). When the degrees of freedom for the bounding box are fixed at four, the model degrees of freedom depend only on the ceiling line. The ceiling line degrees of freedom equal the model degrees of freedom minus 4. That is:

$$df_{\text{ceiling}} = \begin{cases} 2, & \text{linear ceiling line} \\ 2P, & \text{segmented ceiling line} \end{cases} \tag{4.31}$$

where P is the number of peers on the piecewise linear ceiling as defined in Section 4.3.3.

The more degrees of freedom, the more complex the ceiling is. The linear ceiling line has the lowest complexity. This base level is set at 1 (degrees of freedom divided by 2) and represents 1 pair of parameters needed for describing a linear ceiling line (intercept, slope). The complexity of a segmented ceiling line can be described by the number of peers. Each peer can be described by 1 pair of parameters (x-coordinate and y-coordinate). Therefore, complexity is defined as:

$$\mathit{complexity} = \begin{cases} 1, & \text{linear ceiling line} \\ P, & \text{segmented ceiling line} \end{cases} \tag{4.32}$$

where P is the number of peers.

4.5.2 Fit

The model fit metric *fit* (ft) compares a selected ceiling line with the CE-FDH ceiling line. CE-FDH follows the border closely and gives the maximum possible effect size under strict necessity (no cases above the ceiling). By definition, fit of CE-FDH equals 100%. For other selected ceiling lines, fit is usually lower. The fit metric indicates how closely the selected ceiling line follows the pattern of the boundary like the CE-FDH does.[11]

Fit can be expressed as follows:

$$fit = \begin{cases} \frac{d_{\text{ceiling}}}{d_{\text{CE-FDH}}} \times 100\% & \text{if } d_{\text{ceiling}} \leq d_{\text{CE-FDH}} \\ \text{undefined} & \text{if } d_{\text{ceiling}} > d_{\text{CE-FDH}} \end{cases} \tag{4.33}$$

where $d_{ceiling}$ is the effect size of the selected ceiling line and d_{CE-FDH} is the effect size of the CE-FDH ceiling line. Fit is undefined when the effect size of the selected ceiling line exceeds that of the CE-FDH ceiling line, to avoid interpretational issues.

4.5.3 Ceiling accuracy

Model fit metric *ceiling accuracy* (ca) refers to the emptiness of the ceiling zone. It quantifies the degree to which the expected empty corner is indeed empty. A selected ceiling line defines the ceiling zone. With a strict interpretation of necessity, the ceiling zone should not have points, and all points should be in the feasible area. Figure 4.5-left shows an example of two points being in the ceiling zone. The presence of points in the ceiling zone may have several explanations. First, the point can be an observation (case) from an empirical study with measurement error or sampling error (Section 9.8). Second, the selected ceiling line may be too low, resulting in points above it.[12] Third, from the typicality perspective of necessity (Section 2.4.3), the point might represent a rare permissible exception, transforming the expression 'X is *always* necessary for Y' into 'X is *typically* necessary for Y'.

For good model fit nearly all points are in the feasible area. Ceiling accuracy is the percentage of points in the feasible area (on or below the ceiling line):

$$ceiling\,accuracy = \left(1 - \frac{n_{\text{ceiling zone}}}{n}\right) \times 100\% \tag{4.34}$$

where $n_{ceiling\,zone}$ is the number of points in the ceiling zone, and n is the total number of points.[13]

[11] A low fit value (e.g., < 80%) suggests that the selected ceiling line does not closely follow the boundary.

[12] A ceiling line can be selected theoretically or estimated from the data. A *theoretically* selected ceiling line means that the parameters of the ceiling equation are selected a priori, for example, based on previous studies (e.g., in replication studies). It may be that the ceiling line is too low when applied to a new dataset. This would result in having points in the ceiling zone. An *estimated* ceiling line means that the parameters of the ceiling equation are estimated from the current data. For example, when the CR-FDH technique (Section 4.4) is used, normally some points will be in the ceiling zone.

[13] For the dichotomous situation Ragin (1987) suggested at least 80% of points should be in the feasible area for support of necessity. In a prototype version of NCA, Dul et al. (2010) suggested at least 95% should be in the feasible area. A lower ceiling accuracy score than for example 95% classifies several cases as not being compatible with strict necessity, which requires further investigation of these cases or a reconsideration of the selection of the ceiling line.

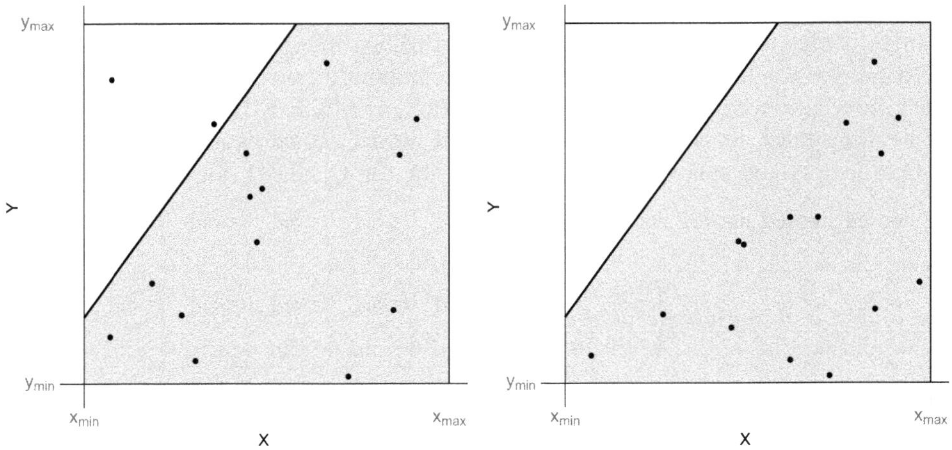

FIGURE 4.5 Estimation of the ceiling line. Left: points in the ceiling zone. Right: no points in the ceiling zone.

4.5.4 Purity – noise and exceptions

A disadvantage of the model fit metric ceiling accuracy is that it does not account for the distance between the point and the ceiling line. Points in the ceiling zone that are farther from the ceiling line present a greater challenge to necessity than points that are closer. The *purity* measure of a point captures the proximity of a point to the ceiling line (Figure 4.6).

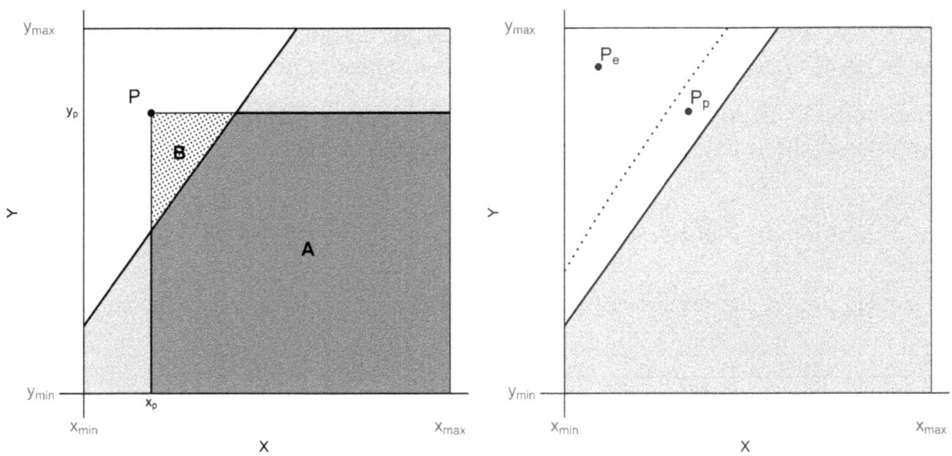

FIGURE 4.6 Purity: capturing the proximity of a point in the ceiling zone to the ceiling line. P = general point in the ceiling zone; P_e = extreme point in the low-purity part of the ceiling zone. P_p = point in the medium-purity part of the ceiling zone. See the text for an explanation of A and B.

Purity is a measure of closeness of a point P in the ceiling zone to the ceiling. Referring to Figure 4.6-left, it is defined as:

$$purity\,(P) = \frac{\text{size of area of F} \cap \text{LR(P)}}{\text{size of area of LR(P)}} \tag{4.35}$$

where F is the feasible area, and LR(P) is the area at the lower-right side of point P within the bounding box. Referring to the linear ceiling line in Figure 4.6-left, the numerator of Equation (4.35) is A and the denominator A + B. Therefore, the purity of a point P for a linear ceiling line is:

$$purity\,(P) = \frac{A}{A+B} \tag{4.36}$$

where the dark-gray area A is the intersection between the feasible area and the LR(P) area, and the dotted area B is the intersection between the ceiling zone and the LR(P) area.

The highest purity value of 1 refers to a point on or below the ceiling line (B = 0, point is not in the ceiling zone). A point above the ceiling line has a purity value < 1. A point in the upper-left corner of the ceiling zone is furthest from the ceiling line, and has the lowest purity value, which equals $1 - d$ where d is the effect size.

Figure 4.6-right shows a dotted line representing points with the same purity value, here 0.9. This line is called the ***iso-purity*** line. For a linear ceiling line the iso-purity line is part of an ellipse[14] and for a stepwise linear function with horizontal and vertical parts it consists of a combination of elliptic and linear segments. The iso-purity line and the ceiling line divide the bounding box into three zones. The *low-purity zone* is the area above the iso-purity line and within the box where the points are far from the ceiling line. Here the purity is low (purity < 0.9 in Figure 4.6-right). The *medium-purity zone* is the area between the iso-purity line and the ceiling line, where the points are close to the ceiling (0.9 ≤ purity < 1 in Figure 4.6-right). The *saturated-purity zone* consists of the feasible area including the ceiling line where purity = 1.

Two model fit metrics are derived from the purity measure: noise and exceptions.

The metric *noise* (*ns*) quantifies uncertainty around the ceiling line due to minor deviations. For good model fit there should be only a limited percentage of points in the *medium-purity zone*.[15]

The model fit metric *exceptions* (*ex*) quantifies violations in the ceiling zone far from the ceiling line, indicating how strict (or tolerant) the necessity claim is. With a deterministic perspective on necessity no cases are allowed in the *low-purity zone*. If a *rare* point exists in the low-purity zone, this point may be considered an exception if measurement error and sampling error are excluded. For a typicality perspective, *rare* exceptions may be permitted in this zone. The proposed model fit metric exceptions captures the number of cases in the low-purity zone.[16]

---

[14]The conic segment arises from calculations with areas.

[15]For example, not more than 5% of points may be tolerable noise for a medium-purity zone corresponding to 0.9 iso-purity (in this zone the points have a purity of 0.9 ≤ purity < 1).

[16]Exceptions should be rare (e.g., ~ 0 in the ceiling zone corresponding to purity < 0.9) to justify a typicality causal perspective. No strict rules exist for what number should be considered as "rare." What matters is that, unlike noise, an exception-case represents a different causal process not captured by necessity, which may be studied separately.

4.5.5 Solidity – support and spread

To ensure high levels of ceiling accuracy and low levels of noise, a high ceiling line well above the points may be selected as illustrated in Figure 4.5-right. However, when the selected ceiling line is far away from points under the ceiling, it may not be informative in the sense that no or only a few points are constrained by the condition, which can make the necessary condition trivial. When more points are close to the ceiling, more points are constrained by the condition and the necessity of the condition becomes more relevant. *Solidity* is a measure of closeness of a point P in the feasible area to the ceiling as illustrated in Figure 4.7.

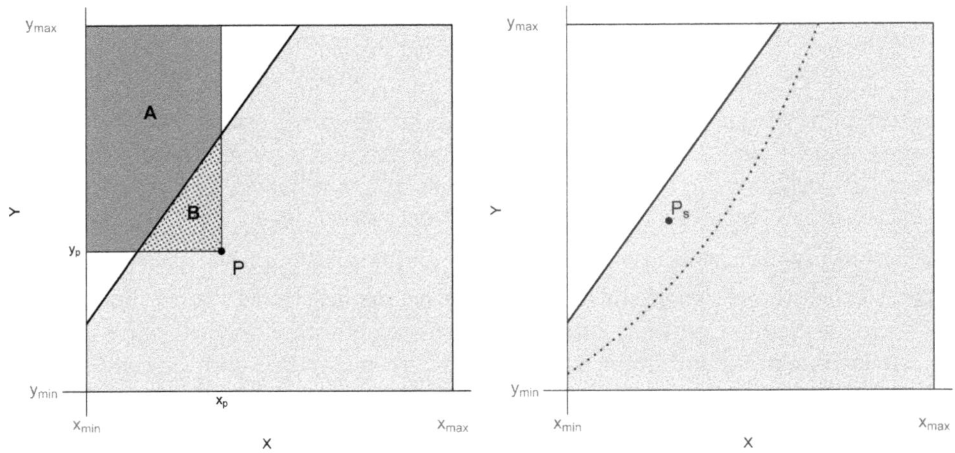

FIGURE 4.7 Solidity: capturing the proximity of a point in the feasible area to the ceiling line. P = general point in the feasible area. P_s = point in the medium-solidity part of the feasible area. See the text for an explanation of A and B.

Referring to Figure 4.7-left, it is defined as:

$$solidity(P) = \frac{\text{size of area of C} \cap \text{UL(P)}}{\text{size of area of UL(P)}} \tag{4.37}$$

where C is the ceiling zone, and UL(P) is the area at the upper-left side of point P within the bounding box. Referring to the linear ceiling line in Figure 4.7-left, the numerator of Equation (4.37) is A and the denominator A + B. Therefore, the solidity equation for a linear ceiling line is:

$$solidity\,(P) = \frac{A}{A+B} \tag{4.38}$$

where the dark-gray area A is the intersection between the ceiling zone and the UL(P) area, and the dotted area B is the intersection between the feasible area and the UL(P) area. The highest solidity value of 1 refers to a point on or above the ceiling line (B = 0, point is not in the feasible area). A point below the ceiling line has a solidity value < 1. A point in the lower-right corner of the feasible area is furthest from the ceiling line, and has the lowest solidity value.

Figure 4.7-right shows a dotted line representing points with the same solidity value, here 0.8. For a linear ceiling line the ***iso-solidity*** line is part of an ellipse. The iso-solidity line and the ceiling line divide the bounding box into three zones. The ***low-solidity zone*** is the area below the iso-solidity line and within the bounding box where the points are far from the ceiling line. Here the solidity is low (<0.8 in Figure 4.7-right). The ***medium-solidity zone*** is the area between the iso-solidity line and the ceiling line, where the points are close to the ceiling. Here the solidity is high, though not perfect ($0.8 \leq$ solidity < 1 in Figure 4.7-right). The ***saturated-solidity zone*** consists of the ceiling zone and the ceiling line where the solidity $= 1$.

Details of the rationale for selecting the specific distance measure for purity and solidity can be found in (Kuik & Dul, 2026).[17]

For good model fit there should be at least a certain percentage of points on the ceiling line or in the medium-solidity zone, which may represent support for necessity. Model fit metric ***support*** (su) quantifies how strongly the data in the feasible area support the ceiling.[18]

While the model fit metric support evaluates the percentage of points on and just below the ceiling to support the ceiling, those points may be clustered and only support certain parts of the line. Preferably all parts of the ceiling are supported. The model fit metric ***spread*** (sp) indicates how evenly the x-positions of the supporting points cover the range $(x_{\text{min}}, x_{\text{max}})$. To compute spread, the number of points on the ceiling line ($satS$) and in the medium-solidity zone ($medS$) are selected, and their X-values (x_i for $i = 1, \ldots, n$) are obtained. For this set of n points the coefficient of variation (CV) is calculated by dividing the standard deviation (sd) by the mean ($\bar{x}$) as follows:

$$\bar{x} = \frac{1}{n} \sum_{i=1}^{n} x_i \tag{4.39}$$

$$sd = \sqrt{\frac{1}{n-1} \sum_{i=1}^{n} (x_i - \bar{x})^2} \tag{4.40}$$

[17] Various other distance measures such as horizontal, vertical, or perpendicular distance to the ceiling line could be used as a proximity measure, but each has limitations. When using horizontal or vertical distance of point P to the ceiling, the variation in the other direction is ignored, which is problematic when the ceiling line is non-linear. Perpendicular distance could solve this but this distance measure loses its orthogonality under affine transformations (e.g., normalization) of the scope. To address these issues, a proximity measure based on the area between the point's horizontal and vertical distances and the ceiling line is used. Assuming that the ceiling line is non-decreasing, this area-based measure has several desirable properties: continuity meaning that it varies smoothly with the point's location, monotonicity meaning that the value increases as the point moves farther from the ceiling line, and invariance meaning that it remains invariant under affine transformations of the scope.

[18] For example, at least 5% of points could be considered acceptable support with a medium-solidity zone corresponding to 0.8 iso-solidity (in this zone the points have a solidity of $0.8 \leq$ solidity ≤ 1.).The proposed suggestion for iso-solidity is less strict than that for iso-purity. The reason is that purity < 1 violates strict necessity. However, solidity < 1 does not violate necessity. Solidity < 1 implies that the single necessary condition is only less informative for sufficiency. If solidity $= 1$ all points are on or above the ceiling, which makes the necessary conditions also sufficient. However, assessing the sufficiency of a single condition is not a goal of NCA, which focuses on single factors being necessary. Low solidity may affect sharpness, see Section 4.5.6.

$$CV = \frac{sd}{\bar{x}} \tag{4.41}$$

To obtain a metric of evenness, the coefficient of variation is transformed as follows:

$$spread = \frac{1}{1 + \frac{sd}{\bar{x}}} \tag{4.42}$$

The spread metric ranges from 0 to 1. High spread values (e.g., ≥ 0.5) are favored over low spread values. Spread is undefined when $\bar{x} = 0$, which can occur when x-values are negative or zero; min-max normalization can avoid this.

4.5.6 Sharpness

The ceiling line serves as a clear boundary that distinguishes the ceiling zone from the feasible area. *Sharpness* (sh) refers to the abrupt change in point density when transitioning between the ceiling zone and the feasible area, and vice versa. Sharpness implies that the density of points in the medium-purity zone (*medP*; the area between the iso-purity line and the ceiling line) excluding the points on the ceiling line, is much lower than the density of points in the medium-solidity zone (*medS*; the area between the iso-solidity line and the ceiling line) excluding the points on the ceiling line, as illustrated in Figure 4.8.

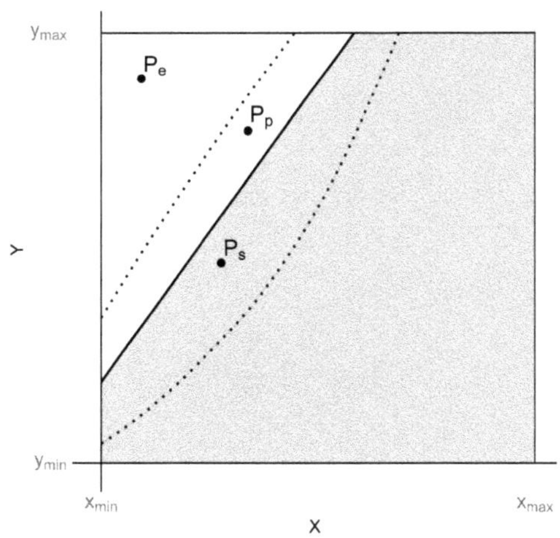

FIGURE 4.8 Point P_s is in the medium-solidity zone of the feasible area between ceiling line and iso-solidity line (dotted). Point P_p is in the medium-purity zone of the ceiling zone between ceiling line and iso-purity line (dotted). Point P_e is in the low-purity (empty) zone of the ceiling zone outside and may be considered an exception.

sharpness is defined as:

$$sharpness = \begin{cases} \frac{medS}{medS+medP} & \text{if } medS + medP \neq 0 \\ \text{NA} & \text{otherwise} \end{cases} \tag{4.43}$$

where *medS* is the number of points in the medium-solidity zone excluding the points on the ceiling line, *medP* is the number of points in the medium-purity zone excluding the

TABLE 4.1 Sharpness values for different combinations of points in the medium-purity zone (rows) and medium-solidity zone (columns). Sharpness > 0.8 is shown in bold.

	0.1	0.2	0.3	0.4	0.5	0.6	0.7	0.8	0.9	1
0.1	0.5	0.67	0.75	0.8	**0.83**	**0.86**	**0.88**	**0.89**	**0.9**	**0.91**
0.2		0.5	0.6	0.67	0.71	0.75	0.78	0.8	**0.82**	**0.83**
0.3			0.5	0.57	0.62	0.67	0.7	0.73	0.75	0.77
0.4				0.5	0.56	0.6	0.64	0.67	0.69	0.71
0.5					0.5	0.55	0.58	0.62	0.64	0.67
0.6						0.5	0.54	0.57	0.6	0.62
0.7							0.5	0.53	0.56	0.59
0.8								0.5	0.53	0.56
0.9									0.5	0.53
1.0										0.5

points on the ceiling line, and NA indicates that a value is not available. Sharpness defines among the cases that lie close to the ceiling (in the ribbon, see below), what fraction are on the "correct" side (below the ceiling) rather than violating it (above the ceiling). Sharpness has a value between 0 and 1. A sharpness value of 0.5 means that there is no distinction in density of points close above and close below the ceiling line. Sharpness above 0.5 indicates that there are more points close below the ceiling line than close above the ceiling line (as intended).

Table 4.1 shows sharpness values for different values of *medP* and *medS*.[19]

4.5.7 NCA ribbon

The NCA ribbon is the combination of the *medium-purity zone* and the *medium-solidity zone*. It represents a type of *uncertainty area* around the ceiling line. It differs from a confidence interval in statistics but serves a similar goal to capture precision. The ribbon is the geometric area calculated from the ceiling line with specified medium-purity and medium-solidity zones (e.g., 0.9-iso and 0.8-iso, respectively), where a specified maximum (purity) or minimum (solidity) number of points (cases, observations) are located. Few points in the medium-purity zone and many points in the medium-solidity zone means high precision; more points in the medium-purity zone and fewer points in the medium-solidity zone means lower precision.

4.5.8 Summary model fit metrics

The eight metrics of NCA's model fit with example guidelines are summarized in Table 4.2. The model fit metrics are part of the broader concept of *credibility*. NCA's credibility for identifying necessary conditions is discussed in Chapter 6.

[19]For good model fit the sharpness should be for example at least 0.8 such that from all points in the ribbon 80% is on the 'correct' side.

TABLE 4.2 Model fit metrics for NCA.

Metric	Description	Symbol	Guideline
Complexity	Half of ceiling degrees of freedom.	cp	Balance.
Fit	Alignment with boundary pattern (% effect size compared to CE-FDH effect size).	ft	High (e.g., $ft \geq 80\%$).
Ceiling accuracy	Percentage of points in the feasible area.	ca	High (e.g., $ca \geq 95\%$).
Noise	Percentage of points in the ceiling zone near the ceiling (medP-zone).	ns	Low (e.g., $ns \leq 5\%$; 0.9 iso-purity).
Exceptions	Number of points in the ceiling zone far from the ceiling (lowP-zone).	ex	Very low (e.g., $ex \sim 0$).
Support	Percentage of points in the feasible area on or near the ceiling (medS-zone).	su	High (e.g., $su \geq 5\%$; 0.8 iso-solidity).
Spread	Evenness of x-positions of points on or near the ceiling (medS-zone).	sp	High (e.g., $sp \geq 0.5$).
Sharpness	Relative density of points in medium-purity (medP) and medium-solidity (medS) zones.	sh	High (e.g., $sh > 0.8$).

4.6 Multiple necessary conditions

Until now the mathematics of necessity has been evaluated by analyzing a single pair of variables: one condition (X) and one outcome (Y). X and Y represent two (related) properties of the phenomenon of interest, and the relationship between these two properties is analyzed using geometry and algebra in the XY-plane. This is NCA's basic approach, which analyzes necessary conditions one by one. This section first formalizes NCA's basic approach of analyzing single necessary conditions. Subsequently, it illustrates this reasoning with an example and gives the arguments for it. Finally, the section shows how several linked necessary conditions can be analyzed as a chain of necessity.

4.6.1 Formalization of single condition approach

NCA intentionally reduces causal complexity by reducing a multidimensional constraint structure to single-condition necessity relations by projection, and assumes that these overall necessity constraints manifest as observable ceilings in the empirical (natural) distribution.

The formalization is as follows.[20] Let $Y \in \mathbb{R}$ denote an outcome variable and let $\mathbf{X} = (X_1, \ldots, X_J)$ denote a vector of J conditions. Assume that the data-generating process implies an upper bound on Y given by a (possibly unknown and complex) function:

$$Y \leq f(\mathbf{X}) \tag{4.44}$$

[20] The specification of the bounding box is tacitly assumed for simplification.

where $\mathbf{X}$ is a vector of conditions and f defines a multidimensional ceiling surface.

NCA does not attempt to model or estimate $f(\mathbf{X})$. Instead, for each condition X_j, NCA considers the projection of the feasible set onto the (X_jY)-plane:

$$(x_j, y) : \exists \mathbf{x}_{-j} \text{ such that } y \leq f(x_j, \mathbf{x}_{-j}). \tag{4.45}$$

This projection induces a bivariate model with a ceiling function:

$$Y \leq f_j(X_j), \tag{4.46}$$

where

$$f_j(x_j) = sup_{\mathbf{x}_{-j}} f(x_j, \mathbf{x}_{-j}), \tag{4.47}$$

and $\mathbf{x}_{-j}$ denotes all conditions except X_j.

The function $f_j(x_j)$ defines the *projected ceiling* for condition X_j. The function $f(x_j, \mathbf{x}_{-j})$ represents the *multidimensional ceiling*, that is, the maximum attainable value of Y for a given combination of all conditions. For a fixed value x_j, the supremum operator $\sup_{\mathbf{x}_{-j}}$ takes the least upper bound of the ceiling $f(x_j, \mathbf{x}_{-j})$ across all admissible values of the remaining conditions. Thus, $f_j(x_j)$ is the highest outcome level that can be achieved at condition level x_j under the most permissive setting of the other conditions. Thus, even if all other conditions take their most lenient possible values, the outcome Y cannot exceed the ceiling value $f_j(x_j)$.

In the context of necessity analysis, this construction ensures that the ceiling function f_j captures the *overall necessity constraint* imposed by X_j: regardless of how the other conditions are configured, the outcome Y cannot exceed $f_j(x_j)$. The empty space above the ceiling is NCA's key interest, not the feasible area below the ceiling. Assuming a well-covered sample, meaning that for each x_j the data include sufficiently permissive realizations of the unobserved conditions so that the upper boundary is observed, the estimated ceiling line is not biased by third variables that are omitted from the analysis. Conversely, insufficient boundary coverage (e.g., due to limited sampling or selection that removes near-boundary cases) may lead to underestimation of the ceiling (the true ceiling is higher) and thus overestimation of the effect size (the true effect size is smaller).

In other words, NCA assumes that the projected ceiling f_j is recoverable from the empirical joint distribution of (X_j, Y), generated under real-world conditions (the *natural distribution*) and incorporating all interactions, dependencies, and selection mechanisms among the conditions. Under this coverage assumption, the observed data exhibit an empty region above $f_j(X_j)$, making the estimated ceiling a candidate for causal interpretation with additional theory.

4.6.2 Illustration

Section 3.3 discusses a necessity theory with multiple necessary conditions (Figure 3.2). For example, Figure 3.2-left shows a theory that includes three properties of the phenomena (three conditions). In mathematical terms, each condition is a dimension of the multidimensional space that models the phenomena. Thus, a model with two conditions X_1 and X_2, and one outcome Y results in a three-dimensional space. The XY-plot becomes a three-dimensional X_1X_2Y-space.

NCA's overall (projected) necessity property is illustrated in Figure 4.9, which shows a volcano mountain where X_1 is the horizontal location in one direction, X_2 the horizontal location in another direction, and Y the vertical elevation.[21]

For example, a point within the mountain can be identified by specific values of X_1, X_2 and Y. The height of the mountain is a *ceiling surface*, which is the set of points representing different mountain heights depending on specific values of X_1, X_2. The ceiling surface is a very complex function. NCA is not interested in modeling or estimating the multidimensional ceiling (maximum value of the outcome for given combinations of values of the conditions). In contrast, NCA takes the projection of a multidimensional ceiling onto each X_jY-plane, resulting in separate ceiling lines for each condition. This projected overall ceiling line in an X_jY-plane represents the global necessity of the single condition X_j for outcome Y and is interpreted independently of other conditions and the rest of the causal structure. It is assumed that this global necessity (overall ceiling line) manifests in the real world and can be observed in the empirical (natural) distribution.

The projections of the three-dimensional ceiling surface of the volcano mountain on the XY-planes result in non-linear ceiling lines (Figure 4.9). When the three-dimensional ceiling surface is linear (e.g., $Y = (X_1 + X_2)/2$, the projections result in linear ceiling lines, as shown in Figure 4.10.

In a multidimensional space, NCA considers one projection at a time and therefore one condition at a time. With several conditions, NCA conducts successive analyses by considering the ceilings in the separate X_jY-planes. For example, with two conditions (X_1 and X_2), two separate analyses are performed: one in the X_1Y-plane and one in the X_2Y-plane. In geometric terms, NCA places a ceiling surface over the data and takes its orthogonal projections onto the X_1Y- and X_2Y-planes. Each condition thus has its own ceiling line, $f_1(X_1)$ and $f_2(X_2)$. Consequently, for a given value $Y = y_c$ two conditions must be satisfied: $X_1 \geq x_{1c}$ AND $X_2 \geq x_{2c}$. Therefore, the maximum possible $Y = y_c$ for given values $X_1 = x_{1c}$, and $X_2 = x_{2c}$ is

$$y_c = min\{f_1(x_{1c}), f_2(x_{2c})\} \tag{4.48}$$

where y_c is a particular outcome value, x_{jc} is the necessary value of the j-th condition, and f_j is the ceiling line of condition X_j.

With more than two conditions, there is a multidimensional ceiling with projections $f_j(X_j)$.

The mathematical description of a necessity AND-combination of multiple necessary conditions is as follows: for achieving an outcome level $Y = y_c$, all conditions must satisfy $X_j \geq x_{jc}$. The maximum possible value of the outcome $Y = y_c$ for given values $X_j = x_{jc}$ is

$$y_c = min_{j=1,\ldots,J}\{f_j(x_{jc})\} \tag{4.49}$$

Here, y_c is a particular outcome value, x_{jc} is the necessary value of the j-th condition, and f_j is the ceiling line associated with condition X_j.

Although NCA performs separate analyses for each condition, these analyses are combined and considered jointly in NCA's bottleneck table (Section 9.11). This allows answering questions such as: "What levels x_{jc} of X_j are necessary for a particular level $Y = y_c$?" and "For given levels x_{jc} of X_j, what is the maximum possible level y_c of Y?"

[21] Data are from the Maungawhau volcano in Auckland, New Zealand and were obtained from the `R` package `volcano`.

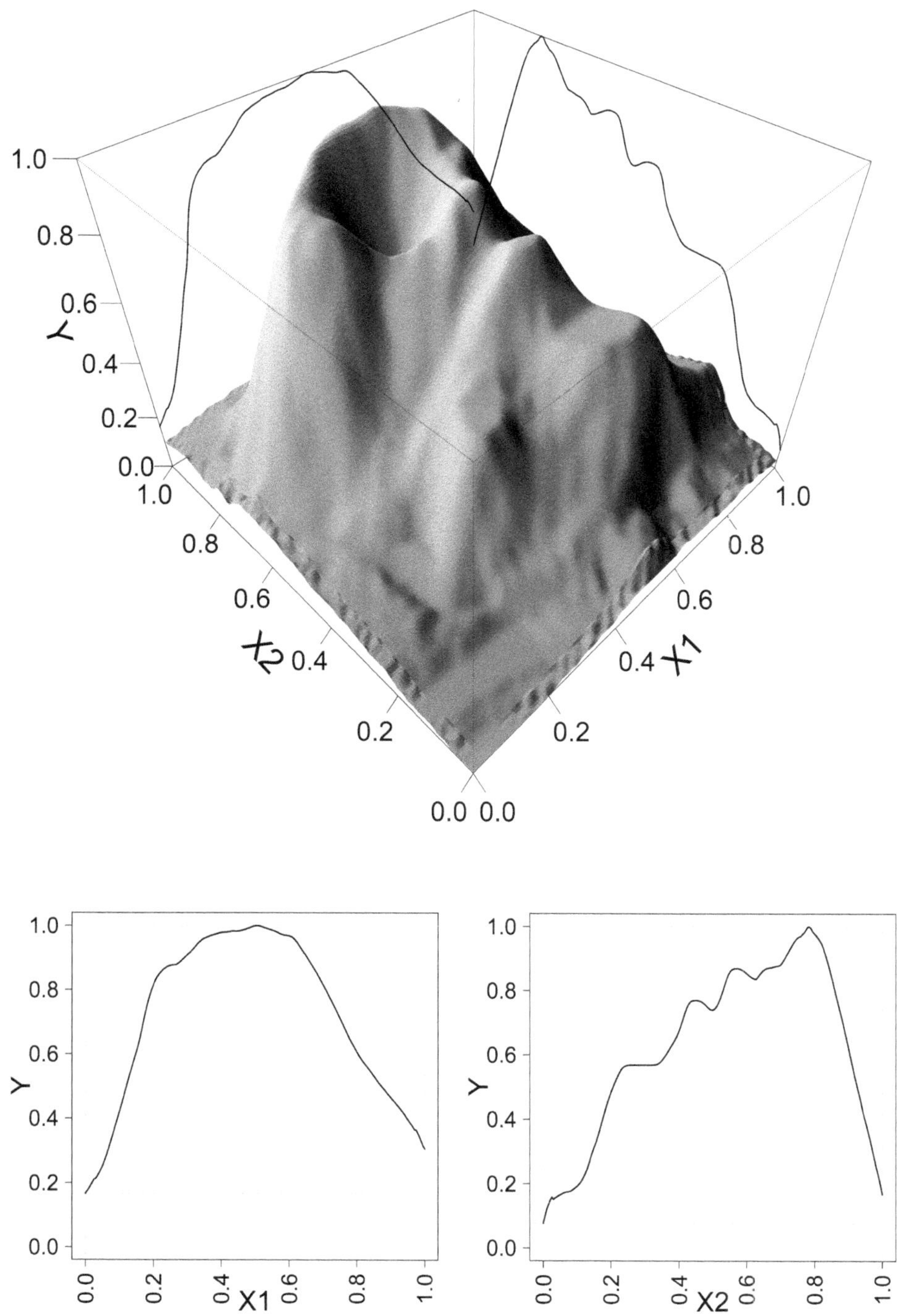

FIGURE 4.9 A nonlinear three-dimensional ceiling (surface) and its projections on the X_1Y and X_2Y-planes (non-linear ceiling lines).

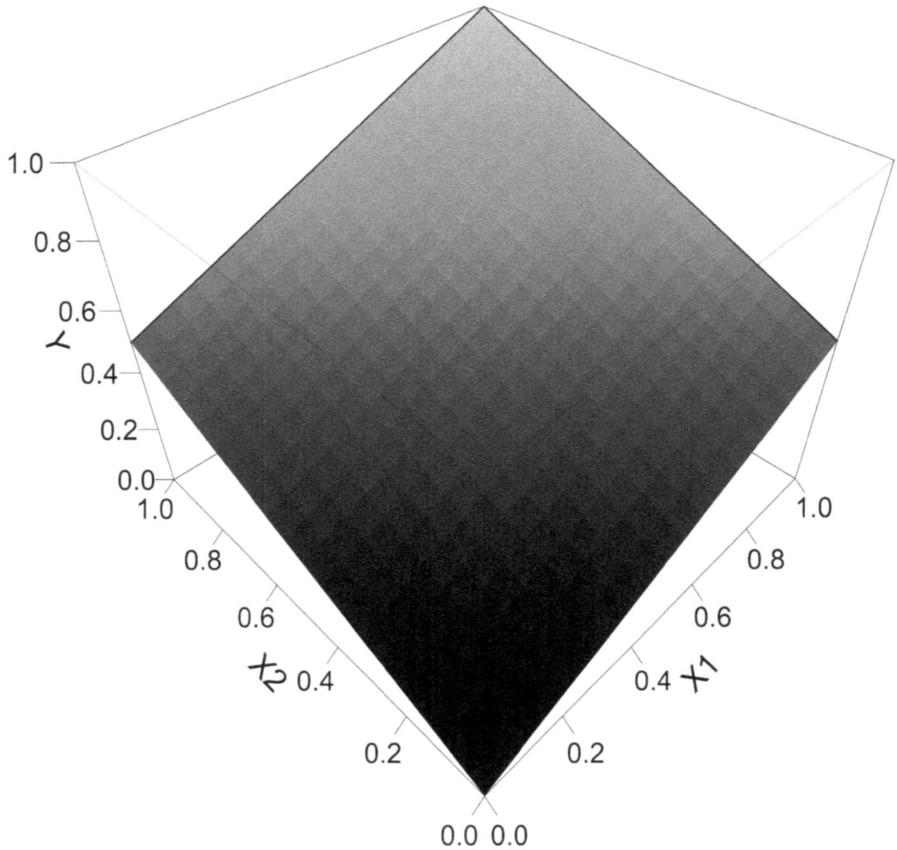

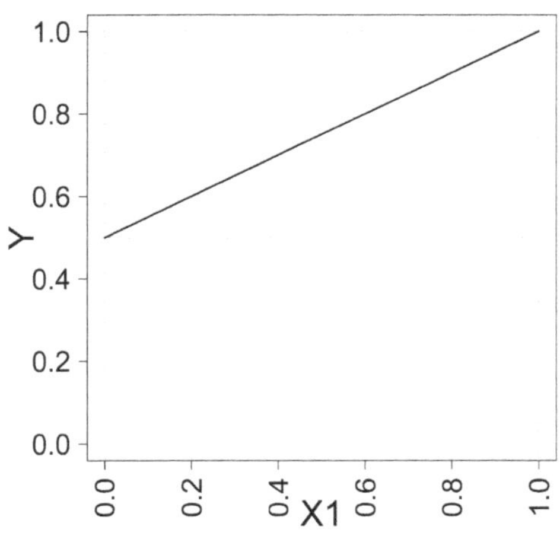

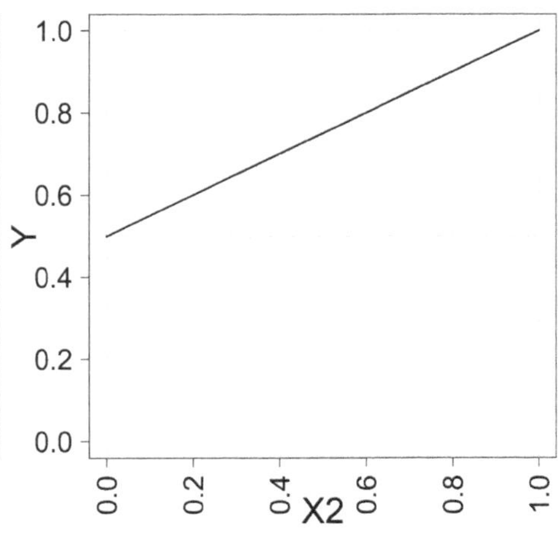

FIGURE 4.10 A linear multivariate three-dimensional ceiling $Y = (X_1 + X_2)/2$ and its projections on the X_1Y- and X_2Y-planes (linear ceiling lines).

4.6.3 Arguments for single condition analyses

There are several reasons why NCA analyzes *projections* of the multidimensional ceiling.

- The *first* reason is the fundamental choice to focus on single factors that are necessary for an outcome. This distinguishes NCA from conventional analyses that study multi-causal phenomena by considering combinations of factors and their joint effects. A multivariate analysis is the realistic option for predicting the presence of an outcome, because often only a combination of factors (and not a single factor) can produce the outcome. In the case of necessity, however, it is realistic and often theoretically sound to propose that a single factor (a necessary condition as a bottleneck) can prevent the outcome from occurring. Therefore, it is possible and useful to study the necessity of single factors for an outcome using projections.

- The *second* reason is that NCA makes statements about single factors that are necessary independently of other factors in the natural distribution. Thus, the global necessity (with an overall ceiling line) of a single factor does not depend on the level of other factors.[22] This allows for a "pure" and straightforward interpretation of necessity: "the factor is necessary" rather than "the factor is necessary depending on other factors." Such pure necessity statements hold independently of other factors. However, the context in which a necessity statement holds is usually not unlimited. The domain in which the necessary condition is assumed to apply must be defined as the *theoretical domain* of the necessity theory (Chapter 3). Such a specification of the theoretical domain must be part of any necessity theory and related formal hypotheses (Chapter 7).

- A *third* reason is that NCA aims to contribute to parsimonious (simple) theories by avoiding complexity that makes theories difficult to understand and less useful. This is a general goal of theory building in applied sciences. NCA provides an elegant way of reducing complexity, particularly in situations where it is hard or impossible to predict the outcome (e.g., when the explained variances of regression models are low or when accurate prediction require very complex models).

- The *fourth* reason is that practical recommendations derived from single necessary conditions are immediately clear and useful: to achieve a target outcome, all necessary conditions must be satisfied; otherwise, failure of this outcome is guaranteed. The absence of a necessary condition cannot be compensated by other factors, including other necessary conditions.

- The *final* reason is that identifying single necessary conditions is more efficient when using multiple analyses of planes (i.e., several ceiling lines) than when conducting a single analysis of a multidimensional space (i.e., one multidimensional ceiling) followed by projection. Modeling and describing a multidimensional ceiling can be complex (e.g., Figure 4.9). Such a modeling corresponds to *frontier analysis* (Section 4.4), which aims to predict the maximum outcome for a given combination of factors. In contrast, NCA describes the maximum possible outcome for a single condition and can therefore focus *directly* on the XY-planes and their ceiling lines.

---

[22]When subsamples are drawn it is possible that the ceiling for the subgroup is lower than the ceiling of the total group, but the ceiling can never be higher. This subgroup ceiling may represent a different necessity theory with a different theoretical domain, see Section 11.3.3 and Chapter 7.

4.6.4 Chain of necessity

Normally, a necessary condition is hypothesized (Chapter 7) and analyzed (Chapter 9) as a necessity relationship between X and Y. When multiple necessary conditions are proposed, several overall necessity relationships between X_j and Y may be hypothesized. In addition to such parallel necessary conditions, a serial system, a chain of necessity conditions, can also be hypothesized. In that case, an overall relationship $X_1 \to Y$ (if present) still holds, but it may be (partly) explained by an indirect chain $X_1 \to \cdots \to Y$.

Section 3.3 discussed, in general terms, a necessity theory consisting of a chain of necessary conditions (Figure 3.3). For example, a certain amount of study time may be necessary for a certain level of knowledge retention (quiz score), which in turn may be necessary for a high exam grade. This section analyzes such a chain algebraically. Numerical examples are provided for a chain with two links.

4.6.4.1 Analysis of chain necessity

A general necessity chain consists of a sequence of necessary conditions: the first condition X_1, intermediate conditions $X_2, X_3, \ldots, X_J$, and the outcome Y:

$$X_1 \to X_2 \to X_3 \to \cdots \to X_J \to Y \tag{4.50}$$

Here, J is the number of necessary conditions (and thus the chain has J links ending in Y). For example, an additional intermediate necessary condition between study hours and quiz score could be active class participation (engagement, asking questions).

For each link, a ceiling function bounds the subsequent variable:

$$X_{j+1} \le f_j(X_j), \quad j = 1, \ldots, J-1, \qquad Y \le f_J(X_J). \tag{4.51}$$

The chain ceiling $f_{\text{chain}}(X_1)$ expresses the implied necessity of X_1 for Y through the chain. It is defined as the composition of the link-specific ceiling functions:

$$Y \le f_J(f_{J-1}(\cdots f_1(X_1)\cdots)) = f_{\text{chain}}(X_1). \tag{4.52}$$

Here, $f_J(X_J)$ is the ceiling function of the last link.

The phrase "a chain is as strong as its weakest link" also applies to a necessity chain. If any link fails to represent a necessity relationship, the chain collapses, and X_1 is not necessary for Y through the chain. This situation can occur, for example, when a link is sufficient but not necessary, or when a link is described only by an average effect or correlation without necessity.

Throughout, all variables are assumed to be min–max normalized, all ceilings are defined within the same bounding box, and all variables are assumed to be able to reach the extremes of that bounding box.

In the linear case with the unit bounding box $[0, 1] \times [0, 1]$, each link is described by a truncated linear ceiling:

$$f_j(X_j) = \min(a_j + b_j X_j,\ 1), \qquad j = 1, \ldots, J. \tag{4.53}$$

Here, a_j is the intercept and b_j is the slope. Assume that each link represents the standard case of necessity (a ceiling line on or above the diagonal in the unit box), i.e., $0 < a_j < 1$ and $1 < a_j + b_j$.

The resulting chained ceiling is again linear:

$$Y \le f_{\text{chain}}(X_1) = a_{\text{chain}} + b_{\text{chain}} X_1. \tag{4.54}$$

The chain intercept and slope are:

$$a_{\text{chain}} = \sum_{j=1}^{J} \left(a_j \prod_{k=j+1}^{J} b_k \right), \tag{4.55}$$

$$b_{\text{chain}} = \prod_{j=1}^{J} b_j, \tag{4.56}$$

with the convention that an empty product equals 1.

For ceilings on or above the diagonal, $a_{\text{chain}} \ge 1 - b_{\text{chain}}$, which implies $d_{\text{chain}} \le \frac{1}{2} b_{\text{chain}}$. In the unit square bounding box $[0,1] \times [0,1]$, the effect size of the (linear) chain ceiling $Y \le a_{\text{chain}} + b_{\text{chain}} X_1$ is:

$$d_{\text{chain}} = \begin{cases} \dfrac{(1 - a_{\text{chain}})^2}{2\, b_{\text{chain}}}, & \text{if } a_{\text{chain}} < 1, \\ 0, & \text{otherwise.} \end{cases} \tag{4.57}$$

As shown in Equation 4.55, each intercept a_j is weighted by the product of downstream slopes $b_{j+1}, \dots, b_J$, and the total intercept a_{chain} accumulates across the chain. As a result, composing multiple necessary conditions can shift the chain ceiling upward relative to a single link with a comparable slope. Consequently, the chain constraint can become less stringent and may even become non-binding. This explains why, even when each link has a positive effect size, the overall chain can yield a null effect of X_1 on Y through the chain.

In other words, for ceilings on or above the diagonal within the bounding box, the "weakest link" principle extends to effect size: the chain effect size is less than or equal to the smallest link effect size.

Since 'a chain is as strong as its weakest link', the effect size d_{chain} is less than or equal to the effect sizes of all links:

$$d_{\text{chain}} \le \min(d_1, d_2, \dots, d_J) \tag{4.58}$$

where d_j is the effect size of the j-th link in the chain.

4.6.4.2 Numeric examples

When the chain consists of two links and assuming ceiling lines above the diagonal, the intercept and slope of the chain necessity function of Equations 4.55 and 4.56, reduce to:

$$a_{\text{chain}} = a_1 b_2 + a_2 \tag{4.59}$$

$$b_{\text{chain}} = b_1 b_2 \tag{4.60}$$

If all variables are min-max normalized to $[0,1]$, the effect size of each link is:

$$d_j = \frac{(1-a_j)^2}{2b_j} \tag{4.61}$$

The equation for effect size d_{chain} of the chain necessity of X_1 for Y is:

$$d_{\text{chain}} = \begin{cases} \frac{(1-a_{\text{chain}})^2}{2b_{\text{chain}}}, & \text{if } a_{\text{chain}} < 1 \\ 0, & \text{otherwise} \end{cases} \tag{4.62}$$

Two numeric examples of a chain consisting of two links (with ceiling lines above the diagonal) illustrate the behavior that the chain is as strong as its weakest link. In the first example, the two ceiling functions are:

$$f_1(X_1) = 0.4 + 2.0 \cdot X_1$$
$$f_2(X_2) = 0.2 + 1.0 \cdot X_2$$

According to Equations (4.60) and (4.59) the intercept and slope for the chain ceiling line are:

$$a_{chain} = 0.6$$
$$b_{chain} = 2.0$$

Following Equation (4.61), the two effect sizes d_1 and d_2 are:

$$d_1 = \frac{(1-0.4)^2}{2\cdot 2.0} = \frac{0.36}{4.0} = 0.09$$
$$d_2 = \frac{(1-0.2)^2}{2\cdot 1.0} = \frac{0.64}{2.0} = 0.32$$

According to Equation (4.62) the chain effect size is:

$$d_{chain} = \frac{(1-0.6)^2}{2\cdot 2.0} = \frac{0.16}{4} = 0.04 \tag{4.63}$$

The three XY-plots are shown in Figure 4.11.

The second example is an example where chain necessity vanishes. Two ceiling functions are:

$$f_1(X_1) = 0.4 + 0.7 \cdot X_1$$
$$f_2(X_2) = 0.7 + 0.8 \cdot X_2$$

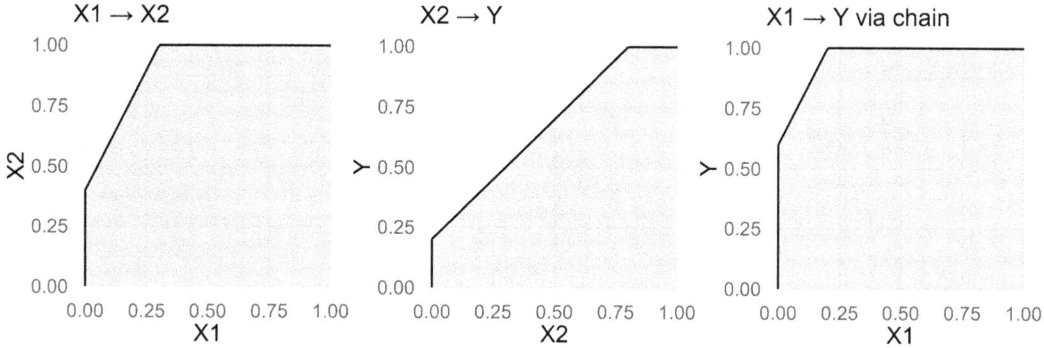

FIGURE 4.11 Example of a necessity chain. The effect sizes of the links are: $X_1 \to X_2 = 0.09$, $X_2 \to Y = 0.32$, $X_1 \to Y = 0.04$.

According to Equations (4.60) and (4.59), the intercept and slope for the composite ceiling lines are:

$$a_{chain} = 1.02$$
$$b_{chain} = 0.56$$

According to Equation (4.61), the two effect sizes d_1 and d_2 are:

$$d_1 = \frac{(1-0.4)^2}{2 \cdot 0.7} = \frac{0.36}{1.4} \approx 0.26$$
$$d_2 = \frac{(1-0.7)^2}{2 \cdot 0.8} = \frac{0.09}{1.6} \approx 0.06$$

Referring to Equation (4.62), the chain effect size is:

$$d_{chain} = 0 \tag{4.64}$$

The three XY-plots are shown in Figure 4.12.

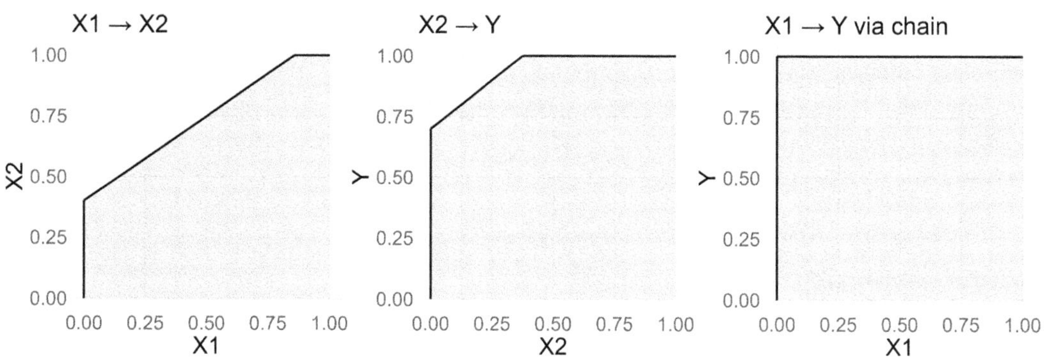

FIGURE 4.12 Example of a necessity chain. The effect sizes of the links are: $X_1 \to X_2 = 0.26$, $X_2 \to Y = 0.06$, $X_1 \to Y = 0$.

The examples illustrate that the chain effect size is smaller than the link effect sizes, and that the chain effect size can be zero while those of the links are not.

5

Statistics

5.1 Summary of this chapter

This chapter discusses several statistical topics relevant to NCA. Although NCA is primarily a (quasi-)deterministic mathematical (geometric) method rather than a statistical method based on probabilities, several statistical approaches and tools are still relevant for NCA. First, the chapter discusses the use of sampling logic in quantitative (large-n) NCA studies (Section 5.2). Second, four specific statistical tools and approaches for NCA are presented. These are a set of permutation tests for testing null hypotheses (Section 5.3), statistical modeling with NCA in particular for Monte Carlo simulation studies (Section 5.4), and the use of these simulations for evaluating the statistical quality of ceiling lines (Section 5.5) and for conducting power analyses (Section 5.6). Finally, the chapter discusses how three common problems in inferential statistics apply to NCA: reverse causality (Section 5.7), measurement error (Section 5.8) and spuriousness (Section 5.9).

5.2 Sampling logic

In large-n quantitative studies, NCA uses a sampling approach that is aligned with common practices in inferential statistics.[1]

For valid statistical inference, the sample should be representative of the population. Representativeness in NCA is especially important with respect to adequate coverage of cases with high values of Y and varying levels of X and combinations of X with other variables. General information on sampling logic in NCA is provided in Dul (2024b) and sampling for NCA's statistical testing is discussed in Section 8.3. Usually, random sampling satisfies the requirements.

5.3 Permutation tests

NCA's set of statistical tests consists of permutation tests to estimate the p-value for evaluating effect sizes and differences between effect sizes. NCA's original test (called

[1] In qualitative (small-n) applications of NCA, such as case studies, purposive case selection can be used without conducting statistical tests, see Section 8.2.3.

DOI: 10.1201/9781003430766-5

"Null" here) evaluates if the observed effect size differs from 0. The other tests are called:

- "Contrast", to evaluate if two conditions for the same Y have different effect sizes,
- "Independent", to evaluate if two independent groups of cases have different effect sizes,
- "Paired", to evaluate if effect sizes at two time points are different.

First, the general principles of the p-value and the permutation test are discussed, followed by a description of each test. The three newer tests are available in the `NCA` software in `R` from version 5.0.0.

5.3.1 General

According to the American Statistical Association (ASA), in statistical null hypothesis testing the p-value is a measure that helps assess the strength of evidence *against a null hypothesis* (Wasserstein & Lazar, 2016). In its 2016 statement on p-values, the ASA discusses what the p-value is, and what it is not. Essentially, the p-value indicates the compatibility of the data with the null hypothesis.

NCA tests the compatibility of the observed necessity effect size, or effect size difference with the null hypothesis of no relationship (the Null test) or no difference in distribution (the Contrast, Independent, and Paired tests). The unique feature of NCA's null hypothesis tests is that it uses the necessity effect size as the test statistic. Note that no null hypothesis test can test an alternative hypothesis. If the null hypothesis is rejected, this does not imply that a specific alternative hypothesis is correct. A small p-value only suggests that the observed effect size (difference) is unlikely under the null hypothesis.

Scientific conclusions about the hypothesis of interest (the alternative hypothesis) should not be based solely on whether a p-value is below a threshold level (e.g., 0.05). For this reason, in concluding that a necessity hypothesis is supported,[2] NCA not only requires a low p-value but also the presence of theoretical support and a practically relevant effect size. These are three minimum requirements for concluding about necessity. Together they protect the analyst against making false positive or false negative conclusions (Chapter 6).

NCA does not make an assumption about the type of distribution of the data (e.g., normality). The permutation test is a valid statistical approach for testing the randomness of effect size (difference) that does not rely on assumptions about the distribution of the data (like normality in t-tests or ANOVA). The permutation test constructs a null distribution empirically by resampling in a specific way from observed data. First, the test statistic, NCA's effect size (difference), is computed using the original data. Next, resamples with the same sample size and the same sets of X- and Y-values are validly created by a specific type of permutation assuming that the null hypothesis of no relationship (or same distribution) applies. By rearranging the data randomly through permutations, any systematic relationship between X and Y, or difference in effect size, is broken. This is done differently for each test (see below). For the resample (permutation), the effect size (difference) is recomputed, and this is repeated many times (e.g., 10,000) to obtain a null distribution of values for the effect size (difference). Finally, to compute the p-value, the observed effect size (difference) is compared with the effect sizes (or effect size differences) obtained under the null distribution. The p-value is the proportion of permuted effect sizes (or effect size differences) under the null that are as extreme or more extreme than the observed effect size (difference).

[2]In the context of statistical testing, this book may use the term "support" of a hypothesis in the meaning of "non-rejection"

The validity of NCA's four permutation tests (Null, Contrast, Independent, Paired) is based on a theorem (Hoeffding, 1952; Kennedy, 1995; Lehmann et al., 2005). The justification of the test rests on the so-called Randomization Hypothesis being true. The Randomization Hypothesis states:

> Under the null hypothesis, the distribution of the dataset D is invariant under the transformation p, that is, for every transformation $p \in \mathcal{P}$, pD and D have the same distribution whenever D has distribution P in H_0 (Lehmann et al., 2005, p. 633).

Here, D denotes the observed dataset (all cases and variables used in the test). The symbol p denotes one admissible randomization operation from the permutation scheme (e.g., a particular shuffling/relabeling/reassignment as defined by the test), and $\mathcal{P}$ denotes the set of all admissible operations of that type. The expression pD denotes the transformed dataset obtained by applying p to D (e.g., shuffling Y across cases while keeping X fixed, swapping within-case labels such as X_1 and X_2, reassigning cases to groups while preserving group sizes, or swapping time labels within a case). Finally, $D \sim P \in H_0$ means that the original dataset D is assumed to be generated from some distribution P that satisfies the null hypothesis H_0; invariance then states that pD is distributed the same way as D under H_0.

In other words, the validity of the permutation test rests on the fact that, under H_0, applying any admissible transformation $p \in \mathcal{P}$ produces a dataset pD that is distributed like the original dataset D. Thus, random draws of p generate valid "null" replicates of the data. Each test simply uses a different admissible transformation set $\mathcal{P}$ (i.e., different permissible relabelings/assignments) corresponding to the hypothesis and the design:

- Null: $H_0 : d_{X \to Y} = 0$.
- Contrast: $H_0 : d_{X_1 \to Y} - d_{X_2 \to Y} = 0$.
- Independent: $H_0 : d_{G_1(X \to Y)} - d_{G_2(X \to Y)} = 0$.
- Paired: $H_0 : d_{T_1(X \to Y)} - d_{T_2(X \to Y)} = 0$.

where d is the effect size, and X_1 and X_2 are two different conditions, G_1 and G_2 two different groups, and T_1 and T_2 two different time points. The differences between the four tests are shown in Table 5.1.

5.3.2 Assumptions for NCA's permutation tests

Beyond representativeness, NCA's permutation tests additionally rely on the assumption of *exchangeability* under the null hypothesis. Exchangeability means that, under H_0, the joint distribution of the observed data is invariant under the admissible permutations (i.e., relabelings/reassignments) specified by the test's permutation scheme. More specifically, for the difference tests, NCA's permutation test assumes that exchangeability holds after a min-max normalization for each group that puts the XY variables in each group on the same (dimensionless) scale, for example a 0–1 scale. Such normalization does not change effect sizes for the two groups involved (Appendix D). In practical terms, this implies that

TABLE 5.1 Comparison of NCA's permutation tests

Test name	Sample(s)	Question	Permutation logic
Null	One	Does X constrain Y at all?	Randomly shuffle Y across cases while keeping X fixed.
Contrast	One	Which condition (X_1, X_2) constrains Y more?	For each case, randomly switch the labels X_1 and X_2 with 50/50 probability.
Independent	Two (Groups)	In which group (G_1, G_2) does X constrain Y more?	Pool both groups (G_1 and G_2), randomly reassign cases to two groups of the same sizes.
Paired	Two (Times)	Does the constraint of X on Y change over time (T_1, T_2)?	For each case, randomly decide whether to switch the time labels (T_1, T_2).

no case has a special status with respect to the tested relationship once the null hypothesis of no necessity (or no difference in effect size) holds. When this condition is met, applying the test-specific permutations (permuting X-Y pairings, swapping within-case condition or time labels, or reassigning group labels) generates valid resamples from the same null distribution.

Representativeness and exchangeability are related but distinct requirements. Representativeness concerns the external validity of the sample (whether conclusions can be generalized to the population), whereas exchangeability concerns the internal validity of the permutation test (whether the permutation distribution correctly reflects H_0). Exchangeability is in general satisfied when the relevant units are independent and identically distributed under H_0, and when the sampling or measurement design does not introduce systematic differences that are not respected by the permutation scheme.

5.3.3 Null test: effect size difference from 0

In the *Null* test, the observed effect size of a single sample is compared with the null of no relationship. This is the standard NCA null test described in detail elsewhere (Dul, Van der Laan, et al., 2020). The test addresses whether X affects Y at all, operationalized as whether effect size $d > 0$. The permutation logic is to generate the null distribution by randomly shuffling Y-values across cases while keeping X-values fixed. This breaks the pairing of X- and Y-values, which is admissible under the null of no relationship.

The transformation set $\mathcal{P}$ (see the Randomization Hypothesis) contains all possible transformations (permutations) of the data under the null; for n cases this corresponds to $n!$ permutations of Y-values across cases (permutations of case labels applied to Y). Under H_0, shuffling Y does not change the joint distribution of the data, so the permuted datasets are valid null replicates. The test statistic is the XY effect size d. The Null test is one-sided.

5.3.4 Contrast test: effect size difference between two conditions

In the *Contrast* test, two observed effect sizes from the same sample are compared with the null of no difference between them. The test addresses which condition, X_1 or X_2, constrains Y more within the same sample. The permutation logic is to generate the null distribution by, for each case, randomly swapping the condition labels 1 and 2 with 0.5 probability.

This preserves each case's observed part of X-values and its Y-value, but randomizes which X-value is labeled X_1 versus X_2, thereby removing any systematic difference between the conditions under the null.

The transformation set $\mathcal{P}$ of admissible permutations under the null consists of all possible within-case label swaps of X_1, X_2 across the n cases (i.e., 2^n possible swap patterns). The null hypothesis is that the joint distribution is invariant to exchanging the labels 1 and 2, implying that X_1 and X_2 constrain Y equally. The test statistic is the difference in necessity effect sizes, $d_{X_1 \to Y} - d_{X_2 \to Y}$. The Contrast test is two-sided.

5.3.5 Independent test: effect size difference between two groups

In the *Independent test*, the observed effect sizes from two independent samples (group G_1 and group G_2) are compared. The test addresses in which group X constrains Y more. The permutation logic is to generate the null distribution by pooling cases from both groups and randomly reassigning them to two groups of the same sizes (i.e., permuting the group labels G_1 and G_2). This removes any systematic group difference under H_0 while preserving the observed (X, Y) pairs.

The transformation set $\mathcal{P}$ of admissible permutations under the null consists of all reallocations of cases to G_1 and G_2 that preserve the original group sizes. The null hypothesis is that the joint distribution of (X, Y) is the same in both groups, so group membership is exchangeable; consequently, reassigning cases to groups does not change the distribution under H_0. The test statistic is the difference in necessity effect sizes, $d_{G_1(X \to Y)} - d_{G_2(X \to Y)}$. The Independent test is a two-sided test.

5.3.6 Paired test: effect size difference between two time points

In the *Paired* test, the observed effect size difference between two dependent (paired) samples (e.g., repeated measures) is compared with the null hypothesis of no effect size difference. In this design, the same cases are measured twice (e.g., at T_1 and T_2 on the same X and Y). The test addresses whether the constraint of X on Y changes over time. The permutation logic is to generate the null distribution by, for each case, randomly deciding whether to swap the time labels T_1 and T_2 (i.e., swapping the paired observations) with probability 0.5.

The transformation set $\mathcal{P}$ of admissible permutations under the null consists of all within-case time-label swaps across the n cases (i.e., 2^n possible swap patterns). The null hypothesis is that the (X, Y) observations at both time points are drawn from the same distribution, so swapping the time labels within cases leaves the joint distribution unchanged. The test statistic is the difference in necessity effect sizes, $d_{T_1(X \to Y)} - d_{T_2(X \to Y)}$. The Paired test is a two-sided test.

5.3.7 Common scope

For comparability of two effect sizes, a common scope should be selected. Depending on the question one tries to answer, this can be the pooled empirical scope (based on observed minimum and maximum scale values), a theoretical scope (e.g., based on theoretical minimum and maximum scale values), or a min-max normalized scope (e.g., [0,1] $\times$ [0,1]) that makes the scales dimensionless and numerically equal.

5.4 Statistical modeling

As mentioned earlier, NCA is primarily a (quasi-)deterministic mathematical (geometric) method rather than a statistical method based on probabilities. For a given XY-pair, the maximum value of Y for a given value of X can be expressed as the inequality:

$$Y \leq f_j(X_j), \qquad X_j \in [x_{j\,\mathrm{min}}, x_{j\,\mathrm{max}}],\ Y \in [y_{\mathrm{min}}, y_{\mathrm{max}}] \tag{5.1}$$

where $f_j(X_j)$ is the ceiling line in the X_jY-plane ($j = 1, \ldots, J$) within the bounding box.

In statistical terms, NCA is a 'semi-parametric' approach that only specifies particular features of the distribution (Greene, 2012, p. 482). In NCA, these specified features are the ceiling lines $f_j(X)$ and the bounding boxes. It is assumed that there are no errors of measurement in variables X and Y. In NCA the distribution under the ceiling is not specified. This is inherent to NCA's goal of analyzing necessary *but not sufficient* conditions: a certain level of X is required for having a certain level of Y but it does not automatically produce that level of Y. In other words, the aim of NCA is not to predict the presence of Y below the ceiling, but to predict the absence of Y above the ceiling.

For some applications with NCA (e.g., Monte Carlo simulations), it is needed to specify the full data generation process (DGP) and to make assumptions about the distribution of data under the ceiling. In that case the analyst must make such an assumption, and Equation (5.1) must be extended with an error term, representing the distribution below the ceiling line:

$$Y = f_j(X) - \varepsilon_j \qquad X_j \in [x_{j\,\mathrm{min}}, x_{j\,\mathrm{max}}],\ Y \in [y_{\mathrm{min}}, y_{\mathrm{max}}] \tag{5.2}$$

where $f_j(X_j)$ is the ceiling line in the X_jY-plane ($j = 1, \ldots, J$) within the bounding box, and ε_j is a non-negative random variable representing the distance of an observation below the ceiling.[3] When the border line is a ceiling line (i.e., when the upper-left or the upper-right corner of the X_jY-plot is empty), ε_j takes non-negative values only. When the border line is a floor line (i.e., when the lower-left or lower-right corner is empty), ε_j takes only non-positive values. Depending on the application, the error term is specified.

Thus, for conducting Monte Carlo simulations with NCA not only the NCA model (Section 4.3) but also the distribution of the data under the ceiling must be specified. This means that the analyst must make choices about:

1. The bounding box (bounded variables).
2. The ceiling line.
3. The distribution of the data.

Examples of bounded variable distributions are the uniform distribution and the truncated normal distribution. The uniform distribution can be considered a distribution with minimum assumptions: only two parameters must be assumed, namely the variable's upper and lower bound. The truncated normal distribution needs two additional parameters (mean

[3] NCA's statistical model differs from a regression model. The regression model is discussed in Section 11.3.1, see Equation (11.1).

and standard deviation). To obtain a bivariate distribution for X and Y while ensuring an empty space above the true ceiling, the selected univariate distributions for X and Y (e.g., both uniform, or one uniform and the other truncated normal) within the bounding box $[0,1] \times [0,1]$ are combined by removing points above the true ceiling line. Figure 5.1 shows typical random samples of n = 100 drawn from the two bivariate population distributions using the `nca_random` function of the `NCA` software in `R`. In Figure 5.1-left both X and Y have uniform-based distributions and in Figure 5.1-right both distributions are truncated normal-based.[4]

In most applications of NCA, the ceiling line is estimated from the data, rather than being determined in advance (e.g., theoretically or based on earlier studies). For estimating the ceiling line with empirical data, the techniques mentioned in Section 4.4 could be used.

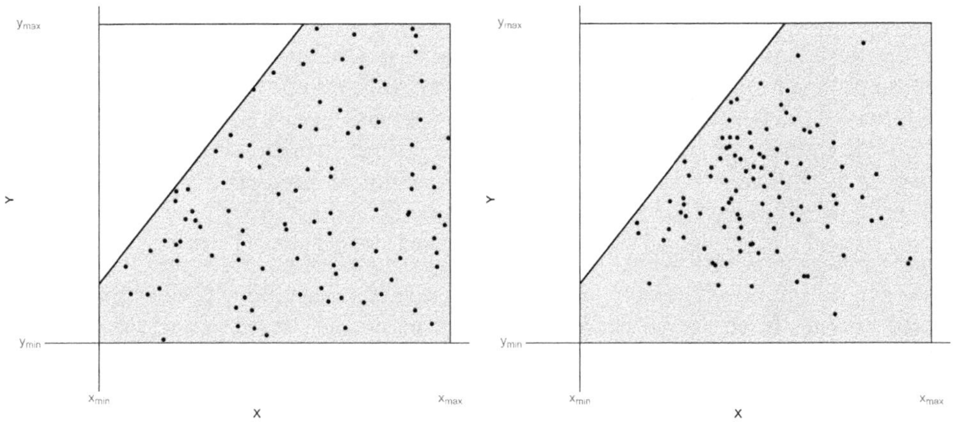

FIGURE 5.1 Random sample (n = 100) from a population with a true linear ceiling line with a uniform-based distribution (Left) and a truncated normal-based distribution (Right).

Note that a necessity relationship cannot be simulated with unbounded distributions that are unbounded in both directions with non-zero probability in every region of the XY-plane, such as the normal distribution. With such variables a necessity effect size cannot exist in the population. The reason is the following. In NCA, a necessity effect is defined by the presence of an empty ceiling zone where no observations occur because they are impossible due to a necessity constraint. This ceiling zone is bounded by the ceiling line and the bounding box. Together, the ceiling line and the bounding box define the feasible area which is the region where observations can occur. If cases are drawn from such distributions, the assumptions needed for a true necessity effect are violated. With an assumption of such distributions, both X and Y can take arbitrarily large (or small) values. As sample size increases, the maximum and minimum values increase, and the observed bounding box expands. There is no natural bound to define a fixed box for the ceiling zone. In a population with such distributions, there is always a non-zero probability that any combination of X, Y occurs, even values far above a supposed ceiling line. Therefore, no region of the space is guaranteed to be empty, and no necessity effect can exist in the population. Because in this situation the ceiling zone is never truly empty in the population, the true necessity effect size that is defined as the limit of the proportion of the bounding box that is unpopulated is exactly zero. Any non-zero effect size seen in a sample is an artifact. Even when the population has

[4]The `nca_random` function in the `NCA` software can be used to generate data from a population with a specified linear ceiling line (intercept and slope) for two types of distributions (uniform-based and truncated normal-based).

no necessity constraint, a finite sample will always have a maximum observed value of X and Y, a finite bounding box (scope), and an empirical ceiling line that appears to limit the upper boundary of observations. This may give the illusion of a ceiling zone. NCA may then calculate a non-zero necessity effect size, but this is due to sampling limits not a true structural constraint.

For example, suppose X and Y are drawn from a bivariate normal distribution $X, Y \sim \mathcal{N}(0,1)$. In a sample of n = 100, the observed range might be approximately [-3, 3] for both X and Y. A ceiling line (e.g., $y = 2 + 0.5x$) may appear to bound the upper edge of the XY-plot. NCA would calculate a non-zero effect size based on the empty region above this line. However, in the full population, values of Y above this line still occur with non-zero probability. As $n \to \infty$, with a fixed theoretical scope, the ceiling zone fills in, causing the effect size to decrease to zero. With an empirical scope the effect size also decreases to zero but extremely slowly.

When NCA calculates an effect size from sample data, it estimates a property of the population. But in the case of such distributions of X and Y, no necessity property exists to estimate. The observed sample statistic does not reflect a meaningful necessity effect in the population. When unbounded variables are used in Monte Carlo simulations it is inherently assumed that necessity does not exist for the phenomenon that is simulated. An analyst's choice to use such a distribution therefore implies the analyst's theoretical expectation that necessity does not exist in the population and that any observed empty space is an artifact. Note that even if a statistically significant empty space would have been detected, NCA does not support necessity if there is no theoretical justification for it (Chapters 6 and 7).

In the remainder of this chapter, two applications of NCA with specification of the full data generation process that include a ceiling line, are presented: the use of Monte Carlo simulation for evaluating the quality of NCA's ceiling lines, and the use of Monte Carlo simulation for conducting a power analysis for NCA. Another application of Monte Carlo simulations for evaluating the statistical credibility of NCA in terms of True Positive Rate (sensitivity) and True Negative Rate (specificity) is presented in Chapter 6 about the credibility of NCA.

5.5 Quality of ceiling lines

When a necessity relationship is present in the population, an empty area in a sample is a manifestation of the necessity effect in the population. To identify this necessity effect, NCA estimates the ceiling line, the bounding box, and the related effect size.

5.5.1 Statistical quality criteria

When analyzing sample data for statistical inference, the calculated statistic (in NCA the ceiling line's effect size) is an *estimator* of the true parameter in the population. An estimator is considered to be a good estimator when for finite (small) samples the estimator is *unbiased* and *efficient*, and for infinite (large) samples the estimator is *consistent*. Unbiasedness indicates that the mean of the estimator for repeated samples approximates the true parameter value; efficiency indicates that the variance of the estimator for repeated samples is small compared to other estimators of the population parameter. Consistency indicates that as sample size increases, the mean of the estimator approaches the true value of the parameter and the variance approaches zero. When the latter applies, the estimator is called *asymptotically unbiased*, even if the estimator is biased in finite (small) samples. An estimator is *asymptotically efficient* if the estimate converges relatively fast to the true population value when the sample size increases.

There are two ways to determine the unbiasedness, efficiency and consistency of estimators: *analytically* and by *simulation*. In the analytic approach the properties are mathematically derived. This is possible when certain assumptions are made. For example, the OLS regression estimator of a linear relationship in the population is 'BLUE' (Best Linear Unbiased Estimator) when the Gauss-Markov assumptions hold (the dependent variable is unbounded, homoskedasticity, error term unrelated to predictors, etc.).

In the simulation approach, a Monte Carlo simulation is conducted by first specifying the true population parameters (variables and distributions). Repeated random samples are drawn from this population, and the estimator is computed for each sample. This yields a sampling distribution of the estimator, which is then compared to the known population parameter.

Both analytic and simulation approaches generally assume an idealized scenario: an infinite population, random sampling, known distributions, and no measurement error. Deviations from these assumptions substantially increase analytic and computational complexity.

Data transformations are not uncommon, for example, to ensure normal distributions or meeting assumptions. However, data transformation may change the nature and content of the theory. In NCA data transformation can change the bounding box, and if not affine, these transformations are likely to change the value of the necessary condition effect size, thereby introducing spurious effect sizes. For example, if X is necessary for Y, then $\log(X)$ may not be necessary for $\log(Y)$ and vice versa (Appendix D). In NCA, the analysis is therefore always conducted with data that represent the theoretical meaning of the concepts in the necessity hypothesis.

5.5.2 Simulation results for NCA's effect size

For NCA, currently no analytic approach to determine bias, efficiency, and consistency exists. However, NCA can rely on simulations to evaluate the quality of its estimates. In this section, Monte Carlo simulation is used for estimating the true effect size in a population using four types of ceiling line estimation techniques. These ceiling lines are the stepwise linear ceiling line CE-FDH, and three linear ceiling lines CR-FDH, C-LP and QR. The QR-ceiling line is not recommended (Section 4.4), but is added for comparison. The theoretical scope (unit box) is selected to make the results comparable.

In Monte Carlo simulation, the full DGP needs to be specified. It is assumed that a high-high direction of necessity exists in the population (high X is necessary for high Y), with a true linear ceiling line $Y = 0.4 + X$. The bounding box is assumed to be the unit box. The corresponding true effect size is 0.18. Next to the specification of the true ceiling line and bounding box, a Monte Carlo simulation also requires a specification of the true distribution of the data under the ceiling line. Three distributions are considered: uniform distribution, symmetric truncated normal distribution with equal means of X and Y (0.5), and left-skewed truncated normal distribution with high mean for X (0.75) and low mean for Y (0.25). The standard deviation of all truncated normal distributions is 0.2.

The simulation draws samples from these predefined population characteristics. The seven sample sizes that are selected vary between 20 and 5000. One hundred samples are drawn per sample size. This is repeated for each form of the ceiling line.[5] Figure 5.2 shows a typical sample selected from each distribution, also showing the four ceiling lines that were tested.

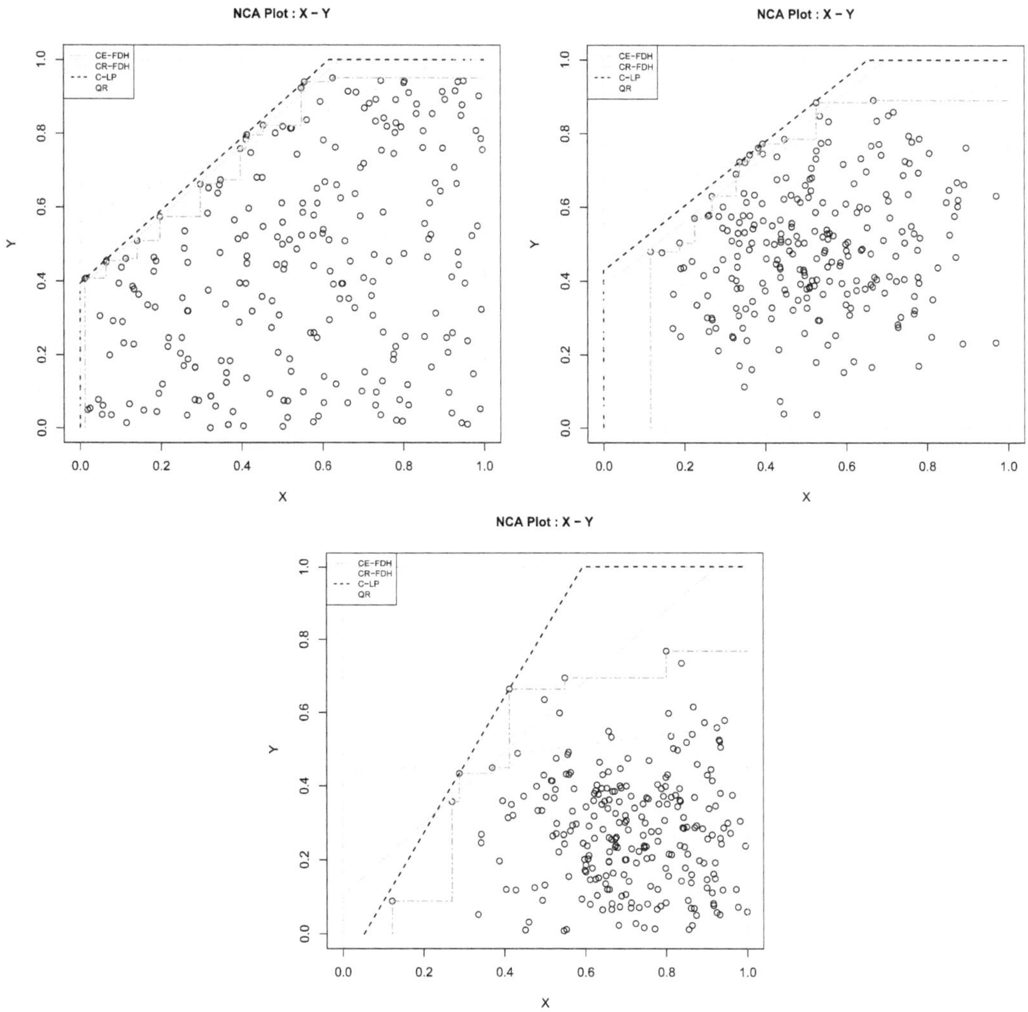

FIGURE 5.2 Random samples with uniform, truncated normal and skewed truncated normal joint distribution with four ceiling lines estimation techniques. True ceiling line = $Y = 0.4 + X$. True effect size is 0.18.

Figures 5.3, 5.4, and 5.5 show the results for the uniform, truncated normal and skewed distributions, respectively. For small samples, the effect sizes are upward biased (the true effect size is smaller) for all estimation techniques. The bias of the estimated effect sizes can be large with the truncated normal and skewed distributions, caused by the limited number of points near the ceiling. These are the only points that are informative for the ceiling line in the population.

When sample size increases, the bias and the variance decrease for all estimation techniques. With increasing sample size, the effect size approaches the true effect size for three ceiling

[5]Additional simulations have been done with different distributions (Massu et al., 2020) but these results are similar and are not reported here.

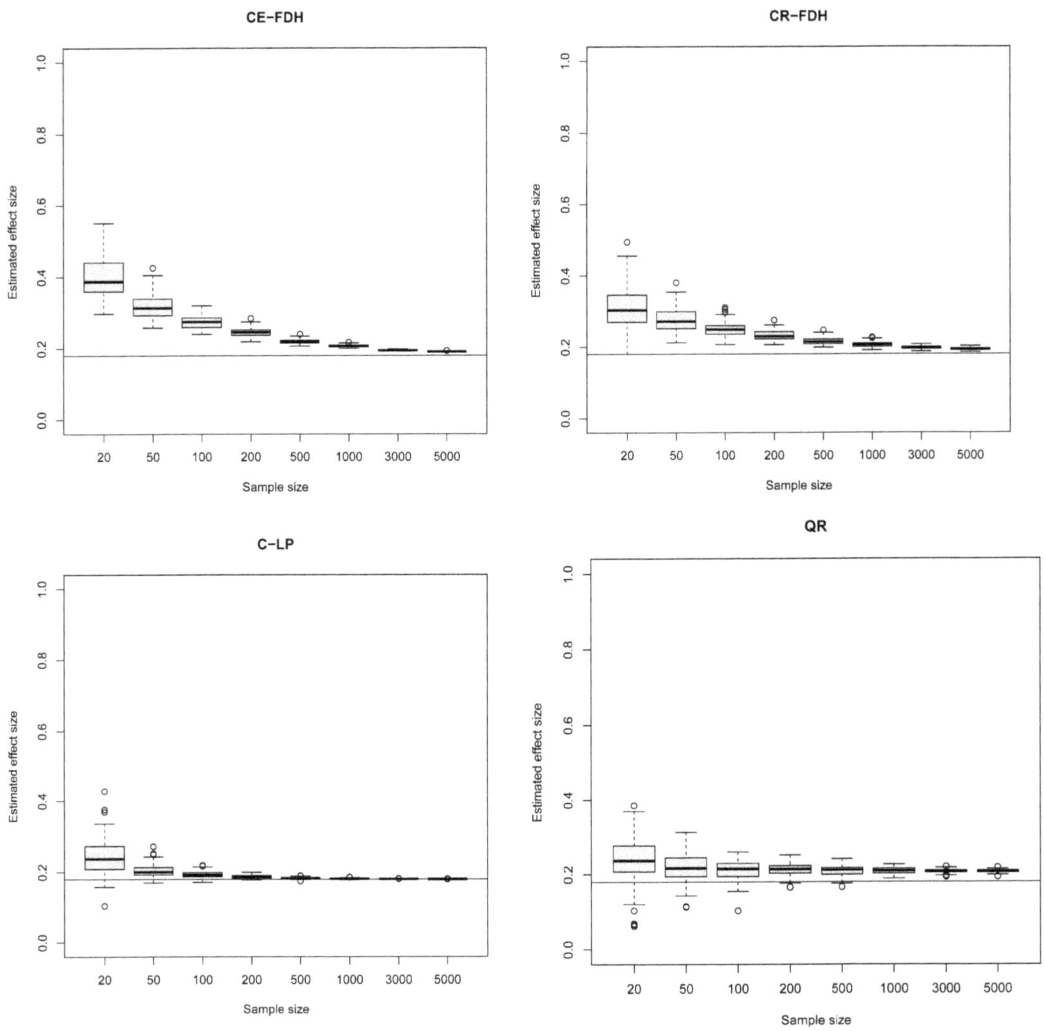

FIGURE 5.3 Monte Carlo simulation results for the effect size of four ceiling line estimation techniques. True ceiling line = $Y = 0.4 + X$. True effect size is 0.18. Uniform distribution.

techniques: CE-FDH, CR-FDH and C-LP. This means that these techniques are asymptotically unbiased and that the three estimators are consistent. Of these ceiling lines, C-LP seems more efficient, reaching the true ceiling line more quickly and with the least variation. With increasing sample size, the QR estimation technique does not approach the true effect size and remains asymptotically biased. Possible reasons are discussed in Section 4.4.

These results only apply to the investigated situations with a true ceiling line that is linear, and data that have no measurement error. Therefore, in this situation the C-LP line may be a preferred technique. Such 'ideal' circumstances may apply in simulation studies, but seldom apply in reality. When the true ceiling line is linear but cases have measurement error, the CR-FDH line may perform better because measurement error usually reduces the effect size (ceiling line moves upwards). When the true ceiling line is not linear, CE-FDH may perform better because this line can better follow the non-linearity of the border

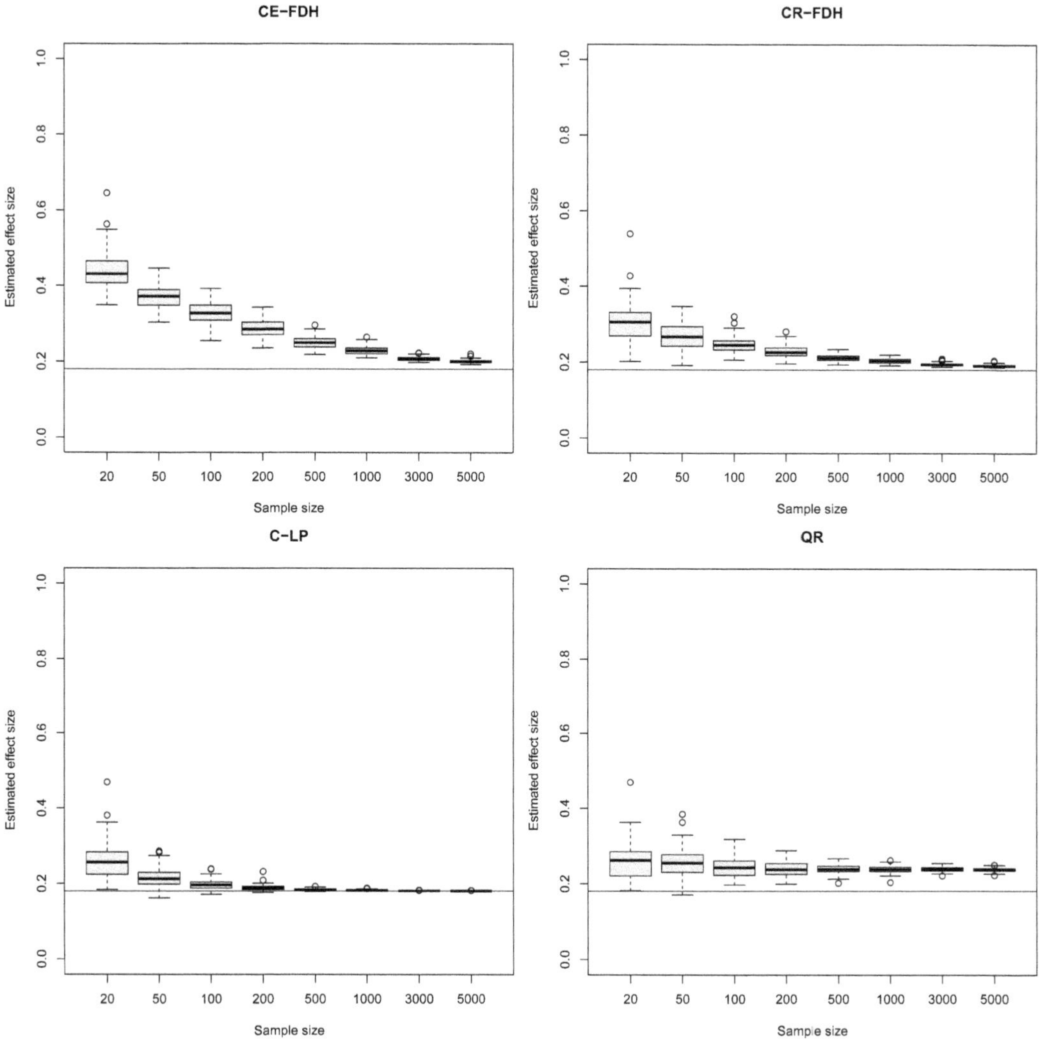

FIGURE 5.4 Monte Carlo simulation results for the effect size of four ceiling line estimation techniques. True ceiling line = $Y = 0.4 + X$. True effect size is 0.18. Truncated normal distribution x-mean = 0.5, y-mean = 0.5.

(Figure 8.1). Further simulations need to be done to clarify the different statistical properties of the ceiling lines under different circumstances. In these simulations the effects of different population parameters (effect size, ceiling slope, ceiling intercept, non-linearity of the ceiling, distribution under the ceiling), and of measurement error (error-in-variable models) on the quality of the estimations could be studied.

5.6 Power analysis

"The power of a test to detect a correct alternative hypothesis is the pre-study probability that the test will reject the test hypothesis (the probability that p will not exceed

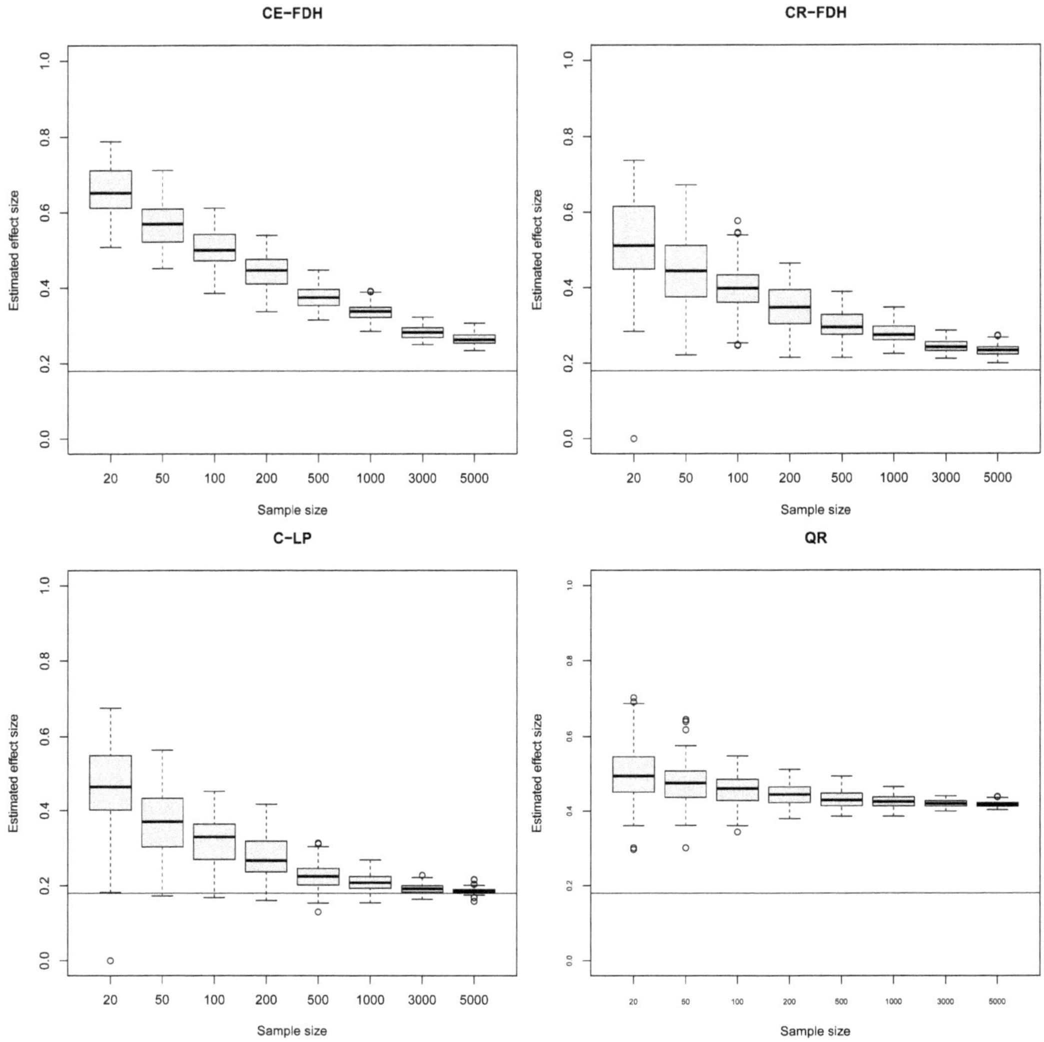

FIGURE 5.5 Monte Carlo simulation results for the effect size of four ceiling line estimation techniques. True ceiling line = $Y = 0.4 + X$. True effect size is 0.18. Skewed truncated normal distribution x-mean = 0.75, y-mean = 0.25.

a pre-specified cut-off such as 0.05)" (Greenland et al., 2016, p. 345). The higher the power of the test, the more sensitive the test is to reject the null hypothesis (and to accept the assumed alternative necessity hypothesis) if necessity exists in the population. Knowledge about the power of NCA's statistical test is useful for planning an NCA study. It can help to establish the minimum required sample size for being able to detect, with a high probability (e.g., 0.8), an existing necessity effect size (difference) that is considered relevant (e.g., for the Null test effect size = 0.10). Note that a post-hoc power calculation as part of the data analysis of a specific study does not make sense (Christogiannis et al., 2022; Lenth, 2001; e.g., Zhang et al., 2019). In a specific study, power does not add new information to the p-value. Because power is a *pre-study* probability, a study with low power (e.g., with a small sample size) can still produce a small p-value and conclude that a necessary condition exists, although the probability of identifying this necessity effect is less than with a high-powered study (e.g., with a large sample size).

Statistical power increases when effect size and sample size increase. For a given expected necessity effect size in the population, the probability of not rejecting the null hypothesis when necessity is true can therefore be reduced by increasing the sample size.

Power of NCA's statistical test can be evaluated by simulation.[6] The simulation presented evaluates the power of NCA's Null test. The simulation includes five population effect sizes ranging from 0.10 (small) to 0.50 (large) (Figure 5.6). The population effect size is the result of a linear population ceiling line with slope = 1 and an intercept that corresponds to the given population effect size.

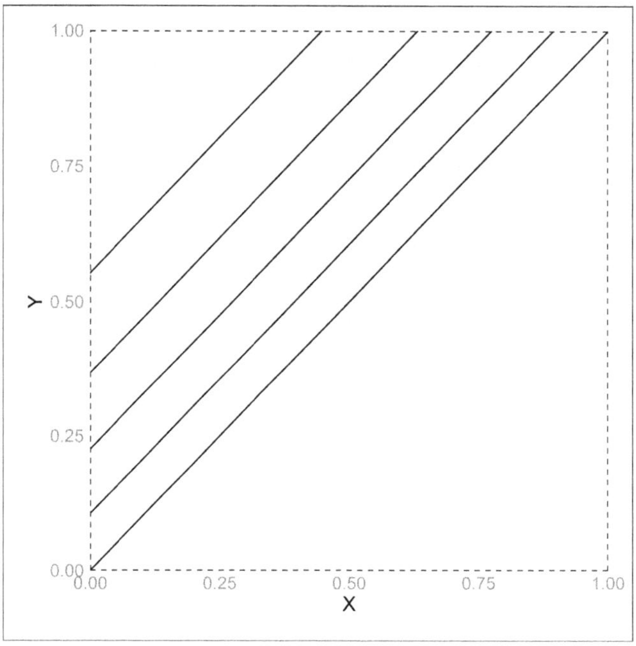

FIGURE 5.6 Five ceiling lines (slope = 1) with corresponding effect sizes of 0.10, 0.20, 0.30, 0.40, and 0.50.

For the simulations, bivariate population distributions are obtained with the procedure as described in Section 5.4. A large number of samples (500) of a given sample size are randomly drawn from the bivariate population distribution. This is done for 2 × 5 × 11 = 100 situations: two bivariate distributions (uniform-based or truncated normal-based), five effect sizes (0.10, 0.20, 0.30, 0.40, 0.50), and 11 sample sizes (5, 10, 20, 30, 40, 50, 100, 200, 500, 1000, 5000). In each situation, NCA's *p*-value is estimated for all 500 samples. The power is calculated as the proportion of samples with estimated *p*-value less than or equal to the selected threshold level of 0.05. This means that if, for example, 400 out of 500 samples have $p < 0.05$, the power of the test in that situation is 0.8. A power value of 0.8 is a commonly used benchmark for a high-powered study.[7]

The relationship between sample size and power for different population necessity effect sizes is shown in Figure 5.7 for the uniform distribution and in Figure 5.8 for the truncated normal distribution. The figures show that when the effect size to be detected is large, the

[6] A pre-study power analysis for the Null test can be done with the nca_power function in the NCA software.

[7] To keep computation time reasonable, the number of permutations to estimate NCA's *p*-value for a given sample is limited to 200, and the number of resamples is set to 500; more accurate power estimations are possible with more permutations and more resamples.

power increases rapidly with increasing sample size. A large necessity effect of about 0.50 can almost always be detected (power ~ 1.0) with a sample size of about 30. However, to obtain a high power of 0.8 for a small effect size of about 0.1, the sample size needs to be doubled (to about 60) with a uniform distribution, and to more than 10 times larger (to about 300) with a truncated normal distribution.

In general, the minimum required sample size to detect a necessary condition is larger for a truncated normal distribution than for a uniform distribution. The reason is that the truncated normal distribution has a lower density of cases in the lower-left and upper-right corners of the XY-plot, such that the empty space in the upper-left corner is less frequently challenged during the permutation test (Figure 5.1). Additional simulations indicate that sample sizes must be larger when the density of observations near the ceiling is smaller, in particular when there are few cases in the upper-right or lower-right corners under the ceiling. Also, larger samples are usually needed when the ceiling line is more horizontal and to a lesser extent when it is more vertical. On the other hand, when sample sizes are large even very small effect sizes (e.g., 0.05) can be detected. Note that a high-powered study only increases the *probability* of detecting a necessity effect when it exists. Low-powered studies (e.g., with a small sample size) can still detect necessity ($p < 0.05$), but have a higher risk of false negative results: concluding that necessity is not supported, when it actually does exist. Note that in small-n studies a dichotomous necessary condition where X and Y can have only two values, can be easily falsified when necessity does not exist in the population using purposive case selection (Section 8.3.2).

5.7 Reverse causality

Reverse causality occurs when the direction of cause and effect between two variables is the opposite of what is assumed or hypothesized. It may invalidate causal interpretations in observational studies. Reverse causality may also be a concern in NCA. When a hypothesis is formulated in NCA, it is expected that X causes Y and not that Y causes X: first X, then Y and not first Y, then X. NCA requires having a theoretical expectation (formal necessity hypothesis, see Chapter 7) that defines the causal direction. For example, if it is hypothesized that the presence or high level of X is necessary for the presence or high level of Y, with X on the horizontal axis and Y on the vertical axis, the upper-left corner of the XY-plot is the *expected empty corner* (Section 3.5 and Chapter 7).

If the temporal order is not guaranteed by the study design (Section 8.2.1), and if it cannot be theoretically excluded, an observed empty area in the upper-left corner could not only represent high-high $X \rightarrow Y$-necessity (presence of high level of X is necessary for presence or high level of Y), but could also represent low-low $Y \rightarrow X$-necessity (absence or low level of Y is necessary for absence or low level of X), thus implying that Y is a necessary cause of X (low-low-necessity). This is illustrated in Figure 5.9 where the hypothesized causal condition is shown on the left and the alternative causal direction on the right. In Figure 5.9-right the XY-plot is rotated 180 degrees in the vertical and horizontal directions, such that the empty corner appears in the lower-right corner. Y is now on the horizontal axis and X on the vertical axis. This visualizes that a high level of condition Y *ensures* (is sufficient for) a certain minimum level of outcome X, and the 'ceiling line' becomes a 'floor line'. If such an effect can be theoretically justified (namely that a *single* cause ensures a certain minimum level of the outcome), the observed empty area in the upper-left corner (Figure 5.9-left) has an alternative causal interpretation: Y is a low-low necessary cause of

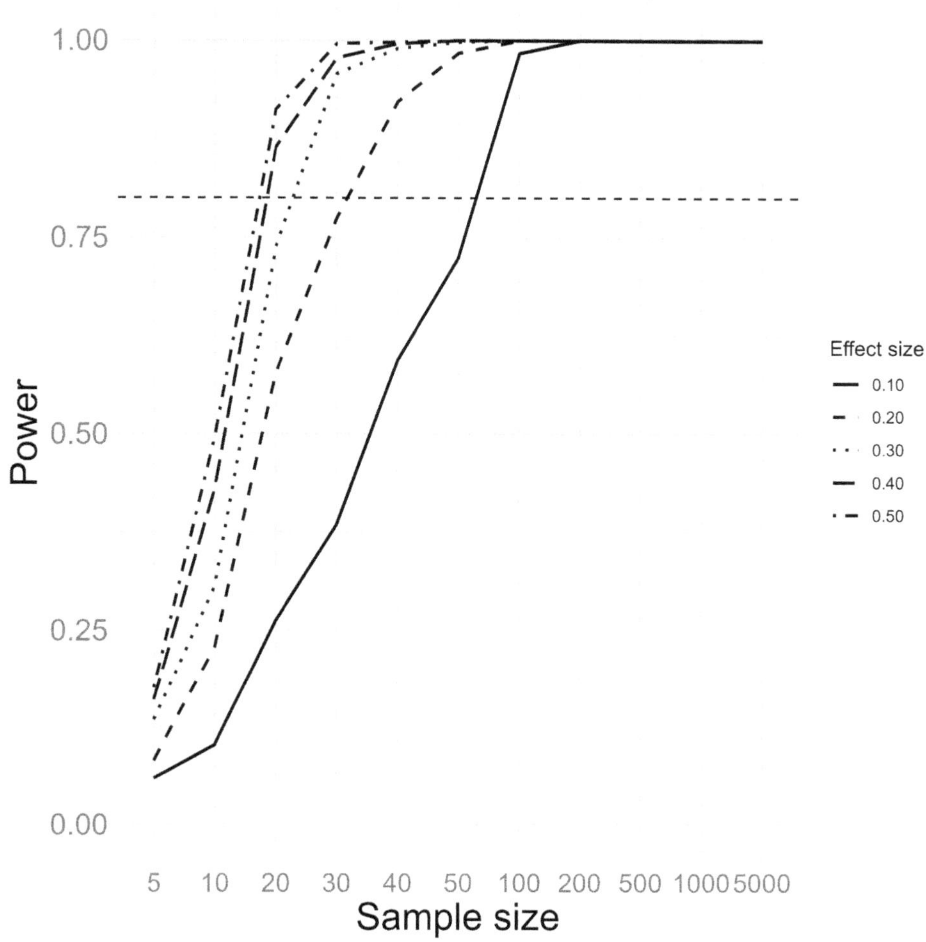

FIGURE 5.7 Power as a function of sample size and effect size for a uniform-based distribution. Curves for effect size: from right to left = from 0.10 to 0.50 effect size.

X: absence of Y is necessary for absence of X. Testing 'forward' necessity causality ($X \rightarrow Y$) in the true presence of reverse causality ($Y \rightarrow X$) may lead to false positive conclusions. However, this risk may be absent or low when reverse necessity causality is unrealistic (e.g., human life does not cause oxygen) or when, in multicausal phenomena, a *single* variable is unlikely to *produce* the outcome at a certain level. When reverse causality is not excluded by study design or theoretically, potential reverse causality should be considered.[8]

5.8 Measurement error

For valid estimations, many methods assume no measurement error in the data; NCA is no different. Measurement error may occur in practice and this affects the findings. In NCA,

[8]The uncertainty about the temporal order between X and Y could be a reason for doing time-lagged study or a necessity experiment. See Section 8.2.

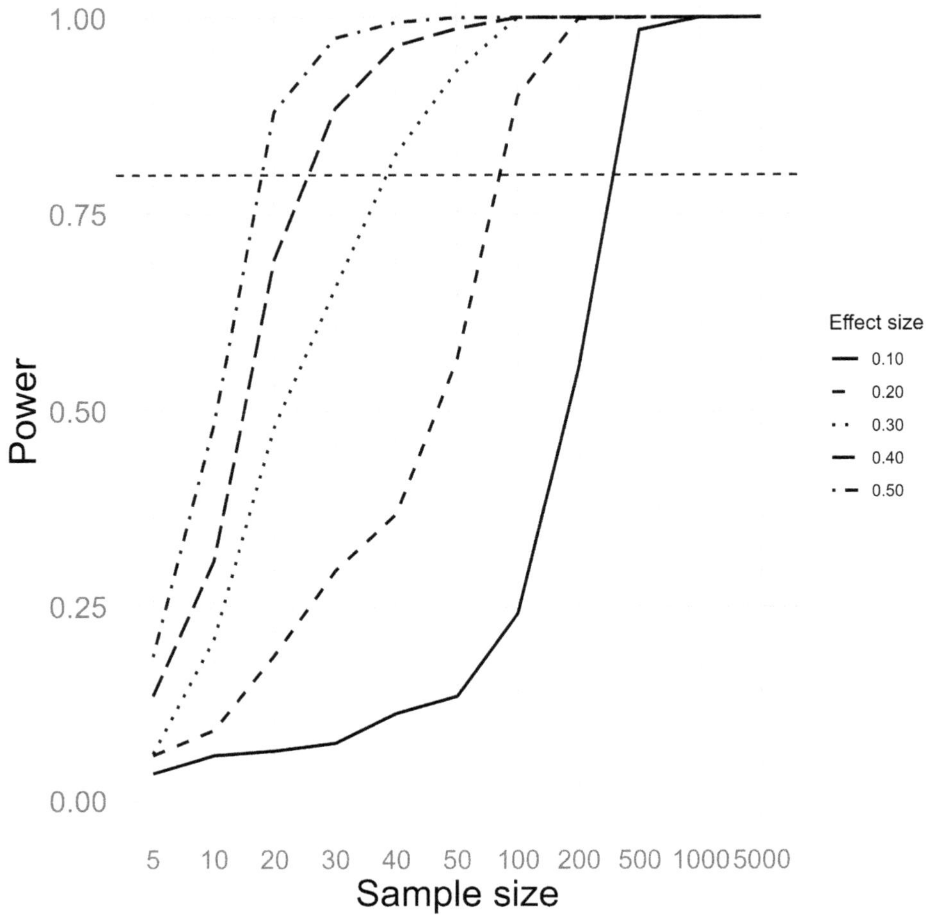

FIGURE 5.8 Power as a function of sample size and effect size for a truncated normal-based distribution. Curves for effect size: from right to left = from 0.10 to 0.50 effect size.

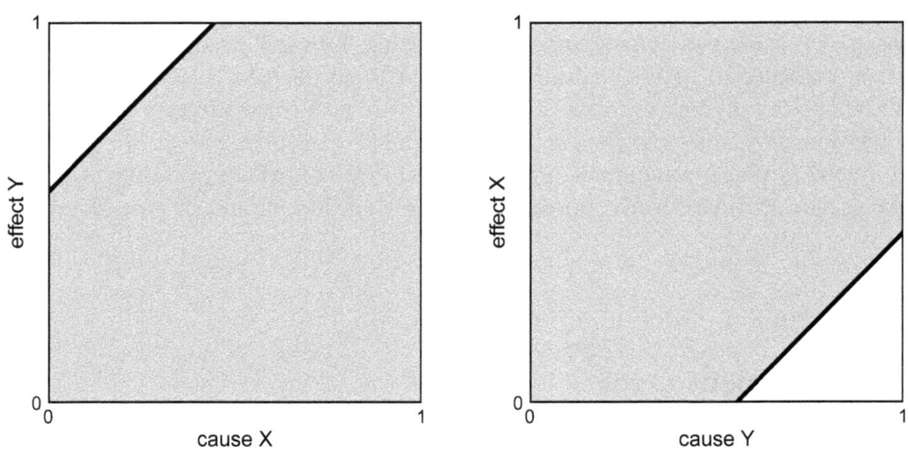

FIGURE 5.9 Reverse causality in NCA. Left: Expected empty area: X causes Y. Right: Possibility of reverse causality: Y causes X.

the effect of measurement error depends on the location of the affected cases. Cases with measurement error that are near the ceiling zone can have a strong influence on the effect size. A case with measurement error may be in the otherwise empty area. This would result in an underestimation of the effect size. Measurement error in cases that are far below the ceiling line generally do not influence the effect size (depending on the size of the measurement error). When the analysis is done with the empirical rather than theoretical scope (Section 4.3), cases with measurement error that are near the borders of the bounding box can affect the size of the box and thus the necessity effect size. Cases with measurement error that can have a large influence on the effect size, are treated in NCA as *potential* ceiling or scope outliers that can be analyzed separately (Section 9.8).

5.9 Spuriousness

An observed relationship is spurious when an alternative causal explanation is more credible, either because it is theoretically more plausible or it fits the empirical evidence better. In NCA, an apparent effect size may be spurious for several reasons.

5.9.1 Lack of theoretical support

The first important reason is the lack of theoretical support: the absence of a convincing necessity hypothesis. During the development of a formal hypothesis, thought experiments are used to make plausible that without X it is not possible to obtain Y. (Chapter 7). Empty spaces in an XY-plot can arise from many reasons other than necessity. The importance of a strong theoretical justification is illustrated by the fact that NCA rejects necessity when such a hypothesis is not available (Section 6.2).

This is also important because NCA lacks a diagnostic signal in the data indicating that a relationship could be spurious. In particular, the observed effect size is insensitive to including or excluding third variables in the analysis.[9] This contrasts with regression analysis, where adding a third variable, such as a confounder that causes both X and Y, as a control variable will often change the regression coefficient of the hypothesized XY relationship (omitted variable bias when the third variable is excluded). This signals that the original XY-relationship may be spurious.[10]

In the context of NCA, a third variable Z that is considered a *sufficiency* cause for X (produces X) and a *necessity* cause for Y, may be a better causal explanation for the observed empty space in the XY-relationship than the necessity of X for Y. If so, the original hypothesis about the XY-relationship may be considered spurious. Considering a third variable Z (provided that theoretical support is available for it) as being sufficient for X and necessary for Y allows conducting a multiple NCA in which both X and Z are

[9]This is also an attractive feature when interpreting observational data causally.

[10]Regression models describe the conditional expectation $E[Y \mid X]$, i.e., the central trend of Y given X. In a two-dimensional XY-plot, a fitted line shows the *unadjusted* bivariate association between X and Y. When potential confounders are included in a multiple regression model, the *conditional* (adjusted) trend may differ from the unadjusted line, often with different intercepts and slopes. The unadjusted relationship is sometimes referred to as *total (bivariate) effect* (including the influence of confounders), whereas the adjusted relationship corresponds to the *net effect*, isolating the association between X and Y after accounting for confounding. The difference between adjusted and unadjusted association can be used as a diagnostic tool for potentially spurious relationships.

evaluated as potentially necessary conditions. However, including the third variable in the analysis does not change the effect size of the XY necessity relationship; instead it can add an explanation of what enables Y.

For example, for driving a non-automatic electric car, having the handle in the Drive position (X) is necessary for the car to drive (Y). Furthermore, a driver (Z) is sufficient for positioning the handle in the Drive position (but not necessary as the handle may already be in the Drive position or another person may place it there). A driver is also necessary for the car to drive (but not sufficient, because X and many other conditions must also be satisfied). NCA would identify both X and Z as necessary conditions if the necessity pattern is observed and theoretical support is available. The estimated effect size for the XY relationship, as estimated from the observed XY data pattern, is not affected by assumptions on the presence or absence of Z. Moreover, Z does not make the data pattern of XY- relationship spurious.

Only when theoretical support for the XY-relationship is lacking or weak can it be helpful to consider an alternative necessary condition Z that is also sufficient for X, to assess theoretically whether the observed XY effect size may be attributable to that causal structure. For example, based on an example presented by Mahoney (2007), suppose the situation that an analyst proposes a weak hypothesis that having no beard (X) is necessary for pregnancy (Y), and observes a statistically significant necessity effect size that meets NCA's criteria for non-rejection. A more credible alternative explanation is that oestrogen (Z) is necessary for pregnancy, and sufficient for having no beard. This theoretical reasoning reveals that the original hypothesis is not defensible. Thus, possible spuriousness in NCA lies not in the data pattern but in the theory attached to it. Claiming that a theoretically justified XY necessity relationship is spurious due to a third variable Z that is sufficient for X and necessary for Y, amounts to proposing a competing *theoretically justified* and *empirically supported* casual explanation in which Z is a sufficient cause for X and a necessary cause for Y and that this alternative explanation is more credible.

In general, it is not easy to find theoretically justified sufficiency relationships involving a single sufficiency cause, except when variables are conceptually very similar. Potential sufficiency relationships of ZX can be analyzed with NCA by analyzing the lower-right corner of the ZX-plot (or the lower-right corner of the XZ-plot when X is considered as a sufficient cause for Z and a necessary cause for Y).

5.9.2 Low-quality data

The second important reason is inadequate data. For estimating necessity from sample data, it is essential that the sample represents the population *of interest*. This target population must be part of the theoretical domain, which is the universe of cases where the theory (formal necessity hypothesis) is expected to hold (also called boundary conditions, scope conditions, range of applicability, contextual constraints, etc.).

When the sample is drawn from a population that is (partly) outside the theoretical domain, or when the sample is selective (sampling a subgroup of the population), the sample is not representative and existing necessity may not be identified; in that case the observed "non-necessity" relationship is spurious (a false negative conclusion). The same holds for samples that are too small to identify an existing necessity relationship in the population (lack of power, see Section 5.6). Similarly, with a sample from outside the theoretical domain, a non-existing necessity relationship may be observed but is spurious (a false positive conclusion).

An additional requirement is that the sample covers all possible combinations of variables, such that cases near the expected empty corner are included. Because population characteristics are normally unknown, adequate coverage can be achieved through random sampling (also needed to meet the requirements for NCA's statistical test) to obtain a large-n sample, or through replication studies with new samples from the same population. If these requirements are not met, the identified empty space (or non-empty space) may be spurious (Section 5.2).

Unreliable or invalid measurement of X or Y (Section 5.8) may also result in spuriousness. The same applies to inappropriate data transformations (e.g., nonlinear transformations) that distort the data pattern (Section 8.5.2).

With a well-specified necessity theory (Chapter 7) and adequate data (Chapter 8), the risk of spurious necessity relationships can be substantially reduced, though never fully eliminated.

6

Credibility

6.1 Summary of this chapter

This chapter evaluates the credibility of NCA to identify necessity from data. In the context of NCA, credibility refers to the trustworthiness of conclusions about necessity from data[1]. The chapter starts by discussing the core criteria to draw conclusions about necessity: there must be a theoretically justified necessity hypothesis, the effect size must be practically relevant (high d-value), and the observed empty space must be unlikely to be compatible with no relationship (low p-value) (Section 6.2). A positive conclusion that necessity is supported requires that all three of these criteria are met. A negative conclusion that necessity does not exist is drawn if any of the three is not satisfied. Next, the chapter evaluates the statistical credibility of NCA using Monte Carlo simulations to analyze how well NCA detects necessity when it is present in the population (True Positive Rate-TPR, sensitivity) and how well NCA avoids detecting necessity when it is absent in the population (True Negative Rate-TNR, specificity) (Section 6.3). Subsequently, the empirical credibility of an NCA study is evaluated focusing on robustness checks (the extent to which the findings are not dependent on the analyst's choices) and model fit metrics (the extent to which the necessity model matches the data) (Section 6.4). Finally, recommendations are given on how to evaluate the credibility of a specific study (Section 6.5).

6.2 Core criteria for necessity

NCA draws a positive conclusion about necessity if three requirements are jointly satisfied: there is theoretical support for necessity (a formal necessity hypothesis), there is a reasonably large effect size (e.g., $d \geq 0.10$), and there is a reasonably small p-value (e.g., $p < 0.05$). NCA draws a negative conclusion about necessity if one or more of these requirements are not met.

To meet the requirement of *theoretical support*, the analyst formulates and theoretically justifies that X and Y are causally related by necessity as outlined in Chapter 7. One implication is that the analyst expects emptiness of a specific corner in the XY-plot. For example, if it is expected that a high level of X is necessary for a high level of Y, no high values of Y are expected when X is low. Therefore, the upper-left corner of the XY-plot is expected to be empty. If no theoretical support for necessity is provided, NCA draws a negative conclusion about necessity, even if effect size and p-value satisfy the criteria.

[1] Credibility in NCA is conceptually unrelated to the same term as used in Bayesian statistics, where it denotes probabilistic confidence in parameter estimates.

DOI: 10.1201/9781003430766-6

To meet the requirement of a *minimum effect size*, NCA estimates the size of the empty space in the expected empty corner of the sample's XY-plot (d-value) and judges whether this size is relevant by comparing it with the selected threshold d-value (e.g., $d \geq 0.10$). If this requirement for a relevant d-value is not met, NCA draws a negative conclusion about necessity even if there is theoretical support and the p-value is below the threshold p-value.

To meet the requirement of a *maximum p-value*, NCA's statistical test is conducted to estimate the p-value of the estimated necessity effect size, and a comparison is made with the selected threshold p-value (e.g., $\alpha = 0.05$). One unique feature of this test is that it uses the size of the expected empty space as the test statistic, thus testing the randomness of this specific emptiness when X and Y are unrelated, rather than being a *general* randomness test (Section 5.3). If the requirement for a small p-value (e.g., $p < 0.05$) is not met, NCA draws a negative conclusion about necessity.[2]

Only if all requirements are met, NCA draws a positive conclusion about necessity in the population (within the common limitations of a study design, and acknowledging the need for replications). If one of them is not met, NCA draws a negative conclusion about necessity in the population. This set of requirements for drawing a conclusion about necessity in the population shows that NCA is more than just a statistical model and that evaluating necessity with just the p-values is not enough. NCA's statistical test and p-value are only one part of the process of making a conclusion about necessity. This fits the commonly accepted view that conclusions about empirically observed effects should not be solely based on the p-value, but also on effect size and theoretical argumentation (see for instance, Cohen, 1994; Cumming, 2013; Greenland & Robins, 1986; Rothman & Greenland, 2005; Wasserstein & Lazar, 2016). Although isolated statistical null hypothesis tests are usually evaluated with concepts like type I error, type II error and power, these concepts are not enough for evaluating the credibility of the NCA approach as a whole, where other non-statistical criteria are also relevant (theoretical support, effect size).

6.3 Statistical credibility

To evaluate the credibility of a binary classification like NCA's positive or negative conclusion about necessity in the population, the True Positive Rate (TPR or sensitivity) and True Negative Rate (TNR or specificity) can be used. Applied to NCA, TPR is the ability of the approach to correctly identify that necessity exists in a population, and TNR is the ability of the approach to correctly identify that necessity is absent in a population. 'Rate' refers to the proportion of randomly drawn samples from that population that were correctly evaluated regarding the presence or absence of necessity.

For testing the credibility of NCA with these evaluation criteria, a Monte Carlo simulation is performed. The simulation evaluates TPR and TNR based on the analysis of samples from two populations: one with the necessary condition (for evaluating TPR), and one without it (for evaluating TNR), and using the three criteria to identify necessity. The simulation assumes that the requirement of 'theoretical support' is satisfied. This implies that it is clear which corner of the sample XY-plot is the expected empty corner and must be analyzed. Here the theoretical expectation is that a high value of X is necessary for a high value of Y, resulting in the upper-left corner as the expected empty corner (corner 1; Section 3.6).

[2]This requirement does not apply in a small-n study with purposeful rather than random selection of cases.

As a reminder, the numbering of the corners is shown in Figure 6.1.

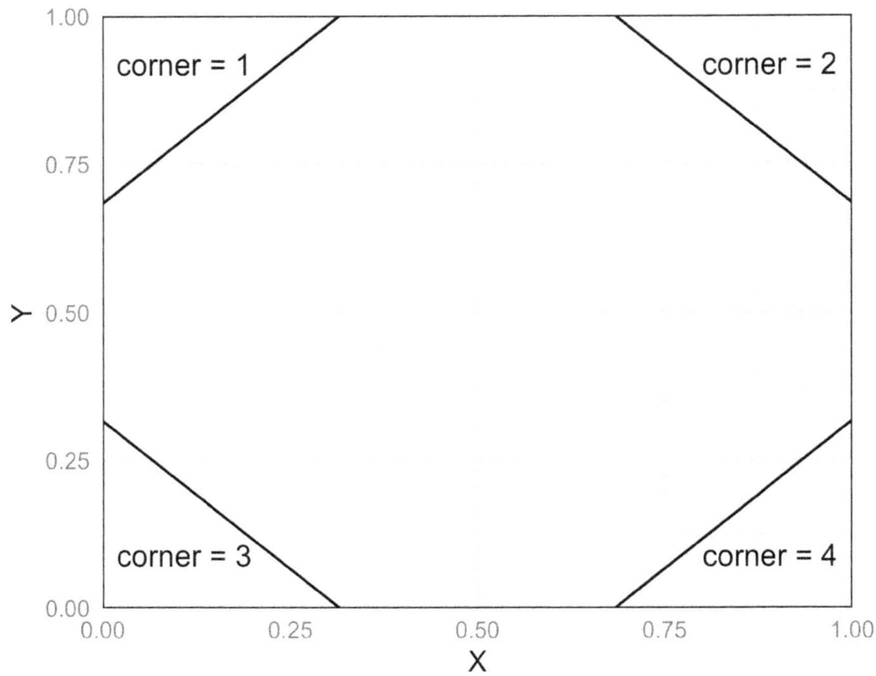

FIGURE 6.1 Numbering of corners in an XY-plot.

For testing the TPR, the simulation must ensure that necessity is present in the upper-left corner of the population XY-plot, so that this corner is empty. Since for testing TPR all samples are drawn from a population where necessity is always present (rather than being present with a certain probability), the False Negative Rate (FNR) equals 1 − TPR. Similarly, for testing TNR, the simulation must ensure that necessity is absent in the upper-left corner in the population XY-plot, thus that this corner is not empty. Since for testing TNR all samples are drawn from a population where necessity is not present, the False Positive Rate (FPR) equals 1 − TNR. Together with the True Positive Rate (TPR), these two measures reflect the model's ability to correctly classify both presence and absence of necessity.

For both tests (TPR and TNR) four scenarios are simulated. In scenario A, X and Y are only related by necessity in the expected empty corner (for testing TPR), or are not related by necessity in the expected empty corner (for testing TNR). In this scenario no other relationships exist between X and Y. In the other scenarios (scenarios B, C, and D) it is assumed that other relationships exist as well. Although an infinite number of types of relationships (bivariate distributions) could exist, here the focus is on other necessity relationships than the one in the upper-left corner. The other relationships produce emptiness in other corners than the corner of interest. However, none of the other relationships are hypothesized or tested. The only goal of additional empty corners in the simulation is to challenge NCA's credibility regarding necessity related to the hypothesized empty corner, when other relationships are also present. The four simulation scenarios about necessity and other relationships in the population are therefore:

A. Necessity (TPR) or non-necessity (TNR) in corner 1. No other relationships.

B. Necessity (TPR) or non-necessity (TNR) in corner 1. Other relationship in corner 2.

C. Necessity (TPR) or non-necessity (TNR) in corner 1. Other relationship in corner 3.

D. Necessity (TPR) or non-necessity (TNR) in corner 1. Other relationship in corner 4.

In all scenarios X and Y are related (TPR) or unrelated (TNR) by necessity in the upper-left corner. In scenario A there is no other relationship. In the other scenarios X and Y are (also) otherwise related because one of the other corners than the expected empty corner of interest is empty[3].

The technical details of the Monte Carlo simulation are as follows. Since NCA requires bounded variables, the bounding box (Section 4.3.1) in the population where cases could exist is defined by the unit box in which X- and Y-values can range from 0 to 1 (scope = 1). The true feasible area of the population (within the bounding box) is the entire unit box when X and Y are unrelated and is a part of the unit box when X and Y have the necessity relationship. The empty corner is assumed to be a triangle consisting of the area between a linear ceiling line (corners 1 and 2) or a linear floor line (corners 3 and 4) and the borders of the unit box. The slopes of the ceiling and floor lines are 1 for corners 1 and 4 and -1 for corners 2 and 3. The sizes of the empty corner range from 0 to a maximum of 0.50. The combination of slope and empty area determines the intercepts of the lines as follows:

corner 1: $a_1 = 1 - \sqrt{2e_1 b_1}$.

corner 2: $a_2 = 1 + |b_2| - \sqrt{2e_2|b_2|}$.

corner 3: $a_3 = \sqrt{2e_3|b_3|}$.

corner 4: $a_4 = 0 - (b_4 - \sqrt{2e_4 b_4})$.

where a is the intercept and b is the slope of the line, and e is the size of the empty space. The subscripts indicate the corresponding corner. For the expected empty corner, the size of the empty space equals the necessity effect size (d) since the scope equals 1. When two opposite corners are empty, the two lines are parallel and when two adjacent corners are empty, the two lines intersect. In the latter situation it is possible that for larger necessity effect sizes and empty space sizes the two lines intersect within the bounding box (Figure 6.2). This reduces the intended necessity effect size, empty space size and scope. In the simulation this situation is avoided to keep the results comparable. Consequently, higher sizes of adjacent empty corners are not evaluated.

It is assumed that within the non-empty area of the bounding box the cases have a uniform distribution (Section 5.4). The `nca_random` function of the NCA software in `R` is used to create such a distribution in the presence of a ceiling line. From this defined population, 1000 random samples are drawn. Subsequently, for each sample the size of the expected empty corner (necessity effect size) is estimated with the CE-FDH ceiling line (with empirical scope). The p-value for each sample is estimated using 1000 random samples from the permutation distribution. The threshold p-value is set at 0.05. This means that necessity is considered not supported (negative conclusion) if the p-value is 0.05 or larger. If the p-value is below 0.05, necessity is considered supported (positive conclusion) only if the other two criteria (theoretical support and effect size) are also met. The theoretical support is assumed

[3]The results remain the same if the simulations are done with a different expected empty corner and corresponding other corners.

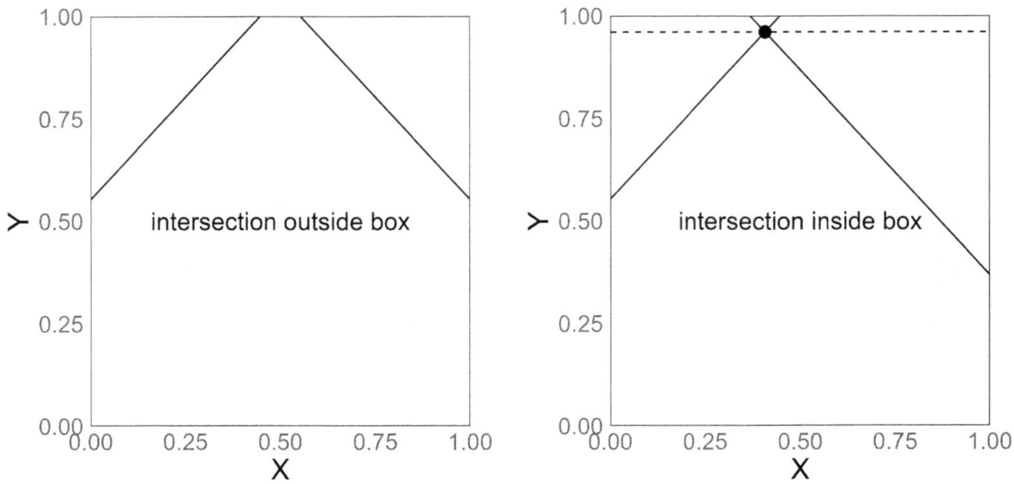

FIGURE 6.2 Intersection of two adjacent empty corners. Left: The lines intersect outside the bounding box. Necessity effect size of corner 1 is 0.1. Size of empty corner 2 is 0.1. Scope = 1. Right: The lines intersect inside the bounding box. Intended necessity effect size is 0.1. Reduced necessity effect size is < 0.1. Intended size of empty corner 2 is 0.2. Reduced size of empty corner 2 is < 0.2. Reduced scope < 1.

to be met. For the effect size the threshold d-value is 0.10. If applicable and possible, the tested sizes of the other empty corner are 0.05, 0.10, 0.20, 0.30, and 0.50.

6.3.1 True Positive Rate (TPR, sensitivity)

Figure 6.3 shows the results of the evaluation of TPR for a true necessity effect size of $d = 0.10$ in corner 1, the expected empty corner (called *necessity corner* in the figure), and with different sizes of other empty corner 2 (Figure 6.3-top-right), other empty corner 3 (Figure 6.3-bottom-left) and other empty corner 4 (Figure 6.3-bottom-right).

In all three graphs, the necessity effect size (the size of the expected empty corner 1) is 0.10, and the size of another empty corner ranges from 0 to 0.50, where possible. A TPR value of 0.80 is added as a benchmark, meaning that for 80 percent of the samples a correct decision is made, namely concluding that necessity is present in the population. This selected value is inspired by the commonly used target of 0.80 for statistical power. NCA draws a *positive* conclusion about necessity if $p < 0.05$ *and* $d \geq 0.10$ (assuming theoretical support).

Scenario A applies when the other corners are *not* empty. The corresponding TPR curve is the same in each TPR graph (empty corner size = 0). This curve is the upper curve in the Top-Right and Bottom-Left graphs, and the lower curve in the Bottom-Right graph. In scenario A, TPR increases with sample size and the benchmark value of 0.80 (80% correct decisions when necessity is present in the population) is reached for a sample size of about 60 (note that the horizontal axis is piecewise-linear: linear between two adjacent ticks but not across ticks).

Table 6.1-top-left shows for scenario A the fraction of samples where $p < 0.05$ (p-based TPR) and the fraction of samples where $d \geq 0.10$ (d-based TPR). Considering both criteria, TPR is the *minimum* of p-based TPR and d-based TPR. It turns out that for scenario

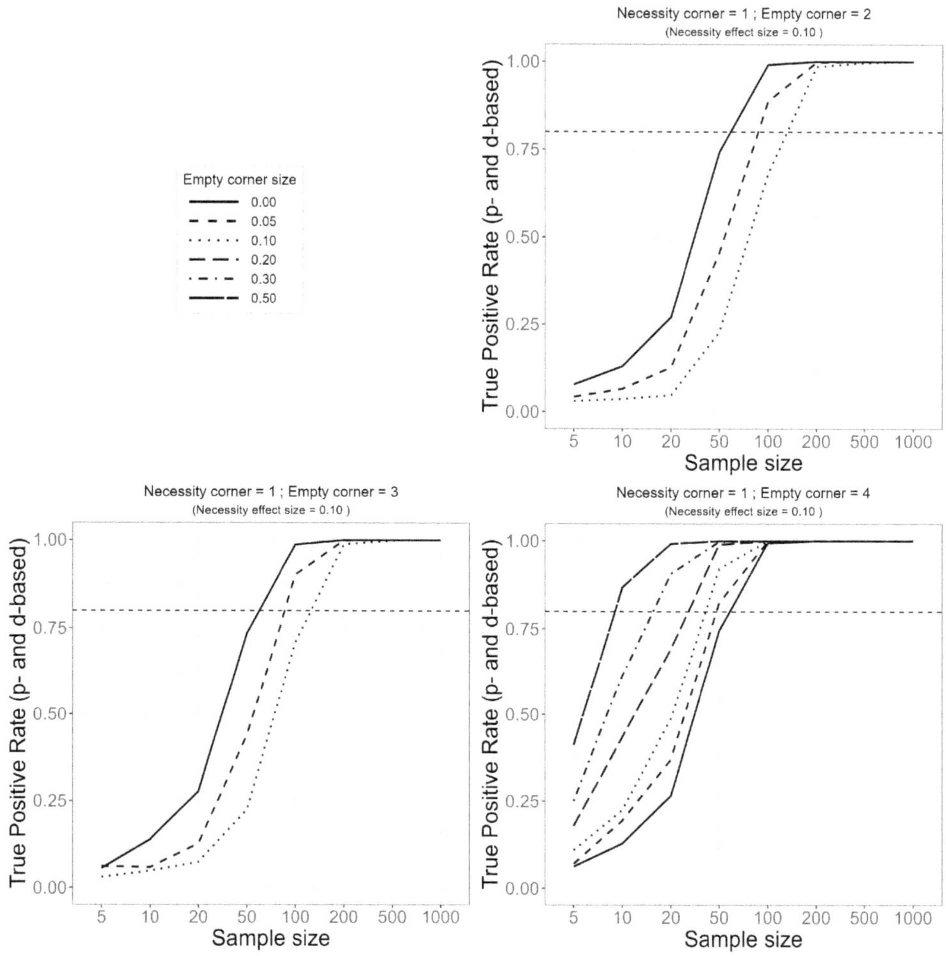

FIGURE 6.3 True Positive Rate (TPR, sensitivity) of the NCA approach for identifying necessity in the population. True necessity effect size in corner 1 = 0.1. Slopes of ceiling/floor lines are 1 (corners 1 and 4) or −1 (corners 2 and 3). Top-Right: other empty corner = 2. Bottom-Left: other empty corner = 3. Bottom-Right: other empty corner = 4.

A, the p-value criterion ($p < 0.05$) is critical for the TPR conclusion. Consequently, TPR corresponds to the power of NCA's statistical test in this simulation scenario.

Scenario B applies when corner 2 (upper-right corner) is also empty (empty corner size = 0.05 or 0.10, see Figure 6.3-top-right). It turns out that the TPR curve shifts to the right with increasing emptiness of corner 2, meaning that for a given sample size the TPR decreases when the emptiness of corner 2 increases. Consequently, higher sample sizes are required to be able to identify necessity in corner 1 if this exists in the population. When the emptiness of corner 2 increases, relatively fewer cases are available to statistically test the emptiness of corner 1. NCA's statistical test uses a permutation procedure (Section 5.3) that creates new cases by allocating Y-values of observed cases randomly to X-values of other observed cases, and random samples are drawn from the obtained permutation distribution (= null distribution). High Y-values of cases allocated to low X-values of other cases can challenge the emptiness of corner 1. If relatively fewer high Y-values are available, the sample size must be larger to have enough high Y-values for the test. For achieving

TABLE 6.1 Fraction of samples that correctly identify necessity: p-based, d-based, TPR (True Positive Rate). Necessity effect size = 0.10 (corner 1 empty) with ceiling slope = 1. Top-Left table: Scenario A – no other corner empty. Top-Right table: Scenario B – corner 2 also empty. Bottom-Left table: Scenario C – corner 3 also empty. Bottom-Right table: Scenario D – corner 4 also empty. Other empty corner size = 0.10.

n	p-based	d-based	TPR	n	p-based	d-based	TPR
5	0.077	0.801	0.077	5	0.030	0.715	0.030
10	0.130	0.890	0.130	10	0.036	0.815	0.036
20	0.271	0.971	0.271	20	0.047	0.895	0.047
50	0.743	0.995	0.743	50	0.228	0.954	0.228
100	0.991	0.999	0.991	100	0.680	0.980	0.680
200	1.000	1.000	1.000	200	0.985	0.993	0.985
500	1.000	1.000	1.000	500	1.000	0.997	0.997
1000	1.000	1.000	1.000	1000	1.000	1.000	1.000

n	p-based	d-based	TPR	n	p-based	d-based	TPR
5	0.030	0.725	0.030	5	0.109	0.843	0.109
10	0.048	0.815	0.048	10	0.224	0.936	0.224
20	0.073	0.889	0.073	20	0.488	0.979	0.488
50	0.224	0.946	0.224	50	0.921	0.995	0.921
100	0.709	0.978	0.709	100	1.000	1.000	1.000
200	0.988	0.994	0.988	200	1.000	1.000	1.000
500	1.000	0.999	0.999	500	1.000	1.000	1.000
1000	1.000	1.000	1.000	1000	1.000	1.000	1.000

a TPR of 0.80, the required sample size is about 75 if the size of the empty corner 2 is 0.05 and about 140 if this size is 0.10. Table 6.1-top-right shows that also in scenario B the p-value criterion ($p < 0.05$) and not the d-value criterion ($d \geq 0.10$) is critical for the TPR conclusion (with one exception for n = 500 where the d-based TPR is smaller than the p-based TPR).

Scenario C (Figure 6.3-bottom-left) applies when corner 3 (lower-left corner) is also empty (empty corner size = 0.05 or 0.10). The graph is very similar to that of scenario B, with some minor differences at lower values of sample size. Also in this scenario higher sample sizes are required to be able to identify necessity in corner 1 when corner 3 is more empty. For achieving a TPR of 0.80, the required sample size is also about 75 if the size of the empty corner 3 is 0.05 and about 140 if this size is 0.10. Also in scenario C the p-value criterion ($p < 0.05$) is dominant.

However, in scenario D the results are very different (Figure 6.3-bottom-right). In this scenario, corner 4 (lower-right corner) is also empty (empty corner size ranges from 0.05 to 0.50). Now the TPR curve shifts to the *left* with increasing emptiness of corner 4, meaning that for a given sample size the TPR credibility of the NCA approach *increases*. This means that *smaller* sample sizes can be used to identify necessity in corner 1. When corner 4 is empty (and corners 2 and 3 are not empty) relatively more cases are available to test the emptiness of corner 1 such that the sample size can be smaller for the permutation test. For achieving a TPR of 0.80 the required sample size can be less than 50 when the corner size is 0.05 and even less than 10 when the corner 4 size is 0.50. However, the possibility of using relatively small sample sizes for identifying necessity has a price in terms of TNR (see below). Also in scenario D the p-value criterion ($p < 0.05$) is critical for the conclusion about TPR (Table 6.1-bottom-right).

Additional TPR simulations with a different necessity effect size (0.20), and different ceiling slopes (0.6, 1.67) in the expected empty corner can be found in Appendix E. These simulations show similar TPR patterns.

6.3.2 True Negative Rate (TNR, specificity)

Figure 6.4 shows the results of the evaluation of TNR.

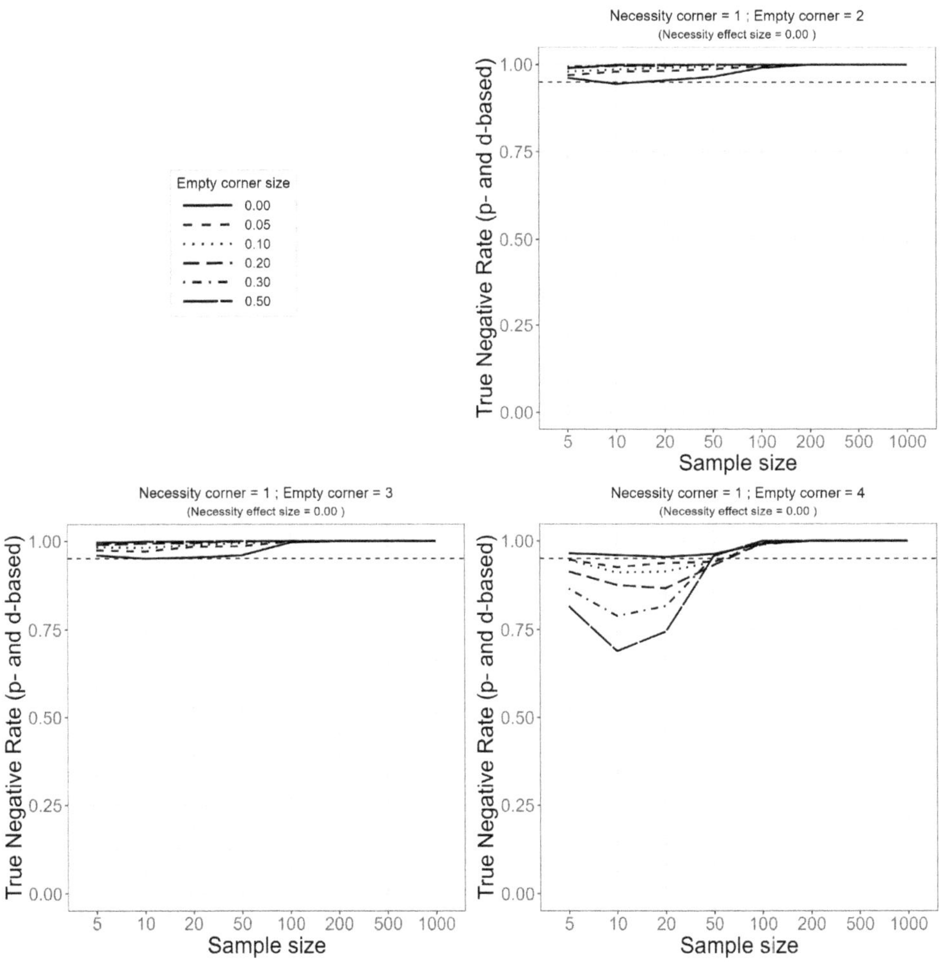

FIGURE 6.4 True Negative Rate (TNR, specificity) of the NCA approach for identifying non-necessity in the population. True necessity effect size in corner 1 = 0 for different sizes of other empty corners. Top-Right: other empty corner = 2. Bottom-Left: other empty corner = 3. Bottom-Right: other empty corner = 4.

In all three graphs (Top-Right, Bottom-Left, Bottom-Right) the expected empty corner is not empty (necessity effect size = 0), while the other corners are empty with sizes ranging from 0 to 0.50. A TNR value of 0.95 is added as a benchmark, meaning that the rate of

TABLE 6.2 Fraction of samples that correctly identify the absence of necessity: p-based, d-based, TNR (True Negative Rate). Corner 1 is tested for necessity (corner 1 not empty). Top-Left table: Scenario A – no other corner empty. Top-Right table: Scenario B – corner 2 empty. Bottom-Left table: Scenario C – corner 3 empty. Bottom-Right table: Scenario D – corner 4 empty. Other empty corner size = 0.10.

n	p-based	d-based	TNR	n	p-based	d-based	TNR
5	0.962	0.413	0.962	5	0.980	0.515	0.980
10	0.945	0.430	0.945	10	0.986	0.577	0.986
20	0.954	0.598	0.954	20	0.991	0.705	0.991
50	0.954	0.877	0.954	50	0.994	0.944	0.994
100	0.953	0.988	0.988	100	0.997	0.998	0.998
200	0.942	1.000	1.000	200	0.997	1.000	1.000
500	0.952	1.000	1.000	500	0.999	1.000	1.000
1000	0.950	1.000	1.000	1000	0.997	1.000	1.000

n	p-based	d-based	TNR	n	p-based	d-based	TNR
5	0.983	0.479	0.983	5	0.946	0.371	0.946
10	0.981	0.539	0.981	10	0.910	0.408	0.910
20	0.989	0.718	0.989	20	0.913	0.596	0.913
50	0.992	0.935	0.992	50	0.917	0.890	0.917
100	0.993	0.994	0.994	100	0.905	0.992	0.992
200	0.998	1.000	1.000	200	0.910	1.000	1.000
500	1.000	1.000	1.000	500	0.896	1.000	1.000
1000	0.999	1.000	1.000	1000	0.918	1.000	1.000

correct decisions confirming the absence of necessity in the population is 0.95. This selected value is inspired by the commonly selected threshold p-value of 0.05 for the false positive rate of a statistical test. NCA draws a *negative* conclusion about necessity if $p \geq 0.05$ *or* $d < 0.10$.

In the TNR evaluation, scenario A applies when not only the expected corner is not empty (necessity effect size = 0) but also the other corners are not empty (empty corner size = 0), such that no relationship exists in the population. The corresponding TNR curves in Figure 6.4 are the same, which are the lowest curves for the Top-Right and Bottom-Left graphs and the highest curve for the Bottom-Right graph. The curve shows that for all sample sizes the TNR is at least 0.95. Up to a sample size of 50, TNR is constant at 0.95 and when the sample size increases the TNR curve moves upward toward 1. The reason is that as long as the p-value is critical, $1 - \text{TNR} = \text{FNR}$ corresponds to the p-value of NCA's statistical test: 5% of the samples drawn from a population where X and Y are unrelated (corresponding to scenario A in the TNR test) are false positives, and 95% are true negatives.

Table 6.2 shows the fraction of samples where $p \geq 0.05$ (p-based TNR) and the fraction of samples where $d < 0.10$ (d-based TNR). Since for each sample NCA draws a *negative* conclusion about necessity if $p \geq 0.05$ *or* $d \leq 0.10$, TNR is the *maximum* of p-based TNR and d-based TNR. Table 6.2-top-left for scenario A indicates that up to sample size 50 the p-value TNR dominates and when the sample size increases, the TNR is based on the d-value. In other words, nearly all samples are correctly evaluated when necessity is absent in the population and other corners are not empty.

Scenario B for TNR (Figure 6.4-top-right) applies when necessity is absent while corner 2 (upper-right corner) is empty (empty corner effect size = 0.05 – 0.50). It turns out that the TNR curves are above the 0.95 benchmark for all sample sizes. Nearly all samples are correctly evaluated when necessity is absent in the population and corner 2 is empty. Table 6.2 top-right shows that for smaller sample sizes the p-value TNR is dominant, whereas for larger sample sizes the d-value TNR dominates, although the differences are small.

The situation for scenario C for TNR (Figure 6.4-bottom-left) when necessity is absent while corner 3 is empty (empty corner effect size = 0.05 to 0.50) is very similar to scenario B. In this scenario, nearly all samples are correctly evaluated when necessity is absent in the population and corner 3 is empty, and the p-value is the primary limiting factor for TNR (Table 6.2-bottom-left).

In scenario D for TNR (Figure 6.4-bottom-right), necessity is absent while corner 4 is empty (empty corner effect size ranges from 0.05 to 0.50). The results are now very different. The TNR curves start below the 0.95 benchmark and with increasing sample size move first downward and then upward and end up above the 0.95 benchmark while moving toward 1. This means that only for a sample size of more than about 50, nearly all samples are correctly evaluated. The pattern of the TNR curve can be explained by the fact that first the p-value is the limiting factor, and subsequently the d-value (Table 6.2-bottom-right).

Similar patterns of the TNR curves can be observed when the distribution in the feasible space is based on a normal rather than uniform distribution of the cases (Appendix E).

If the d-value were not included as decision criterion for necessity, TNR would have been considerably lower for scenario D. This is shown in Figure 6.5 where the TNR decision is only based on the p-value. Figure 6.5-bottom-right shows that TNR decreases with increasing sample size and increasing size of empty corner 4.[4]

Because NCA bases its judgement about necessity not only on the p-value but also on the d-value (and on the availability of theoretical support), the methodology has credibility regarding both TPR and TNR, and the analyst is reasonably protected against making both false positive and false negative conclusions.

6.4 Empirical credibility

The previous results show that it is possible to identify necessary conditions when they exist, and to correctly determine when they do not exist. Nevertheless, even when the criteria for necessity (theoretical support, large d-value, low p-value) are satisfied, the TPR and TNR results also show that there is still a possibility that necessary conditions are not identified when they exist, or that they are identified when they do not exist. This can also happen when limitations exist regarding the quality of model specification (Section 4.3), study design, sampling, and measurement (Chapter 8), or data analysis (Chapter 9). To further evaluate the credibility of an NCA finding, several empirical credibility approaches may be considered.

[4]All simulations show again that when scenario A applies in Figures 6.4 and E.4 with the curves for other corner size = 0 and in Table 6.2-top-left, and when the p-value is dominant, NCA's statistical test is valid (Dul, Van der Laan, et al., 2020).

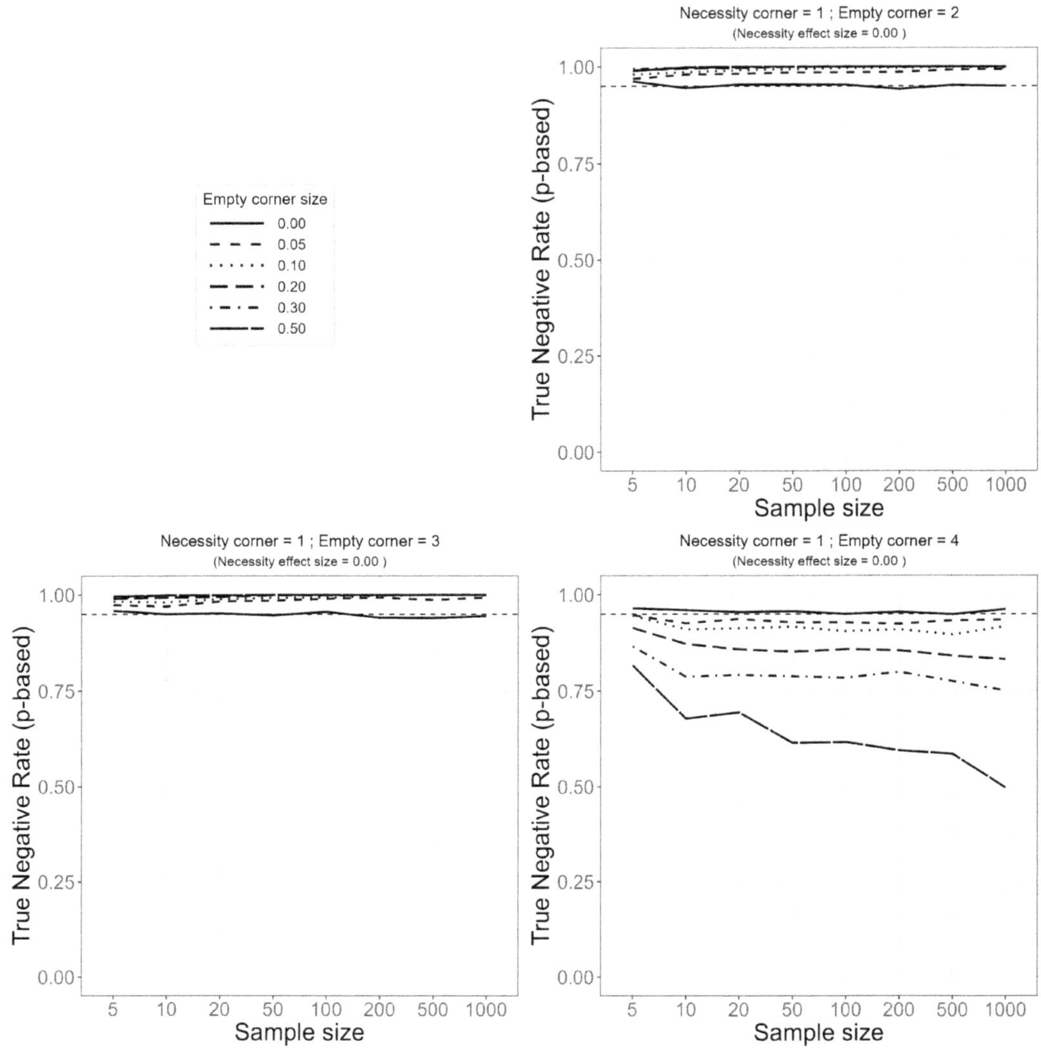

FIGURE 6.5 True Negative Rate (TNR, specificity) of the NCA approach for identifying necessity in the population with only the p-value criterion. True necessity effect size in corner $1 = 0$ for different other empty corner sizes. Top-Right: empty corner $= 2$. Bottom-Left: empty corner $= 3$. Bottom-Right: empty corner $= 4$.

6.4.1 Robustness

When conducting an empirical study, the analyst must make various theoretical and methodological choices to obtain results. Often, other plausible choices could also have been made. In a robustness check, the sensitivity of the results to other plausible choices is evaluated. The check focuses on the sensitivity of the main estimates and the analyst's main conclusion.

In NCA, the effect size and the p-value are two main estimates for concluding whether a necessity relationship is credible. These estimates can be sensitive to several choices. Some choices are outside the realm of NCA, for example choices related to the data that are used as input to NCA (see Chapter 8 about study design, sampling, and measurement) or the threshold level of statistical significance. Other choices are NCA-specific, such as the choice of the ceiling line (Section 4.3.2) and the bounding box (Section 4.3), the threshold level of the necessity effect size (Section 4.4.1), and the handling of NCA-relevant outliers (Section 9.8). The credibility of support for necessity increases when the results are robust. Section 9.12 provides further details of NCA's robustness checks.

6.4.2 Model fit

Section 4.5 introduced eight model fit metrics: complexity, fit, ceiling accuracy, noise, exceptions, support, spread, and sharpness. The credibility of the evidence for necessity increases when the model better fits the data.

6.5 Recommendations for evaluating credibility

The three core criteria for establishing necessity (theoretical support, practical relevance in terms of large effect size, and statistical relevance in terms of small p-value), the robustness check requirement, and the eight model-fit metrics are summarized in Table 6.3. The table provides examples of numerical guidelines, but these should not be interpreted as strict decision rules. Ultimately, credibility is a matter of informed judgment and optimization by the analyst and other stakeholders after considering all credibility aspects. In an empirical study, it is rare for all credibility aspects to be fully satisfied.

In the following Part II of the book, the principles of NCA as discussed in Part I are applied to empirical studies.

TABLE 6.3 Credibility aspects of NCA.

Credibility aspect	Goal	Description	Symbol	Guideline
Theoretical support	Identifies expected empty corner. Allows interpretation of findings.	Necessity hypothesis.	H1	Specify formal necessity hypothes
Practical relevance	Ensures relevance of findings.	Effect size.	d	High (e.g., $d \geq 0.1$
Statistical significance	Tests whether the observed effect could arise by chance (permutation test).	p-value.	p	Low (e.g., $p < 0.0$
Robustness	Checks sensitivity of findings.	Repeated analysis.	Δd, Δp	Report.
Model fit – complexity	Quantifies model complexity used to balance parsimony against accuracy.	Half of ceiling degrees of freedom.	cp	Balance.
Model fit – fit	Quantifies how close the chosen ceiling line is to the standard ceiling line (CE-FDH).	Alignment with boundary pattern (% effect size compared to CE-FDH effect size).	ft	High (e.g., $ft \geq$ 80%).
Model fit – ceiling accuracy	Quantifies the degree to which the expected empty corner is indeed empty.	Percentage of points in the feasible area.	ca	High (e.g., $ca \geq$ 95%).
Model fit – noise	Quantifies uncertainty around the ceiling line due to minor deviations.	Percentage of points in the ceiling zone near the ceiling (medP-zone).	ns	Low (e.g., $ns \leq 5$% 0.9 iso-purity).
Model fit – exceptions	Quantifies violations in the ceiling zone far from the ceiling line, indicating how strict (or tolerant) the necessity claim is.	Number of points in the ceiling zone far from the ceiling (lowP-zone).	ex	Very low (e.g., ex ~0).
Model fit – support	Quantifies how strongly the data in the feasible area support the ceiling line.	Percentage of points in the feasible area on or near the ceiling (medS-zone).	su	High (e.g., $su \geq 5$ 0.8 iso-solidity).
Model fit – spread	Quantifies the extent to which observations in the feasible area near the ceiling are dispersed, rather than clustered along the ceiling line.	Evenness of x-positions of points on or near the ceiling (medS-zone).	sp	High (e.g., $sp \geq 0.$
Model fit – sharpness	Quantifies how abrupt the transition in the data is across the boundary between ceiling zone and feasible area.	Relative density of points in medium-purity (medP) and medium-solidity (medS) zones.	sh	High (e.g., $sh > 0.$

Part II

Application

Part II of the book demonstrates how to apply the principles of NCA in an empirical study, whether in research or in a practice setting. The primary aim of an NCA study is to identify conditions X (e.g., actionable factors) that enable or constrain an outcome Y, which may be either desirable (e.g., performance) or undesirable (e.g., disease). Identifying necessary conditions rather than average contributing factors provides novel insights in research and opens new possibilities for effective action in practice. Once the specific goal of the empirical study is defined, the NCA method involves formulating a formal necessity hypothesis (X is necessary for Y) and testing this hypothesis using empirical data.

After establishing the goal, the application of NCA consists of four stages:

1. Formulate the necessary condition hypothesis.
2. Collect the data.
3. Analyze the data.
4. Report the results.

The first chapter (Chapter 7) discusses the steps that are needed for developing a formal necessity hypothesis. The next chapter (Chapter 8) discusses data collection in terms of the selection of study design (observational study, longitudinal study, case study, experimental study), sampling and selection of cases, measurement (getting scores for X and Y), and the dataset. This is followed by Chapter 9, which discusses NCA's data analysis. Chapter 10 provides guidance on reporting an NCA study. Chapter 11 illustrates the use of NCA in multimethod studies (NCA with regression and NCA with QCA) and Chapter 12 describes the application of NCA in practice. This part ends with a summary and personal reflections.

7

Hypothesis

7.1 Summary of this chapter

This chapter presents the development of a formal necessary condition hypothesis. It is part of a necessity theory and should be testable, non-trivial, plausible, and accompanied by a causal explanation. First, it is explained why NCA adopts a hypothesis formulation-and-testing approach to answer research and practice questions (Section 7.2). The next section (Section 7.3) clarifies why the hypothesis development process differs from usual practice (e.g., lack of past research on necessary conditions). This is followed in Section 7.4 by an overview of knowledge sources that can be used to develop necessity hypotheses (explicit and tacit knowledge from academia and practice). Subsequently, the three steps for developing a formal hypothesis are presented, including decision tools to support this process (Section 7.5). Step 1 establishes a preliminary necessity theory that consists of selecting a "best guess" necessity hypothesis from available sources of information that is embedded in theory. Step 2 uses thought experiments to challenge the hypothesis and, if needed, refine it. Step 3 deepens the causal explanation, and formulates the final formal hypothesis to be tested with empirical data.

7.2 Hypothesis formulation and testing

After the theoretical or practical goal is set, and it is decided to apply NCA, often the question of interest is: "Is X necessary for Y"? or "Which factors are necessary for Y"?

NCA advocates starting the study by providing a preliminary answer to the question before collecting empirical data. Such an answer can be based on existing sources of knowledge. NCA is an approach in which the hypothesis is first formulated and subsequently tested with empirical data. In this deductive approach, the hypothesis can have three states: 'preliminary', 'formal' and 'tested'. A *preliminary necessity hypothesis* is a 'best guess' answer to the above questions based on existing sources of knowledge and, for example, abductive reasoning. A *formal necessity hypothesis* is a deepened preliminary hypothesis that is embedded in theory and that was challenged by thought experiments regarding testability, triviality, plausibility, and completeness. A *tested necessity hypothesis* is a formal hypothesis that was subjected to tests using empirical data, where the outcome of the test is that the hypothesis is rejected or not-rejected (supported) by the observed evidence. In this book, a hypothesis is considered supported or plausible when it is not rejected after being challenged empirically. Rejection may arise from the absence of a plausible causal explanation, a very small necessity effect size, or a large p-value. However, support for, or plausibility of, a necessary condition does not automatically establish its truth. Greater confidence requires,

DOI: 10.1201/9781003430766-7

for example, support from replication studies. Even then, alternative causal explanations for the observed empty corner are possible (Section 2.7). Ultimately, any conclusion about a hypothesis involves human judgment, as discussed in Sections 2.7 and 6.5. Such a judgment may persist only briefly or endure for a long time until rejection.

As discussed below, a formal hypothesis specifies the expected necessity relationship, provides a causal explanation, and specifies the domain where it is expected to hold, among others. This is also helpful in different subsequent stages of an empirical study:

1. Study design: It helps to focus the study based on existing knowledge (Section 8.2).
2. Sampling or case selection: It specifies from which theoretical domain cases must be sampled or selected (Section 8.3).
3. Results: It satisfies NCA's requirement for theoretical support for a decision about (non-)rejection of the hypothesis (Sections 6.2 and 9.10).
4. Discussion: It helps to interpret why the hypothesis is rejected or not, and what the theoretical and practical consequences are (Chapters 10 and 12).

In principle, NCA can also be used in a hypothesis-building (inductive) approach. In situations where the phenomenon of interest is novel or poorly understood and existing knowledge is not available, empirical data may be collected to extract a preliminary necessity hypothesis from data (exploratory or theory-building study). Such a situation is further discussed in Section 8.3.2. However, this book focuses on using empirical data only for hypothesis-testing and not for exploration/hypothesis-building for two reasons:

1. It is normally possible to formulate a preliminary necessity hypothesis from existing knowledge (Section 7.3) without collecting new empirical data.
2. To obtain an empirically tested hypothesis, it is more efficient to formulate a hypothesis first, even if existing knowledge is scarce, followed by one round of data collection for testing, compared to first having a round of data collection for empirically observing a preliminary hypothesis, followed by a second round of data collection with new data for empirically testing this hypothesis. Without testing the hypothesis with new data, the hypothesis remains untested, even if it is formalized and refined afterwards. According to a basic principle of the scientific methodology, using the same data for building and testing does not qualify for a proper empirical test of the hypothesis. This 'predesignation' principle was articulated in the Neyman–Pearson framework of hypothesis testing (Neyman & Pearson, 1933) and has an analogue in modern machine learning, where one dataset is used for training and another for testing. In other words, "It is never OK to snoop at the data before formulating a hypothesis—at least not if the same data are to be used in testing that hypothesis" (Mayo, 1996, p. 295).

As theories in the social sciences are usually semantic (expressed in words), a necessity hypothesis is formulated qualitatively as X is necessary for Y without specifying the exact levels of X and Y, other than the possibility to specify presence/high level or absence/low level of X and Y (Chapter 3). Therefore, in this chapter only the *in kind* dichotomous version of the necessary condition (necessity-in-kind, NiK) is considered, where the presence/absence of X is necessary for the presence/absence of Y (Figure 3.6).

7.3 Why developing necessity hypotheses is different

Developing a necessity hypothesis differs from developing a conventional probabilistic sufficiency hypothesis. Conventional hypothesis development draws on a wealth of existing knowledge about average effects. Reviewing established theories, causal explanations, and prior empirical studies is usually a central part of this process. In such cases, a literature review provides the main justification for the hypothesis, demonstrating that it is plausible, theoretically grounded, and worth (re)testing with empirical data.

In contrast, the current body of knowledge about necessity is still limited. Theoretical frameworks and empirical studies specifically addressing necessity are not yet widespread, and the traditional literature on probabilistic sufficiency offers little direct guidance. That literature hints at important factors that potentially could be necessary conditions, but usually provides no evidence. It may also contain indirect hints or isolated statements about necessity logic and causality. Although this can serve as inspiration or secondary support for developing necessity hypotheses, more is needed to develop well-grounded necessity hypotheses.

Because a cumulative body of theory and evidence on necessity causality is often lacking, developing necessity hypotheses requires an approach that relies more heavily on the analyst's creativity, critical reasoning, and substantive expertise. By integrating solid knowledge of necessity logic and necessity causal reasoning with substantive domain knowledge, and by engaging in discussions with peers, analysts can formulate convincing and meaningful necessity hypotheses. It is often surprising how quickly a relevant preliminary necessity hypothesis can emerge once this reasoning process is clear.

Whereas developing an average-effect hypothesis generally involves dedicating most effort to the literature review, developing a necessity hypothesis centers on conceptual reasoning and logical argumentation. The goal is to construct a coherent narrative that justifies why a given condition must be present for the outcome to occur, and why the outcome will not occur if the condition is absent, and no compensation is possible.

Generally, such a narrative starts with identifying an important factor (known or presumed) that contributes to a desired outcome but is not the only factor that helps to produce the outcome. Conventional literature often establishes that such a factor is important on average.[1] The next step is crucial for necessity reasoning: it formulates the expectation that the presence or a high level of the outcome will disappear if the condition is absent or is reduced, corresponding to necessity logic: if not X, then not Y. The narrative then explains why this disappearance occurs, for example, through what mechanisms the absence of X undermines Y. This is followed by an argument that other factors cannot compensate for this absence: no substitute can replace the missing necessary condition; the outcome will not be possible without it. Finally, the narrative specifies the domain of applicability. This is the boundary within which the necessity claim holds. It clarifies whether it applies universally (e.g., oxygen is necessary for human life) or only within a specific group or context (e.g., being pregnant is necessary for having a child).

For example, it may be assumed that class attendance is necessary for school performance. A minimal narrative could be formulated as follows: *Class attendance is widely recognized as an important determinant of school performance. Meta-analyses show that students who attend classes more regularly tend to achieve higher grades on average. Attendance is not*

[1] Factors that are not important on average may still be necessary (Table 11.5).

the only factor: intelligence, motivation, prior preparation, and study habits also matter. However, class attendance is crucial, as it provides the behavioral foundation through which learning processes operate effectively. If class attendance is withdrawn, school performance deteriorates, a pattern consistent with necessity logic. This disappearance occurs for several reasons. Students who skip classes miss essential explanations, examples, and discussions on which exams and assignments depend. Absence removes opportunities for immediate correction and reinforcement, allowing misconceptions to persist. Reduced attendance may also weaken motivation and belonging, leading to further reductions in performance. Other factors cannot replace the absence of class attendance. Digital resources or peer notes can alleviate short-term gaps but do not replicate the real-time cognitive and social engagement that in-class participation provides. This necessity relationship applies primarily to students in traditional, face-to-face courses where class sessions deliver core content and interactive learning is integral to success. It may not apply universally, for example, in fully online, asynchronous courses where students can compensate for absence through other structured forms of engagement. Within the conventional classroom teaching format, however, attendance remains a necessary condition for school performance.[2]

This narrative defines not only the hypothesis (Class attendance is necessary for high school performance), but also gives a causal explanation (students miss essential learning elements, lack of interactions and motivation, no compensation possible by digital resources) and defines the theoretical domain (conventional classroom teaching).

While creativity is central for formulating the narrative, existing information sources can help generate, sharpen and justify hypotheses about necessary conditions. This chapter provides a detailed plan and practical tools for this process. While it may appear complex at first, it soon becomes almost automatic once one becomes familiar with the reasoning steps involved.

7.4 Knowledge sources for developing necessity hypotheses

The development of a formal necessity hypothesis begins with the selection of a preliminary hypothesis, based on existing knowledge. Various sources such as academic literature, practical insights, and expert judgment can inform this selection. To ensure the relevance of the necessity hypothesis for both theory and practice, knowledge from scholarly research, real-world experience, and the expertise of the analyst can be combined. Since formulating a necessity hypothesis is primarily a creative process it can rely both on explicit knowledge (documented knowledge from academia and practice) and on tacit knowledge (undocumented expert knowledge from scholars and practitioners). This process may be iterative and involve people with different knowledge backgrounds. The result is a defensible claim made by the analyst (and others), which requires a test with empirical data to be (non-)rejected. Or, according to Popper:

[2] Elements of such a reasoning can be found in Tynan et al. (2020). In this study the necessity-in-kind analysis showed that class attendance is necessary for high grades ($d = 0.28$, $p < 0.001$), and the subsequent necessity-in-degree analysis found that students needed to attend at least 50% of classes to achieve a grade of 90% or higher.

> "[Great scientists] are men[3] of bold ideas, but highly critical of their own ideas; they try to find whether their ideas are right by trying first to find whether they are not perhaps wrong" (K. R. Popper, 1985, p. 119)

This section elaborates on the two sources of information for selecting a preliminary necessity hypothesis: explicit knowledge and tacit knowledge.

7.4.1 Explicit knowledge

Explicit knowledge is documented in the academic literature (e.g., books, journal articles) and practice literature (e.g., best practice documents, manuals). Since practice is more focused on 'doing' than on 'documenting', less explicit knowledge is available from practice than from academia. Practice documents could contain know-how and statements that refer to necessity using terms like 'requirements', 'must have', 'critical success factors', or 'bottlenecks', etc. This section focuses on academic literature since it contains much information that can be used to formulate a preliminary necessity hypothesis. Four approaches for identifying them can be distinguished:

1. Sometimes the academic literature includes explicit *necessity theories*, whose necessity hypotheses may need to be (re)tested.
2. The literature may also contain *statements* referring to necessity.
3. The literature contains *causal models*, which are formal models primarily developed for probabilistic sufficiency causality, but elements may be useful for necessity causality as well.
4. The literature contains *causal narratives* for describing probabilistic sufficiency relationships or more general causes that contribute to the outcome that can inspire necessity reasoning.

These sources of explicit scholarly knowledge can be used to derive a *preliminary necessity hypothesis*. Each source is discussed below in more detail.

7.4.1.1 Necessity theories

As discussed in Section 3.4, some theories explicitly state necessity relationships. Such theories can be established or emerging, and may be 'pure' necessity theories consisting only of necessity relationships, or 'embedded' necessity theories consisting of a combination of necessity relationship and other relationships. As the examples in Section 3.4 show, often, necessity theories have not been (extensively) tested empirically. (Re)testing selected necessity relationships from these theories can provide valuable insights into their empirical support and generalizability.

[3] In Popper's time, the terms 'man' and 'mankind' were commonly used in academic writing to refer to humans in general, not just males.

TABLE 7.1 Equivalent phrases for 'X is necessary for Y'. Left: Enablers. Right: Constraints.

Enablers (The presence of X ...)	Constraints (The absence of X ...)
X is necessary for Y	X constrains Y
X is needed for Y	X limits Y
X is critical for Y	X blocks Y
X is crucial for Y	X bounds Y
X is essential for Y	X restricts Y
X is key for Y	X stops Y
X is fundamental for Y	X inhibits Y
X is pivotal for Y	X hinders Y
X is imperative for Y	X prevents Y
X is indispensable for Y	X impedes Y
X is a prerequisite for Y	X disables Y
X is a requirement for Y	X disallows Y
X is a conditio sine qua non for Y	X is a barrier for Y
X is a pre-condition for Y	X is a bottleneck for Y
X allows Y	X is a hurdle for Y
X enables Y	Without X there cannot be Y
X permits Y	No X then no Y
There must be X to have Y	
X is a must-have for Y	
Y requires X	
Y needs X	
Y demands X	

7.4.1.2 Necessity statements

Phrases using terms like 'necessary' or 'necessary but not sufficient' are very common in the academic literature (and in practice). Although several statements may use the word necessity loosely as a synonym for important, many refer explicitly to necessity logic and causality, in particular when it is stated 'for what' the condition is necessary. For example, in the field of supply chain management, Bokrantz & Dul (2023) systematically reviewed necessity statements in major journals and identified 157 statements that can be considered theoretical necessity claims. Most of these statements have not been tested at all, or not been tested properly for necessity (e.g., using regression analysis). A similar result was found by Richter & Hauff (2022) when analyzing necessity statements in the field of international business. The literature also uses phrases that are directly related to necessity, without using the words 'necessity' or 'necessary'. Examples of such phrases are listed in Table 7.1. Several phrases reflect that the necessary condition must be present to have the outcome (enablers); others reflect that the absence of the necessary conditions guarantees the absence of the outcome (constraints). Existing necessity or equivalent statements can be reviewed, and selected statements can be candidates for a preliminary necessity hypothesis.

7.4.1.3 Causal models

The academic literature contains numerous causal models based on probabilistic sufficiency relationships between variables. Such models are generally evaluated through regression-based methods. All or some of the relationships in the models could be revisited with a necessity perspective on the relationship. Revisiting the relationship with a new causal perspective is *not* a replication or robustness test of the existing model. It yields a new necessity model or theory that provides new insights.

An example of a probabilistic sufficiency model is shown in Figure 7.1 about nine factors that contribute to academic achievement. While these factors have only been modeled as having an average effect on the outcome and were tested with regression-based studies, some studies in the literature refer to these factors as critical (e.g., X. Ma & Wang, 2001), in the meaning of 'highly influential'. This model could be revisited from the perspective of necessity causality thereby re-interpreting the word critical as necessary.

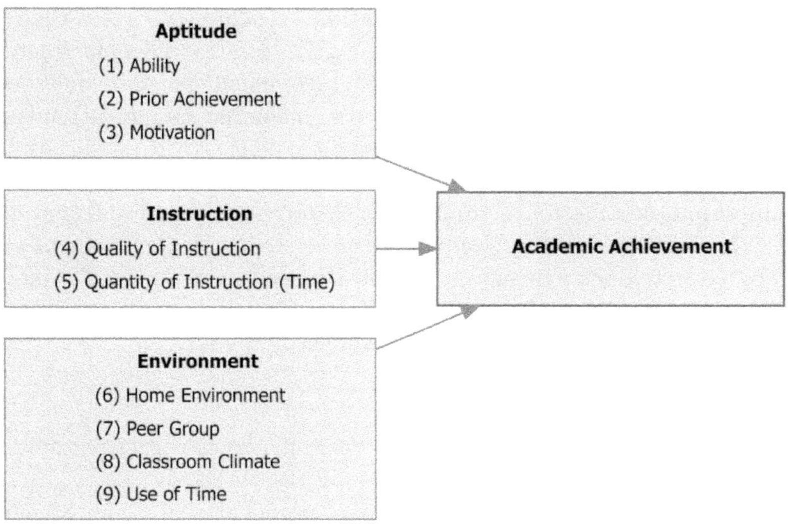

FIGURE 7.1 Example of a causal model (originally for describing probabilistic sufficiency) for academic achievement with 9 factors that contribute to academic achievement. After: Walberg (1984).

It is also possible that two concepts that were not (directly) related in a probabilistic sufficiency-based model, are related in a necessity model. The original role of a concept (e.g., independent, dependent, mediator, moderator, control) is not relevant for formulating a new role for the concept in a different causal model (condition or outcome in a necessity model).

An additional possibility is to evaluate suggestions for future studies that are made in a study based on average effects. It is not uncommon that a publication about a model suggests to apply NCA in future studies to (also) investigate the phenomena from a necessity perspective. Likewise, literature reviews of studies with regression-based causal models and thus probabilistic sufficiency perspectives often suggest exploring the phenomenon with NCA and thus with a necessity perspective as well to further enrich the state of knowledge. Chapter 1 gives examples of publications that suggest using NCA in future research.

7.4.1.4 Causal narratives

When proposing or justifying hypotheses, academic publications often use causal narratives: stories with explanations that connect causes to effects. For example, in describing probabilistic sufficiency relations, the narrative outlines the mechanism through which a cause contributes to an effect. These causal mechanisms can be critically evaluated by

assessing whether the absence of a particular factor disrupts the mechanism. By systematically
analyzing causal terminology (Table 7.1), and identifying factors that could break the mechanism (potential necessary conditions) it may be possible to formulate a preliminary necessity hypothesis. For example, Hausknecht et al. (2008) argue that organizational commitment has a negative effect on absenteeism as follows:

> "[...] [O]rganizational commitment is defined as a collective sense of affective or emotional attachment to an organization [...]. Employees are thought to develop organizational attachments via experiences in the unit where they work [...]. For example, employees of high-commitment units may engage in more community maintenance behaviors, including regular attendance at work. Thus, high-commitment work units are likely to be associated with stricter attendance norms [...]. At the individual level, meta-analytic evidence has supported the negative relationship between commitment and absenteeism [...]. In view of these findings, and the premise that high commitment reflects a strong collective attachment to organizational values and goals, including attendance at work, we expect that at the unit level of analysis, organizational commitment is negatively related to absenteeism" (Hausknecht et al., 2008, p. 1225)

Although in this quote the relationship between organizational commitment and absenteeism is clearly described as a probabilistic sufficiency relationship, such a causal narrative could be approached from the perspective of necessity as well. One might wonder if the absence of organizational commitment would stop low absenteeism. In this example, the authors themselves explicitly suggest this by stating:

> "When the level of analysis is the work unit, organizational commitment may be a necessary but insufficient condition for low absenteeism" (Hausknecht et al., 2008, p. 1225)

---

and after conducting a regression analysis they conclude (although this cannot be supported by a regression-based analysis):

> "Taken together, the findings [...] show that high organizational commitment is necessary to avoid high levels of absenteeism" (Hausknecht et al., 2008, p. 1239)

Explicit statements about necessity on causal reasoning in the literature may support the necessity argument, even though necessity was not tested in the original study (Section 7.4.1.2).

7.4.2 Tacit knowledge

Undocumented expert knowledge can be used to formulate a preliminary necessity hypothesis. Experts often possess theoretical or practical insights that can be interpreted in terms of necessity, in particular when these insights are described with phrases from Table 7.1. This approach is especially useful in fields where explicit knowledge is scarce, while experts believe necessity relationships exist.

Two types of experts can be distinguished: academic experts (scholars) and practitioner experts (practitioners). *Scholars* are persons with deep theoretical or empirical knowledge developed through research. Drawing upon their expertise, comprehensive understanding of the academic literature, and possibly practical experience, they may identify potential necessary conditions either directly by recognizing explicit hypotheses, or indirectly, by recognizing important factors for an outcome that could also be interpreted as necessary conditions. *Practitioners* are professionals with hands-on experience in a field outside academia (e.g., consultants, managers, employees, engineers, policy-makers, educators, designers). While they may not articulate their knowledge as hypotheses, they often operate based on an implicit "theory in use" (Dul & Hak, 2008), a solid belief grounded in practice and experience that explains why a certain factor causes a certain effect. These informal theories may include necessity logic. This is not unlikely as practitioners commonly focus on factors that are required for success ('critical success factors'), or on fail factors that prevent success by identifying and addressing 'bottlenecks'. Their practical insights can be systematically translated into preliminary necessity hypotheses.

An array of qualitative methods exists to capture experts' tacit knowledge, ranging from informal conversations, in-depth interviews, and creative group sessions to more structured approaches such as think-aloud protocols, and Delphi studies. However, tacit knowledge has a complex status in academia. It is often valued implicitly (many scholars rely on it), but devalued explicitly when used as evidence or justification, because it does not fit the norms of transparency, replicability, and formal theorizing that define "scientific" knowledge. If a hypothesis is (partly) developed based on tacit practitioner knowledge, the approach used for obtaining this knowledge can be reported and it can be justified that such knowledge is not anecdotal, but accumulated through extensive experience with the practice, system, or process, thus empirically grounded.

7.5 Developing a formal necessity hypothesis

A formal necessity hypothesis has five characteristics:

Theory-grounded: the hypothesis is embedded in a necessity theory. According to Chapter 3, a theory is defined when four elements are defined: focal unit, concepts (condition and outcome), (direction of) the hypothesis[4], and theoretical domain.

[4]In the context of formulating a formal hypothesis for testing, the term 'hypothesis' instead of 'proposition' (Chapter 3) is used to describe the relationship between the theory's concepts.

Testable. A hypothesis is empirically testable if the concepts X and Y can be operationalized into measurable variables.

Non-trivial. A hypothesis is non-trivial when both the absence of the condition and presence of the outcome are possible, otherwise it is trivial.

Plausible. A hypothesis is plausible when virtually no cases are reasonably expected in the empty corner.

Complete. A hypothesis is complete when a plausible explanation for *why* condition X is necessary for outcome Y is available.

The development process of a formal necessity hypothesis consists of three steps that are shown in Figures 7.2, 7.4, and 7.6, respectively.

In the first step (Figure 7.2), a preliminary necessity hypothesis is selected based on the available sources of information (Section 7.4), and is embedded in a *preliminary necessity theory* by defining focal unit, concepts, (direction of) the hypothesis, and theoretical domain. Concepts are defined such that they are measurable and the hypothesis becomes testable. After this step the hypothesis is *theory-grounded* and *testable.*

In the second step (Figure 7.4) a *thought experiment* is conducted in which the preliminary necessity theory is challenged and possibly revised. After this step the hypothesis is also *non-trivial* and *plausible.*

In the third step (Figure 7.6) a detailed *causal explanation* for the hypothesis is given (Chapter 2). This final step makes the hypothesis *complete.* The formal hypothesis is now ready to be tested with empirical data.

Below, each step is discussed in more detail and illustrated with an example of the preliminary hypothesis that 'An internet connection is necessary for attending an online meeting'.

7.5.1 Step 1: Define a preliminary necessity theory

The first step transforms the preliminary *hypothesis* into a preliminary necessity *theory.* This is a theory that has measurable concepts, a hypothesis describing the expected necessity relationship between the concepts, and a description of the theory's focal unit and theoretical domain (Chapter 3). However, this theory is not yet challenged in a thought experiment, which is the topic of Step 2. Figure 7.2 shows the hypothesis development tool for Step 1 of the development process.

To begin, a preliminary hypothesis is formulated based on existing knowledge (Section 7.4). This hypothesis is then elaborated into a preliminary necessity theory by precisely defining the four constituent elements of a necessity theory:

- Precise definition of *focal unit.* The focal unit is the entity of the theory to which the theory applies (e.g., person, team, country, project, etc.).

Sometimes, defining the focal unit is simple. For example, the Theory of Planned Behavior (TPB, Ajzen, 1991) is a theory about individual behavior. The focal unit is evident: 'individual' or 'person'. Other times, the focal unit may not be immediately clear, for example, when theorizing the relationship between parent height and child height (Galton, 1886).[5] The focal unit is neither parent nor child but 'parent-child pair'. The focal unit is expressed as a singular noun, not a plural noun. Specification is important because cases to be sampled

[5]This example is discussed in Section 11.3.

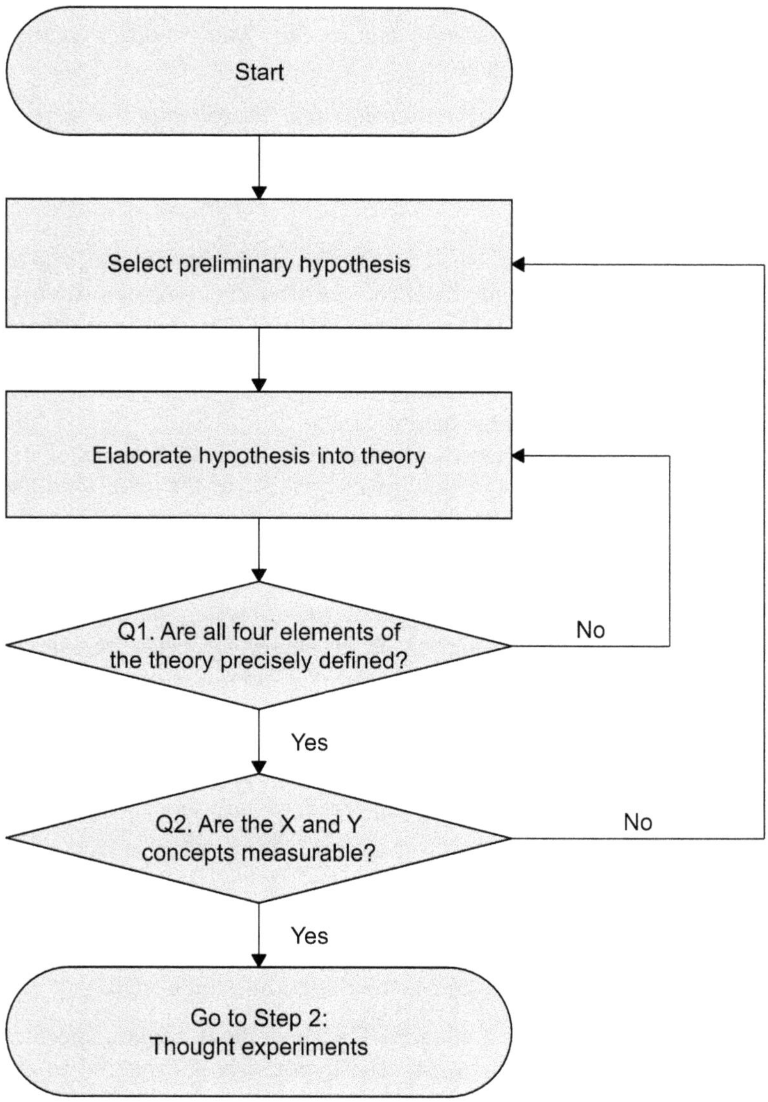

FIGURE 7.2 Hypothesis development tool for Step 1 of the development process of a formal necessity hypothesis: Defining a preliminary necessity theory.

or selected for testing the hypothesis must be specific units/examples of the focal unit (e.g., a person, a parent-child pair). Additionally, the concepts of the theory are characteristics of the focal unit.

- Precise definition of the *concepts*. The concepts are the varying characteristics that are relevant for the theory. For a necessity theory, the concepts are condition X and outcome Y. The outcome often represents something either desirable or undesirable. Desirable outcomes include, health, well-being, performance, innovation, success, etc. Undesirable outcomes may be sickness, risk, disease, failure, etc. The condition is the concept that must be present to have the outcome.

In the TPB example the condition is *Intention for the behavior* (of a person), and the outcome is *Performance of the behavior* (of a person). In the body height example the condition is *Parent body height* (of a parent-child pair) and the outcome is *Child body height* (of a parent-child pair).

- Precise specification of the *hypothesis*. Once the concepts are defined, the hypothesis can be formulated as X *is necessary for* Y. If the condition is desirable, the hypothesis could be formulated using one of the enabling terms of Table 7.1 to indicate that the necessary condition supports the outcome. If the condition is undesirable, the hypothesis could be formulated using one of the constraining terms of Table 7.1 to indicate that the absence of the necessary condition prevents the outcome.
  Usually, theories in the social sciences are semantic, such that levels of its concepts are only expressed in binary terms and qualitatively (e.g., in terms of absence/presence, or low/high), without giving exact levels. For example, the presence of X is necessary for the presence of Y, or high X is necessary for high Y. This is the implicit meaning of X is necessary for Y. The conceptual model of this hypothesis is shown in Figure 3.6-top-left. This figure shows a specification of the direction of necessity as + nc + or high-high (Section 3.5): presence/high X is necessary for presence/high Y. Other possible directions are absence/low X is necessary for presence/high Y (- nc +; low-high), presence/high X is necessary for absence/low Y (+ nc -; high-low), and absence/low X is necessary for absence/low Y (- nc -; low-low) (see the other conceptual models in Figure 3.6). The specified direction of the hypothesis defines the *expected empty corner*: which corner of an XY-plot is expected to be empty if the hypothesis holds (Figures 3.7 and 3.8).

For the TPB example, the preliminary hypothesis is: the presence of *intention for the behavior* is necessary for the presence of *performance of the specific behavior.* For the body height example, the preliminary hypothesis is: a high *parent body height* is necessary for a high *child body height.* In both hypotheses, the direction is high-high.

- Precise definition of the *theoretical domain.* The theoretical domain specifies the universe of units (cases) of the focal unit to which the hypothesis is expected to apply. This universe can be defined in various ways, such as temporally (e.g., a specific time period), geographically (e.g., a region, country, or global context), demographically (e.g., age group, profession), or conceptually (e.g., type of organization or system). The specified theoretical domain determines which cases must be sampled (from a population in the theoretical domain) or selected (directly from the theoretical domain).

For the TPB example, the theoretical domain is 'All adults in the world'. For the body height example, the theoretical domain is 'All adult parent-child pairs in the world, when children have reached adulthood'.

After the preliminary hypothesis is embedded in theory, the first question (Q1) of Figure 7.2 checks if precise definitions are available for four elements of the theory: focal unit, concepts, hypothesis and theoretical domain.

Q1: Are all four elements of the necessity theory precisely defined? When all elements of a necessity theory are specified, the hypothesis is considered to be embedded in theory.

Q2: Are the X and Y concepts measurable? This question checks whether the concepts can be transformed into measurable variables. When concepts cannot be measured directly, substitute measurements (proxies) can be considered, such as measuring an informant's perception or belief about an otherwise unmeasurable behavior or event, although this approach may compromise measurement validity. NCA accepts any meaningful, valid, and reliable scores of X and Y as input (Chapter 8). When the X and Y concepts of the necessity theory can be operationalized into measurable variables, the hypothesis is considered testable.

7.5.1.1 Example Step 1

The hypothesis development tool for Step 1 (Figure 7.2) can be applied to the example about the necessity of an Internet connection for attending an online meeting. First the preliminary necessity hypothesis that is formulated based on expert knowledge is selected.

Select preliminary hypothesis: Internet connection is necessary for Online meeting attendance.

Next, the hypothesis is embedded in theory.

Elaborate hypothesis into theory:

Q1: Are all four elements of the necessity theory precisely defined?

- *Focal unit*: 'Person'. A person may or may not have an Internet connection and the person may or may not attend an online meeting.
- *Condition concept* X: Internet connection, defined as 'the communication pathway between a device such as a computer or smartphone and the global network known as the Internet'. This concept can be absent (no Internet connection) or present (Internet connection).
- *Outcome concept:* Y: Online meeting attendance. This concept is defined as 'being part of a real-time, multimedia interaction among multiple participants'. This concept can also be absent (no Online meeting attendance) or present (Online meeting attendance).
- *Direction of the hypothesis*: the *presence* of an Internet connection is necessary for the *presence* of an Online meeting attendance (high-high).
- *Theoretical domain*: All persons in the world

Answer → yes.

Q2: Are the concepts Internet connection (X) and Online meeting attendance (Y) measurable? The presence or absence of a person's Internet connection, and the presence or absence of a person's online meeting attendance can be measured in different ways such as by analyzing the platform and network information, user activity logs, observations, questionnaires, and other data sources. This makes the hypothesis testable → yes.

Figure 7.3 shows the conceptual model and the expected empty corner for this preliminary theory. The preliminary hypothesis predicts that all persons that do not have an Internet connection also do not attend an online meeting, and that all persons that attend an online meeting have an Internet connection.

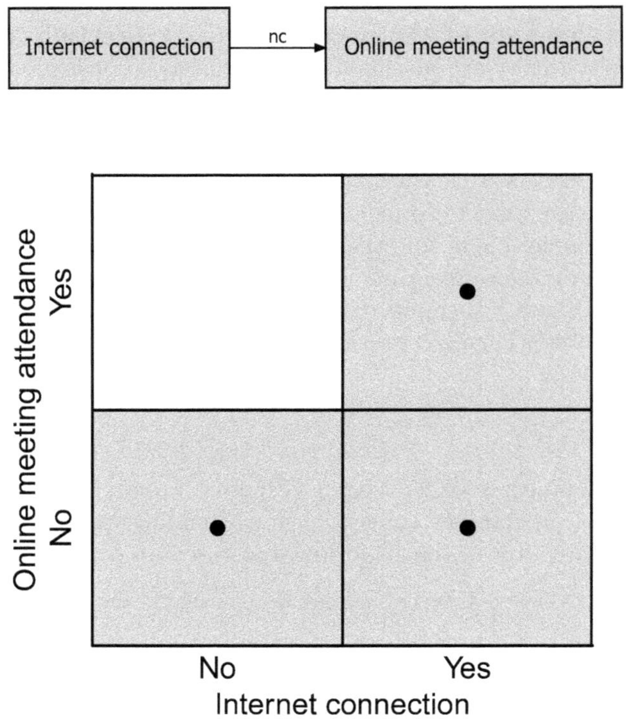

FIGURE 7.3 Conceptual model and XY-table with expected empty corner when the necessity proposition: having an Internet connection is necessary for Online meeting attendance.

7.5.2 Step 2: Conduct thought experiments

The second step of the development process of a formal necessity hypothesis consists of conducting thought experiments with the preliminary theory including its hypothesis, which is now elaborated into theory and is testable. Figure 7.4 shows Step 2 of the development process.

A *thought experiment* in NCA is a mental evaluation to check the validity of the necessity theory and its hypothesis. Since a necessity hypothesis has only one X and one Y, the prediction of the necessity theory is straightforward: the outcome is absent if the condition is absent, and the condition is present if the outcome is present. Therefore it is usually easy to evaluate the hypothesis mentally. The basic idea is to evaluate if the expected empty corner is indeed empty. Assuming that presence/high X is necessary for presence/high Y, the expected empty corner is the upper-left corner.

A thought experiment can help to check if the preliminary theory may be trivial or implausible. Testing trivial or implausible hypotheses with empirical data is not interesting and a waste of resources. Furthermore, if the thought experiment shows that many counterexamples exist (the expected empty space is not empty) the hypothesis should be rejected. With a few counterexamples, further evaluation is needed and the theory may be refined by redefining the concepts or the theoretical domain, such that the (revised) hypothesis of

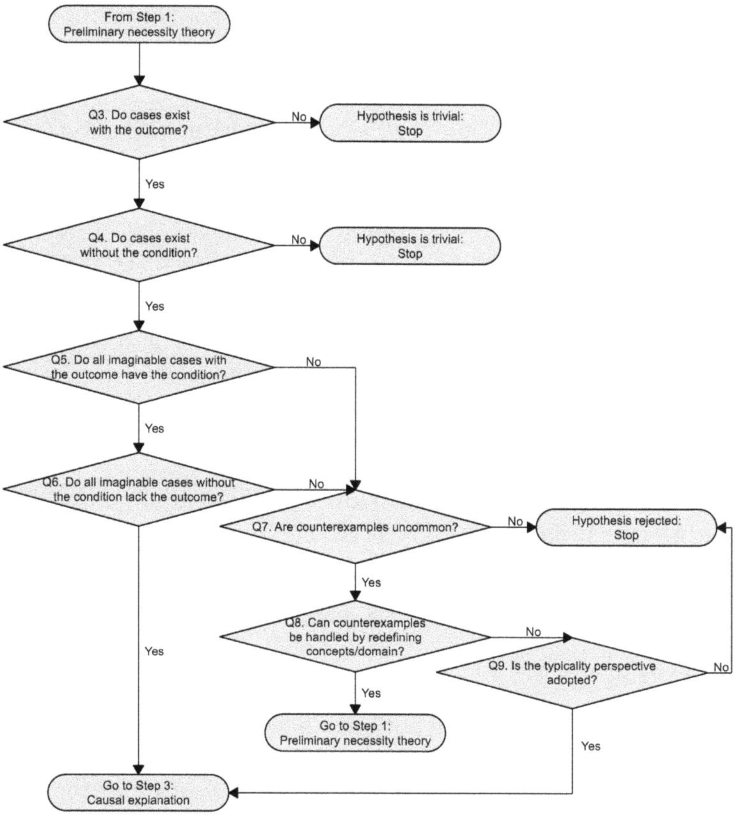

FIGURE 7.4 Step 2 of the development process of a formal necessity theory: using thought experiments to challenge the preliminary theory.

the revised theory is plausible. Three thought experiments can be done (references to the questions in Figure 7.4 are given in brackets):

- Thought experiment for non-triviality (Q3, Q4).
- Thought experiment for plausibility (Q5, Q6).
- Thought experiment for counterexamples and exceptions (Q7, Q8, Q9).

7.5.2.1 Non-triviality

A trivial necessity hypothesis is a hypothesis that is expected to hold, but is not informative because the condition or the outcome is nearly (constant) in the situation of interest (i.e., in the theoretical domain where the hypothesis is supposed to hold). Triviality applies when the condition is virtually always present, or the outcome is virtually always absent. The first type of triviality where the condition is constant (always present) and the outcome varies, is depicted in Figure 3.11-bottom-left. An example is the hypothesis that oxygen (X) is necessary for attending an online meeting (Y). This hypothesis holds because when an online meeting is held (Y is present), oxygen (X) is present and needed for the participants to have the meeting. The hypothesis is trivial because oxygen is virtually never absent. The second type of triviality where the outcome is constant (always absent) and the condition varies, is depicted in Figure 3.11-top-left. An example is the hypothesis that having good food during lifetime (X) is necessary for a human to reach age 125. The necessity claim of this hypothesis is reasonable, but it is trivial because reaching age 125 is rare.

The situation of a constant condition that is always absent and an outcome that varies (Figure 3.11-bottom-right) and the situation of a constant outcome that is always present and a condition that varies (Figure 3.11-top-right) both result in a clear rejection because the outcome can be present when the condition is absent. In these situations the necessary condition hypothesis is immediately rejected, not because it is trivial but because it is not necessary.

Therefore, there are two questions to be raised and answered in a thought experiment for checking triviality of the theory and its hypothesis (Figure 7.4):

Q3. Do cases exist with the outcome? This question checks if the outcome is virtually always absent. If not, the theory is trivial.

Q4. Do cases exist without the condition? This question checks if the condition is virtually always present. If not, the theory is trivial.

If either of these questions is answered negatively, the theory and its hypothesis can be considered trivial and not worth testing with empirical data.

7.5.2.2 Plausibility

A plausible hypothesis is a non-trivial hypothesis (both the condition and outcome can vary) that cannot be rejected in a thought experiment. The goal of this thought experiment is to consider condition and outcome together to find supportive cases for the hypothesis, and in particular possible counterexamples that challenge the hypothesis. This thought experiment for plausibility resembles an empirical test of a necessity hypothesis using a small number of cases (Section 8.3.2).

To find supportive cases, two questions can be asked:

Q5. Do all imaginable cases with the outcome have the condition? If the answer yes, this is an indication that the hypothesis may hold.

Q6. Do all imaginable cases without the condition lack the outcome? If the answer yes, this is another indication that the hypothesis may hold.

If the answer to both questions is affirmative, all cases from the theoretical domain that can be imagined are *supportive cases* for the hypothesis. When only supportive cases and no counterexamples can be imagined, the hypothesis is plausible. If the answer to one or both questions is negative, *counterexamples* are identified. These counterexamples need further evaluation before a decision can be made about the plausibility of the hypothesis.

7.5.2.3 Counterexamples and exceptions

When counterexamples outnumber supportive cases, it is clear that the hypothesis needs to be rejected. However, when there is a limited number of counterexamples compared to supportive cases an additional evaluation is needed. Therefore, when counterexamples exist, the following question is answered first:

Q7: Are counterexamples uncommon? If counterexamples are common, the theory should be rejected. If counterexamples are uncommon, they need further evaluation before the analyst makes a decision to reject or keep the hypothesis, possibly after refinement of the theory or adoption of the typicality perspective (see below).

The further analysis consists of considering the possibility of incorporating or excluding counterexamples in the theory by redefining the condition X, the outcome Y, or the theoretical domain. *Broadening the condition X* or *narrowing the outcome Y* can make counterexamples become supportive cases within the redefined theory. *Narrowing the theoretical domain* excludes the counterexamples from the redefined theory.[6]

Therefore, the analyst has three options to handle counterexamples: (1) broadening the condition, (2) narrowing the outcome, and (3) narrowing the theoretical domain. It is the analyst's choice which option to adopt. Option 1 broadens the ways in which the condition can allow the outcome. Often, the broader condition is a higher-order concept that integrates the original concept and the counterexample concept. By broadening the condition, counterexamples that previously displayed an absence of the condition now display a presence of the condition, and become supportive cases under the refined theory. However, this risks making the condition too broad and weakening its explanatory power: almost any facilitating mechanism could then be considered part of the condition. Option 2 narrows the outcome to a more specific form, reducing the scope of what is being explained. With the redefined outcome concept, the exception becomes a supportive case that does not violate the hypothesis. However, this risks excluding cases that are theoretically relevant. Option 3 restricts the theoretical domain by focusing only on a subset of cases where necessity holds. With a new definition of the theoretical domain, the exception should not have been selected for testing the hypothesis because the claim of the hypothesis is only valid in a defined theoretical domain. While this approach preserves the original concepts and hypothesis, it reduces the generalizability of the theory. Nevertheless, it seems reasonable to adopt option 3 since a theoretical domain is seldom homogeneous and universal. In option 3, the original hypothesis is retained but its theoretical domain is narrowed. This strategy preserves the preliminary hypothesis while specifying the scope of its application more precisely, provided that no known counterexamples exist.

If any of the options for handling counterexamples is selected, it is assumed that the redefined theory holds for all imaginable cases selected from the theoretical domain.[7]

If counterexamples exist the following question needs to be answered:

Q8. Can counterexamples be handled by redefining concepts/domain? If the answer is yes, the theory is adjusted, so that there are no more counterexamples. After the counterexamples are handled by redefining the necessity theory (concepts or domain) it is assumed that the hypothesis holds for all imaginable cases of the theoretical domain, such that the testing of the hypothesis can start with a deterministic view on necessity (Section 2.4.1).

If the answer to Q8 is no, it may be that (some) counterexamples remain. The analyst now has two options to proceed:

---

[6] Narrowing the condition X weakens the theory because it applies to fewer cases from reality. Broadening the outcome Y and broadening the theoretical domain adds new counterexamples.

[7] It is, of course, possible that during empirical testing, counterexamples emerge that were not anticipated during the thought experiment. In the NCA method, counterexamples are empirically defined as uncommon cases in the expected empty space, far from the ceiling, see Section 4.5.4. Such findings result in an empirical rejection of deterministic necessity. After reporting this result, the analyst may then reflect on whether the original deterministic interpretation of necessity was overly strict. However, adopting the typicality perspective after the results are known, and suggesting that this causal perspective was adopted before the results were known, should be avoided (HARKING, see Hollenbeck & Wright, 2017).

1. Adopting the *deterministic perspective* of necessity, single counterexamples reject the hypothesis (Section 2.4.1).
2. Adopting the *typicality perspective* of necessity. The counterexamples are *rare* and cannot be explained or handled. They are considered *exceptions* that are not driven by necessity and do not reject the necessity hypothesis (Section 2.4.3). If counterexamples are not rare, the hypothesis should be rejected. Note the difference between exceptions and noise (Section 4.5.4).

Adopting either perspective is a matter of analyst judgment. The *deterministic perspective* may be adopted when strict logical necessity is required without exceptions, when the theoretical domain contains few cases, or when logical rigor is prioritized. The *typicality* perspective may be adopted when it is recognized that rare exceptions can exist that cannot be explained with necessity logic.

Therefore, if counterexamples remain the following question needs to be answered:

Q9. Is the typicality perspective adopted? If the answer is no, the theory's hypothesis is rejected; if yes, it is plausible and (rare) exceptions are allowed.

7.5.2.4 Example Step 2

The development tool for Step 2 (Figure 7.4) can be applied to the example about the necessity of an Internet connection for attending an online meeting. First, the preliminary necessity theory from Step 1 is selected. The first question for the triviality check is:

Q3. Do people exist who attend an online meeting? In 2021, about 500 million people daily had online meetings. Since then the number of daily users presumably has considerably increased → yes.

The second question of the triviality check is:

Q4. Do people exist without an Internet connection? In 2025, about one-third of the world population had no access to the Internet → yes.

Since both questions are answered affirmatively, the hypothesis is non-trivial for the selected theoretical domain (all people in the world). To check for the plausibility of the hypothesis, another two questions are raised:

Q5. Do all imaginable persons attending an online meeting have an Internet connection?

Q6. Do all imaginable persons without Internet connection not attend an online meeting? The answer to both questions is no. It is possible that persons attend an online meeting at home or in an organization in a Local Area Network (LAN). These persons attend an online meeting without being connected to the Internet. Both questions Q5 and Q6 are therefore answered negatively, indicating that counterexamples exist. Trying to find counterexamples, even if they seem uncommon, is an essential part of conducting a thought experiment to test plausibility of a necessity hypothesis → no.

Since counterexamples exist, the next step is to evaluate if they are uncommon:

Q7: Are people with LAN connections uncommon? For the online meeting example, the number of persons attending an online meeting with a LAN connection compared to the number of persons attending an online meeting with an Internet connection is relatively small. Attending an online meeting with a LAN connection is relatively uncommon, suggesting further analysis of these cases is needed to refine the theory → yes.

Q8: Can counterexamples be handled by redefining concepts/domain? The three options to handling counterexamples are:

Broadening the condition means that the original definition of Internet connection (the communication pathway of devices such as computer or smartphone to the global network known as Internet) is broadened such that it incorporates LAN networks. The concept *Internet connection* could be broadened to the concept *Digital network connection* defined as the communication pathway of devices such as computer or smartphone to a digital network. 'Digital network' now includes Internet and LAN.

Narrowing the outcome means that the original definition of the outcome Y (Online meeting attendance: being part of a real-time, multimedia interaction among multiple participants) is narrowed such that it incorporates the exception. For example, *Online meeting attendance* could be narrowed to *Public online meeting attendance* defined as being part of a real-time, multimedia interaction among multiple participants from the general public, excluding private online meetings (that use LAN) from the definition. For the cases that were previously counterexamples, the condition 'Internet connection' is absent (though LAN is present), and the redefined outcome concept 'Public online meeting attendance' is also absent (because the LAN meeting is considered a private meeting).

The third way to handle an exception is to *exclude counterexamples* from the theory by *narrowing the theoretical domain*. In the online meeting example, narrowing the theoretical domain means that the original definition, 'all persons in the world', would be revised such that it excludes people that use a LAN network in a private meeting. For example, the domain *All persons in the world* could be narrowed to *All persons who are geographically distant from each other*, which would exclude those in a local gathering (home, organization) from the theoretical domain.

In summary, for the example, the three possible ways to handle the exception by redefining concepts/domain are:

Option 1 (broadening condition):

Hypothesis: Digital network connection is necessary for Online meeting attendance. *Focal unit*: Person. *Condition* X: **Digital network connection: the communication pathway of devices such as computer or smartphone to a digital network**. *Outcome* Y: Online meeting attendance: real-time, multimedia interaction among multiple participants. *Theoretical domain*: All persons in the world.

Option 2 (narrowing the outcome):

Hypothesis: Internet connection is necessary for Public online meeting attendance. *Focal unit*: Person. *Condition* X: Internet connection: the communication pathway of devices such as computer or smartphone to the global network known as Internet. *Outcome* Y: **Public online meeting attendance defined as a real-time, multimedia interaction among multiple participants from the general public**. *Theoretical domain*: All persons in the world.

Option 3 (narrowing the theoretical domain):

Hypothesis: Internet connection is necessary for Online meeting attendance. *Focal unit*: Person. *Condition* X: Internet connection: the communication pathway of devices such as computer or smartphone to the global network known as Internet. *Outcome* Y: Online meeting attendance: real-time, multimedia interaction among multiple participants. *Theoretical domain*: **All persons who are geographically distant from each other**.

It is supposed that the second option is selected for further development in Step 3 (Section 7.5.3.1) → yes.

7.5.3 Step 3: Provide causal explanation

In the previous step, the preliminary necessity theory including its hypothesis was either rejected or accepted during a thought experiment, possibly after revision of concepts or theoretical domain, or by adopting the typicality perspective. If accepted, the next step is to explain *why* X is necessary for Y. This involves developing a causal explanation. During the formulation of the preliminary necessity theory including its hypothesis and during the process of conducting thought experiments, the analyst implicitly used a causal reasoning about necessity. In the last step before empirical testing of the hypothesis, the causal reasoning is explicitly articulated and further deepened. Since causality is a human interpretation rather than a directly observable phenomenon (Chapter 2), the analyst must construct a plausible and coherent narrative to justify the proposed necessity-based causal link.

Thus, describing the causal explanation is a creative process that can be based on insights obtained during the thought experiments, and can be enriched with causal reasoning from the academic literature, from practice, and from the analyst's own experience (Sections 7.4.1 and 7.4.2).

The causal narrative can be inspired by answering three questions that are positioned in the XY-plot shown in Figure 7.5:

1. Explaining why X is an *enabler* for Y (why the presence of X allows the presence of Y).
2. Explaining why X is a *constraint* for Y (why the absence of X leads to the absence of Y).
3. Explaining why there is *no substitute* for the absence of X (why no other paths without X can lead to the outcome Y).

The first question builds on conventional causal reasoning, exploring how one factor allows an outcome and in this way helps to produce it. The second and third, however, go beyond standard approaches: they introduce the need to reason counterfactually and to explore the uniqueness of X in the causal structure. These perspectives encourage creative and rigorous thinking about necessity and irreplaceability.

One core idea of necessity causality is that the absence of X causes the absence of Y. More precisely, if the necessity cause is reduced or removed, the present or high outcome should disappear. This thought experiment resembles a true 'loss of function', see the necessity experiment discussed in Section 8.2.4.

The three guiding questions are part of Step 3 of the hypothesis development tool (Figure 7.6).

Q10. Is it explained why X enables Y? The accepted hypothesis from the thought experiments is first considered from the perspective that X is an enabler for Y. To explain why a condition *enables* an outcome, the starting point is the *presence* of X. The explanation then focuses on how this presence can plausibly contribute to producing the outcome, possibly through a chain of necessity (Sections 3.3 and 4.6.4). Common narratives about

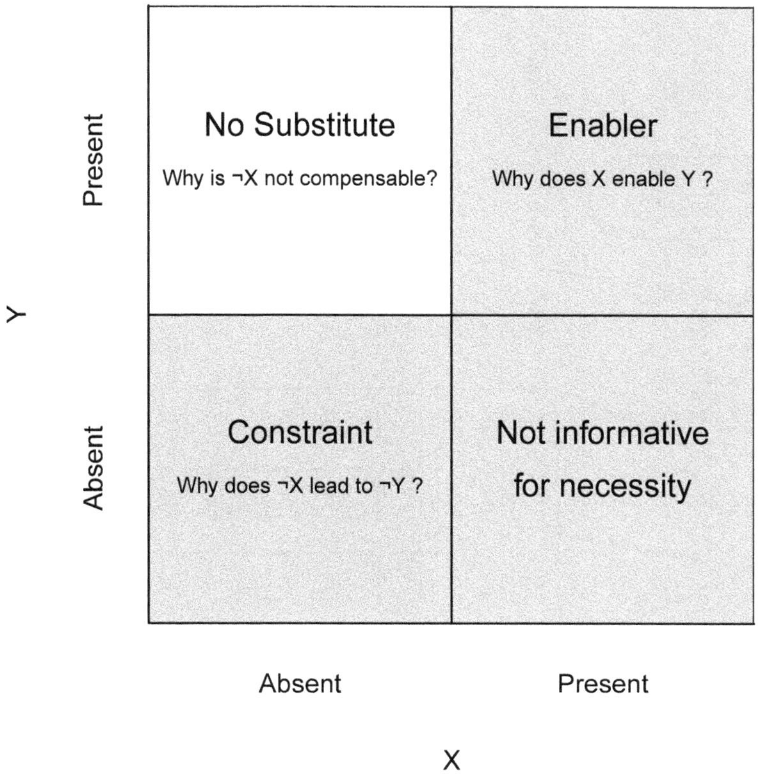

FIGURE 7.5 Questions for explanation why X is necessary for Y. $\neg X$ and $\neg Y$ mean absence of X and Y.

the causal mechanisms through which X contributes to Y can support this explanation. While such narratives are used to describe probabilistic sufficiency where the presence of X increases the likelihood of Y, the same reasoning can be used to illustrate X's role as a contributor: when X is present, Y *may* occur, but it is not guaranteed; Y may still be absent. The contributor becomes only an enabler when also the next two questions apply: if X is absent, Y will not occur.

Q11 Is it explained why X constrains Y? Now X is considered from the perspective of being a constraint for Y. To explain why a condition *constrains* an outcome, the starting point is the *absence* of X. The explanation then focuses on how this absence disrupts the causal mechanism through which X contributes to the outcome as described in the answer to Q10, thereby preventing the outcome from occurring.

Q12. Is it explained why there is no substitute for the absence of X? But even if X is not present, it may be that its absence is compensable by another factor or path. Therefore, the hypothesis is considered from the perspective of the possibility of substitutes. To explain why there is *no substitute* for the absence of a condition, the starting point is again the *absence* of X. The explanation then focuses on whether any alternative path (mechanism) could plausibly exist to still produce the outcome. If no viable substitutes exist, the absence of X guarantees the absence of Y. If an alternative path is discovered that was not identified in Step 2, the analyst may return to Q7 of Step 2.

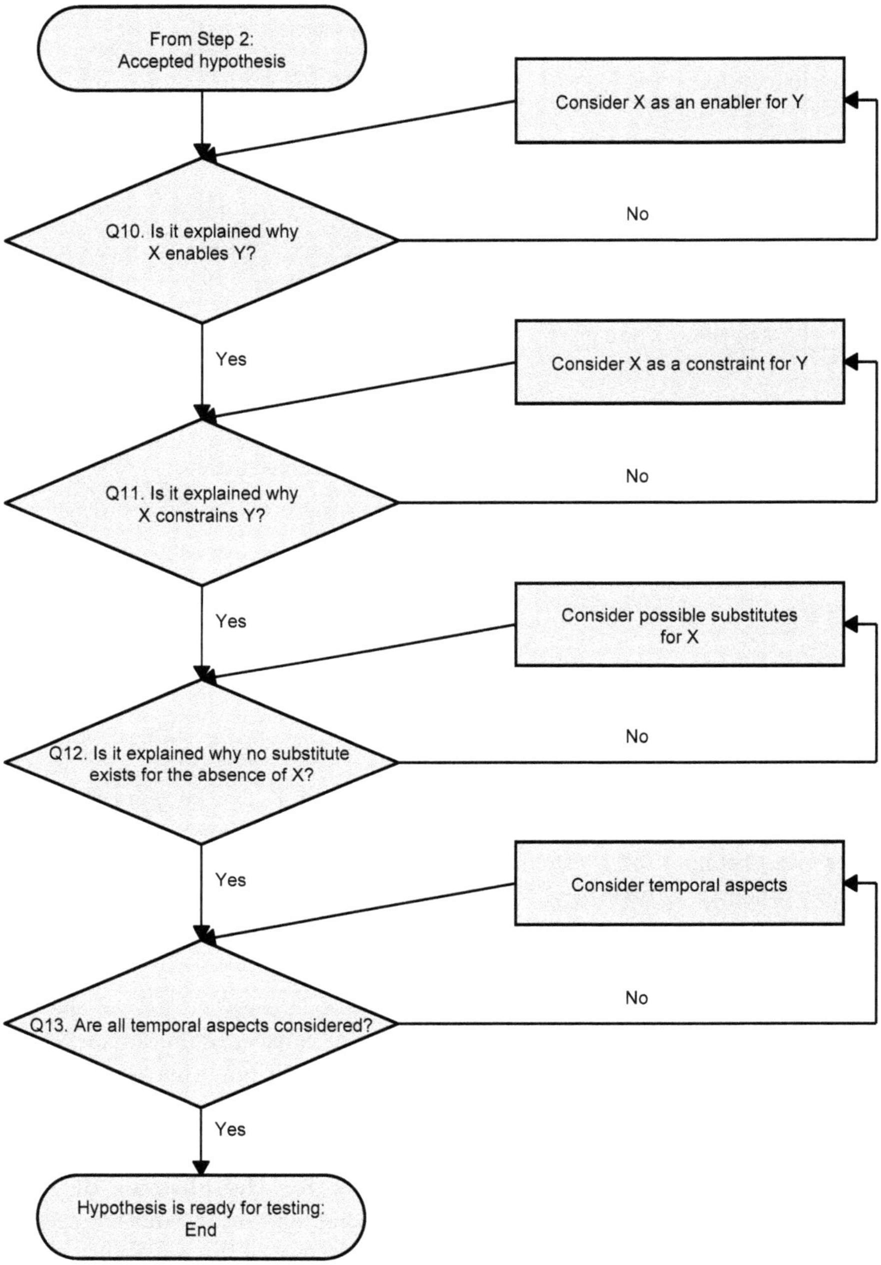

FIGURE 7.6 Hypothesis development tool for Step 3 of the development process of a formal necessity hypothesis: Adding a causal explanation.

Q13. Are temporal aspects considered? If Q10-Q12 are answered affirmatively, the three components (enabler, constraint, substitute) provide the basis for a narrative for explaining why X is a necessary cause for Y. In addition to the explanation that X is essential to produce Y, and that its absence cannot be substituted, also three temporal aspects need to be considered. First, one fundamental requirement for establishing causality is *temporal order*: X must precede Y. If this sequence is not self-evident, the expected temporal ordering should be explicitly stated and justified. Doing so helps to address the risk of reverse causality (Section 5.7). Second, it is important to clarify whether X is necessary for the *onset* of Y or for its *continuation*. For instance, a match is necessary to ignite a fire (onset), but not to keep it burning (continuation). In contrast, oxygen is necessary for both the onset and continuation of the fire. This distinction can be incorporated into the causal explanation or into the definition of the outcome concept itself. Third, the necessity of X for Y may be *time-dependent*. That is, X may only be necessary during certain time periods. For example, in the past, attention to sustainability may not have been necessary for a company's success, whereas it may be essential today. This time-sensitive aspect of necessity can be included either in the causal narrative or in the specification of the theoretical domain.

After providing the temporal aspects, the hypothesis is complete. The formal hypothesis is now fully specified and is ready for empirical testing:

Hypothesis: X is necessary for Y. (add direction if needed) *Focal unit* = ... *Concept X* = ... *Concept Y* = ... *Theoretical domain* = ... *Causal explanation* = ...

7.5.3.1 Example Step 3

The hypothesis development tool for Step 3 (Figure 7.6) can be applied to the example about the necessity of an Internet connection for attending an online meeting. The accepted hypothesis after the thought experiments is 'Internet connection is necessary for Public online meeting attendance'.

Q10. Is it explained why Internet connection enables Online meeting attendance? First, it is evaluated if an Internet connection can be considered as an enabler for Public online meeting attendance. This enabling role can be explained as follows: the presence of a communication pathway makes it possible for a participant to connect to a platform and to communicate remotely. The platform provides the necessary technical infrastructure for video, audio, and chat functionalities that allow a public online meeting to occur[8] → yes.

Q11. Is it explained why Internet connection constrains Public online meeting attendance? Next, it is considered if an Internet connection is a constraint for a public online meeting. The constraining role of an Internet connection can be explained as follows. Without a communication pathway it is also not possible to connect to a digital platform, such that participants are unable to connect remotely. The essential infrastructure for real-time communication is missing. As a result, the conditions required for a public online meeting are not met, and the person cannot attend the meeting. The absence of the communication pathway thus blocks the occurrence of the outcome → yes.

[8] However, the mere presence of the Internet connection does not guarantee that the meeting will be attended. Other conditions must also be met, such as the participant's willingness to join, availability, and the scheduling of the meeting itself. Thus, the Internet connection enables the possibility of the outcome (a public online meeting), but it does not ensure its occurrence. Ensuring the outcome would mean that corner 4 is also empty. However, the hypothesis is only about necessity, and does not make a statement whether or not the condition is also sufficient. For the hypothesized necessity, corner 4 is not relevant.

Q12. Is it explained why there is no substitute for the absence of X? Next, it is considered if, when a digital platform is absent, other technologies for example, email or satellite communication could serve as substitutes. Email is an asynchronous communication tool that may support collaboration in a broader sense, but does not enable the real-time interaction required for a public online meeting. Satellite communication may offer connectivity for specific groups or in remote areas, but does not facilitate general public online meetings. If no substitute exists in the theoretical domain, with the absence of an Internet connection the outcome cannot be achieved, confirming that X is irreplaceable in this context $\rightarrow$ yes.

Q13. Are temporal aspects considered? An online meeting cannot begin without a working Internet connection already in place. Thus, the Internet connection must exist before or at least at the onset of the meeting. The Internet connection is necessary both for the onset of the online meeting (to initiate the call or session), and the continuation of the meeting (to maintain the live connection throughout). The necessity of an Internet connection for an online meeting is time-dependent in a technological sense: Before Internet-based communication tools existed (e.g., before the 1990s), an Internet connection was not necessary; online meetings did not exist as a concept, so only after 1990 the hypothesis is non-trivial. In the future, alternative technologies might make traditional Internet connections obsolete and become substitutes $\rightarrow$ yes.

If all questions are answered affirmatively, the hypothesis is now a formal necessity hypothesis, and is ready to be tested with empirical data.

8

Data

8.1 Summary of this chapter

This chapter describes the steps between the hypothesis and data analysis that are needed to obtain empirical data for testing a formal necessity hypothesis. The hypothesis discussed in the previous chapter (Chapter 7) guides which data should be collected. The hypothesis has precise definitions of (1) the focal unit, (2) the concepts X and Y, (3) the direction of necessity and a causal explanation, and (4) the theoretical domain where the hypothesis is supposed to hold. Empirical data refer to observed x- and y-scores (values or levels) of variables that represent the concepts X and Y. These scores are measured on cases selected from the theoretical domain. To collect high-quality empirical data for testing the necessity hypothesis, three key decisions must be made: (1) selection of study design, which is the type of study (observational study, longitudinal study, case study, or experimental study discussed in Section 8.2), (2) sampling or selection of cases from the theoretical domain: random sampling and purposive case selection (Section 8.3), and (3) conducting the measurement to obtain valid and reliable x- and y-scores (e.g., qualitative or quantitative data; archival or new data) for each case (Section 8.4). Section 8.5 discusses the format of a useful dataset.

While NCA does not impose new requirements on data collection and treats empirical data as input to NCA rather than as part of NCA, some choices must be tailored to fit the logic of necessity. This applies, for example, to the study design when a necessity experiment is done (Section 8.2.4.1), or to the use of purposive case selection in a small-n qualitative study (Section 8.3.2).

8.2 Study design

Study design refers to the type of study that is done to test the hypothesis. This book distinguishes between the *observational study*, which is a study that analyzes a large number of cases in the natural environment (e.g., large-n quantitative study), the *longitudinal study*, which is an observational study where cases are studied at different time points, the *case study*, which is an observational study with a small number of cases in the natural environment (e.g., small-n qualitative study), and the *experimental study*, which is a study in which the condition X is manipulated and the outcome Y is observed. This section discusses each of these four study designs in the context of NCA.

DOI: 10.1201/9781003430766-8

8.2.1 Observational study

The most common design of an NCA study is the observational study. The analyst observes the phenomena of interest in the natural setting without manipulating X and Y. Such studies are often cross-sectional, meaning that X and Y are observed at a single point in time. The observational study is usually associated with a large-n *quantitative study* where many cases are sampled (Section 8.3.1), and scores of X and Y are numeric (Section 8.4.1).

The popularity of this design in NCA can be explained not only by the popularity of quantitative studies in general, but also because the method is not vulnerable to omitted variable bias. This is one reason in regression-based studies for conducting randomized experiments (see Section 11.3.1). The absence of omitted variable bias allows findings from an observational NCA study to be readily interpreted causally without taking special measures to control for other variables, assuming that an adequate sample (Section 5.2) and a solid necessity theory (Chapter 7) are available. Thus, the observational study is particularly useful when the hypothesis is well-developed with a plausible causal explanation that also specifies the temporal aspect (first X then Y) with a proper sample that adequately covers the population of interest. NCA does not impose new requirements for designing a good observational study.

8.2.2 Longitudinal study

In this book, a longitudinal study (common name in statistics) and a panel study (common name in econometrics) are treated as synonyms, and the term longitudinal study is used. A longitudinal study is an observational study in which the same cases are followed over time and X and Y are measured at different time points.

When a necessity hypothesis includes an expectation about differences of necessity over time, or when the analyst is unsure about the possibility of reverse causality (see Section 5.7), a longitudinal study may be appropriate. It may be that the analyst expects that necessity exists at one time point but not at another or that the necessity between time points differs in strength (effect size).

Three types of longitudinal study are discussed here. First, it may be that data at different time points are available, but that the analyst is not interested in time per se. Then these data can be pooled together, resulting in a design that corresponds to the single cross-sectional observational study (e.g., Jaiswal & Zane, 2022b). Second, if the analyst is interested in time trends, NCA can be applied for each time point separately. This corresponds to a multiple cross-sectional observational study. Third, if the analyst wants to avoid a reverse causality interpretation of the findings, X-scores are selected from a point in time before the Y-score is selected. Such a time-lagged study ensures that the condition X existed before the occurrence of the outcome Y. NCA does not impose new requirements for designing a good longitudinal study.

Below, the longitudinal study is illustrated with an example of the hypothesis that a country's economic prosperity is necessary for a country's average life expectancy:

- *Focal unit*: Country.
- *Condition concept* X: Economic prosperity.
- *Outcome concept:* Y: Life Expectancy.

- *Hypothesis*: A high level of Economic prosperity is necessary for a high level of Life expectancy (high-high).
- *Theoretical domain*: All countries in the world.

Economic prosperity captured by log GDP per capita and Life expectancy by Life expectancy at birth, expressed in years. The data are archival data from the World Bank. The data span a period of 64 years between 1960 and 2023, for 211 countries. Data are available for 11,020 country-years.

8.2.2.1 Pooled data

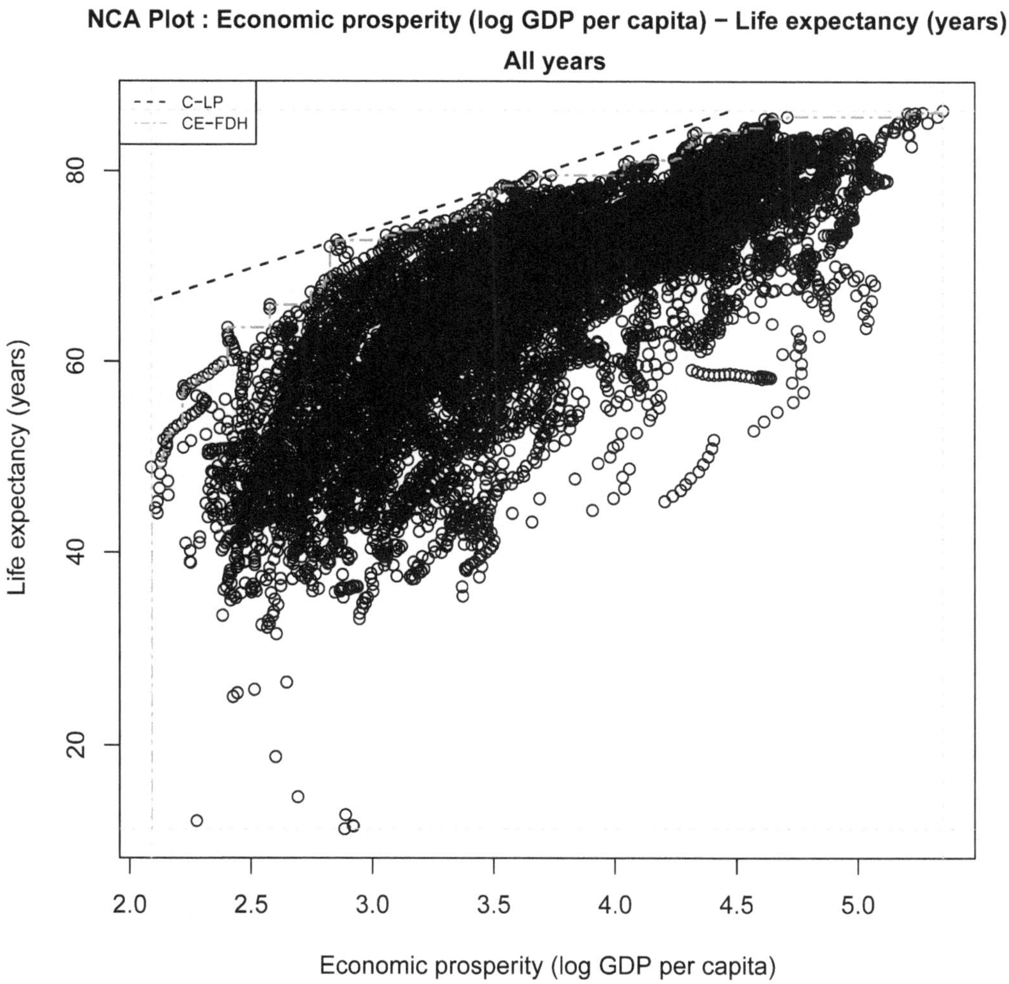

FIGURE 8.1 XY-plot of Economic prosperity and Life expectancy with two ceiling lines (Pooled World Bank data from 1960 to 2023 for all countries/years).

In the first illustration, the data are pooled to test the hypothesis. Figure 8.1 shows the XY-plot. Each data point is a country-year combination. The plot shows a clear empty space in the upper-left corner as expected. This suggests that a *necessity* relationship exists between Economic prosperity and Life expectancy[1]. This indicates that a certain

[1]The NCA analysis is done with the C-LP ceiling line (Section 4.4), which is a linear ceiling line that

level of life expectancy is not possible without a certain level of economic prosperity. For example, for a country's life expectancy of 75 years, an economic prosperity of about 3 (10^3 = 1,000 equivalent US dollars) is necessary. Such a level of economic prosperity is not sufficient as many cases with an Economic prosperity level of 3 have a lower level of Life expectancy[2].

8.2.2.2 Time trends

In the second illustration, the data for the different time points are used to test expected time trends regarding the necessity of high Economic prosperity for high Life expectancy[3]. Table 8.1 shows the NCA parameters for each separate year.[4]

TABLE 8.1 NCA parameters per year.

Year	Ceiling intercept	Ceiling slope	Effect size	*p*-value
1960	2.2	19.8	0.19	0.001
1961	18.8	15.0	0.18	0.001
1962	28.8	11.6	0.19	0.001
1963	27.5	12.2	0.19	0.001
1964	27.7	12.3	0.18	0.001
1965	38.2	8.7	0.21	0.001
1966	37.6	9.0	0.20	0.001
1967	41.4	7.7	0.22	0.001
1968	42.3	7.5	0.22	0.001
1969	43.3	7.2	0.22	0.001
1970	35.4	10.0	0.19	0.001
1971	36.5	9.6	0.19	0.001
1972	53.5	4.8	0.20	0.001
1973	56.3	4.2	0.19	0.001
1974	58.3	3.8	0.19	0.001
1975	38.9	9.0	0.19	0.001
1976	40.3	8.7	0.18	0.001
1977	51.1	5.7	0.19	0.001
1978	52.2	5.5	0.18	0.001
1979	52.6	5.5	0.18	0.001
1980	56.1	4.6	0.17	0.001
1981	45.5	7.6	0.17	0.001

has no observations above it. The effect sizes for the pooled data with the two ceiling techniques are 0.10 ($p < 0.001$) and 0.14 ($p < 0.001$), respectively.

[2]There is also a well-known positive *average* relationship (imaginary line through the middle of the data), but this average trend is not considered here. Furthermore, the lower-right corner is empty as well. This suggests that a low level of Economic prosperity is necessary for a low level of Life expectancy. However, this low-low necessity relationship was not claimed in the hypothesis that is tested here. The necessity of low X for low Y can be evaluated with the `corner = 4` argument in the `nca_analysis` function of the `NCA` software (Appendix B)

[3]The time trend expectation can be included in the causal explanation of the hypothesis, see Section 7.5.3.

[4]Only the C-LP ceiling technique is used. For making the NCA results comparable per year, the theoretical scope for each year is the scope of the pooled data (based on the absolute minima and maxima of empirically observed Economic prosperity and Life expectancy in the data).

1982	45.7	7.6	0.17	0.001
1983	45.0	7.8	0.16	0.001
1984	46.9	7.4	0.16	0.001
1985	47.6	7.3	0.16	0.001
1986	49.1	6.9	0.16	0.001
1987	50.7	6.5	0.15	0.001
1988	51.2	6.5	0.15	0.001
1989	50.2	6.8	0.15	0.001
1990	56.5	5.2	0.14	0.001
1991	58.6	4.7	0.14	0.001
1992	59.1	4.7	0.13	0.001
1993	58.3	4.9	0.13	0.001
1994	57.4	5.2	0.13	0.001
1995	56.7	5.4	0.13	0.001
1996	56.1	5.6	0.13	0.001
1997	54.7	5.9	0.13	0.001
1998	57.1	5.4	0.12	0.001
1999	56.4	5.6	0.12	0.001
2000	56.0	5.7	0.12	0.001
2001	55.8	5.8	0.12	0.001
2002	56.6	5.6	0.12	0.001
2003	57.5	5.5	0.11	0.001
2004	56.7	5.7	0.11	0.001
2005	57.0	5.7	0.11	0.001
2006	57.9	5.5	0.11	0.001
2007	59.1	5.3	0.10	0.001
2008	31.9	13.1	0.11	0.001
2009	59.0	5.5	0.09	0.001
2010	58.0	5.7	0.10	0.001
2011	57.1	5.9	0.10	0.001
2012	55.5	6.3	0.10	0.001
2013	54.4	6.6	0.10	0.001
2014	54.4	6.6	0.10	0.001
2015	55.8	6.3	0.10	0.001
2016	57.0	6.0	0.10	0.001
2017	57.0	6.0	0.10	0.001
2018	57.5	6.0	0.09	0.001
2019	57.8	5.9	0.09	0.001
2020	49.7	7.8	0.11	0.001
2021	51.3	7.4	0.11	0.001
2022	51.3	7.5	0.10	0.001
2023	48.9	8.2	0.10	0.001

The interpretation of time trends focuses on the change of necessity effect size, ceiling line intercept and ceiling line slope over time. Table 8.1 and Figure 8.2-top-left show the clear trend that necessity effect size reduces over time. This indicates that, over the years, Economic prosperity has become less of a bottleneck for high Life expectancy. Figures 8.2-

top-right and 8.2-bottom show that this decrease of effect size is due to the increase of the intercept (upward moving ceiling line) and decrease of the slope (ceiling line flattens).

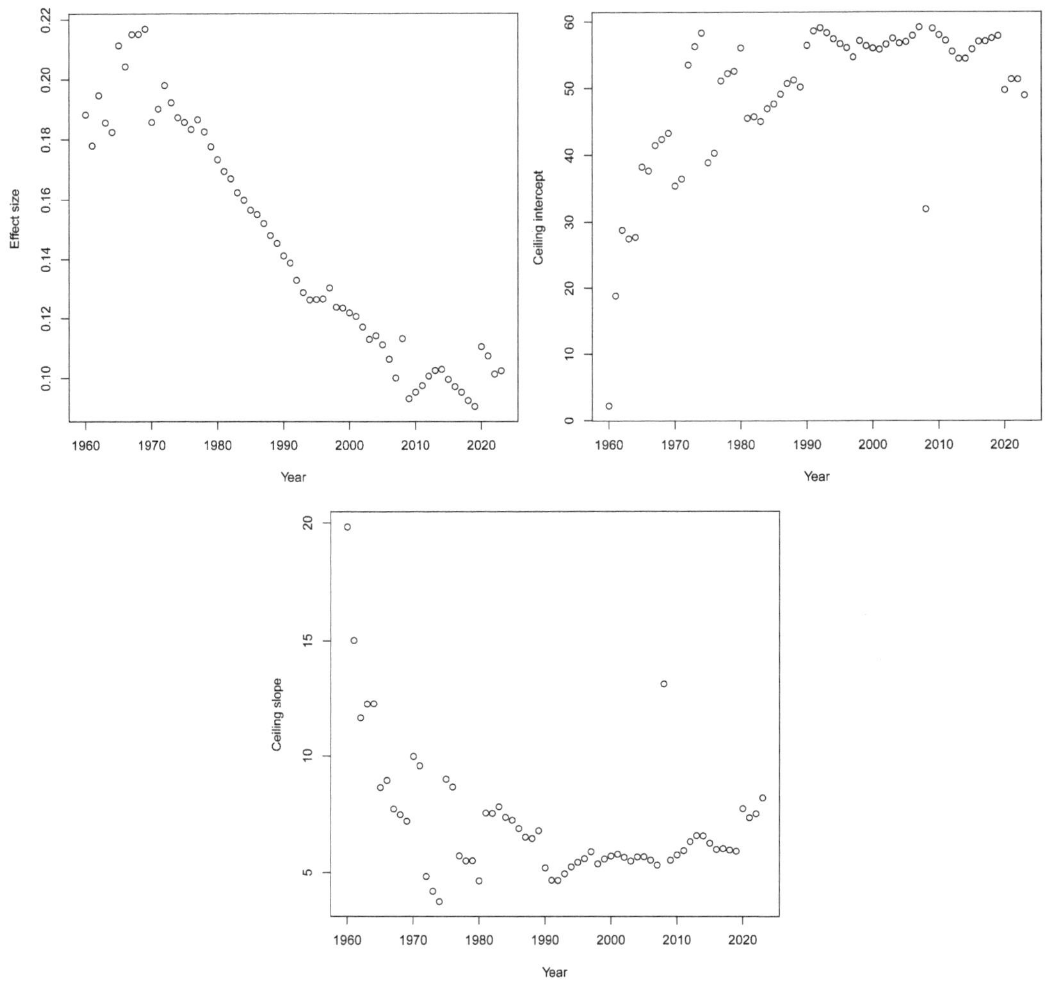

FIGURE 8.2 Time trends of NCA parameters for the necessity of Economic prosperity for Life expectancy. Top-Left: necessity effect size. Top-Right: ceiling intercept. Bottom: ceiling slope.

Figure 8.3 compares the XY-plots for the first (1960) and last (2023) year in the dataset and shows that the effect size in 1960 was higher than in 2023.

NCA's statistical difference test for paired observations (Section 5.3.6) shows that the 0.09 difference between the effect sizes of 1960 and 2023 (0.19 and 0.10, respectively) is statistically significant ($p < 0.001$).

In 1960 the maximum life expectancy for an Economic prosperity of 3 ($10^3 = 1000$ equivalent US dollars) was about 60 years, and in 2023 it was about 75 years. In 2023 compared to 1960, the countries are more clustered near the ceiling, indicating that the difference in Life expectancy between countries diminished, but that a hard border of maximum possible Life expectancy for a given Economic prosperity level remains.

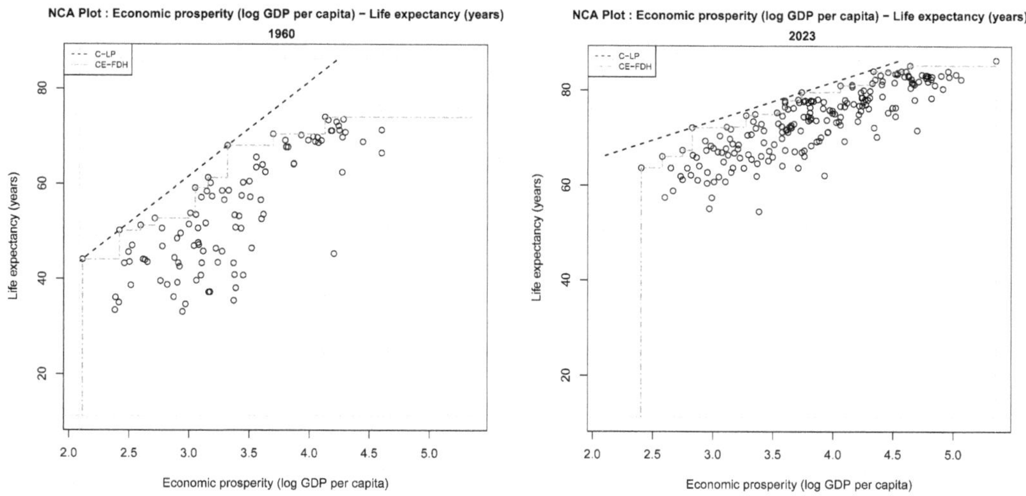

FIGURE 8.3 *XY*-plot of Economic prosperity and Life expectancy for 1960 (Left) and 2023 (Right).

Socio-economic policy practitioners could make new interpretations from these necessity time-trends. For example, in the past, economic prosperity was a major bottleneck for high life expectancy. Currently, most countries are economically prosperous enough for having the possibility of a decent life expectancy of 75 years; if these countries do not reach that level of life expectancy, other factors than economic prosperity are the reason for it. If the trend continues, in the future Economic prosperity may not be a limiting factor for high Life expectancy, because the empty space in the upper-left corner disappears.

8.2.2.3 Time-lagged study

The observed empty space in the upper-left corner could also be the results of reverse causality. In reverse causality Y existed before X. An empty space in the upper-left corner could then also indicate that a low level of Y is necessary for a low level of X. This would mean that a low level of Life expectancy is necessary (Y) for a low level of Economic prosperity X. When forward necessity makes sense, the argument for reverse causality is often weaker but cannot be excluded. With a time-lagged study, reverse causality is less likely when an empty space in the upper-left corner is observed while the X-score is obtained at time point (T_1) before the time point (T_2) of the Y-score. The lag should be long enough for Y to develop in the presence of X and other causal factors. For example, by using a 5-year time-lag the analysis with Economic prosperity in 2018 and Life expectancy in 2023 has a similar empty space in the upper-left corner (Figure 8.4).

The differences between Figures 8.4 and 8.3-right are minor.

8.2.3 Case study

The case study investigates the phenomena in a small number of cases in their natural environment. Such a design is associated with a small-n qualitative study that allows a rich description of the phenomena. In such a study, relationships are often described in words rather than numerically.

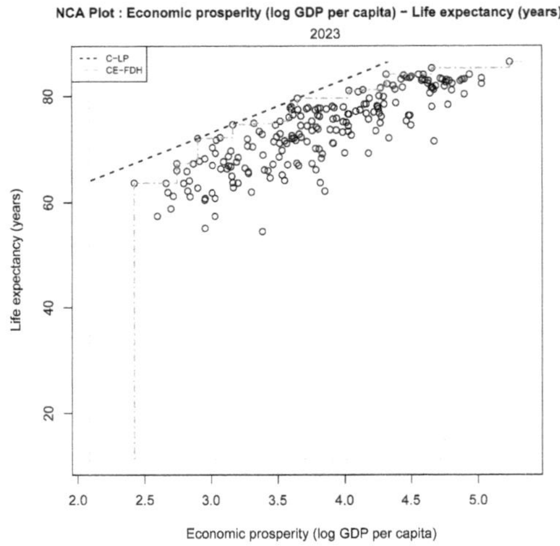

FIGURE 8.4 *XY*-plot of Economic prosperity in 2018 and Life expectancy in 2023.

NCA can be applied to a case study design in two ways. First, NCA's methodology provides the causal logic to describe phenomena. Case studies are often used to explore complex causal relationships within specific contexts, and NCA complements this by introducing the idea that some factors may be necessary (but not sufficient) for an outcome to occur. In this way, NCA helps case investigators to move beyond traditional probabilistic sufficiency cause-effect explanations toward identifying the minimal requirements that must be present for a phenomenon to manifest.

Second, NCA's data analytic method provides the tools to analyze qualitative data within or across cases. In multiple-case designs, coded evidence from interviews, documents, or observations can be transformed into ordinal or categorical indicators of conditions and outcomes. These can then be used as input scores for NCA to identify which conditions consistently act as enablers or bottlenecks across cases. The results can be interpreted to explain how and why these conditions function as necessary conditions.

If a necessity hypothesis already exists, the case study design can be used to test this hypothesis with the small number of cases. Case studies explicitly designed for testing necessity hypotheses are rare, but implicitly many studies do test such a hypothesis given the way the cases are selected (e.g., Fujita & Kusano, 2020; Harding et al., 2002, see Section 8.3.2 for more details). However, testing a necessity hypothesis as part of an existing case study design, commonly aimed at exploration and theory building, offers a new opportunity for many qualitative studies (e.g., Novales et al., 2025).

When testing a necessity hypothesis with a small number of cases selected from the theoretical domain (deductive case study), a rich description of the context is not required, although it is not objectionable either. The only requirement is that the qualitative data can be scored into levels of the condition and the outcome (e.g., in terms of low, medium, high). Because necessity is independent of other variables, scoring the concepts X and Y is enough for testing the hypothesis. Given the small number of cases and their non-random selection, an estimation of p-values is not feasible. However, the reduced data requirements often allow the analyst to dive deeper into the cases for better understanding the necessity mechanisms, or selecting more than a few cases, bringing the design closer to an observational study (Section 8.2.1). Case selection for testing necessity hypotheses is discussed in Section 8.3.

8.2.4 Experimental study

In an experimental study design, X is first manipulated and afterwards the effect on Y is observed, such that the required causal direction is ensured: first X then Y. This excludes the possibility of reverse causality that Y occurs before X (Section 5.7). Whereas uncertainty about the causal direction could be handled with a time-lagged longitudinal study, where the condition is measured at T_1 and the outcome at T_2, such a study can only detect necessity if the outcome has developed during this time period. To have more control, the necessity experiment is an alternative approach for establishing the causal direction when the analyst is uncertain about reverse causality. With this goal in mind, the experiment is useful for testing a necessity hypothesis in the situation that reverse necessity is theoretically plausible.

The experimental setup for a necessity experiment differs from the traditional probabilistic sufficiency experiment in several ways. Whereas in a common experiment X is manipulated (increased) to produce the *presence* of Y on average, in a necessity experiment X is manipulated (decreased) to produce *absence* of Y (assuming a high-high hypothesis or more precisely: assuming that for a desired level of $y = y_c$, x must be equal to or larger than x_c).

The necessity experiment starts with cases *with* high level of the outcome ($y \geq y_c$) and *with* high level of condition ($x \geq x_c$), then the condition is *removed or reduced* ($x < x_c$), and it is observed if the outcome has *disappeared or reduced* ($y < y_c$). This contrasts the traditional "average treatment effect" experiment in which the research starts with cases with low outcome ($y < y_c$) and with low condition ($x < x_c$), then the condition is *added or increased* ($x \geq x_c$), and it is observed if the outcome has *appeared or increased* ($y \geq y_c$) .

The necessity experiment is conducted as follows:

1. Obtain a *representative sample* from the population.
2. Check the expected empty corner.
3. Select the entire sample or part of it.
4. Divide this sample randomly into two groups (treatment and control group).
5. Remove ($x \to 0$) or reduce ($x \to low$) the condition in the treatment group.
6. Observe: (1) whether the effect size does not disappear; and (2) whether the outcome reduction in the treatment group is larger than the outcome reduction in the control group. If both are true, necessity is supported.

This necessity experiment ensures the causal direction (first X, then Y). The randomization between groups ensures that the group constitutions are similar, and that possible time effects (e.g., effect of other disappearing necessary conditions) are controlled for. The possible disappearance/reduction of the outcome due to other necessary conditions happens in both groups; the treatment group must have more outcome disappearance/reduction than the control group.

Below several types of necessity experiments depending on whether X and Y are dichotomous or continuous are discussed.

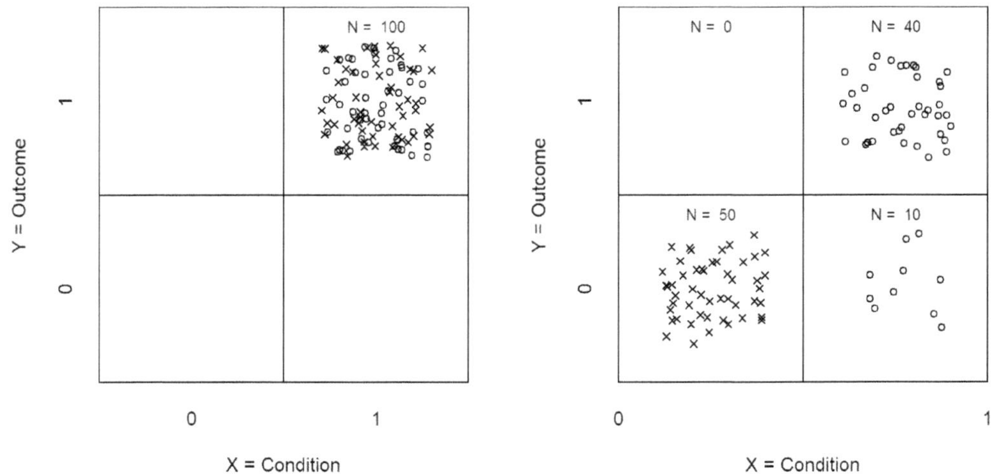

FIGURE 8.5 XY-table of the necessity experiment before (Left) and after (Right) manipulation. X = Condition (0 = no; 1 = yes). Y = Outcome (0 = no; 1 = yes). ∘ = control group. × = treatment group.

8.2.4.1 Necessity experiment: dichotomous X and dichotomous Y

For dichotomous variables the experiment to evaluate a high-high necessity hypothesis the steps are conducted as follows:[5]

1. Obtain a *representative sample* from the population.
2. Observe that *expected corner* is empty (corner 1).
3. Select the cases of corner 2. The experiment is done with these cases. In Figure 8.5-left it is assumed that 100 cases are available for the experiment.
4. *Randomly assign* these cases into two groups:
   - Treatment group (symbol ×).
   - Control group (symbol ∘).
5. *Apply the treatment*:
   - Treatment group: Remove the condition: $x \to 0$.
   - Control group: Keep the condition: $x = 1$.
6. *Observe* the outcome Y in both groups (Figure 8.5-right):
   - Treatment group: number of cases that move to corner 1.
   - Treatment group: number of cases that move to corner 3.
   - Control group: number of cases that move to corner 4.
   - Control group: number of cases that remain in corner 2.

[5] Convention in this book for corner numbers: 1 = upper-left; 2 = upper-right; 3 = lower-left; 4 = lower-right.

If all treatment cases have moved from corner 2 to corner 3, but not to corner 1, and if not all control cases have moved from corner 2 to corner 4 (because another necessary condition disappeared during the manipulation) there is evidence that X is necessary for Y. If treatment cases have moved from corner 2 to corner 1, this indicates that Y can occur without X and that X is not necessary for Y. If control cases have moved from corner 2 to corner 4, this suggests that there is a time effect. This is not a problem for concluding that X is necessary for Y as long as not all control cases have moved to corner 4.[6]

8.2.4.2 Necessity experiment: dichotomous X and continuous Y

Figure 8.6 illustrates the situation when the condition is dichotomous and the outcome is continuous. Before the manipulation the condition is present ($x = 1$). The y-values include the *maximum value* of the outcome ($y = 1$). Necessity is supported when after the manipulation ($x \to 0$) the maximum value in the treatment group is reduced to a value of $y = y_c$, resulting in an empty upper-left corner of the XY-plot. The necessary condition can be formulated as $x = 1$ is necessary for $y \geq y_c$. The necessity effect size is the empty area divided by the scope. In this example the effect size = 0.18.

8.2.4.3 Necessity experiment: continuous X and dichotomous Y

Figure 8.7 illustrates the situation when the condition is continuous and the outcome is dichotomous. Before the manipulation, the condition is present: $x \geq x_{high}$, where x_{high} is a value above the presumed necessity level $x = x_c$. The treatment reduces the value of x of all treatment cases. Some cases may end below $x = x_{high}$ but above $x = x_c$. Since the bottleneck value x_c is not reached, these cases have $y = 1$. Other treatment cases have new x-values below the necessity value $x = x_c$. The latter group of cases will then end up in the lower-left corner with $y = 0$. Necessity is supported when after manipulation no cases with $y = 1$ exist with $x < x_c$. In other words, all cases with $x < x_c$ should be located in the lower-left corner. The necessary condition can be formulated as $x \geq x_c$ is necessary for $y = 1$.

8.2.4.4 Necessity experiment: continuous X and continuous Y

In a necessity experiment with a continuous condition and a continuous outcome, the starting point is the existence of an empty area in the upper-left corner of the XY-plot (Figure 8.8-left). The manipulation consists of reducing the level of the condition in the treatment group but not in the control group. Necessity is supported when after the manipulation no cases enter the empty zone. This is shown in Figure 8.8-right, where the necessity effect size before manipulation (0.25) stays intact after manipulation.

---

[6]Note that an odds ratio (OR) is often used to estimate the average effect when X and Y are dichotomous. The odds ratio summarizes how much more likely (in terms of odds) the outcome is to be absent in the treatment group (where X is removed) compared to the control group (where X is retained). The mathematical formula for the odds ratio is $OR = \frac{c3/c1}{c4/c2} = \frac{c3 \cdot c2}{c1 \cdot c4}$ where $c1$, $c2$, $c3$, and $c4$ refer to the number of cases in corner 1, corner 2, corner 3, and corner 4, respectively. After the treatment, $c2$ is the number of cases in the control group with outcome success, $c4$ is the number of cases in the control group with outcome failure (spontaneous decay), $c1$ is the number of cases in the treatment group with outcome success (success remains after removing X), and $c3$ is the number of cases in the treatment group with outcome failure. With an observed data pattern of an empty space in corner 1, the odds ratio is infinite, regardless of the distribution of the cases in the control group: $(50/0)/(10/40) = \infty$. This situation is called "perfect separation" and is considered a problem for average effect statistical analysis. A common advice is to remove the X from further analysis. However, X is a very important predictor of the outcome: the perfect predictor ensures absence of the outcome when X is absent: a necessary condition.

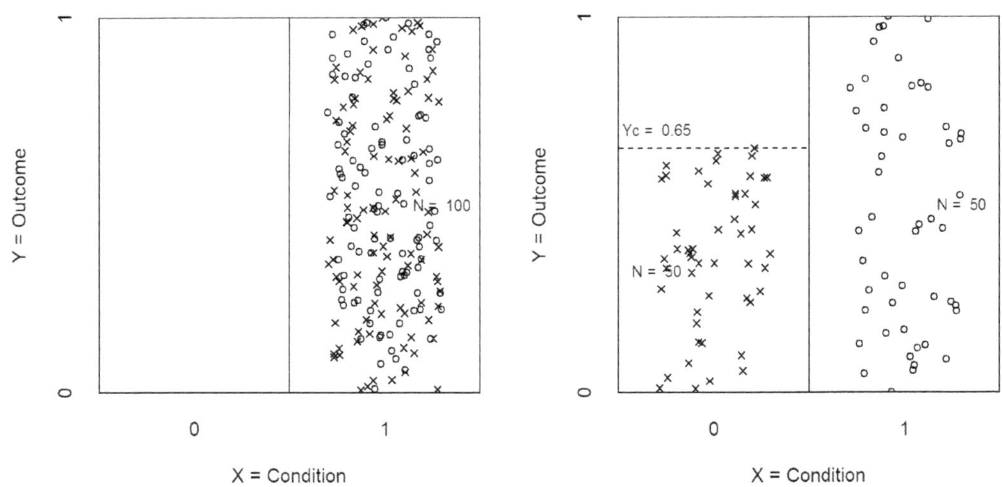

FIGURE 8.6 XY-plot of the necessity experiment with a continuous outcome and a dichotomous condition before (Left) and after (Right) manipulation. X = Condition (0 = no, 1 = yes). Y = Outcome between 0 and 1. ∘ = control group. × = treatment group. y_c is the ceiling value of Y for $x = 0$.

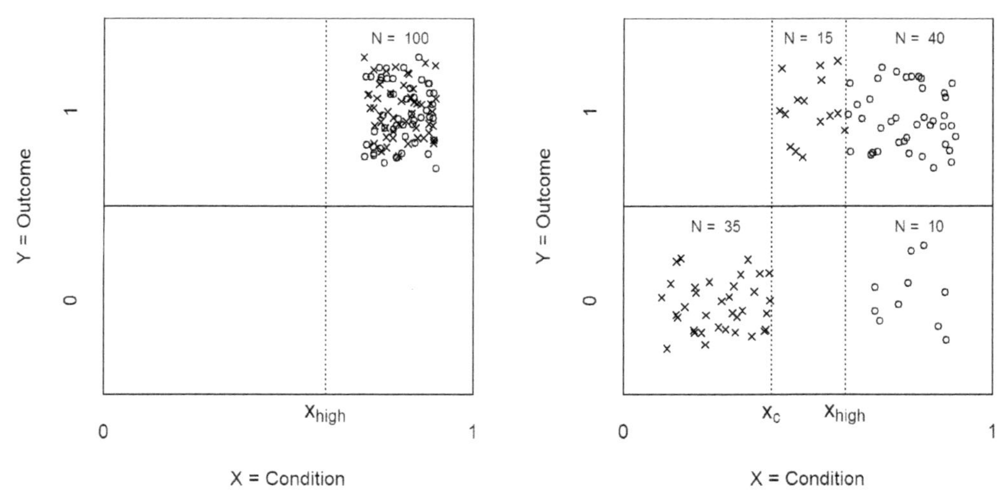

FIGURE 8.7 XY-plot of the necessity experiment with a continuous condition and a dichotomous outcome before (Left) and after (Right) manipulation. X = Condition (0 = no, 1 = yes). Y = Outcome between 0 and 1. ∘ = control group. × = treatment group. x_c is ceiling value of x-value for $y = 1$. x_{high} is the minimum x-value for selecting cases with $y = 1$.

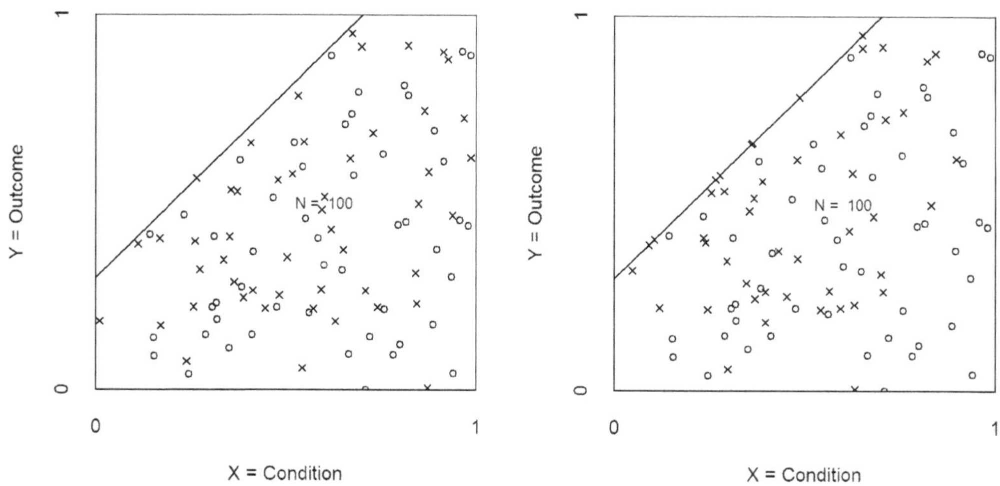

FIGURE 8.8 *XY*-plot of a necessity experiment with a continuous condition and outcome before (Left) and after (Right) manipulation (reducing X). Effect size remains after manipulation. ∘ = control group. × = treatment group.

However, when cases enter the empty zone, the necessity is reduced (smaller effect size) or rejected (no empty space). Figure 8.9-right shows an example of a reduction of the necessity effect size from 0.25 to 0.15.

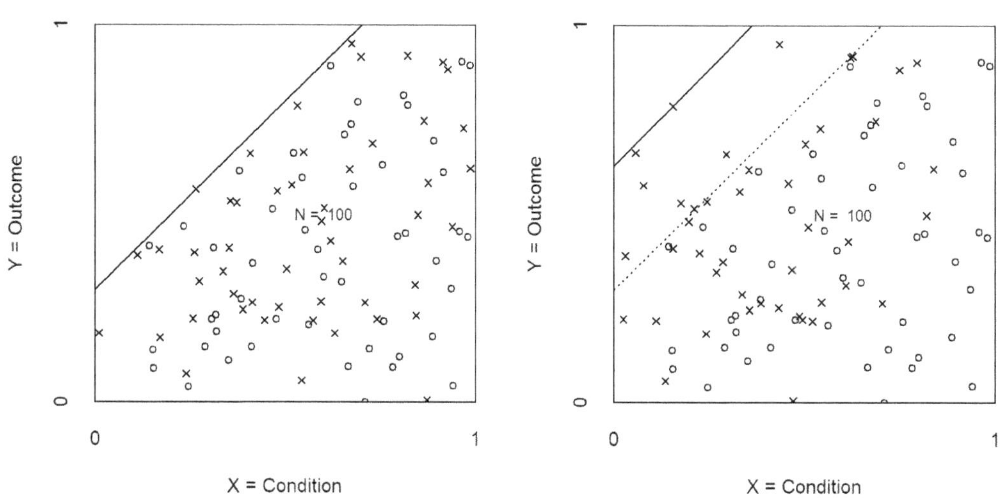

FIGURE 8.9 *XY*-plot of a necessity experiment with continuous conditions and outcome before (Left) and after (Right) manipulation (reducing X). Effect size reduces after manipulation. ∘ = control group. × = treatment group.

Figure 8.10 displays an example of an experiment in which a necessity effect disappears after experimental manipulation. In such cases, the necessity hypothesis should be rejected.

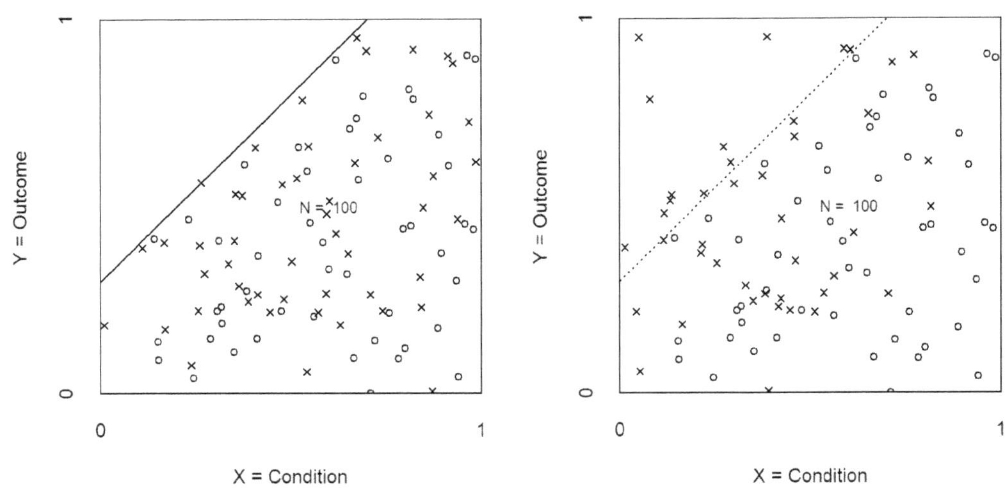

FIGURE 8.10 *XY*-plot of an experiment to challenge necessity with continuous condition and outcome before (Left) and after (Right) manipulation of reducing X. Effect size disappears after manipulation. ∘ = control group. × = treatment group.

In situations with a poorly covered sample, before manipulation the necessary condition may not be visible as an empty space in the upper-left corner because only cases are selected where the conditions are satisfied. After manipulation, the new scope (empirical or theoretical) is extended and a hidden empty space becomes visible. (Figure 8.11).

8.3 Sampling and selection of cases

After the choice of the type of study design, cases must be obtained from the theoretical domain where the necessity hypothesis is supposed to hold. The cases are used to test the hypothesis. In general it is a good strategy to select as many cases as possible. However, necessity theory may be broadly applicable in the theoretical domain consisting of many cases. If not all cases can be selected, a thoughtful selection must be done. NCA offers two types of case selection: large-n sampling for a quantitative study and small-n case selection for a qualitative study.

8.3.1 Large-n sampling

The goal of large-n sampling for a quantitative NCA study is to have a sample that is representative of a defined population from the theoretical domain. Usually, different populations where the cases have some relevant common characteristic (e.g., from the same country) could be selected. Since the theoretical domain may not be homogeneous, different populations could have different effect sizes, although the hypothesis claims that overall

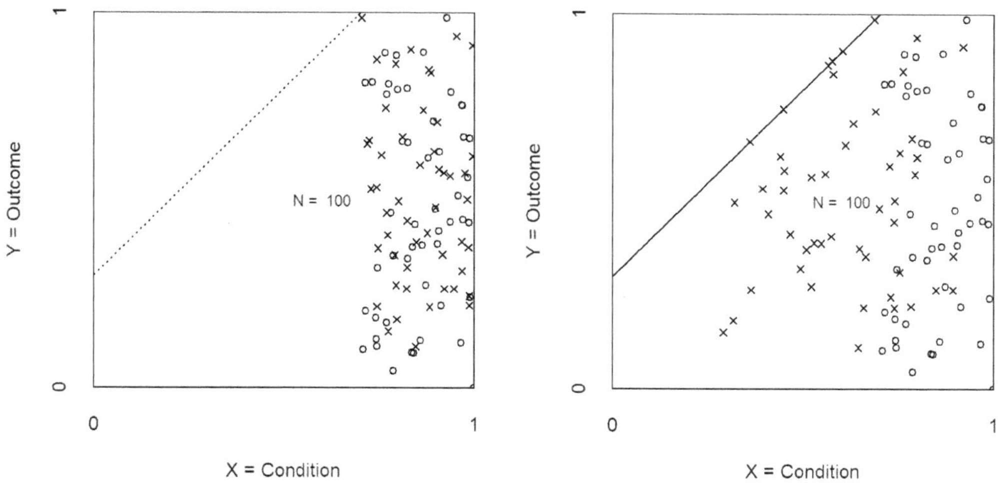

FIGURE 8.11 XY-plot of a necessity experiment with a continuous condition and outcome before (Left) and after (Right) manipulation (reducing X). Effect size appears after manipulation. ∘ = control group. × = treatment group.

necessity exists for all cases. The selection of the population itself is not guided by statistical principles and can often be done for convenience. However, after the population is selected from the theoretical domain, *random sampling* of a large sample helps to ensure that the population is well-covered in the sample and that findings based on the sample can be statistically generalized to the population (inferential statistics). A possible approach to achieve this is to have a sampling frame (list of cases) of the population and to select cases randomly from this list. This ideal sampling approach contrasts with the more common approach of *convenience sampling* where cases are selected from the population for the analyst's convenience, for example because the analyst has easy access to the cases. With such a sampling procedure, the sample may not be representative of the population, which can lead to biased effect size estimates and invalid p-values. The decision on the size of the sample depends on available resources and other practical factors. The minimum required sample size can be guided by the expected power of the study. Power refers to the minimum required sample size such that an existing relevant necessity effect size can be detected with high probability (Section 5.6).

8.3.2 Small-n case selection

It is possible to test a necessity hypothesis with a small number of cases, even with one case (e.g., Dul & Hak, 2008). In qualitative studies, usually a small number of cases is selected with a certain purpose in mind.[7] For testing a necessity hypothesis, the purpose of sampling is to have cases where the outcome is present ($Y = 1$) or the condition is absent ($X = 0$). After selecting such cases, it is observed if the condition is present ($X = 1$) or the outcome is absent ($Y = 0$), respectively. If the condition is absent in cases where the

[7]For example *typical cases* that represent a wider population of cases, *deviant cases* that are unusual, or *maximum variation cases* to ensure maximum variation of a certain characteristic.

outcome is present, or the outcome is present in cases where the condition is absent, the necessary condition hypothesis is rejected for these cases. This type of purposive sampling in small-n studies is possible with a deterministic view on necessity (no exceptions) and when the conditions and the outcome have dichotomous scores (e.g., absent/present, low/high). Since according to the hypothesis necessity applies to each single case in the theoretical domain, a single case can be selected for the test. If necessity is rejected in that single case, the deterministic version of the necessity hypothesis is rejected. However, support for necessity based on only one or a few cases is often not convincing for a conclusion about the entire theoretical domain. Other cases that falsify the necessary condition hypothesis may not have been selected. Following replication logic, the confidence in a hypothesis increases when more cases do not falsify the hypothesis.

Purposive selection of cases for testing a necessity hypothesis was used implicitly in a study by Fujita & Kusano (2020) of highly controversial visits by Japanese Prime Ministers (PM's) to the Yasukuni Shrine. Three necessary conditions were expected: a conservative ruling party, a government enjoying high popularity, and Japan's perception of a Chinese threat. They tested these hypotheses by selecting all five cases where the outcome was present ($Y = 1$, PM's who visited the shrine) from all 22 cabinets between 1986 and 2014. They found that the three conditions were present in all cases ($X = 1$) suggesting support for necessity, at least no rejection. Additionally, they selected the 17 cases where at least one potential necessary condition was absent ($X = 0$) and found that in these cases no visit occurred ($Y = 0$), giving further support for necessity.

In another small-n qualitative study, Harding et al. (2002) studied rampage school shootings in the USA. They first identified five potential necessary conditions: gun availability, cultural script (a model why the shooting solves a problem), perceived marginal social position, personal trauma, and failure of a social support system.[8] Subsequently, they tested the necessary conditions with two cases of school shooting ($Y = 1$) and found support for necessity (non-rejection) because the five conditions were present in these cases ($X = 1$).

Note that this approach of selecting a small number of cases for testing a necessity hypothesis is used in Chapter 7 as part of thought experiments to evaluate a preliminary necessity hypothesis based on existing knowledge (Section 7.5.2).

8.3.3 Replication

Since the necessity hypothesis is assumed to hold for all cases in the entire theoretical domain and the samples or selected cases that are used for testing the hypothesis are only a subset of the theoretical domain, replication studies with other cases are needed. Replication is re-testing the hypothesis with new data (Köhler & Cortina, 2023). A single shot study is not conclusive about a hypothesis in the entire theoretical domain. A next study should select cases from other parts of the theoretical domain, i.e., other samples from the sample population and samples from other populations for large-n sampling, or other cases for small-n case selection. Although a hypothesis can never be proven correct, the confidence in its correctness increases if repeated tests with different samples/cases fail to falsify the hypothesis.

[8] They identified these five potential necessary conditions not by using existing knowledge (Section 7.2) but by first conducting an inductive case study. The authors selected two cases where the outcome was present (school shooting) and observed common factors. The two cases were different from the cases used for testing the hypothesis. Identifying "important" factors in this way has a long tradition in explorative and theory-building case study designs. For example, Znaniecki (1934) introduced 'analytic induction' for finding causal relationships. This method identifies common characteristics in cases where the outcome is present, thus identifies potential necessary conditions (Katz, 2001; W. S. Robinson, 1951).

8.4 Measurement

After the cases are sampled/selected, the X and Y characteristics of each case must be measured. In NCA the scores are often numeric (numbers, quantitative data) but words or letters (qualitative data) can also be used to represent levels of concepts (e.g., high or low). The goal of measurement is to obtain valid and reliable scores of the concepts X (the conditions) and Y (the outcome) that are defined in the necessity hypothesis. Scores are valid when they represent the value of a concept (measuring what you want to measure) and reliable when they are consistent under identical circumstances.

8.4.1 Quantitative data

Quantitative data are data expressed by numbers where the numbers are scores (values or levels) with a meaningful order and meaningful distance. Scores can have two levels (dichotomous, e.g., 0 and 1), a finite number of levels (discrete, e.g., 1,2,3,4,5) or an infinite number of levels (continuous, e.g., 1.8, 9.546, 306.22).

NCA can be used with any type of quantitative data. It is possible that a *single indicator* represents an X or Y concept. Then this indicator score is used to score the concept. For example, in a questionnaire study where a subject or informant is asked to score conditions and outcomes with seven-point Likert scales, the scores are one of seven possible (discrete) scores.

It is also possible that a *construct* that is built from several indicators is used to represent the concept of interest. Separate indicator scores are then summed, averaged or otherwise combined (aggregated), for example for scoring a formative index. It is also possible to combine several indicator scores statistically, for example by using the factor scores resulting from factor analysis, or construct scores of latent variables resulting from a measurement model.

8.4.2 Qualitative data

Qualitative data are data expressed by names, letters or numbers without a quantitative meaning. For example, gender can be expressed with names (male, female, other), letters (m, f, o) or numbers (1, 2, 3). Qualitative data are discrete and usually have no order between the scores (e.g., names of people) or if there is an order, the distances between the scores are unspecified (e.g., low, medium, high). Although with qualitative data a quantitative NCA (with the `NCA` software for R) is not possible, it is still possible to apply NCA in a qualitative way using visual inspection of the XY-table or plot.

When X and Y of cases are scored qualitatively, for example X = A,B, or C, and Y = no or yes, the corresponding XY-table has 6 cells where a case has a position in one of the cells. The hypothesis may be that B is necessary for yes. The analyst can inspect which X-score can reach that Y-score. When the logical order of the condition is absent, the position of the empty space can be in any upper cell (left corner, middle or right; assuming the hypotheses that high X is necessary for high Y). When a logical order exists between the qualitative X-scores, the empty space is in the upper-left corner. In both cases the size of the empty space is arbitrary and no meaningful effect size can be calculated. However, when Y is quantitative, the distance between the highest observed Y-score and the next-highest observed Y score could be an indication of the constraint of X on Y.

8.4.3 Set membership data

NCA can be used with set membership scores. This is often done when NCA is combined with QCA (Section 11.5). In QCA 'raw' variable scores are 'calibrated' into set membership scores with values between 0 and 1. A set membership score indicates to what extent a case belongs to the set of cases that have a given characteristic. In crisp set QCA the scores are dichotomous (0 or 1), and in fuzzy set QCA the scores are discrete or continuous between 0 and 1.

8.5 Dataset

After the measurements are done, the X- and Y-scores are integrated into a dataset. An appropriate format is having cases as rows and conditions and outcome as columns, where the first row (header) consists of the names of the conditions and outcome, and the first column the names of the cases.

Table 8.2 shows the dataset of the example used in Chapter 9 that is part of the `NCA` software.

8.5.1 Archival data

NCA needs a dataset with X- and Y-values as input to empirically test the hypothesis that X is necessary for Y. Often, new data on X and Y are gathered for conducting such a test. This requires a decision about the study design, the sampling/selection of cases, and the measurement of X and Y. The quality of NCA depends on the quality of conducting these steps. However, it is also possible that an existing dataset is used. It may be that data on X and Y have already been tested for other purposes in previous studies. For example, a large number of datasets have been used with the goal of testing probabilistic sufficiency relationships (e.g., average effect of X on Y using regression analysis). If existing datasets contain X- and Y-scores of cases that are part of the theoretical domain, and if study design, sampling/case selection and measurement meet the quality criteria, the datasets from earlier studies may be useful for testing the necessity hypothesis. Given the trends toward open science, relevant datasets for testing necessity hypotheses are available, for example the World Bank dataset used in Section 8.2.2. When using archival data, the origins of the data should be evaluated and acknowledged, possible adaptations of X- and Y-scores should be justified and reported, and relevant earlier studies that use the same data should be evaluated and referenced to ensure accumulation of knowledge (Dul et al., 2024). In a multimethod study that combines NCA with other methods (Chapter 11), data that were collected for the other method may be re-used for NCA.

8.5.2 Data transformation

In quantitative (statistical) studies, data are often transformed before the actual analysis is done. For example, in machine learning, cluster analysis and multiple regression analysis, data are often standardized. The sample mean is subtracted from the original score and divided by the sample standard deviation to obtain the z-score: $X_{new} = (X - X_{mean})/SD$. This allows, for example, a comparison of the regression coefficients of the variables, independently of the variable scale. Min-max normalization is another type of data transformation: $X_{new} = (X - X_{min})/(X_{max} - X_{min})$. Such normalized variables have scores

TABLE 8.2 An example dataset that is part of the NCA software in R.

	Individualism	Risk taking	Innovation performance
Australia	90	84	50.9
Austria	55	65	52.4
Belgium	75	41	75.1
Canada	80	87	81.4
Czech Rep	58	61	14.5
Denmark	74	112	116.3
Finland	63	76	173.1
France	71	49	77.6
Germany	67	70	109.5
Greece	35	23	12.0
Hungary	80	53	5.4
Ireland	70	100	62.3
Italy	76	60	19.7
Japan	46	43	171.6
Mexico	30	53	1.2
Netherlands	80	82	68.7
New Zealand	79	86	14.9
Norway	69	85	75.1
Poland	60	42	3.5
Portugal	27	31	11.1
Slovak Rep	52	84	3.5
South Korea	18	50	42.3
Spain	51	49	17.3
Sweden	71	106	184.9
Switzerland	68	77	149.7
Turkey	37	50	1.4
UK	89	100	79.4
USA	91	89	214.4

between 0 and 1. Transforming original scores into z-scores or min-max normalized scores are both examples of a *linear* transformation.

In principle, NCA does not need data transformation. The effect size is already normalized between 0 and 1 and allows comparison of effect sizes of different variables. Also, NCA does not make assumptions about the distribution of the data. Furthermore, interpretation of the results with original data is often easier than the interpretation of the results with transformed data as the original data may have a more direct meaning than the transformed data. If, nevertheless, the analyst wishes to transform data, a linear transformation (or in general an affine transformation) does not affect NCA's estimation of the effect size or the p-value (Appendix D).

However, a *non-linear* data transformation as is often done in quantitative (statistical) studies affects NCA's estimation of effect size and p-value and is normally not acceptable for NCA. The transformation may change the cases that are used to establish the ceiling line, and this will affect the effect size. This can happen, for example, when data with a skewed distributed are log-transformed to obtain more normally distributed data. This is relevant when a statistical approach assumes normally distributed data, but this is not a requirement in NCA. Calibration in QCA is another example of a non-linear transformation. Original variable scores ('raw scores') are transformed into set membership scores, often using a logistic function ('s-curve'). Calibration is needed in QCA as a set-theoretic method, but not in NCA, that can also handle raw scores.

Non-linear transformations should be avoided in NCA, unless there is a good reason for it. Such a good reason can be that a non-linear transformed original score is the valid representation of the concept of interest (X or Y in the necessity hypothesis). For example, the concept of a country's economic prosperity may be expressed as the log-transformed GDP (Section 8.2.2.1). In QCA, the calibration process can justify the non-linear transformation of the original data. However, the general advice is to refrain from non-linear data transformation when the original data represent the concepts of interest, and only to transform data non-linearly when this is needed to validly capture the concept of interest.

9

Data analysis

9.1 Summary of this chapter

In the previous chapters, the formal hypothesis was developed (Chapter 7) and the dataset with scores of condition(s) X and outcome Y became available (Chapter 8). This chapter focuses on testing the hypothesis quantitatively by analyzing the data using the `NCA` software in `R`. First, the example that is used to illustrate the analysis is presented (Section 9.2). After the hypotheses are formulated, the steps for testing them with the data are: (1) Model specification (Section 9.3); (2) Selection of model evaluation criteria (Section 9.4); (3) Visual inspection of the XY-plot (Section 9.5); (4) Ceiling line estimation (Section 9.6); (5) Effect size and p-value estimation (Section 9.7); (6) Outlier analysis (Section 9.8); (7) Model fit metrics calculation (Section 9.9); (8) Necessity-in-kind decision (Section 9.10); (9) Bottleneck analysis for necessity-in-degree (Section 9.11); (10) Robustness checks to evaluate the sensitivity of the findings for choices made in steps (1), (2), and (6) (Section 9.12).

9.2 Illustrative example

In this chapter, the elements of NCA's data analysis are illustrated with the example called `nca.example` that is part of the `NCA` software (Appendix B). Before any data analysis with NCA can start, a *formal* necessity hypothesis (Chapter 7) and an adequate dataset (Chapter 8) must be available.

The *formal necessity hypothesis* for the illustrative example is as follows:

- *Hypothesis*: Individualism is necessary for Innovation performance (high-high).
- *Focal unit* = Country.
- *Concept X* = Individualism: the extent to which a country's people feel independent, as opposed to being interdependent as members of larger wholes.
- *Concept Y* = Innovation: the country's transformation of knowledge into new products, processes, and services.
- *Theoretical domain* = All countries in the world.
- *Causal explanation* = (Summary) Enabler: In highly individualistic societies, individuals have greater freedom to experiment and propose disruptive ideas, which contributes to producing innovative products and services. Constraint: Without enough individualism, potential innovators lack the autonomy and psychological safety to transform

DOI: 10.1201/9781003430766-9

ideas into innovations, such that high levels of a country's innovation are not achievable. No substitute: R&D infrastructure, financial resources, or government policies or other known contributing factors to innovation cannot compensate for the absence of an individualistic mindset. Temporal aspects: Individualism must be present before innovation is possible, must stay present to maintain innovation and is needed in all time periods.

The example's *data* consists of archival data (Section 8.5.1), collected from two different sources. Data on cultural dimensions were obtained from Hofstede's (1980) country cultures study. The value for a country's Risk taking was obtained by subtracting the country's Uncertainty avoidance from 135. Data on Innovation performance are from a study by Gans & Stern (2003) on innovation differences between countries. Both studies are large-n observational studies. By combining the datasets, a convenience sample of 28 countries with information on the three variables was obtained. Scores for the cultural variables are aggregated questionnaire item scores (for details see Hofstede, 1980). Scores for Innovation performance equal the country's innovation index (Gans & Stern, 2003). The data and how they were obtained have the following characteristics:

- *Study design* = observational study.
- *Sampling* = convenience sampling (two different sources); n = 28 countries.
- *Measurement* = Individualism by questionnaires; Innovation performance by document analysis.
- *Dataset* = `nca.example` (part of the `NCA` software for `R`).

9.3 Model specification

NCA's model specification consists of specifying the bounding box and the ceiling line (Section 4.3).

9.3.1 Bounding box specification

NCA assumes that X and Y are bounded variables ($x_{min}, x_{max}, y_{min}, y_{max}$) (Section 4.3). The bounds define the bounding box where observations could be possible given the extreme values of X and Y. Two types of bounding boxes can be distinguished.

The bounding box with a *theoretical scope* is defined by theoretical bounds of X and Y.[1] It may be that theoretical limits exist, such as for human body temperature ($27°C \leq T \leq 44°C$). The theoretical bounds may be known or approximated (e.g., from previous empirical studies). It is also possible that X and Y are measured on scales with known minimum and maximum values, such as Likert scales or percentage scales.

The bounding box with an *empirical scope* is defined by the minimum and maximum values of X and Y that are observed (in the dataset). This option is often selected when theoretical bounds are unknown. It can also be selected to avoid overestimation of the effect size.

In a situation where groups of similar cases are compared (e.g., measured at different time points in a longitudinal study, resamples from a population in simulation study), the

[1] The scope is the area of the bounding box.

theoretical or scale minima and maxima or the observed minimum and maximum values of the pooled dataset should be selected as the common theoretical scope of each single group for comparability (Sections 5.3.7 and 8.2.2.2).

In the illustrative example, the empirical scope is selected. The scales used for collecting the data do not have minimum and maximum values and no strong theoretical rationale exists for selecting other extreme values for X and Y than those observed in the dataset.

9.3.2 Ceiling line specification

Specifying the ceiling line involves choosing the *location* (relative to an empty corner) and the *form*.

The *location* depends on the direction of the hypothesis (Section 3.5), which defines the *expected empty corner*. The proposed hypothesis of the illustrative example is a high-high hypothesis where presence/high value of X is necessary for presence/high value of Y. Therefore, the expected empty corner is the upper-left corner of the XY-plot (corner 1).

Next to the location, also the *form* of the ceiling line needs to be specified (Sections 4.3.2 and 4.4). NCA assumes that the ceiling line is non-decreasing. A *linear* ceiling line (e.g., CR-FDH or C-LP) may be specified when the border between the area with and without cases is theoretically assumed to be linear, which is the most common assumption. The CR-FDH ceiling allows some noise above the ceiling such that the ceiling zone is not perfectly empty near the ceiling line (see model fit metric *noise*, Section 4.5.4). This line may be selected when noise exists in the data such as measurement error. The C-LP line may be selected when the data are obtained from high-quality measurement approaches, or are perfect as for example in simulated data. This line may also be selected to ensure that the linear ceiling line has no cases in the ceiling zone. The default CR-FDH ceiling is often selected when the condition and outcome have many levels (e.g., are continuous).

For modeling purposes, a linear ceiling line may be attractive (Sections 4.3.2.3 and 4.4.1.3). For a given bounding box, it requires only two parameters such as ceiling intercept and ceiling slope. These parameters may be theoretically decided (e.g., based on earlier data or other knowledge) or empirically estimated with the data. The latter is most commonly used.[2] Although estimation of a ceiling line and effect size are not needed when the ceiling line and scope are predefined, the other steps of the data analysis are still relevant.

A *segmented linear* (non-decreasing) ceiling line (e.g., CE-FDH or CE-VRS) may be specified when the border is non-linear (but non-decreasing). The CE-FDH ceiling is a piecewise linear function where peers (dominant points) are connected by line segments (Sections 4.3.2.1 and 4.4.1.1). CE-FDH is a stepwise linear ceiling line that connects peers with horizontal and vertical line segments. The CE-VRS is a concave piecewise linear ceiling with linear segments that connect a selection of the CE-FDH peers such that the ceiling is concave (Sections 4.3.2.2 and 4.4.1.2). When estimated from data, both lines do not have points in the ceiling zone. CE-FDH may also be selected when the condition or outcome have few levels (e.g., are discrete) or when the ceiling line is irregular or non-linear. When the non-linearity is concave, the CE-VRS line may be an option. The default segmented ceiling line is CE-FDH.

---

[2]Note that in fsQCA the 'ceiling line' is always predefined as the diagonal of the bounding box $[0,1]\times[0,1]$, with ceiling intercept $= 0$ and ceiling slope $= 1$. In NCA it is also possible to pre-define the ceiling line. This option is integrated in NCA's software package for R from version 5.0.0.

For modeling purposes, a piecewise linear ceiling line may be attractive for an irregular border between feasible area and ceiling zone because it can better follow the non-linearity (though non-decreasing) of the border than a linear ceiling line. It is less attractive when the number of line segments is large. This requires a large number of parameters to specify the model, which makes the model complex (see the model fit metric *complexity* Section 4.5.1) and has the risk of overfitting (Section 9.9.1).

Analysts should critically reflect on the choice of ceiling line prior to conducting the analysis, and aim to select a ceiling line that is expected to best suit the study (Section 5.5). The most commonly used ceiling lines are the default lines CE-FDH for non-linear ceilings and discrete variables, and CR-FDH for linear ceilings and continuous variables.

In the illustrative example, the ceiling line is specified in two steps. Since the X- and Y-data are continuous and it is not expected a priori that the border is non-linear, the linear default ceiling line (CR-FDH) is provisionally selected. After exploration of outliers (Section 9.8) and model fit (Section 9.9), a better ceiling line may be selected (if available) to test the hypothesis.

9.4 Selection of evaluation criteria

NCA applies three main criteria for a binary decision to reject or non-reject (support) a necessary condition hypothesis: adequate theoretical support, a sufficiently large effect size, and a low p-value.[3] For non-rejection all three criteria must be satisfied: theoretical support, a large effect size (d) and a small p-value (Section 6.2). If any of these criteria is not satisfied, the hypothesis is rejected. The first requirement can be satisfied if a formal necessity hypothesis is available that gives the required theoretical support. The process of developing a formal necessity hypothesis is discussed in Chapter 7.[4] The other two criteria are discussed below.

9.4.1 Effect size threshold

Generally, the larger the effect size, the more practically relevant the necessary condition. For the binary decision about the hypothesis, a threshold level must be selected. Ideally, the selection is based on practical considerations: the minimum level of d that is considered relevant for practice. Since such a consideration may be difficult, analysts often select the arbitrary benchmark suggested in Dul (2016b): $d = 0.10$. However, in situations where a weak constraint can have significant consequences, a lower threshold level may be selected; when consequences are costly or critical, a higher threshold may be needed.

In the illustrative example, the selected effect size threshold is $d = 0.10$.

[3] For small-n case studies with purposive selection of cases rather than random sampling (Section 8.3.2), the p-value is not informative, and the selection is only based on effect size and theoretical support.

[4] Note that just formulating a necessity hypothesis without challenging it with thought experiments, without specifying the theoretical domain, or without providing a causal explanation for necessity does not qualify as a *formal* necessity hypothesis.

9.4.2 *p*-value threshold

The *p*-value is meant to reduce the risk of false positive conclusions: concluding that a non-existing necessary condition exists. A high *p*-value suggests that the observed data are compatible with the null of no relationship (Section 5.3). For this purpose, a threshold *p*-value (α) is selected to evaluate if the observed effect size is statistically significant, meaning that the data are not compatible with the null hypothesis (the data may be compatible with an alternative hypothesis). The threshold *p*-value must be small enough to make it plausible that the observed effect size is not compatible with the situation of no effect between X and Y. Often, $\alpha = 0.05$ is selected. For exploration, a higher threshold level ($\alpha = 0.10$) is sometimes applied. For claims of new discoveries, Benjamin et al. (2018) suggest using $\alpha = 0.005$.

In the illustrative example, the selected *p*-value threshold is $\alpha = 0.05$, which is commonly used.

9.4.3 Selection of target outcome

For the practical application of NCA results, it is (optionally) possible to specify the target outcome y_c, which is a specific level of Y that is considered relevant for practical action. The model estimates the corresponding target condition level x_c that is necessary for achieving the target outcome. If necessity-in-kind (NiK) is supported and a bottleneck analysis of necessity-in-degree (NiD) is warranted (see below Sections 9.10 and 9.11), NCA identifies the condition target level required to make the desired target outcome possible. A *bottleneck case* is a case that does not satisfy the necessary level x_c for achieving the target level y_c. Bottleneck cases may be the focus for practical action based on NCA results (Chapter 12).

If the outcome is *desirable* (e.g., performance), the target outcome is the minimum performance level that must be achieved. If the outcome is *undesirable* (e.g., disease), the target outcome is the maximum level of the condition below which the undesired target outcome will not occur. These "rules" apply to all cases.

In the illustrative example, the outcome is desired (innovation performance). The selected target outcome is "very good innovation performance", defined as the level corresponding to the 90th percentile in the dataset, that is, being part of the top 10% performers.

9.5 Visual inspection of the XY-plot

After hypothesis formulation, data collection, model specification and selection of evaluation criteria, the starting point for NCA's data analysis is a visual inspection of the XY-plot. This is a qualitative examination of the data pattern. By visual inspection the following questions can be answered:

- Is the expected corner of the XY-plot empty?
- Is the selected ceiling line appropriate (or if the ceiling line is not specified: what is an appropriate ceiling line)?
- Which cases are possibly outliers?
- What is the data pattern in the rest of plot?

An *XY*-plot for the illustrative example can be produced using the NCA software for R as follows:[5]

```
# Load the NCA package and the example
library(NCA)
data(nca.example)

# Run the NCA analysis
model <- nca_analysis(nca.example, 1, 3, ceilings = "cr_fdh")
nca_output(model, summaries = FALSE)
```

With this code the *XY*-plot is displayed where X = Individualism and Y = Innovation performance. All 28 cases are shown as dots (Figure 9.1-left). The plot also shows the selected ceiling line (CR-FDH).[6] Figure 9.1-right shows the same plot and now with annotations for the visual inspection.[7]

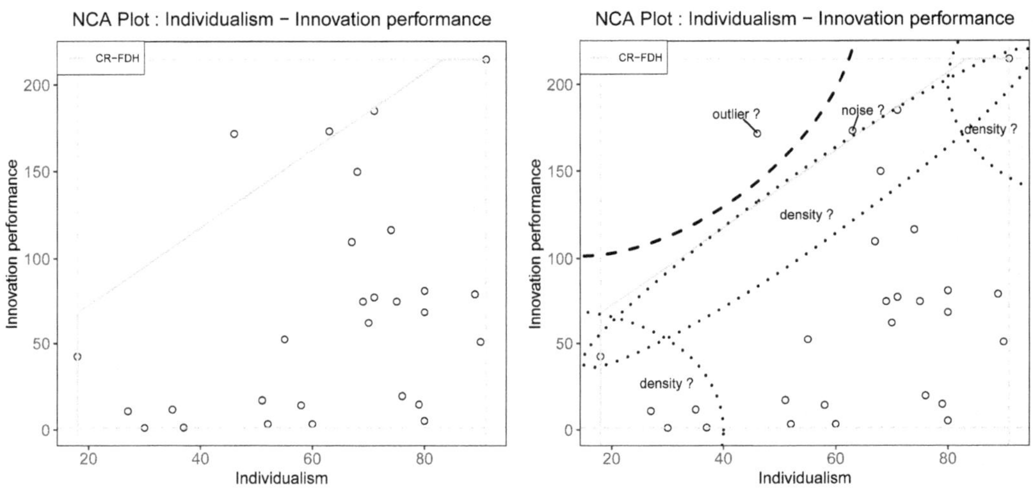

FIGURE 9.1 Visual inspection of the *XY*-plot of the illustrative example. Left: output from the NCA in R package. Right: same output with annotations.

9.5.1 Is the expected corner empty?

According to the hypothesis, a high level of X is necessary for a high level of Y such that the upper-left corner is expected to be empty. Figure 9.1 shows, generally, that cases with a low value of X have a low value of Y, and that cases with a high value of Y have a high value of X. The upper-left corner is indeed empty. A non-empty expected empty corner with cases near the geometrical corner would have indicated that the hypothesis should be rejected, unless only a few cases are located there without cases in the surrounding empty area; such cases could be outliers and would need further inspection. One case in Figure 9.1 could be a candidate outlier, see Section 9.5.3.

[5]See Appendix B for a description and demonstration of the NCA software for R.

[6]It may be that the ceiling line was not yet selected during the model specification (Section 9.3). Then the two default ceiling lines or another line may be displayed (by using the `ceilings` argument in the `nca_analysis` function of the NCA software) to select the 'best' ceiling line after visual inspection.

[7]The annotations are not produced by the NCA software but added manually for illustration.

9.5.2 Is the selected ceiling line appropriate?

In the illustrative example the border is assumed to be linear, and the CR-FDH ceiling line is (provisionally) selected during the model specification. However, the *XY*-plot shows that the border is somewhat irregular. This suggests that representing the border by a linear ceiling may not be optimal, causing model fit issues. This will be evaluated in more detail using model fit metrics (Section 9.9).

9.5.3 Which cases are possible outliers?

Outliers are cases that are relatively 'far away' from other cases. NCA defines an outlier as a case that has a large influence on the effect size when removed (Section 9.8). This means that cases that help to produce the ceiling line (peers) and cases that define the empirical bounding box (observed minimum and maximum values of X and Y) are potential outliers. Isolated potential outliers may be identified by visual inspection of the *XY*-plot and may give rise to conducting a formal outlier analysis (Section 9.8).

In the illustrative example, one point is located in the apparent ceiling zone relatively far away from other points and from the ceiling line (Figure 9.1). This point is a candidate for being an outlier case, and an outlier analysis is therefore warranted. The other points in the ceiling zone (above the ceiling) also contradict strict necessity, but may not be considered an outlier but rather noise because the point is closer to other points and to the apparent ceiling line. Noise may, for example be due to measurement error, and can be captured with the model fit metric *noise* (Section 4.5.4).

9.5.4 What is the data pattern in the rest of the plot?

Although NCA focuses on the analysis of the expected empty corner and the ceiling line, the distribution of cases in the rest of the *XY*-plot can also be informative for the credibility of necessity (Chapter 6). In particular, the density of cases near the ceiling line and density of cases in adjacent corners to the empty corner (upper-right and lower-left) may contain information that is relevant for necessity.

9.5.4.1 Density of cases near the ceiling line

Cases just below the ceiling line 'support' the ceiling line. With no or very few cases just below the ceiling line, the line may not be informative. The informativeness of the ceiling line is captured with NCA's model fit metric called *support* (Section 4.5.5).[8]

The cases near the ceiling should not be clustered but preferably spread along the ceiling line. The spread of the ceiling line is captured with NCA's model fit metric *spread* (Section 4.5.5). In the illustrative example, the cases near the ceiling line do not seem to be clustered and are reasonably spread.

Theoretically, the ceiling line represents a sharp border between the area with and the area without cases. This means that the densities just below and just above the ceiling should be different with considerably more cases just below the ceiling compared to just above the ceiling. This feature of the ceiling line is captured in NCA's *sharpness* metric (Section 4.5.6).

---

[8]Note that cases near the ceiling line can be considered as best cases or innovators (assuming that the outcome is desirable, and the condition is an effort). For a given level of the condition they have the maximum possible outcome compared to other cases.

In the illustrative example, few cases are present just below the ceiling and the densities just below and just above the ceiling are not very different, such that the ceiling is not a sharp border.

9.5.4.2 Density of cases in adjacent corners

The presence of many cases in the lower-left corner, and/or in the upper-right corner, may be supportive for the non-randomness of the necessary condition. If only a few cases are present in these corners, the emptiness of the upper-left corner could be a random effect. To mimic the null-distribution where X and Y are unrelated, NCA's permutation test re-samples from the observed sample while shuffling labels of observations (see NCA's null hypothesis test, Section 5.3.3). To challenge randomness, it is important that enough cases are in the upper-right or lower-left corners. If these cases do not exist, the p-value will be high, resulting in a rejection of the necessity hypothesis, despite the presence of an expected empty corner and possibly despite presence of a necessary condition in the population (false negative conclusion).

In the illustrative example, it appears that a reasonable fraction of cases is present in the lower-left corner, suggesting that the p-value may be small such that the necessity explanation of the observed empty corner is not rejected. Section 9.7 presents the formal evaluation of the effect size regarding statistical significance.

9.6 Ceiling line estimation

During model specification (or after visual inspection of the XY-plot) the location and form of the ceiling line are specified. When the specific ceiling parameters (Section 4.3.3) are not yet available (the most common situation) NCA estimates these parameters. For a linear ceiling line these are the intercept and the slope, for a segmented linear ceiling line these are the coordinates of the peers. For most types of ceiling lines (CE-FDH, CR-FDH, CE-VRS, CR-VRS, C-LP) this is done by first estimating the Free Disposal Hull (Section 4.4). The CE-FDH ceiling line corresponds to this line, whereas the other ceiling lines are derived from this line by using all or a selection of peers or otherwise based on the peers (C-LP) as explained in Section 4.3.2.

In the illustrative example, the CR-FDH ceiling line is selected (Figure 9.1-left), but without specifying its parameters (ceiling line intercept and ceiling line slope). The ceiling line parameters can be estimated with the `NCA` software for `R` as follows:

```
library(NCA)
data(nca.example)

# Run the NCA analysis with the specified ceiling line
model <- nca_analysis(nca.example, 1, 3, ceilings = "cr_fdh")

# Extract intercept & slope
Intercept <- nca_extract(model = model,
                         x = "Individualism",
                         ceiling = "cr_fdh",
                         param = "Intercept")
```

```
Slope     <- nca_extract(model = model,
                         x = "Individualism",
                         ceiling = "cr_fdh",
                         param = "Slope")

# Print the ceiling line equation
cat(
  "Ceiling line equation (CR-FDH):  Y =",
  round(Intercept, 3), "+",
  round(Slope, 3), "* X", "\n"
)

## Ceiling line equation (CR-FDH):  Y = 28.353 + 2.23 * X
```

The estimated ceiling intercept is 28.353 and the estimated ceiling slope is 2.23.

9.7 Effect size and *p*-value estimation

The effect size (Section 4.4.1) can be calculated from the ceiling zone (Section 4.3.2) and the scope (Section 4.3). The ceiling zone is calculated from the (estimated) parameters of the ceiling line. The scope is calculated from the observed or theoretically determined minimum and maximum values of X and Y. The effect size is the test statistic for NCA's permutation test to estimate the p-value (Section 5.3).

For the illustrative example, the calculations and estimations can be done with the `NCA` software for `R`. The number of repetitions for obtaining the permutation distribution for estimating the p-value is set at 10,000. This number generally balances accuracy and computation time.

```
library(NCA)
data(nca.example)

# Run the NCA analysis with the specified ceiling line
# 10,000 repetitions for estimating the p-value
set.seed(123) # for reproducibility of the estimated p-value
model <- nca_analysis(nca.example, 1, 3,
                      ceilings = "cr_fdh",
                      test.rep = 10000)

## Setting up parallelization, this might take a few seconds...
##      Do test for  :  cr_fdh - IndividualismDone test for:
##        cr_fdh - Individualism

# Show the results
print(model)

##
## --------------------------------------------------------------------------

## Effect size(s):
```

```
##                 cr_fdh p
## Individualism 0.31   0.168
## ----------------------------------------------------------------------
```

During the computation a message "Do test for: ..." appears when the analysis is running, which changes into "Done" when the analysis is finished. The results show that the effect size $d = 0.31$, which is greater than the selected threshold level of $d = 0.10$. However, this result is not statistically significant given the estimated p-value of 0.168 for the selected threshold level of $\alpha = 0.05$. Therefore, the initial conclusion is that the hypothesis is rejected.[9]

9.8 Outlier analysis

Including or excluding outliers from a data analysis can change the results. This applies to any data analysis method, and also to NCA. In NCA, an outlier is an 'influential' case that has a large influence on the necessity effect size when removed.

9.8.1 Outlier types

Because the effect size is calculated by dividing a numerator (*ceiling zone*) by a denominator (*scope*), and an outlier is defined by its influence on this division (effect size), two types of outliers can exist in NCA: *ceiling outliers* and *scope outliers.*

A *ceiling outlier* is a case that helps to determine the position of the ceiling line and that largely changes the effect size if it is removed from the dataset. Since the peers determine the ceiling line, they are potential ceiling outliers. When a ceiling outlier is removed, it usually increases the ceiling zone and thus *increases* the necessity effect size. A *scope outlier* is a case that determines the empirical scope and that largely changes the effect size if it is removed. When a scope outlier is removed, it usually decreases the scope and thus *increases* the effect size. When a case is both a ceiling outlier and a scope outlier, it determines not only the position of the ceiling but also the size of the scope. When such a case is removed, it simultaneously decreases the ceiling zone and the scope. The net effect is often a *decrease* of the necessity effect size.

Duplicate cases are cases with exactly the same X- and Y-values. They are possible when the variable scores are discrete, for example when measured with a Likert scale. When duplicate cases determine the ceiling line or scope, the effect size will not change when only one of the duplicates is removed. Therefore, single cases that have a duplicate have no influence on the effect size when removed, and cannot be outliers. A set of duplicate cases can be outliers and can be removed as a set. Cases that are below the ceiling line and have no extreme values of X or Y cannot be outliers.

9.8.2 Outlier identification

There are no scientific rules for deciding if a case is an outlier. The decision depends primarily on a judgment by the analyst. For a first impression of possible outliers, the analyst may visually inspect the XY-plot as discussed in Section 9.5.3. In the literature, rules of thumb

[9] Note that a rejection of a necessity hypothesis (if robust) can be an important result: the condition may be important, but is not necessary, and thus compensable if absent or low.

are used for identifying outliers. For example, a variable score is an outlier score 'if it deviates more than three standard deviations from the mean', or 'if 1.5 interquartile ranges is above the third quartile or below the first quartile'. These pragmatic rules address only one variable (X or Y) at a time. Such single variable rules of thumb may be used for identifying potential scope outliers (unusual minimum and maximum values of condition or outcome), but they are not suitable for the identification of potential ceiling outliers. The reason is that ceiling outliers depend on the *combination* of X- and Y-values. For example, a case with a non-extreme Y-value may be a regular case when it has a large X-value, because it does not define the ceiling. Then removing the case does not change the effect size. However, when X is small, the case may define the ceiling; removing the case may change the effect size considerably. This means that both X and Y must be considered when identifying ceiling outliers in NCA.

All potential ceiling outliers and potential scope outliers can be evaluated one by one by calculating the absolute and relative effect size changes when the case is removed (with replacement). Such an outlier analysis can be done with the `NCA` software for `R` using the `nca_outliers` function. The output is a list of potential outliers, ranked by how much the effect size decreases when the outlier is removed.[10] For the illustrative example, the outlier analysis can be done as follows:

```
library(NCA)
data(nca.example)

# Run the NCA outlier analysis
outliers <- nca_outliers (nca.example, 1, 3, ceiling = 'cr_fdh')

# Show the results
print(outliers)

##      outliers eff.or eff.nw dif.abs dif.rel ceiling scope
## 1 Japan         0.31   0.44    0.13    44.0       X
## 2 USA           0.31   0.22   -0.08   -26.8       X     X
## 3 South Korea   0.31   0.36    0.06    18.2       X     X
## 4 Finland       0.31   0.31    0.00     1.4       X
## 5 Sweden        0.31   0.30    0.00    -0.6       X
## 6 Mexico        0.31   0.31    0.00     0.1             X
```

The table shows in the first column the case names of the potential outliers. The second column displays the original effect size when no cases are removed (*eff.or*) and the third column displays the new effect size when the outlier is removed (*eff.nw*). The fourth column shows the absolute difference (*dif.abs*) between the new and the original effect sizes, and the fifth column the relative difference (*dif.rel*) between the new and the original effect sizes expressed as a percentage. A negative sign means that the new effect size is smaller than the original effect size.

The output shows six potential outliers. The new effect sizes range from 0.22 to 0.44, absolute differences range from −0.08 to +0.13, and relative differences range from −26.8% to +44%. Three outliers are just ceiling outliers (Japan, Finland, Sweden), one is just a scope outlier (Mexico), and two cases are both ceiling and scope outlier (USA and South Korea). South Korea defines the ceiling line and the minimum X of the scope. USA defines the ceiling line

[10]Note that for brevity *potential outliers* are called *outliers* in printed and plotted output. The analyst judges if a potential outlier is a true outlier or not.

and maximum X and maximum Y of the scope. Mexico does not define the ceiling line but defines the minimum Y.

For an inspection of the cases in the XY-plot, the `plotly = TRUE` argument in the `nca_outliers` function can be used to display an interactive plot with highlighted potential outliers (the static version is shown in Figure 9.2).

```
library(NCA)
data(nca.example)

nca_outliers (nca.example, 1, 3, ceiling = 'cr_fdh' , plotly = TRUE)
```

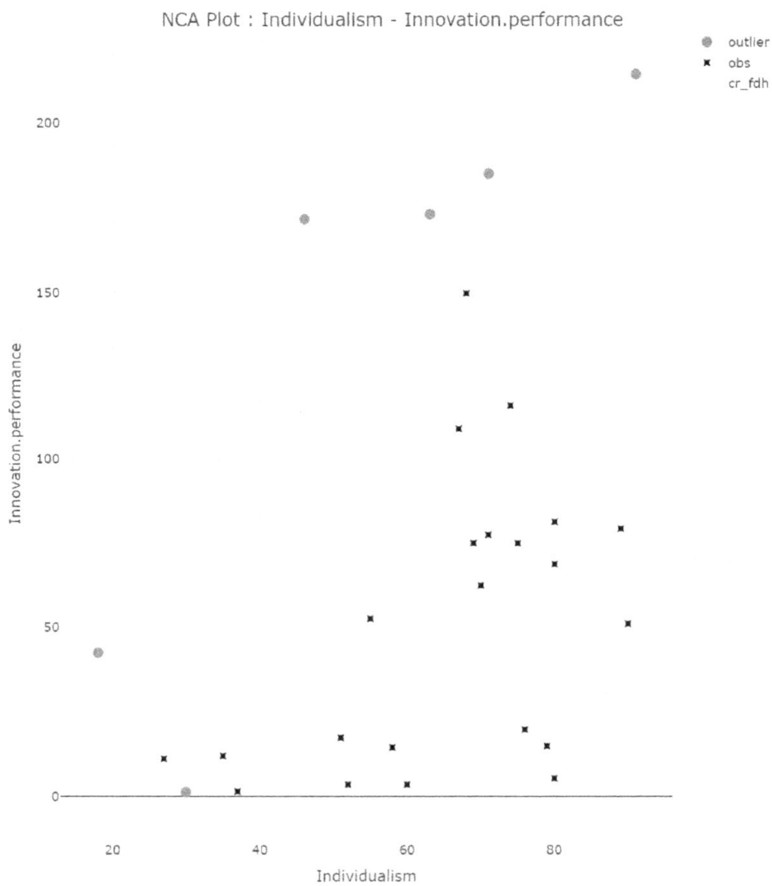

FIGURE 9.2 Plotly plot with potential ceiling and scope outliers distinguished from other observations.

From the three potential *ceiling outliers* that are not also a scope outlier (Japan, Finland, Sweden) and are in the center of the plot (Figure 9.2), two cases increase the effect size when removed, and one decreases it. The relative effect size differences when cases are removed range from −0.6% to 44.0%.[11] The potential *scope outlier* that is not also a ceiling outlier (Mexico) increases the effect size when removed, but its influence is marginal. The two potential *ceiling outliers* that are also potential *scope outliers*, USA and South Korea,

[11] In the software, small changes in effect size can be ignored. By default, combinations that change the effect size by less than 0.01 are left out.

change the effect size when removed with relative effect size differences of −26.8% and 18.2%, respectively.

All other cases are not potential outliers because they have no influence on the effect size when removed. Note that after a potential outlier is removed and the effect size difference is calculated, the case is added again to the dataset before another potential outlier is evaluated (removing with replacement).

In the `nca_outlier` function it is also possible to evaluate two or more potential outliers at once by specifying k = number of potential outliers. This evaluates the change of effect size when k potential outliers are removed as a group (and replaced as a group). Evaluating multiple outliers is useful in the following situations:[12]

- Existence of duplicate cases that are far from the other cases.
- Existence of a cluster of cases that is far from the other cases.
- Existence of multiple single cases that are not near each other, but far from the other cases.

Sometimes, removing more outliers does not lead to a larger change in effect size compared to removing just one or two of them. When the argument `condensed = TRUE` is used, these larger combinations are normally hidden from the results. However, a combination is still shown if it produces a larger effect size change than any other combination that does not include one of the cases in that combination. In other words, even with `condensed = TRUE`, a combination remains visible when it adds meaningful new information that cannot be explained by smaller or unrelated combinations.

The remaining combinations are sorted by how much they change the effect size (from largest to smallest). This is the absolute value of the difference between the original effect size and the new effect size. If several combinations cause the same change, smaller combinations are listed first. Only the top combinations are shown, up to the user-specified limit (default = 25).

For $k > 2$ the procedure repeats with a next round of identifying new outliers. In general, all possible $\binom{n}{k} = \frac{n!}{k!(n-k)!}$ outlier combinations are evaluated, where k is the combination size and n is the total number of outliers found so far.

In the illustrative example, the results for $k = 2$ potential outliers are:

```
library(NCA)
data(nca.example)

outliers <- nca_outliers (nca.example, 1, 3, ceiling = 'cr_fdh', k = 2,
                          condensed = TRUE)
print(outliers)

##                    outliers eff.or eff.nw dif.abs dif.rel ceiling scope
## 1   Japan - South Korea       0.31    0.51    0.20    65.6       X     X
```

[12]For $k > 1$ the number of combinations of potential outlier cases can be large. The selection of combinations of two potential outliers is as follows. First, all potential single-case outliers (ceiling and scope outliers) in the original dataset are identified (List1). Next, all cases from List1 are removed from the dataset and new single-case outliers are identified in the remaining dataset (List2). Then both lists are combined into one set of outliers. The effect size changes of all combinations of two outliers are evaluated by comparing the effect size without the pair with the effect size with the pair (the original dataset).

```
## 2  Japan - Finland           0.31  0.50   0.19   63.1    X
## 3  Japan - Sweden            0.31  0.47   0.16   52.0    X
## 4  Japan - Mexico            0.31  0.44   0.14   44.1    X   X
## 5  Japan                     0.31  0.44   0.13   44.0    X
## 6  USA - Sweden              0.31  0.19  -0.12  -39.4    X   X
## 7  USA - Finland             0.31  0.22  -0.09  -28.7    X   X
## 8  USA                       0.31  0.22  -0.08  -26.8    X   X
## 9  USA - Mexico              0.31  0.22  -0.08  -26.7    X   X
## 10 South Korea - Finland     0.31  0.38   0.07   23.6    X   X
## 11 Japan - USA               0.31  0.37   0.06   20.5    X   X
## 12 South Korea - Sweden      0.31  0.37   0.06   19.7    X   X
## 13 South Korea - Mexico      0.31  0.36   0.06   18.3    X   X
## 14 South Korea               0.31  0.36   0.06   18.2    X   X
## 15 Japan - Austria           0.31  0.36   0.05   17.0    X
## 16 South Korea - Portugal    0.31  0.35   0.05   15.0    X   X
## 17 USA - South Korea         0.31  0.28  -0.03   -9.3    X   X
## 18 South Korea - Greece      0.31  0.31   0.01    2.3    X   X
## 19 Finland - Mexico          0.31  0.31   0.00    1.5    X   X
## 20 Finland                   0.31  0.31   0.00    1.4    X
## 21 Finland - Sweden          0.31  0.31   0.00    1.1    X
## 22 Sweden                    0.31  0.30   0.00   -0.6    X
## 23 Sweden - Mexico           0.31  0.31   0.00   -0.5    X   X
## 24 Mexico                    0.31  0.31   0.00    0.1        X

## # Not showing 33 possible outliers
```

The output shows 24 combinations of potential outliers with the largest influence on the effect size when a combination of two outliers or a single outlier is removed (see *dif.abs* and *dif.rel* for the absolute and relative differences, respectively). A potential outlier combination consists always of at least one single outlier. Removing Japan in combination with South Korea, Finland, Sweden or Mexico, increases the effect size change. When Japan is removed in combination with other countries, the change of the effect is less than when only Japan is removed.[13]

From the example it appears that Japan is the most serious potential single outlier.

9.8.3 Outlier decision approach

After potential outliers are identified, the next question is what to do with them. NCA suggests a two-step approach for making a decision about an outlier. In the first step, it is evaluated whether the outlier is caused by a sampling or case-selection error, or a measurement error. In the second step, it is decided if an 'outlier for unknown reasons' should be part of the phenomenon (using the deterministic perspective on necessity) or if it should be considered an *exception*.

In the first step of NCA's outlier decision approach, the potential outlier case is selected. Next, it is evaluated if the case has sampling or case selection error. This refers to a case not being part of the theoretical domain of the *formal* necessity hypothesis. For example, the case may be a large company, whereas the necessity theory applies to small companies only. An outlier case due to sampling or case selection error should be removed from further analysis. If the case is not an outlier because of sampling or selection error, it should be

[13]Note that for $k > 1$ the `plotly` XY-plot only shows the single potential outliers as for $k = 1$.

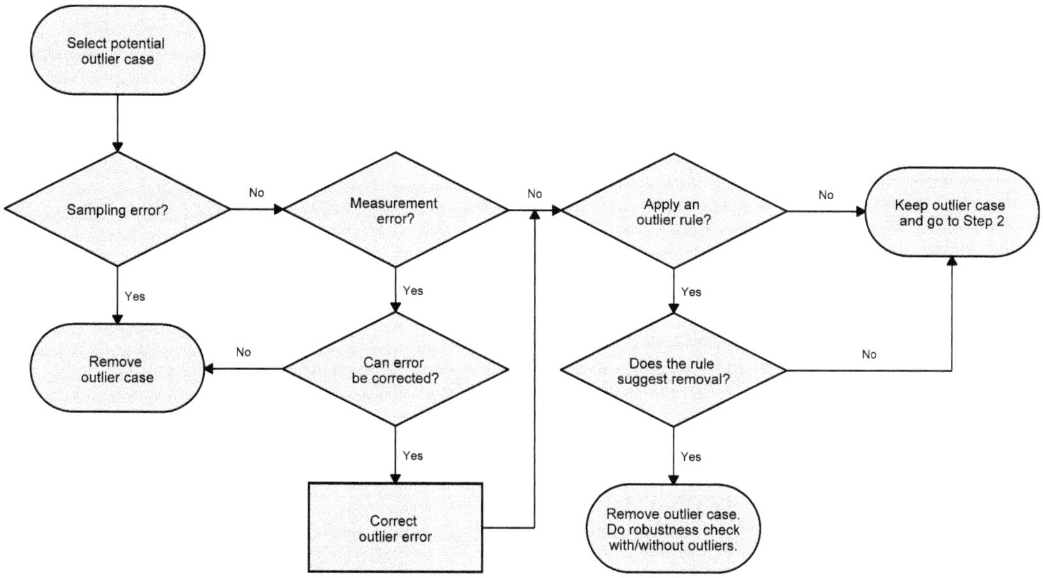

FIGURE 9.3 Outlier decision tree for Step 1 of the outlier decision approach.

evaluated if the case may have measurement error. The condition or the outcome may have been incorrectly scored, which could happen for a variety of reasons. If the case has measurement error and this error can be corrected, the outlier case becomes a regular case and is kept for the analyses.

If the measurement error cannot be corrected, two situations may apply. If the measurement error occurs in the outcome score, the case should be removed from further analysis of all necessary conditions (if there are multiple necessary conditions). If the measurement error occurs in the condition score, the case should be removed from further analysis of that condition, but the analyst must decide to keep it or remove it for the analysis of other conditions. This depends on the analyst's judgement if the case is generally problematic, or only regarding the specific condition where it was identified as an outlier case (the case may not be an outlier case for other conditions).

If the outlier is not caused by sampling/selection error or measurement error or no information about sampling error or measurement error is available, the outlier may be considered an 'outlier for unknown reasons'. The analyst may then decide to apply an outlier rule that is justifiable and convincing. A mechanistic outlier rule (e.g., based on percentages) to decide whether the case should be kept or removed is usually not desirable. Such rules are arbitrary but their consequences on the results are usually large. Therefore, applying a mechanistic outlier rule is generally not advised. If it is used anyway and outliers are removed mechanistically, it should be reported and justified, and a robustness check should be done comparing the results with and without applying the outlier rule (Section 9.12). Note that concluding that necessity is rejected because of presence of outliers may be an important result. The condition may on average influence the outcome, but apparently it is not essential as cases exist that can compensate the absence or a low value of the condition. The outliers that reject necessity may be best or innovative cases that are able to achieve a desired outcome without or with a low level of the condition. If the outcome is undesirable, outlier cases may need different action to reduce the outcome than standard actions focused on the condition. In other words, outliers may be informative rather than being errors. *Rejecting* necessity due to outliers and discussing outliers may be a more valuable result than *accepting* necessity after removal of outliers. If no outlier rule is applied (the preferred approach), the potential outlier case is kept, and the next step of NCA's outlier decision approach begins.

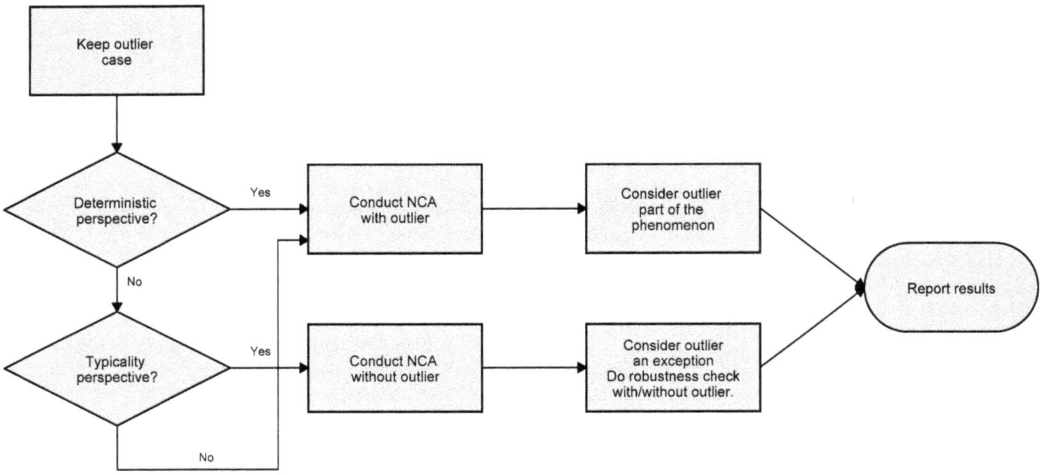

FIGURE 9.4 Outlier decision tree for Step 2 of the outlier decision approach.

The decision tree for the second step is illustrated in Figure 9.4. This picks up from the decision to keep the outlier case because the case has no known error, and no outlier rule is applied. To continue the analysis two perspectives are distinguished for necessity causality: the *deterministic perspective*, and the *typicality perspective* (Section 2.4.3 and Dul, 2024a). The deterministic perspective, in principle, does not allow points above the ceiling. However, some noise may be tolerated, namely when the case is close to the ceiling in the medium purity ceiling zone (Section 4.5.4). If this perspective is adopted the analysis continues with the outlier.

The typicality perspective, in principle, allows outlier cases that are further away from the ceiling, but only if they are rare. These outliers are called *exceptions*. Accepting the typicality perspective means accepting that the outlier is not an error but behaves differently than other cases with another (unknown) explanation than necessity. Thus, the typicality perspective can only be adopted if the case:

- is a rare exception.
- is (presumably) not an error.
- is isolated and far from the ceiling (not noise).
- behaves differently than other cases for unknown reasons.

If the typicality perspective is adopted, the analysis continues without the outlier, because it does not represent the necessity phenomenon that is studied. It can, however, be studied separately because of the unusual characteristic. Since a case without an error is removed from the analysis, a robustness check should be done with and without the exception to show how it affects the results.

If the typicality perspective is not adopted, the analysis continues with the outlier.[14]

[14]Note that the typicality perspective can only be used when *rare* exceptions exist. Having many points

In general, the handling of outliers (keep or remove) relies on a series of decisions made by the analyst. It is advised to perform robustness checks to evaluate the sensitivity of the results to specific outlier decisions and to report the results (Section 9.12).

In the illustrative example, the outlier approach is applied as follows. The case Japan is the only potential outlier to be considered. The first step of the outlier decision approach (Figure 9.3) is the evaluation of whether Japan could be an outlier case because of sampling/selection error. The theoretical domain of the hypothesis was defined as "all countries in the world" (Section 9.2). Japan is one of these countries, so sampling/selection error does not apply.

Next, the question is whether Japan's scores for X and/or Y could have measurement error. This seems unlikely as these scores are obtained from a reputable archival dataset that has been widely used elsewhere, and discussions of measurement error specifically for Japan have not emerged before. The next consideration is to apply an outlier rule for Japan.

Since a mechanistic outlier rule is not preferred, and Japan is a ceiling outlier for which no generally accepted outlier rules are available, the result of the first step of the outlier decision approach is to keep Japan as outlier case for further consideration.

In the second step of the outlier approach (Figure 9.4) it is decided to consider Japan as an exception for unknown reasons. One speculation may be that Japan's high level of innovation is focused more on incremental innovation of systems than on disruptive start-up style of innovation based on high-risk and high failure entrepreneurship of individuals. Despite the relatively small number of cases in the dataset (28) the illustrative example considers Japan a rare exception. This means that the full analysis must be re-done to obtain an estimate of the main necessity effect without Japan, and that Japan is considered an exception, to be discussed separately in the report. Additionally, a robustness check should be conducted and reported, comparing the results with and without the exception.

In the following analysis, Japan is removed from the `nca.example` dataset, followed by re-doing the data analysis:

```
library(NCA)
data(nca.example)

nca.exampleN <- nca.example[-14,] #exclude Japan
set.seed(123) # for replicability of estimated  p-value
modelN <- nca_analysis(nca.exampleN, 1,3,
                       ceilings = 'cr_fdh',
                       test.rep = 10000)

## Setting up parallelization, this might take a few seconds...
##       Do test for  :  cr_fdh - IndividualismDone test for:
##        cr_fdh - Individualism

modelN
```

far above the ceiling or having points that are not isolated from the rest of the points, do not qualify for adopting the typicality perspective. The typicality perspective is only meant for a very limited number of isolated points that represent another (unknown) explanation than necessity. Typicality is about "rare exceptions" (Goldstein, 2012) and applied to NCA's deterministic causal logic (Chapter 2), deterministic necessity becomes "deterministic with exceptions" (Dul, 2024a), not probabilistic necessity. Also see Section 7.5.2.3 on how to deal with exceptions during the development of a formal necessity hypothesis.

```
##
## --------------------------------------------------------------------------

## Effect size(s):
##                cr_fdh p
## Individualism 0.44   0.025
## --------------------------------------------------------------------------
```

The results show that the effect size $d = 0.44$ and the p-value $= 0.025$. Both values meet the selected threshold levels of $d = 0.10$ and $\alpha = 0.05$ such that the necessity hypothesis is not rejected (see, however, Section 9.9 about model fit and Section 9.12 about robustness checks).

To compare the results with and without the exceptions, it is possible to extract the XY-plot of this analysis that is done without the exception, and then to add the exception back to show its original position, and to include the original ceiling line as well using the following script.

```
library(NCA)
library(ggplot2)
data(nca.example)
# Coordinates of the exception
x <- nca.example["Japan", "Individualism"]
y <- nca.example["Japan", "Innovation performance"]

# Intercept and slope of original ceiling line
model <- nca_analysis(nca.example, 1, 3, ceilings = "cr_fdh")
model
a <- nca_extract(model, x = "Individualism",
                 ceiling = "cr_fdh",
                 param = "Intercept")
b <- nca_extract(model, x = "Individualism",
                 ceiling = "cr_fdh",
                 param = "Slope")

#2 NCA without exception
modelN <- nca_analysis(
  nca.example[row.names(nca.example) != "Japan", ], 1, 3,
  ceilings = "cr_fdh", custom = c(a,b) )
modelN

# XY plot for comparison
out <- nca_output(modelN, plots = TRUE, summaries = FALSE)
p <- out$plots[["Individualism"]] +
  geom_point(aes(x = x, y = y), shape = 1, colour = "blue") +
  geom_point(aes(x = x, y = y), shape = 4, colour = "red",
             size = 2, stroke = 1.2)
print(p)
```

Figure 9.5 shows the XY-plot with the outlier case, the original ceiling line and the new ceiling line. The above procedures are for one outlier at a time. The judgment becomes more complex when the analyst considers removing more than one outlier one by one or at the

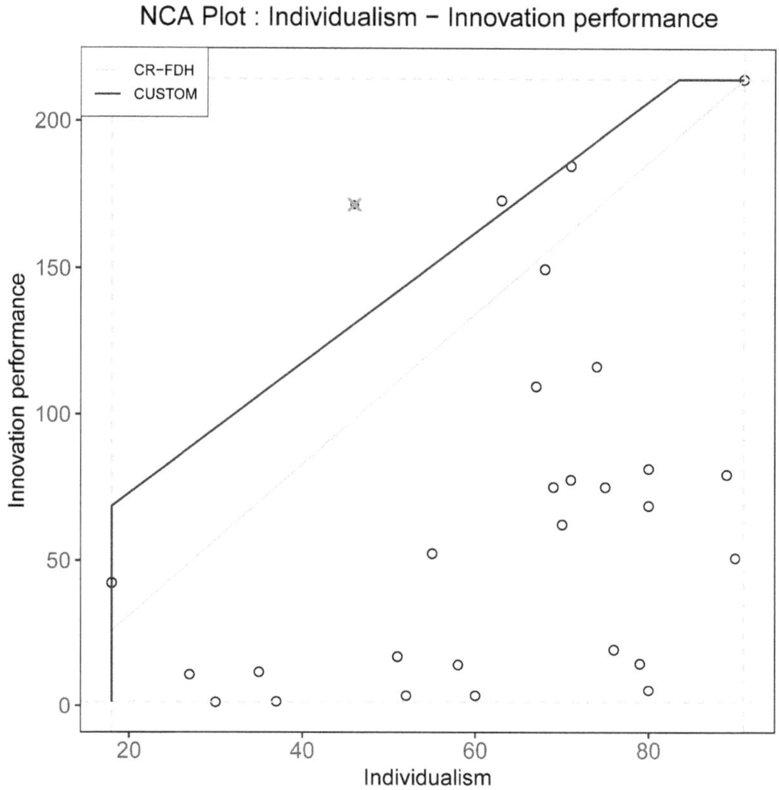

FIGURE 9.5 XY-plot of illustrative example with one exception (crossed out). Light line: ceiling line after removal of the exception. Dark line: ceiling line before removal of the exception.

same time. For example, after removing the first outlier, a case that was originally not part of the first set of single potential outliers, may be identified as a new potential outlier, etc. Generally, removing outliers is not advised unless the analyst has very good reasons for it.

9.9 Model fit

In Chapter 4, specific model fit metrics for NCA were introduced to evaluate how well the chosen necessity model (bounding box and ceiling line) represents the necessity relationship suggested by the data. Since NCA's model is mathematical and not statistical (no error term), common statistical model fit metrics such as R^2 or likelihood-based fit metrics are not applicable. To illustrate the use of NCA's model fit metrics, they are applied to the illustrative example. In the `NCA` software for `R`, the values of the model fit metrics can be obtained using `summaries = TRUE` in the `nca_output` function as follows:

```
library(NCA)
data(nca.example)
```

```
nca.exampleN <- nca.example[-14,] #exclude Japan
set.seed(123) # for replicability of estimated  p-value
modelN <- nca_analysis(nca.exampleN, 1,3,
                       ceilings = 'cr_fdh',
                       test.rep = 10000)

## Setting up parallelization, this might take a few seconds...
##         Do test for  :  cr_fdh - IndividualismDone test for:
##           cr_fdh - Individualism

nca_output(modelN, plots = FALSE, summaries = TRUE)

##
## --------------------------------------------------------------------------

## --------------------------------------------------------------------------
##
## Number of observations    27
## Scope                   15563.6
## Xmin                       18.0
## Xmax                       91.0
## Ymin                        1.2
## Ymax                      214.4
##
##                            cr_fdh
## Ceiling zone              6872.010
## Effect size                  0.442
## # above                      3
## Slope                        2.580
## Intercept                  -20.338
## p-value                      0.025
## p-accuracy                   0.003
##
## Complexity                   1
## Fit                        80.1%
## Ceiling accuracy           88.9%  (24)
## Noise                      11.1%   (3)
## Exceptions                  0  %   (0)
## Support                    14.8%   (4)
## Spread                       0.42
## Sharpness                    0.50
##
## Abs. ineff.              1819.580
## Rel. ineff.                11.691
## Condition ineff.             0.014
## Outcome ineff.             11.679
```

9.9.1 Complexity

Complexity refers to the ceiling degrees of freedom. Linear ceiling lines (C-LP, CR-FDH, CR-VRS) have fixed, low complexity ($cp = 1$), whereas segmented lines, such as CE-FDH (stepwise linear) and CE-VRS (concave linear) have higher complexity which depends on the number of peers that define the ceiling. Complexity is a model fit metric that needs to be balanced between a low score for good generalizability but possibly poor boundary fit, and a high score for good boundary fit but poor generalizability and greater susceptibility to small measurement errors (i.e., risk of overfitting).

For the illustrative example, the complexity is 1 because the selected ceiling line is a linear ceiling line (CR-FDH).

9.9.2 Fit

As discussed in Section 4.5.2, *fit* is the closeness of a chosen ceiling line's effect size to the CE-FDH benchmark (100%), expressed as a percentage. This metric indicates how well a ceiling line follows the data border. Low values (e.g., $< 80\%$) suggest poor fit with the irregularity of the boundary.

For the illustrative example, the fit is 80.1%.

9.9.3 Ceiling accuracy

Ceiling accuracy refers to the number of points on or below the ceiling line (Section 4.5.3). High ceiling accuracy means the ceiling line correctly separates the ceiling zone from the feasible area. If there are no points above the ceiling, the ceiling accuracy is 100%. A ceiling accuracy below for example 95% suggests that too many cases exist above the ceiling, and that the ceiling line may be too low.

In the illustrative example, the ceiling accuracy is 88.9%.

9.9.4 Noise

As discussed in Section 4.5.4, *noise* refers to points just above the ceiling that violate strict necessity. The medium-purity zone is the area of tolerable noise if the percentage of points is low, for example $< 5\%$.

Figure 9.6 shows the *NCA ribbon* in the *XY*-plot of the illustrative example. The upper part of the ribbon represents iso-purity $= 0.9$. It shows that three points are in the medium-purity zone. With $N = 27$ this implies that noise is 11.1%.

9.9.5 Exceptions

Section 4.5.4 discusses the role of *exceptions* in NCA. An exception is a rare case that seriously violates necessity. Exceptions are located in the low-purity zone that is far from the ceiling. No points are allowed in the low-purity zone for deterministic necessity. If a rare point is in the low-purity zone it could be an exception that behaves differently than the entire group and needs to be discussed separately. Figure 9.6 shows that in the illustrative example there are no points in the low-purity zone, suggesting that there are no exceptions. The case (Japan) that was classified as exception in the outlier analysis is removed from the dataset. This choice is evaluated in the robustness check below (Section 9.12).

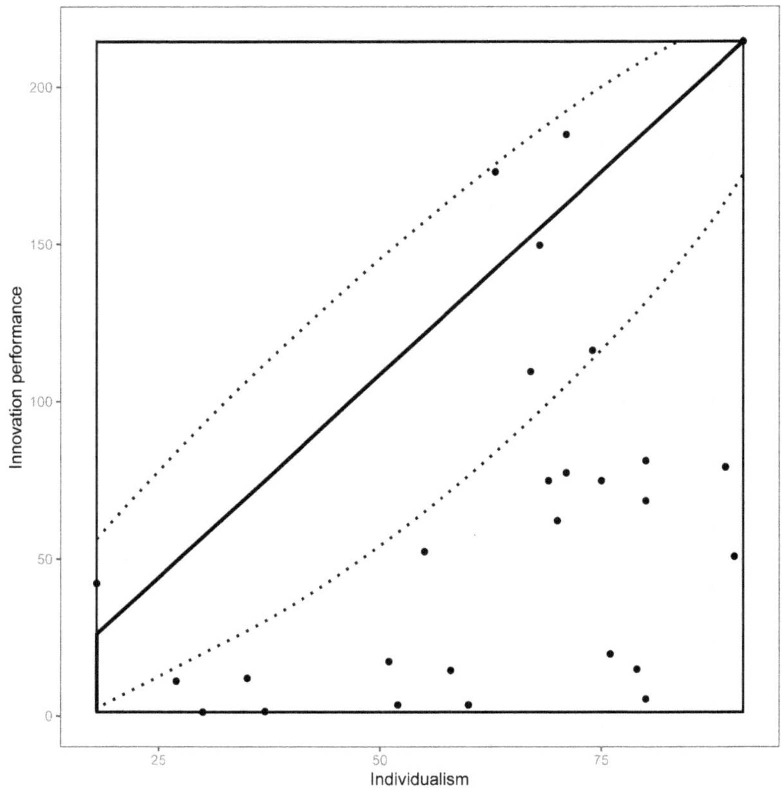

FIGURE 9.6 The NCA ribbon in the XY-plot of the illustrative example (without outlier) with the preliminary CR-FDH ceiling line. Upper dotted line: iso-purity = 0.9. Lower dotted line: iso-solidity = 0.8.

9.9.6 Support

Support quantifies how strongly the data in the feasible area support the ceiling line (Section 4.5.5). The points on the ceiling line and in the medium-solidity zone serve this purpose. For good support, at least for example 5% of cases in the feasible area should be located on or near the ceiling line in the medium-solidity zone.

In the illustrative example, one point is on the ceiling line and three points are below it in the medium-solidity zone. This implies that (14.8%) is in the ceiling line or in the medium-solidity zone (*support* = 14.8%).

9.9.7 Spread

As discussed in Section 4.5.5, *spread* quantifies the extent to which observations in the feasible area near the ceiling are dispersed, rather than clustered along the ceiling line. Spread is 1 if all points are spaced uniformly, and is close to 0 if the points are clustered. A spread value of for example > 0.5 may be adequate.

In the illustrative example, the spread is 0.42.

9.9.8 Sharpness

Sharpness quantifies how abrupt the transition in the data is across the boundary between ceiling zone and feasible area (Section 4.5.6). If there are substantially more points in the medium-solidity zone than in the medium-purity zone, the ceiling is sharp (for example, if $sharpness > 0.8$).

In the illustrative example, in both the medium-purity zone and in the medium-solidity zone the number of points is 3. Consequently, $sharpness = 0.5$, meaning that the ceiling line does not sharply separate the feasible area from the ceiling zone.

9.9.9 Evaluation of model fit

In summary, the model fit parameters with the CR-FDH ceiling line are:

$$\begin{aligned} Complexity &= 1 \\ Fit &= 80.1\% \\ Ceiling\ accuracy &= 88.9\% \\ Noise &= 11.1\% \\ Exceptions &= 0 \\ Support &= 14.8\% \\ Spread &= 0.42 \\ Sharpness &= 0.50 \end{aligned}$$

In all, the model fit of the illustrative example suggests that, although necessity is not rejected based on the effect size and the p-value (and the availability of theoretical support), its support from the perspective of model fit is not strong. Although after removing Japan, no further outliers exist in the low-purity zone of the ceiling zone, relatively many points are just above the ceiling in the medium-purity zone, as indicated by the metrics *ceiling accuracy* (too low) and *noise* (too high). Also, the ceiling line does not provide a clear separation between feasible area and ceiling zone, as indicated by the *sharpness* metric (too low). This suggests that an alternative linear ceiling line that is placed somewhat higher would give a better model fit. The C-LP ceiling line could be a candidate (Section 4.4). Therefore, the provisional model specification using the default CR-FDH ceiling line may be replaced by a model with a C-LP ceiling line to improve model fit:

```
library(NCA)
data(nca.example)
nca.exampleN <- nca.example[-14,]

set.seed(123)
modelFINAL <- nca_analysis(nca.exampleN, 1,3,
                           ceilings = 'c_lp',
                           test.rep =10000)

## Setting up parallelization, this might take a few seconds...
##      Do test for  :  c_lp - IndividualismDone test for:
##        c_lp - Individualism
```

```
nca_output(modelFINAL, summaries = TRUE)
```

```
##
## --------------------------------------------------------------------------
## --------------------------------------------------------------------------
##
## Number of observations    27
## Scope                  15563.6
## Xmin                      18.0
## Xmax                      91.0
## Ymin                       1.2
## Ymax                     214.4
##
##                      c_lp
## Ceiling zone     5094.910
## Effect size         0.327
## # above             0
## Slope               2.907
## Intercept         -10.020
## p-value             0.034
## p-accuracy          0.004
##
## Complexity          1
## Fit               59.4%
## Ceiling accuracy  100  %
## Noise               0  %  (0)
## Exceptions          0  %  (0)
## Support           18.5%  (5)
## Spread             0.41
## Sharpness          1.00
##
## Abs. ineff.      5373.780
## Rel. ineff.        34.528
## Condition ineff.   18.892
## Outcome ineff.     19.278
```

Figure 9.7 shows the XY-plot with the C-LP ceiling line.

The results show that the effect size $d = 0.33$ is greater than the selected threshold level of $d = 0.10$ and is statistically significant given the estimated p-value of 0.034 and selected threshold level of $p = 0.05$. Therefore, the initial conclusion that the hypothesis is supported remains.

Figure 9.8 shows the C-LP ceiling line with the NCA ribbon.

The new model fit parameters for the C-LP ceiling line are (old values between parentheses):

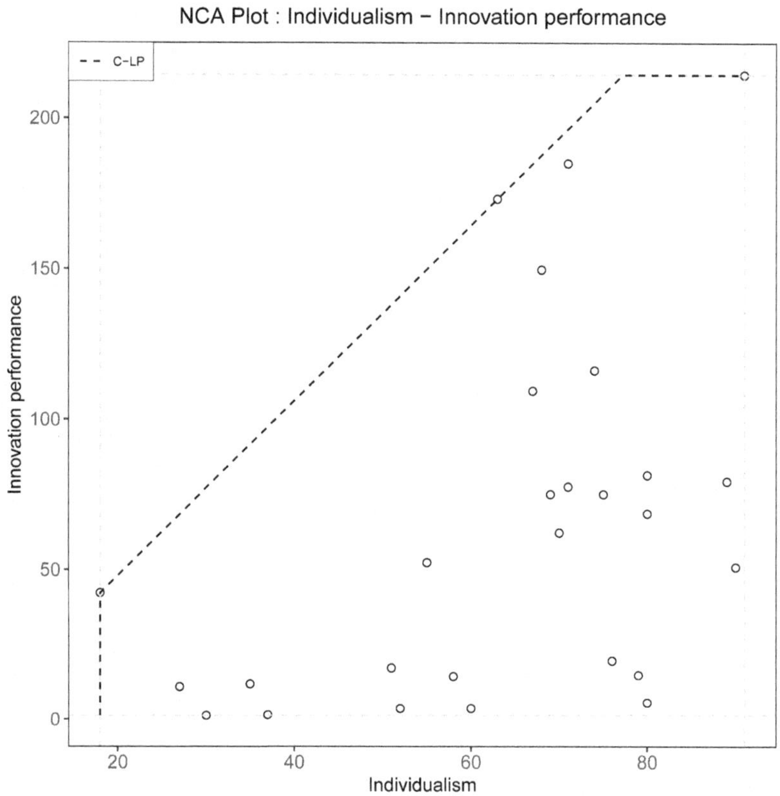

FIGURE 9.7 *XY*-plot of final results of illustrative example without exception and with the C-LP ceiling line.

$$\begin{aligned} Complexity &= 1(1) \\ Fit &= 59.4\%(80.1\%) \\ Ceiling\ accuracy &= 100\%(88.9\%) \\ Noise &= 0\%(11.1\%) \\ Exceptions &= 0(0) \\ Support &= 18.5\%(14.8\%) \\ Spread &= 0.41(0.42) \\ Sharpness &= 1.00(0.50) \end{aligned}$$

All but one model fit metrics are the same or have improved. The *fit* metric has decreased to 59.4%. This indicates that the C-LP line poorly fits the irregular form of the border between the feasible area and the empty space.

The example shows that assessing and achieving model fit is an act of balancing between several indicators of fit, and that it is unlikely that all indicators will reach an optimal level. The analyst must make certain choices that should be justified and reported for transparency.

In the illustrative example the analysis with the C-LP model is selected to evaluate the hypothesis.

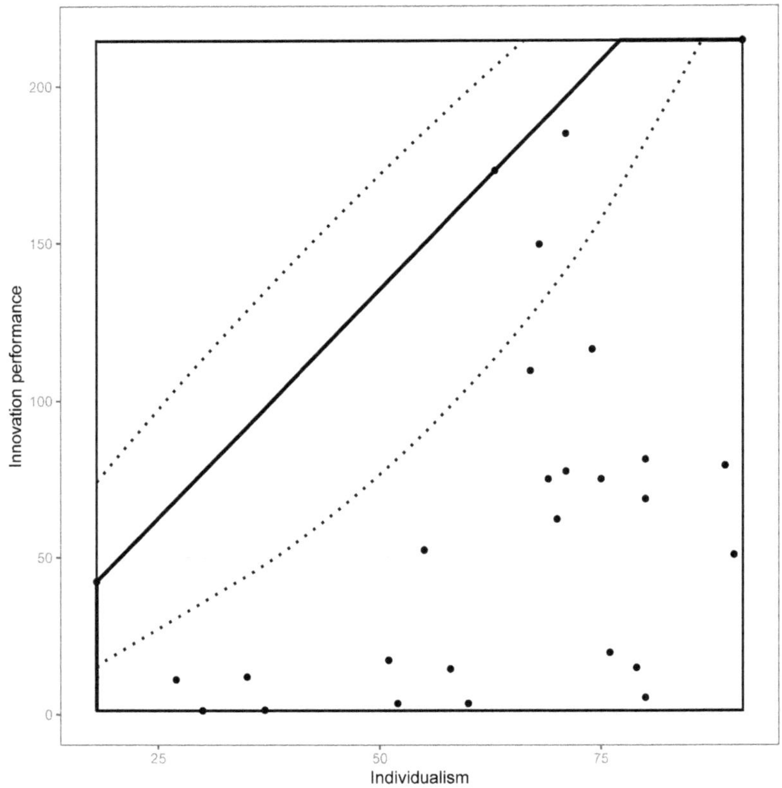

FIGURE 9.8 The NCA ribbon in the XY-plot of the illustrative example (without outlier) with the final C-LP ceiling line. Upper dotted line: iso-purity = 0.9. Lower dotted line: iso-solidity = 0.8.

9.10 Conclusion about necessity-in-kind

After the analysis, a binary conclusion can be made about the hypothesis. The hypothesis is supported (not rejected) if the observed effect size is equal to or above the effect size threshold, *and* the p-value is below the p-value threshold, *and* there is theoretical support (the formal hypothesis). The hypothesis is rejected if the observed effect size is below the effect size threshold, *or* the p-value is equal to or above the p-value threshold, *or* there is no theoretical support (no formal hypothesis).

Given these choices made for the illustrative example, the hypothesis is supported because there is:

- theoretical support (Section 9.2);
- practical significance: the effect size $d = 0.33$ is at least $d = 0.10$;
- statistical significance: the p-value $p = 0.034$ is less than $p = 0.05$.

9.11 Bottleneck analysis: necessity-in-degree

For additional insights, a supported necessary condition *in kind* can also be formulated quantitatively as a necessary condition *in degree*: *level* x of X is necessary for *level* y of Y.

In the illustrative example a target outcome corresponding to 90th percentile was selected: a performance level that is so high that it can only be achieved by 10 percent of the cases.

The target condition that is required for the target outcome can be directly observed from the XY-plot in which the ceiling line and the target outcome are drawn. Figure 9.9 shows the target outcome level for the illustrative example as a horizontal line $y = 159.0$, which corresponds to the selected 90th percentile level. This line intersects with the ceiling line at point $x = 58.17$.[15]

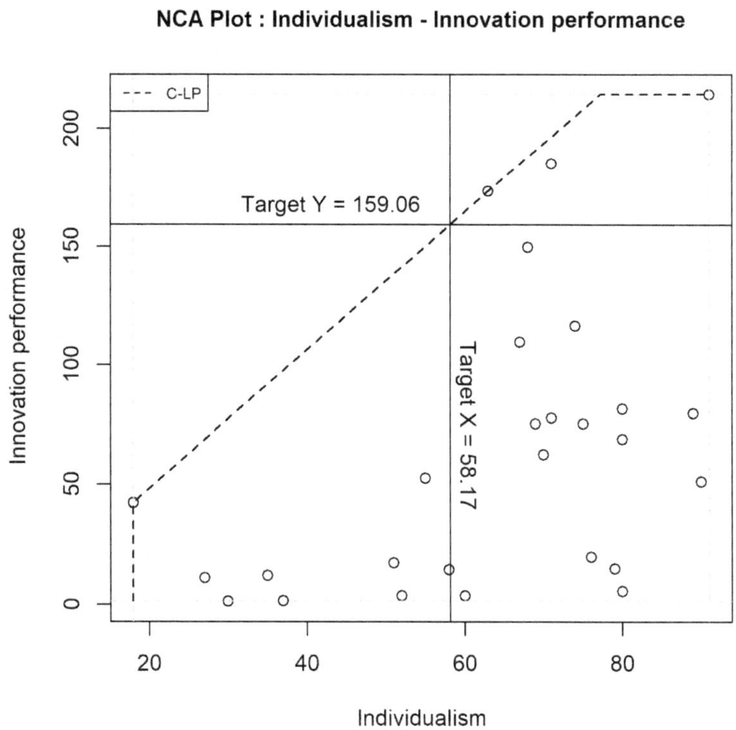

FIGURE 9.9 The final XY-plot of the illustrative example showing the target outcome for the 90th percentile and the corresponding required target level of the condition.

The *bottleneck table* is a helpful tool for evaluating necessary conditions in degree for any target level of the outcome. The bottleneck table is the tabular representation of the ceiling line, in which the first column represents the outcome and the next columns represent the

[15]The target condition for a given target outcome can be computed from the inverse of the ceiling line. For a linear ceiling line like the C-LP, the inverse can be computed from the intercept and slope of the ceiling line according to Equation (4.16). The bottleneck table presented in Section 9.11.2 shows how to obtain the value of the target condition for a target outcome value expressed in percentiles.

necessary conditions. The values in the first column are the levels of Y, and the values in the next columns are the levels of X, corresponding to the ceiling line. By reading the bottleneck table row by row from left to right, the required threshold level(s) of the condition(s) X for a particular level of Y can be evaluated.

The bottleneck table can be produced with the `NCA` software using the argument `bottlenecks = 'TRUE'` in the `nca_output` function. For the illustrative example the bottleneck table can be produced as follows:

```
library(NCA)
data(nca.example)
nca.exampleN <- nca.example[-14,]

modelFINAL <- nca_analysis(nca.exampleN, 1,3, ceilings = 'c_lp')
nca_output (modelFINAL, summaries = FALSE, plots = FALSE,
            bottlenecks = TRUE)
```

```
##
## --------------------------------------------------------------------------------

## Bottleneck C-LP (cutoff = 0)

## Y Innovation performance (percentage.range)

## 1 Individualism          (percentage.range)

## --------------------------------------------------------------------------------
## Y       1
## 0       NN
## 10      NN
## 20      0.7
## 30      10.8
## 40      20.8
## 50      30.9
## 60      40.9
## 70      51.0
## 80      61.0
## 90      71.1
## 100     81.1

##
```

In the default bottleneck table, the values of the outcome and the condition are expressed as 'percentage range' with 11 levels between 0% and 100%, corresponding to steps of 10%. The 'percentage range' format of the bottleneck table means that the values of X and Y are linearly transformed by min-max normalization (with 0% and 100% bounds) to obtain values between 0 and 100 percent. 0% corresponds to the minimum value, 100% to the maximum value, and 50% to the mid-value between these two extremes. This transformation facilitates the comparison of X- and Y-values within and between studies. However, as illustrated below, other formats to represent X and Y may be more informative for a specific study.

The bottleneck table of the illustrative example shows that up to target outcome of 20%, X is not necessary for Y ('NN'). For higher levels of the target outcome, X is necessary.

For a target outcome of 50%, at least $x = 30.9\%$ is needed, and for 90%, at least $x = 71.1\%$ is needed.

To obtain a more detailed picture, the size of the steps can be adjusted, for example to 5 using the `step.size` argument in the `nca_analysis` function:

```
library(NCA)
data(nca.example)
nca.exampleN <- nca.example[-14,]

modelFINAL <- nca_analysis(nca.exampleN, 1,3, ceilings = 'c_lp',
                           step.size = 5)
nca_output (modelFINAL, summaries = FALSE, plots = FALSE,
            bottlenecks = TRUE)
```

```
##
## --------------------------------------------------------------------------
## Bottleneck C-LP (cutoff = 0)
## Y Innovation performance (percentage.range)
## 1 Individualism          (percentage.range)
## --------------------------------------------------------------------------
## Y       1
## 0       NN
## 5       NN
## 10      NN
## 15      NN
## 20      0.7
## 25      5.7
## 30      10.8
## 35      15.8
## 40      20.8
## 45      25.8
## 50      30.9
## 55      35.9
## 60      40.9
## 65      45.9
## 70      51.0
## 75      56.0
## 80      61.0
## 85      66.0
## 90      71.1
## 95      76.1
## 100     81.1
##
```

9.11.1 Levels expressed as actual values

The levels of X and Y in the bottleneck table can also be expressed as actual values. Actual values are the values of X and Y as they are in the original dataset (and in the XY-plot). This way of expressing the levels can help to compare the bottleneck table results with the original data and with the XY-plot results. Expressing the levels of X with actual values can be done with the argument `bottleneck.x = 'actual'` in the `nca_analysis` function, and the actual values of Y can be shown in the bottleneck table with the argument `bottleneck.y = 'actual'` as follows:

```
library(NCA)
data(nca.example)
nca.exampleN <- nca.example[-14,]

modelFINAL <- nca_analysis(nca.exampleN, 1,3, ceilings = 'c_lp',
                           bottleneck.y = "actual",
                           bottleneck.x = "actual")
nca_output (modelFINAL, plots = FALSE, summaries = FALSE,
            bottlenecks = TRUE)
```

```
##
## --------------------------------------------------------------------------

## Bottleneck C-LP (cutoff = 0)

## Y Innovation performance (actual)

## 1 Individualism          (actual)

## --------------------------------------------------------------------------
## Y       1
## 1.20    NN
## 22.52   NN
## 43.84   18.530
## 65.16   25.865
## 86.48   33.200
## 107.80  40.534
## 129.12  47.869
## 150.44  55.204
## 171.76  62.539
## 193.08  69.874
## 214.40  77.209

##
```

While the default number of steps remains the same, the scores in which X and Y are expressed has been changed to actual values.

For practical purposes, specific values for Y can be selected, for example, values that correspond to the scale for scoring Y, or specific target values of interest. The `steps` argument in the `nca_analysis` function can be used for this purpose. For the illustrative example, the step values may correspond to for example the minimum and maximum Y-values and multiples of 10, as follows:

```
library(NCA)
data(nca.example)
nca.exampleN <- nca.example[-14,]

min_y <- min(nca.exampleN$`Innovation performance`)
max_y <- max(nca.exampleN$`Innovation performance`)
modelFINAL <- nca_analysis(nca.exampleN, 1,3, ceilings = 'c_lp',
                           bottleneck.y = "actual",
                           bottleneck.x = "actual",
                           steps = c(min_y, seq(10, 210, by = 10),max_y))
nca_output (modelFINAL, plots = FALSE, summaries = FALSE,
            bottlenecks = TRUE)
```

```
##
## ---------------------------------------------------------------------------
## Bottleneck C-LP (cutoff = 0)
## Y Innovation performance (actual)
## 1 Individualism          (actual)
## ---------------------------------------------------------------------------
## Y       1
## 1.2     NN
## 10.0    NN
## 20.0    NN
## 30.0    NN
## 40.0    NN
## 50.0    20.649
## 60.0    24.089
## 70.0    27.530
## 80.0    30.970
## 90.0    34.411
## 100.0   37.851
## 110.0   41.291
## 120.0   44.732
## 130.0   48.172
## 140.0   51.612
## 150.0   55.053
## 160.0   58.493
## 170.0   61.933
## 180.0   65.374
## 190.0   68.814
## 200.0   72.255
## 210.0   75.695
## 214.4   77.209
##
```

9.11.2 Levels expressed as percentiles

The levels of X and Y can also be expressed as percentiles. The percentile is a score where a certain percentage of scores fall below that score. For example, the 90th percentile of X is a score of X that is so high that 90% of the observed X-scores fall below that X-score, and thus only 10% above it. Similarly, the 5th percentile of X is a score of X that is so low that 5% of the observed X-scores fall below that X-score, and 95% above it.

The X- and Y-values in the bottleneck table can be expressed by percentiles by using the argument `bottleneck.x = 'percentile'` and/or `bottleneck.y = 'percentile'` in the `nca_analysis` function.

For a given target level of Y that is expressed in percentiles (e.g., the 90th percentile target outcome as pre-defined in the illustrative example), the actual bottleneck value of X can be obtained as follows, which corresponds to the above approach to find the target condition:[16]

```
library(NCA)
data(nca.example)
nca.exampleN <- nca.example[-14,]

modelFINAL <- nca_analysis(nca.exampleN, 1,3, ceilings = 'c_lp',
                           bottleneck.y = "percentile",
                           bottleneck.x = "actual",
                           steps = c(0, 90))
nca_output (modelFINAL, plots = FALSE, summaries = FALSE,
            bottlenecks = TRUE)
```

```
##
## --------------------------------------------------------------------------

## Bottleneck C-LP (cutoff = 0)

## Y Innovation performance (percentile)

## 1 Individualism          (actual)

## --------------------------------------------------------------------------
## Y      1
## 0     NN
## 90    58.170

##
```

A useful approach is to express Y-values as actual values and X-values as percentiles, as follows:

```
library(NCA)
data(nca.example)
nca.exampleN <- nca.example[-14,]

min_y <- min(nca.exampleN$`Innovation performance`)
max_y <- max(nca.exampleN$`Innovation performance`)
```

[16]Note that the `steps` argument in the `nca_analysis` function interprets a single number as the number of steps to be displayed in the bottleneck table, and multiple numbers as specific Y-values to be displayed. For displaying a specific y-value, it must be accompanied by another (e.g., arbitrary) y-value.

```
modelFINAL <- nca_analysis(nca.exampleN, 1,3, ceilings = 'c_lp',
                           bottleneck.y = "actual",
                           bottleneck.x = "percentile")
nca_output (modelFINAL, plots = FALSE, summaries = FALSE,
            bottlenecks = TRUE)
```

```
##
## --------------------------------------------------------------------------
## Bottleneck C-LP (cutoff = 0)
## Y Innovation performance (actual)
## 1 Individualism          (percentile)
## --------------------------------------------------------------------------
## Y        1
## 1.20     NN (0)
## 22.52    NN (0)
## 43.84    3.7 (1)
## 65.16    3.7 (1)
## 86.48    11.1 (3)
## 107.80   18.5 (5)
## 129.12   18.5 (5)
## 150.44   29.6 (8)
## 171.76   37.0 (10)
## 193.08   51.9 (14)
## 214.40   74.1 (20)
##
```

In this format, the bottleneck table can be interpreted as follows. The percentile level of X corresponds to the percentage of cases that are *unable* to reach the threshold level of X that is required for the target level of Y (in the same row). The number between brackets corresponds to the number of cases that were not able to achieve the target outcome. For example, in the row with target level $Y = 150.44$, the percentile score of X is 29.6. This means that 29.6% of the cases (= 8 cases) were not able to achieve the required level of X for the target level of Y.

Thus, percentile scores of X provide an indication of the importance of the bottleneck by revealing how many cases in the sample were unable to reach the required level of the necessary condition for a particular level of the outcome.

9.11.2.1 Levels expressed as percentage of maximum

The final way of expressing the levels of X and Y is less commonly used: the percentage of their maximum levels. It indicates how far the score is from the maximum score. The bottleneck table can be expressed with percentage of maximum by using the arguments `bottleneck.x = 'percentage.max'` or `bottleneck.y = 'percentage.max'` in the `nca_analysis` function.

9.11.3 Multiple necessary conditions

The bottleneck table can include multiple necessary conditions for which necessity-in-kind has been established (Section 4.6). Conditions that are not necessary (e.g., because theoretical support is missing, the effect size is too small, or the p-value is too large), should not be included in the table. For the illustrative example, the bottleneck table can be extended with the condition Risk taking. Theoretical arguments suggest that Risk taking is necessary for Innovation performance, and the criteria for effect size and p-value are also satisfied according to the analysis below. For consistency, the same form of the ceiling line (C-LP), type of scope (empirical), and evaluation criteria ($d \geq 0.10$ and $p < 0.05$) are used. Again, the outlier that was detected earlier is excluded from the analysis:

```
library(NCA)
data(nca.example)
nca.exampleN <- nca.example[-14,]

set.seed(123)
modelFINAL <- nca_analysis(nca.exampleN,
                           c(1,2),3,
                           ceilings = 'c_lp', test.rep = 10000,
                           bottleneck.y = "actual",
                           bottleneck.x = "percentile",
                           steps = c(min_y, seq(10, 210, by = 10),max_y))
print(modelFINAL)
nca_output (modelFINAL, summaries = FALSE,
            bottlenecks = TRUE)
```

```
##
## --------------------------------------------------------------------------------
## Effect size(s):
##              c_lp p
## Individualism 0.33 0.034
## Risk taking   0.33 0.002
## --------------------------------------------------------------------------------
```

```
##
## --------------------------------------------------------------------------------
## Bottleneck C-LP (cutoff = 0)
## Y Innovation performance (actual)
## 1 Individualism          (percentile)
## 2 Risk taking            (percentile)
## --------------------------------------------------------------------------------
## Y       1         2
## 10.0    NN (0)    NN (0)
## 20.0    NN (0)    NN (0)
## 30.0    NN (0)    3.7 (1)
## 40.0    NN (0)    3.7 (1)
## 50.0    3.7 (1)   7.4 (2)
## 60.0    3.7 (1)   7.4 (2)
## 70.0    7.4 (2)   7.4 (2)
## 80.0    11.1 (3)  14.8 (4)
## 90.0    11.1 (3)  14.8 (4)
```

```
## 100.0    18.5 (5)  22.2 (6)
## 110.0    18.5 (5)  37.0 (10)
## 120.0    18.5 (5)  37.0 (10)
## 130.0    18.5 (5)  37.0 (10)
## 140.0    22.2 (6)  44.4 (12)
## 150.0    29.6 (8)  48.1 (13)
## 160.0    33.3 (9)  51.9 (14)
## 170.0    37.0 (10) 51.9 (14)
## 180.0    40.7 (11) 59.3 (16)
## 190.0    48.1 (13) 59.3 (16)
## 200.0    63.0 (17) 70.4 (19)
## 210.0    70.4 (19) 81.5 (22)
## 214.4    74.1 (20) 81.5 (22)
```

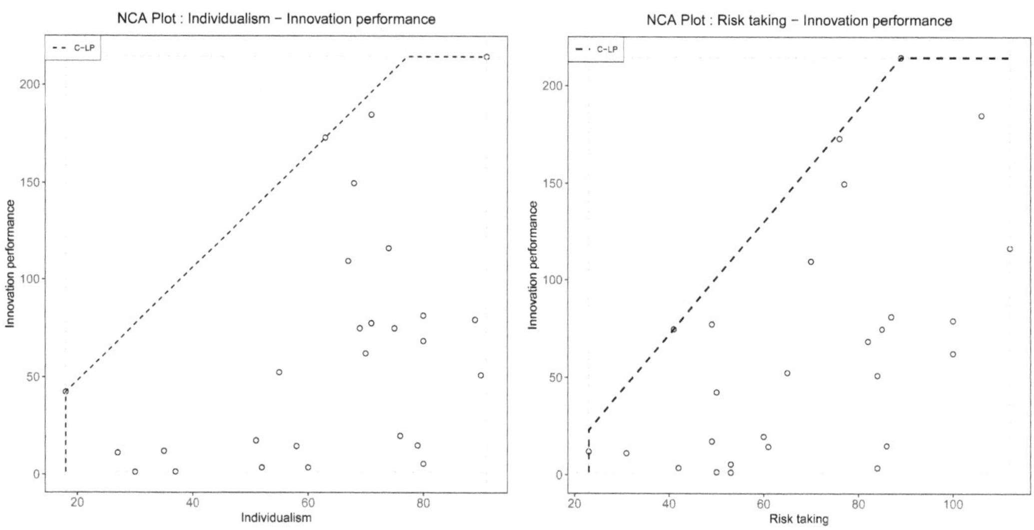

FIGURE 9.10 XY-plots of final results of illustrative example including a second necessary condition.

The results show that the ceiling lines and the effect sizes are similar for the two conditions. Risk taking is necessary with an effect size of $d = 0.33$ and a p-value of $p = 0.002$. The bottleneck table for the two conditions shows that different values of Individualism and Risk taking are needed to make the outcome possible. For example, for $Y = 30$, Individualism is not necessary, but Risk taking is necessary. However, this is not a very important bottleneck because only 3.7% (1 case) cannot achieve the required level of X for this target level of Y. To be able to achieve a target level of for example $Y = 170$, both Individualism and Risk taking are necessary. This target outcome cannot be achieved by 37.0% of cases (10 cases) for Individualism, and 51.9% of cases (14 cases) for Risk taking. If for a given case the required level of one of the two conditions is not achieved, the corresponding level of the outcome will not be reachable.

9.11.4 Interpretation of the bottleneck table for other corners

The above sections refer to analyzing the upper-left corner in the XY-plot, corresponding to the situation that the presence or a high value of X is necessary for the presence or a high value of Y. As indicated above, in the situation that X is expressed as 'percentage range', 'actual value' or 'percentage maximum' the bottleneck table shows the minimum required level of X for a given value of Y. When X is expressed as 'percentiles', the bottleneck table shows the percentage of cases that are unable to reach the required level of X for a given value of Y.

However, the interpretation of the bottleneck table is different when other corners than the upper-left corner are analyzed (Figure 3.8).

9.11.4.1 Bottleneck table for corner = 2

For corner = 2, the upper-right corner is empty suggesting that the absence or low value of X is necessary for the presence or high value of Y. This means that for a given value of Y, the value of X must be equal to or *lower* than the threshold value according to the ceiling line. When X is expressed as 'percentage range', 'actual value' or 'percentage maximum', the bottleneck table shows the *maximum* required level of X for a given value of Y. In other words, for each row in the bottleneck table (representing the target level of Y), the corresponding level of X must be *at most* the level mentioned in the bottleneck table. When X is expressed as 'percentiles', the bottleneck table still shows the percentage and number of cases that are unable to reach the required level of X for a given value of Y.

9.11.4.2 Bottleneck table for corner = 3

For corner = 3, the lower-left corner is empty suggesting that the presence or high value of X is necessary for the absence or low value of Y. This means that for a given value of Y, the value of X must be equal to or *higher* than the threshold value according to the 'ceiling line', which is now a floor line. When X is expressed as 'percentage range', 'actual value' or 'percentage maximum' the bottleneck table shows the *minimum* required level of X for a given value of Y. In other words, for each row in the bottleneck table (representing the desired level of Y), the corresponding level of X must be *at least* the level mentioned in the bottleneck table. However, the first column in the bottleneck table (representing Y) is now reversed. The first row has a high Y-value and the last row has the low value. When percentage-based scores of Y are used (i.e., percentiles, percentages of the range, and percentages of the maximum), a low percentage now corresponds to a high level of Y and a high percentage to a low level. When X is expressed as 'percentiles', the bottleneck table still shows the percentage and number of cases that are unable to reach the required level of X for a given value of Y.

9.11.4.3 Bottleneck table with corner = 4

For corner = 4, the lower-right corner is empty suggesting that the absence or low value of X is necessary for the absence or low value of Y. This means that for a given value of Y, the value of X must be equal to or *lower* than the threshold value according to the 'ceiling line', which is now a floor line. When X is expressed as 'percentage range', 'actual value' or 'percentage maximum' the bottleneck table shows the *maximum* required level of X for a given value of Y. For each row in the bottleneck table (representing the desired level of Y), the corresponding level of X must be *at most* the level mentioned in the bottleneck table. The first column in the bottleneck table (representing Y) is again reversed. The first row has a high Y-value and the last row has the low values as the desired outcome is low. Again, when percentages are used for Y (percentile or percentage of maximum) a low percentage corresponds to a high level of Y and a high percentage to a low level of Y, and when X is expressed as 'percentiles', the bottleneck table still shows the percentage and number of cases that are unable to reach the required level of X for a given value of Y.

9.11.5 NN and NA in the bottleneck table

'NN' (Not Necessary) in the bottleneck table means that X is not necessary for Y for the particular level of Y. With any value of X it is possible to achieve the particular level of Y.

'NA' (Not Applicable) in the bottleneck table is meant to warn that it is impossible to compute a value for X. There are two possible reasons: the first of which occurs more often than the second:

1. The maximum possible value of the condition for the particular level of Y according to the ceiling line is lower than the actually observed maximum value. This can happen for example when a CR ceiling (which is a trend line) intersects the right bound ($X = x_{max}$) of the bounding box. If this happens, the analyst can either explain why NA appears in the bottleneck table, or can change NA into the highest observed level of X. The latter can be done with the argument `cutoff = 1` in the `nca_analysis` function.
2. In a bottleneck table with multiple conditions, one case determines the y_{max}-value and that case has a missing value for the condition with the NA (but not for another condition). When all cases are complete (have no missing values) or when at least one case with a complete observation (x,y_{max}) for the given condition exists, NA will not appear. The analyst can choose to explain why NA appears, or to exclude the incomplete case from the bottleneck table analysis (and accept that the y_{max} in the bottleneck table does not correspond to the actually observed y_{max}).

9.12 Robustness checks

When conducting an empirical study, the analyst must make a variety of theoretical and methodological choices to obtain results. For example, as shown above the choice of the form of the ceiling line was not obvious. Often, plausible alternative choices can be made for several aspects of the analysis. To evaluate the sensitivity of the results to the analyst's choices, robustness checks can be conducted. Typically, the checks focus on the sensitivity of the main estimates and the study's main conclusion (e.g., rejection or non-rejection of the hypothesis based on effect size and p-value).[17]

In NCA, the effect size and the p-value are two main estimates for concluding whether a necessity relationship is credible. These estimates can be sensitive to several choices. Some choices may not be specific for NCA, for example choices related to the data that are used as input to NCA (study design, sampling, measurement, Section 8.2) or the statistical significance level (α, Section 9.4.2). Other choices are NCA-specific, such as the choice of the ceiling line (e.g., CE-FDH or CR-FDH in the model specification, Section 9.3), the choice of the scope (empirical or theoretical, Section 9.3.1), the threshold level of the necessity effect size (e.g., 0.10 or different, Section 9.4.1), the target outcome (Section 9.4.3) and the handling of NCA-relevant outliers (inclusion/ exclusion, Section 9.8).

[17] For example, in regression analysis the regression coefficient and its p-value are the main estimates that can be sensitive to the selection of the functional form of the regression equation (e.g., linear or non-linear), or the inclusion or exclusion of control variables in the regression model (X. Lu & White, 2014). Another example of an analyst's choice is the threshold p-value (e.g., 0.05 or 0.01).

This section discusses several choices that are candidates for a robustness check in NCA. These choices refer to model specification (ceiling line and scope, Section 9.3), the selection of evaluation criteria for the hypothesis (threshold levels for d, and p-value, level of the target outcome, Section 9.4), and the handling of outliers (Section 9.8). The analyst selects only relevant robustness checks. It makes no sense to do a robustness check for outliers if the visual inspections of the XY-plots clearly indicate that there are no outliers, or that a certain ceiling line is clearly the preferred choice for theoretical or empirical reasons.

For the illustrative example, the effect of the analyst's choices on the conclusion is evaluated. The choices made in the illustrative example for the final analysis were:

- *Ceiling line* = C-LP
- *Scope* = empirical
- *Effect size threshold* = 0.10
- *p-value threshold* = 0.05
- *Target outcome*: 90th percentile
- *Outliers*: one outlier removed (considered an exception; application of the typicality necessity).

9.12.1 Different ceiling line choice

In the illustrative example, the final choice was made to conduct the primary analysis with the C-LP ceiling line based on the evaluation of model fit (Section 9.9.9). One of the weaknesses of this selection is the low *fit* metric, that suggests that the line does not follow the irregular borderline well. The CE-FDH ceiling line is a reasonable alternative ceiling line that performs better regarding this aspect, although this line is more complex and may risk overfitting (Section 9.9.1). The analysis with this ceiling line is as follows:

```
library (NCA)
data(nca.example)
nca.exampleN <- nca.example[-14,]
#ceiling line change
set.seed(123)
modelN2 <- nca_analysis(nca.exampleN, 1,3, ceilings = 'ce_fdh', test.rep = 10000)
##Setting up parallelization, this might take a few seconds...
modelN2

nca_output(modelN2, summaries = FALSE)

##
## --------------------------------------------------------------------------------
## Effect size(s):
##   ce_fdh   p
## Individualism  0.55   0.006
## --------------------------------------------------------------------------------
```

The results with the alternative ceiling line show that the effect size $d = 0.55$ is larger than the original effect size $d = 0.33$ and that the p-value is smaller (0.006 compared to 0.034). Both ceiling lines produce the same conclusion that necessity is supported, which suggests a robust result. However, given the low fit metric, the bottleneck table analysis would yield different results.

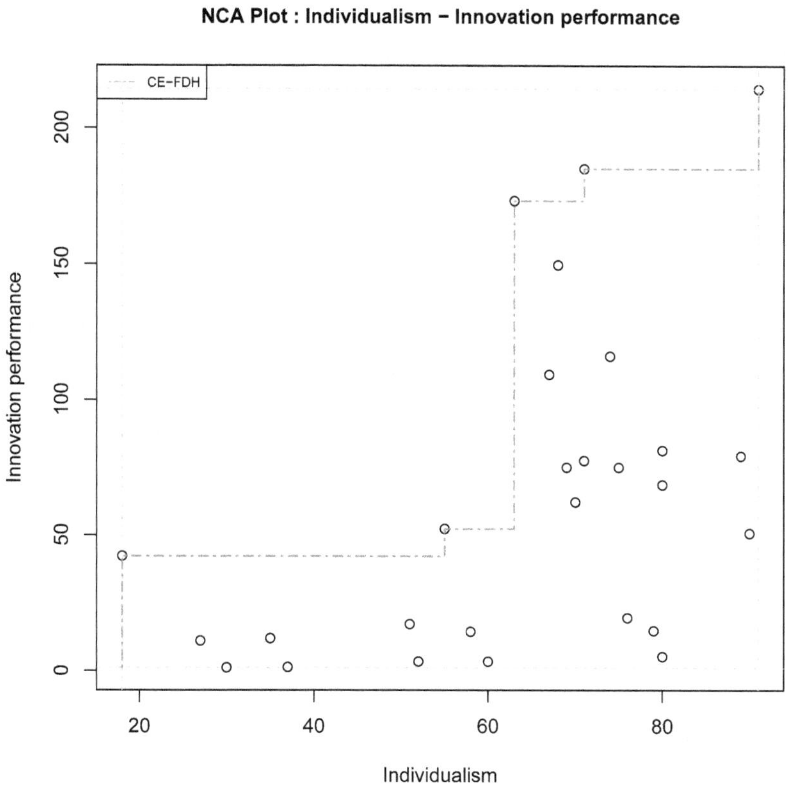

FIGURE 9.11 Ceiling robustness check for the illustrative example: Results with CE-FDH ceiling line.

9.12.2 Different choice of the scope

In the illustrative example the choice was made to use the empirical scope. The reason was that this scope usually avoids overestimation of the effect size, and that the variables have no natural upper or lower bound (unlike, for example, Likert scales or percentage scales). However, it may be possible to assume lower and upper limits for the variables. For example, the minimum and maximum observed values for Individualism in all countries are 6 and 91, respectively (Hofstede, 1980), whereas the minimum and maximum values in the analyzed dataset of 28 countries are 18 and 91, respectively. Based on this, the theoretical bounds for Individualism could be set at [6, 91]. According to Gans and Stern (2003), Innovation performance ranged between 1.2 and 214.4 in the year 2000. They predicted that, by 2005, the maximum innovation performance would be 258. Assuming that Innovation performance cannot be negative, the minimum theoretical value could be set at 0. This means that the theoretical bounds for Innovation performance could be set at [0, 258], resulting in a relevant theoretical scope of [6,91] $\times$ [0,258].

This can be analyzed with the following `R`-code:

```
library (NCA)
data(nca.example)
nca.exampleN <- nca.example[-14,]

set.seed(123)
```

```
modelN3 <- nca_analysis(nca.exampleN, 1,3,
                        ceilings = 'c_lp',
                        test.rep = 10000,
                        scope = c(6,91,0,258))

## Setting up parallelization, this might take a few seconds...
##     Do test for  :  c_lp - IndividualismDone test for:
##       c_lp - Individualism

modelN3

##
## --------------------------------------------------------------------------
##               c_lp p
## Individualism 0.49 0.031
## --------------------------------------------------------------------------

nca_output(modelN3, summaries = FALSE)
```

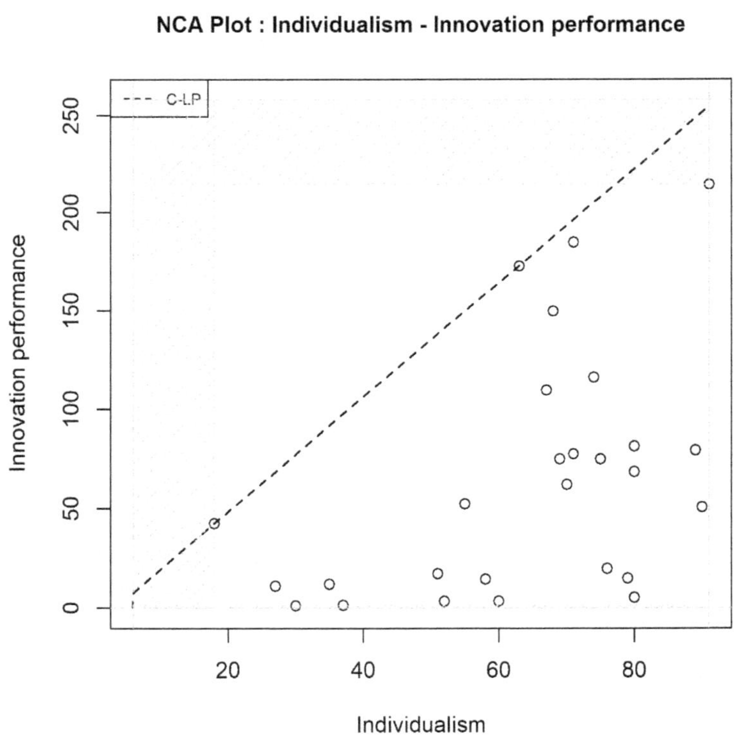

FIGURE 9.12 Scope robustness check for the illustrative example: Comparing the results with the empirical and with the theoretical scope (difference is shaded).

The check reveals that an analysis with the larger theoretical scope yields a larger effect size. In the current example, this does not change the main conclusion.

9.12.3 Different choice of the threshold level of the effect size

The analyst could also have made another choice for the effect size threshold level. If the threshold level is changed from the common 0.10 value ("small effect") to a larger value of 0.20 ("medium effect"), the conclusion for the illustrative example about the hypothesis would not change, as the original effect size $d = 0.33$. The robustness check of effect size threshold suggests a robust result.

9.12.4 Different choice of the threshold level of the p-value

The analyst could also have made another choice for the p-value threshold level (α). If the threshold level is changed from the common 0.05 value to a stricter level of $p = 0.01$, the conclusion in the above example about the hypothesis would change. This robustness check indicates that the statistical significance of the result is fragile.

9.12.5 Different outlier choice

The fourth robustness check is the outlier robustness check. NCA results can be sensitive to the handling of outliers as they, by definition, influence the effect size estimate (Section 9.8). An earlier decision to include potential outliers could be challenged by excluding one or more potential outliers with the largest influence on the effect size if removed. Similarly, an earlier decision to exclude one or more outliers with a relatively large influence on the effect size, could be challenged by including them, or by excluding additional outliers.

In the illustrative example, one single outlier was removed from the analysis. This decision is challenged in two ways: (1) the removed outlier (Japan) is included again in the analysis, (2) an additional outlier that makes a pair with Japan is also excluded (Finland).

The `R`-code for this robustness check is as follows:

```
library (NCA)
data(nca.example)
nca.exampleN <- nca.example[-14,]

#outlier change 1 (include the earlier removed outlier)
set.seed(123)
modelN4 <- nca_analysis(nca.example, 1,3, ceilings = 'c_lp', test.rep = 10000)

## Setting up parallelization, this might take a few seconds...
##     Do test for  :  c_lp - IndividualismDone test for:
##       c_lp - Individualism

modelN4

##
## --------------------------------------------------------------------------

##              c_lp p
## Individualism 0.16 0.319
## --------------------------------------------------------------------------

nca_output(modelN4, summaries = FALSE)
```

```
modelN4b <- nca_analysis(nca.example, 1,3, ceilings = "c_lp",
                         bottleneck.y = 'percentile',
                         bottleneck.x = 'percentile',
                         steps = c(85,90,95))
nca_output(modelN4b, summaries = FALSE, bottlenecks = T)

##
## --------------------------------------------------------------------------------
## --------------------------------------------------------------------------------
## Y       1
## 85     3.6 (1)
## 90     21.4 (6)
## 95     32.1 (9)

#outlier change 2 (remove an additional outlier)
nca_outliers(nca.example, 1,3,
             ceiling = 'c_lp',
             k = 2,
             condensed = TRUE) #Japan and Finland

## Setting up parallelization, this might take a few seconds...

##                    outliers eff.or eff.nw dif.abs dif.rel ceiling scope
## 1  Japan - Finland             0.16   0.35    0.19   117.2       X
## 2  Japan - South Korea         0.16   0.34    0.17   106.7       X     X
## 3  Japan - Mexico              0.16   0.33    0.16   101.2       X     X
## 4  Japan                       0.16   0.33    0.16   101.0       X
## 5  USA - South Korea           0.16   0.04   -0.12   -72.7       X     X
## 6  USA                         0.16   0.06   -0.11   -65.3       X     X
## 7  USA - Mexico                0.16   0.06   -0.11   -65.3       X     X
## 8  USA - Australia             0.16   0.06   -0.11   -64.8       X     X
## 9  Japan - USA                 0.16   0.26    0.10    62.4       X     X
## 10 South Korea - Portugal      0.16   0.14   -0.03   -16.4       X     X
## 11 South Korea                 0.16   0.14   -0.02   -12.3       X     X
## 12 South Korea - Mexico        0.16   0.14   -0.02   -12.2       X     X
## 13 USA - Sweden                0.16   0.15   -0.01    -8.1       X     X
## 14 USA - Finland               0.16   0.17    0.00     2.2       X     X
## 15 Mexico - Turkey             0.16   0.16    0.00     1.1       X     X
## 16 Mexico                      0.16   0.16    0.00     0.1       X     X

nca.exampleN2 <- nca.example[-c(7,14),]

set.seed(123)
modelN5 <- nca_analysis(nca.exampleN2, 1,3,
                          ceilings = 'c_lp',
                          test.rep = 10000)

## Setting up parallelization, this might take a few seconds...
##  Do test for  :  c_lp - IndividualismDone test for:
##  c_lp - Individualism
```

```
modelN5

##
## --------------------------------------------------------------------------
## 			c_lp p
## Individualism 0.35 0.026
## --------------------------------------------------------------------------

nca_output(modelN5, summaries = FALSE)
modelN5b <- nca_analysis(nca.exampleN2, 1,3, ceilings = "c_lp",
                         bottleneck.y = 'percentile',
                         bottleneck.x = 'percentile',
                         steps = c(85,90,95))
nca_output(modelN5b, summaries = FALSE, bottlenecks = T)

##
## --------------------------------------------------------------------------
## --------------------------------------------------------------------------
## Y      1
## 85     19.2 (5)
## 90     23.1 (6)
## 95     42.3 (11)
```

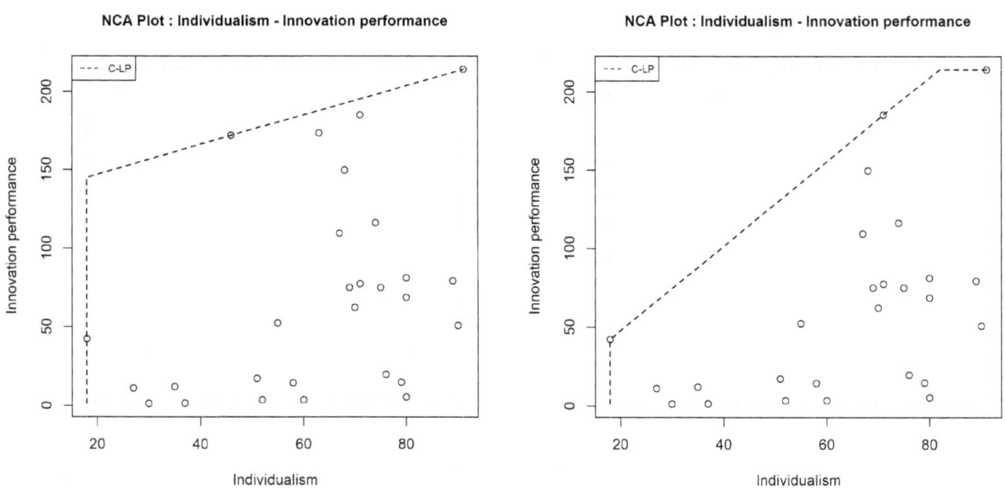

FIGURE 9.13 Outlier robustness check for the illustrative example: Results with including the outlier (Left), and excluding an additional outlier (Right).

The results of these outlier changes are shown in Figure 9.13. The first outlier change (re-introducing the outlier, Figure 9.13-left) reduces the effect size from 0.33 to 0.16 and produces a statistically non-significant result ($p = 0.319$). The second outlier change (removing a second outlier, Figure 9.13-right) increases the effect size slightly to 0.35 while keeping the results significant ($p = 0.026$).

The outlier robustness check indicates that the results are sensitive to outliers.

9.12.6 Different choice of the target outcome

The last robustness check explores the stability of the bottleneck analysis in terms of the percentage of cases that cannot achieve the target outcome because the condition is not satisfied. With the 90th percentile specified as the target outcome, one could for example check how the results change if the target outcome is set somewhat lower (85th percentile) or higher (95th percentile).

```
library(NCA)
data(nca.example)
nca.exampleN <- nca.example[-14,]

modelN <- nca_analysis(nca.exampleN, 1,3, ceilings = "c_lp",
                       bottleneck.y = 'percentile',
                       bottleneck.x = 'percentile',
                       steps = c(85,90,95))

nca_output (modelN, plots = FALSE, summaries = FALSE, bottlenecks = TRUE)
```

```
##
## --------------------------------------------------------------------------------

## --------------------------------------------------------------------------------
## Y        1
## 85     18.5 (5)
## 90     33.3 (9)
## 95     40.7 (11)
```

Figure 9.14 shows that for the original target outcome (90th percentile) 33.3% of the cases (9 cases) did not achieve the required minimum level of the condition. For a target outcome of 85th percentile this percentage reduces to 18.5% (5 cases) and for 95th percentile this percentage increases to 40.7% (11 cases). The percentage of bottleneck cases changes considerably with minor changes of the target outcome.

9.12.7 Robustness table

The results of the robustness checks can be summarized in an NCA robustness table (Table 9.1) where the original NCA results are compared with the results when alternative feasible choices are made. The first row shows the original results, and the next rows display the results under alternative choices. To check robustness of *necessity-in-kind*, the effect size, the *p*-value, and the conclusion about necessity (assuming theoretical support) are presented. To check the robustness of *necessity-in-degree*, the percentage (%) and number (n) of bottleneck cases that are unable to achieve the required level of the condition for a selected target level of the outcome are presented.

The robustness checks indicate that the original necessity-in-kind result is sensitive to the exclusion/inclusion of the identified single outlier (Japan). When this outlier is added again to the analysis, the effect size reduces and the statistical significance disappears. Considering a rare single outlier as an exception according to the typicality necessity approach is possible, as the other robustness checks generally show more stable results, with effect sizes well above the threshold level of 0.10. The statistical significance is somewhat fragile as changing the significance threshold level from 0.05 to 0.01 makes the results statistically insignificant. This may be related to the relatively small sample size.

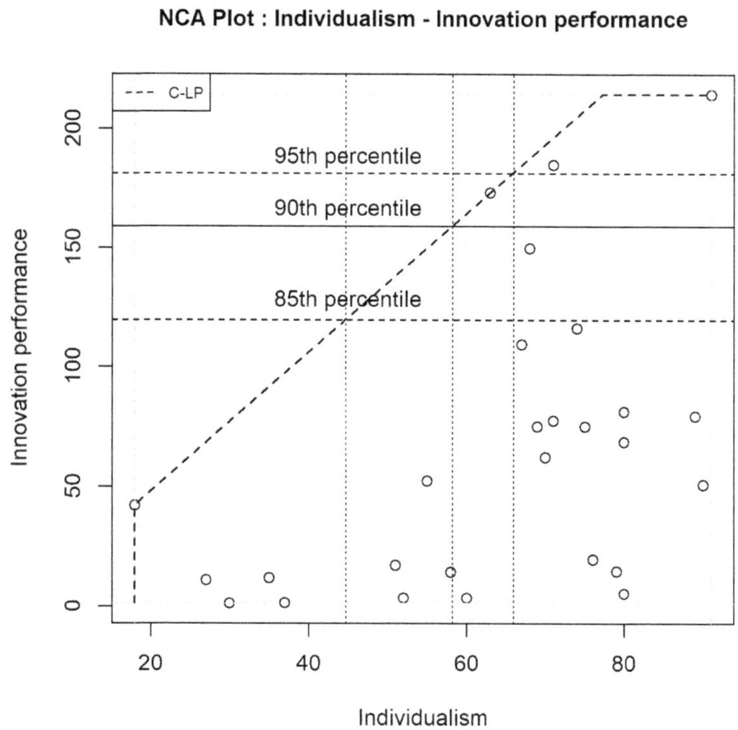

FIGURE 9.14 Target outcome robustness check. The effect of changing the target outcome from 90th percentile to 85th and 95th percentile.

TABLE 9.1 NCA robustness table for the illustrative example.

Robustness check	Effect size	*p*-value	Necessity	Bottlenecks(%)	Bottlenecks(n)
Original	0.33	0.034	yes	33.3	9
Ceiling change	0.55	0.006	yes	37.0	10
Scope change	0.49	0.031	yes	33.3	9
d threshold change	0.33	0.034	yes	33.3	9
p threshold change	0.33	0.034	no	33.3	9
Outlier change 1	0.16	0.319	no	21.4	6
Outlier change 2	0.35	0.026	yes	23.1	6
Target lower	0.33	0.034	yes	18.5	5
Target higher	0.33	0.034	yes	40.7	11

The robustness checks regarding the necessity-in-degree results show that the number of bottleneck cases reduces when outliers are handled differently, or when the selected level of the target outcome is somewhat reduced. This suggests that necessity-in-degree results are fragile, which can be explained by the irregular border between the feasible area and the ceiling zone.

Before the original results can be generalized, replications with other samples are needed. Generalization should only be made when a consistent trend emerges across multiple studies. This is a general principle that applies to any single empirical study, including an NCA study.

9.13 Meta-analysis

Given the growth of NCA publications (see Figure C.1), it can become a realistic option to conduct a meta-analysis of NCA studies for a specific topic. The general aim of a meta-analysis is to systematically combine results from multiple independent studies that address the same research question. A common goal is to estimate an overall effect size across studies by combining results from different samples. This gives a better estimate of the 'true' effect size in an underlying population from which the samples are assumed to be drawn. In regression analysis, this is done by estimating an overall mean effect size using weighting of the effect sizes of the different samples (e.g., based on sample size). In other words, a weighted average is used as an estimate of the overall effect size. The assumption is made that the effect sizes can be compared (e.g., in each study the XY variables are measured on the same scale), and that there is no measurement error.

However, this approach does not apply to NCA because the sample necessity effect size is usually larger than the true effect size (Section 5.5.2), such that the weighted average of the sample effect sizes is not a proper representation of the true effect size. Instead, with the same assumptions as in a regular regression-based meta-analysis (no measurement error, comparable effect sizes), and an additional assumption that the scopes of the separate studies are the same (empirically or by setting the same theoretical scope for each study), the minimum of the sample effect sizes may be the best representation of the true effect size. Alternatively, for estimating the overall ceiling, all samples could be combined into one master sample (as they are assumed coming from the same population) and then the effect size of the combined sample could be the estimation of the true effect size. Such meta-analysis is called an "individual participant meta-analysis" (IPD meta-analysis) and assumes that raw data are available. If it cannot be assumed that the samples are from the same (underlying) population, then differences in sample effect sizes could be explained by the different characteristics of the population. If absence of measurement error cannot be assumed, then an NCA outlier-analysis may possibly find cases with measurement error that could have partly caused sample effect size differences (and could be deleted).

10

Reporting

10.1 Summary of this chapter

Previous chapters explained how to develop a formal necessity hypothesis (Chapter 7), and how to subsequently test it using empirical data (Chapter 8) and NCA's data analysis approach (Chapter 9). This chapter describes how to report an NCA study. The goal is to report insights from the study, how these were obtained, and how they contribute to theoretical understanding and practical application. The chapter focuses on a written report for an academic audience (e.g., a journal article), but it can also be the basis for other reporting formats (e.g., presentations) and other audiences (e.g., practitioners, the general public). First, Section 10.2 distinguishes between several types of NCA publications (empirical, theoretical, methodological introductions, and practical) and provides examples of publications and recommendations. Second, Section 10.3 follows the common format of an academic publication (introduction – theory/hypotheses – methods – results – discussion) to present recommendations for conducting and reporting a high-quality NCA study. This section includes the SCoRe checklist for authors, editors, reviewers, and others to evaluate the quality of an NCA study and its reporting.

10.2 Types of NCA publications

NCA publications can be categorized into empirical, theoretical, methodological introductions, and practical publications. In the remainder of this chapter, it is assumed that the reader of the publication is a scholar or an engaged practitioner who is interested in NCA or in the substantive topic to which NCA is applied. While readers may not be experts on NCA, it is likely that they are familiar with the substantive topic.

10.2.1 Empirical publication

An empirical NCA study applies NCA's methodology (e.g., necessity logic) and NCA's method (e.g., for data analysis) to data that represent reality. The goal is to obtain a better understanding of a specific phenomenon of interest. In an academic setting, the intention is to contribute to the theoretical understanding of the phenomenon. In a practical setting, the intention is to contribute to solving a practical problem related to the phenomenon. A high-quality empirical NCA publication often has two main parts. The first part uses a *necessity causal perspective* for reframing current understanding of the phenomenon by distinguishing between essential (necessary) and contributing (substitutable) factors. The formulation of one or more formal necessity hypotheses is a central part of it. This implies that a formal hypothesis is formulated before the data analysis is conducted. While it is possible to conduct NCA in an inductive/exploratory manner (i.e., without a predefined

DOI: 10.1201/9781003430766-10

hypothesis), this book focuses on the deductive approach for reasons mentioned in Chapter 7. The second part is the *empirical testing* of the hypotheses by providing evidence that supports or refutes the expected necessity relationship.

In an academic setting, the empirical publication is intended to provide a *contribution* to the academic knowledge and literature. The report of the study could contribute to a 'conversation' about a certain topic in the academic literature (e.g., Campbell & Aguilera, 2022; Lange & Pfarrer, 2017).[1] The conversation starts with a *common ground*. This refers to an ongoing conversation about the problem/topic in the primary literature (i.e., what do we know?), and to which NCA contributes new insights. After that, an interesting *complication* is highlighted. This could, for example, be a missing element or an inconsistency in the academic conversation (i.e., what don't we know?). Afterwards, it is explained why this complication is a *concern* (i.e., so what?), and it is reported how the *course of action* (i.e., NCA) addresses this complication. Finally, the *contribution* is described by showing how and why the new insights are now part of the ongoing conversation. For an empirical publication, the narrative to report the insights could focus on developing and testing a necessity hypothesis from scratch, or developing and testing an existing suggestion for a necessity relationship. Tables 10.1 and 10.2 provide examples of how such contributions could be formulated, respectively.

TABLE 10.1 Example of the formulation of the contribution of an empirical NCA publication that develops and tests a new necessity hypothesis.

Topic
Common ground (What do we know?)
Several contributing factors to the outcome of interest are identified in the primary literature that describes a complex multicausal phenomenon.
Complication (What don't we know?)
We do not know which factors are essential.
Concern (So what?)
If the essential factor is not known, we do not know if it is present or not. If it is not present the outcome will not occur no matter the other factors. Acting on other factors than the bottleneck factor is a waste of resources.
Course of action
We analyze the phenomenon from the causal perspective of necessity by developing a formal necessity hypothesis using the NCA methodology and testing it with the NCA method.
Contribution
The phenomenon is now described in more detail by distinguishing between factors that contribute to the outcome and can be substituted (as commonly done), and essential factors that must be in place.

[1]The ideas presented here are in line with typical approaches for presenting theoretical ideas using the 5 C's of theory development: Constructs, Connections, Contingencies, Causal mechanisms, and Contributions (Lange & Pfarrer, 2017). However, this is only one way of structuring an academic argument about NCA. Academic writing is, after all, a matter of preference/style.

TABLE 10.2 Example of the formulation of the contribution of an empirical NCA publication that further develops and tests an existing necessity hypothesis.

Topic
Common Ground (What do we know?) A necessity relationship is suggested in the primary literature to describe a phenomenon.
Complication (What don't we know?) But necessity logic was not used when empirically testing that relationship, so we do not know if it is a true necessity relationship.
Concern (So what?) Whether or not the relationship is a necessity relationship makes a difference for theoretical understanding and practical action. If necessary, the factor is essential in theory and practice; if not, the factor 'only' contributes and can be substituted.
Course of action We develop the proposed necessity relationship toward a formal necessity hypothesis using the NCA methodology and test it with the NCA method to ensure theory-method fit.
Contribution There is now empirical evidence to support or reject the claim of a necessity relationship.

In an empirical NCA publication, the following sections (or similar) are often present:

- Introduction: introduction of necessity logic/theory and description of the contribution.
- Theory/hypotheses: development of (one or more) formal necessity hypotheses using NCA's methodology for necessity causal logic.
- Methods: description of the data analysis approach using the NCA method.
- Results: presentation of NCA's results including the conclusion about (non-)support of the hypotheses.
- Discussion: description of the potential importance of the (non-)support of the hypotheses for theory and practice.

Appendix C provides an overview of empirical journal articles until the end of 2025 that apply NCA. These publications are presented here to illustrate in which fields NCA has been applied. They do not serve as good examples of high-level applications of NCA. Detailed recommendations for conducting and reporting an empirical NCA study at a high level are presented in Section 10.3.

10.2.2 Theoretical publication

In some cases, the theoretical implications of using necessity theory according to the NCA methodology to describe a phenomenon may be complicated enough to warrant a separate publication. In a *theoretical* NCA publication, the focus can be on formulating necessity theory or defining concepts. The main contribution is the use of necessity logic. In such a publication, attention may be paid to NCA's specific characteristics of this logic, such as conditional logic with a causal assumption, necessity-in-degree, or the typicality necessity perspective. Unlike in an empirical publication, a theoretical publication does not include an analysis of empirical data.

Few examples exist of theoretical publications with NCA (Table 10.3).

TABLE 10.3 Examples of theoretical journal articles with NCA.

Year	Author(s)	Journal/Book	Main contribution
2020	Swab, Sherlock, Markin and Dibrell	Family Business Review	Clarify and advance understanding of the Socio-emotional Wealth (SEW) construct in family business research.
2022	Andrevski and Miller	Academy of Management Review	Show that forbearance is conceptually different from counteraction, word response, and involuntary nonresponse and is a deliberate, strategic choice.

10.2.3 Methodological introduction publication

The goal of a *methodological introduction publication* is to introduce NCA in a specific field by explaining its methodology and method.

Since the introduction of NCA in 2016, publications have introduced the approach, or part of it, in specific fields, often illustrated with examples from that field. Examples of these publications are shown in Table 10.4.

TABLE 10.4 Examples of publications with methodological introductions of NCA.

Year	Author(s)	Journal/Book	Main contribution
2019	Tóth, Dul and Li	Annals of Tourism Research	Introduction of NCA in tourism
2020	Tynan, Credé and Harms	Learning and Individual Differences	Introduction of NCA in education
2020	Dul, Karwowski and Kaufman	book chapter (Edward Elgar)	Introduction of NCA in creativity
2021	Dul, Hauff and Tóth	book chapter (Edward Elgar)	Introduction of NCA in marketing
2021	Hauff, Guerci, Dul and van Rhee	Human Resource Management Journal	Introduction of NCA in HRM
2022	Richter and Hauff	Journal of World Business	Introduction of NCA in international business

2022	Greco, Guilera, Maldonado-Murciano, Gómez-Benito and Barrios	International Journal of Environmental Research and Public Health	Introduction of NCA in public health
2023	Linder, Moulick and Lechner	Entrepreneurship Theory and Practice	Introduction of NCA in entrepreneurship
2023	Bokrantz and Dul	Journal of Supply Chain Management	Introduction of NCA in supply chain management
2023	Lee, Dul and Tóth	book chapter (Emerald)	Introduction of NCA in tourism and hospitality
2023	Lee and Lu	Journal of Hospitality & Tourism Research	Introducing necessity theorizing in hospitality and tourism research
2026	Conde	Journal of Marketing Analytics	Introduction of NCA in sales
2026	Marchetti, Koster, Dul	Clinical Psychology Review	Introduction NCA in clinical psychology

As yet, only a small selection of fields has been covered by such publications, while NCA has been applied in many fields and could be applied in even more. This means there is room for more publications introducing NCA to a specific field. Introducing NCA in a new specific field has particular added value if it can be explained that:

- Necessity logic is particularly useful for the field in general or for specific *topics and challenges* (e.g., a list of potential topics/challenges that can benefit from NCA).
- *Necessity thinking* already (often implicitly) exists in the literature of the field (e.g., a list of necessity statements from the field).
- NCA is *different* from conventional methodologies and methods (e.g., by explaining the logic and the steps of NCA using an example from the field).
- NCA brings new *insights* to the field (e.g., an example of applying NCA to a topic from the field).

10.2.4 Practical publication

A final category of publications is relatively scarce. In the first 10 years of NCA, the focus was on the development of the methodology and the introduction in the academic world. Although academic publications often stress the practical relevance and potential of NCA, its uptake in practice has only just started. Early adopters, innovators and pioneers may have used NCA in their practical settings, but such activities usually do not lead to public publications. Yet, experiences from the use of NCA in practice may help to further develop the method and yield valuable insights for both theory and practice. Some possibilities of using NCA in practice are discussed in Chapter 12.

10.3 SCoRe checklist and recommendations

The publications listed in Appendix C are examples of adoptions of NCA during the first decade of NCA's existence. Applying an innovative methodological approach is courageous, but it can also be risky. "When an emerging method's standards of good practice are not yet broadly known, an early and fast adoption of the method without deep understanding of its details may result in incorrect interpretations that depart from the original ones and may result in invalid conclusions" (Dul, 2022, p. 1). The advancement of NCA and the availability of tools for applying the method appropriately can help move NCA into the next phase. In light of this, the SCoRe checklist and its recommendations are presented to support application of the method at the highest possible level.

The checklist and recommendations are meant to stimulate high-quality *empirical* NCA studies and their reporting. The SCoRe checklist is a tool for authors, peers, reviewers, editors, and readers in general to evaluate the quality of the execution and reporting of an NCA study. The acronym refers to the three themes covered in the checklist: **S**trengthening (theoretical rigor), **Co**nducting (data and analysis quality), and **Re**porting (transparency). The checklist can help to identify study parts that are well developed, and those that need improvement.

The checklist is accompanied by best-practice guidelines. For each checklist item, recommendations are provided that can be used by analysts and authors for conducting their NCA study and for their publication about it. SCoRe can also help reviewers and others to provide constructive feedback.

The checklist and recommendations can be used for any empirical publication that uses NCA, whether NCA is the only method or is combined with other methods. In Sections 11.4.2 and 11.6.5, additional recommendations are given for applying NCA in combination with SEM and QCA, respectively.

SCoRe builds on several earlier works with basic guidelines for conducting and reporting NCA studies (Dul et al., 2023), recent developments in NCA reasoning (Dul, 2024a), necessity theorizing (Bokrantz & Dul, 2023), sampling for NCA (Dul, 2024b), the use of archival data in NCA (Dul et al., 2024), and the combined application of NCA and QCA (Dul, 2022) or PLS-SEM (Hauff et al., 2024; Richter, Hauff, Ringle, et al., 2023). Additionally, it incorporates topics newly introduced in this book, such as the formulation of a formal necessity hypothesis (Chapter 7), the use of robustness checks (Section 9.12) and updated recommendations for the combined application of NCA with SEM (Section 11.4) and NCA with QCA (Section 11.5). SCoRe is based on current knowledge, but as NCA and its application are continuously developing, the checklist may be updated in the future.

Following the common structure of an empirical publication, the checklist and its recommendations have six parts: Introduction/General, Theory/Hypotheses, Methods – data, Methods – data analysis, Results, Discussion. Each part discusses 5-10 items to be addressed for a high-level empirical NCA publication. The items have different priority levels. Some are essential ('must-haves') that must be addressed in any NCA publication. If any of the must-haves is missing, the quality of the publication should be seriously questioned. Others are important ('should-haves') to increase the quality of the publication, and yet other items are the 'nice-to-haves'. When addressing all items, the publication meets the highest quality standard from the perspective of applying NCA, necessary for publishing in a high-ranking journal. Some items can be easily addressed, others need more reflection. For each item, requirements, recommendations, and sources for further reading are provided.

Below the checklist items are presented, followed by the recommendations.[2]

10.3.1 Introduction/General

Table 10.5 is the checklist for the Introduction/General part of applying and presenting NCA. The introduction should clearly define why NCA is used, describe it correctly, distinguish it from other causal methods, and highlight its role in the study.

TABLE 10.5 SCoRe checklist – Questions for Introduction/General.

Item	Priority	Question	Check
1	Must-have	Are the goals and contributions of applying NCA explicitly stated?	☐
2	Must-have	When referring to necessity, are only words that correctly describe a necessity relationship used?	☐
3	Must-have	Is it explained that NCA is an approach that combines a specific causal logic (methodology) and the related data analysis technique (method)?	☐
4	Must-have	Is a proper comparison made between NCA's causal perspective and conventional causal perspectives (e.g., probabilistic sufficiency, configurational sufficiency)?	☐
5	Must-have	In multimethod studies, is NCA described and used as an approach with a specific perspective on causality and data analysis, and not merely as a robustness check or an add-on to other methods (and vice versa)?	☐
6	Nice-to-have	If the necessity analysis is a prominent part of the publication, do the title and abstract reflect this necessity perspective/NCA?	☐

The following recommendations help to achieve the goals of the checklist items:

1. State the primary and (possibly) secondary goals and contributions of the empirical NCA study. The study may primarily contribute to theory development by, for example, developing and testing new necessity hypotheses, developing an existing necessity claim further toward a formal necessity hypothesis and testing it with NCA, or replicating results of previously tested necessity hypotheses. Secondary goals could be methodological (e.g., introducing NCA to a new field/phenomenon), or practical (e.g., using results of NCA for practically meaningful recommendations/insights).
   See this book: Chapter 7, Section 10.2.

[2]An interactive version, and an AI-assisted version of the SCoRe checklist can be accessed via the book-page https://jandul.github.io/NCA/.

Suggested articles: Bokrantz & Dul (2023); Dul (2024a); Hollenbeck & Wright (2017); Köhler & Cortina (2023); Bergh et al. (2022).

2. When referring to necessity relationships, use appropriate terminology. In practice, this means using equivalents/derivatives of the statement that X is necessary for Y. Avoid ambiguous statements (X causes Y, X correlates with Y, X is associated with Y, X affects Y), or incorrect (sufficiency-based) statements such as X produces Y, X drives Y, X increases/decreases Y, etc.
   See this book: Table 7.1.

3. Refer to NCA as a methodological approach that combines the theoretical perspective of necessity causality (rather than common probabilistic or configurational causality) with a corresponding data analysis method, ensuring theory-method fit. Emphasize that NCA entails more than just a data analysis or statistical method.
   See this book: Chapters 1, 2, 5.
   Suggested article: Dul (2024a).

4. When, in the context of an NCA study, a comparison is made between NCA's causal perspective and the causal perspective inferred from conventional regression-based/statistical approaches, use the term probabilistic sufficiency for the latter, and not just sufficiency. The term sufficiency suggests determinism (as applied in QCA's configurational sufficiency logic). This recommendation is often violated in studies that combine NCA with SEM. When NCA is compared with QCA's necessity analysis, use the term necessity analysis of QCA for the latter (and not NCA) and explain the differences between the necessity analysis of QCA (only necessity-in-kind) and that of NCA (also necessity-in-degree).
   See this book: Sections 2.3, 2.4.
   Suggested articles: Dul (2024a); Dul (2016a); Vis & Dul (2018); Dul (2022).

5. In multimethod studies, use and refer to NCA as an approach with a specific perspective on causality and data analysis, and do not incorrectly refer to it as a robustness check or add-on to other methods (and vice versa). This recommendation is often violated in studies that combine NCA with SEM.
   See this book: Sections 2.7, 11.4, 11.5.
   Suggested article: Dul (2024a).

6. If important parts of the publication and/or its conclusions are based on necessity logic and NCA, ensure that this is reflected in the title and abstract by using words that reflect necessity logic or NCA.
   See this book: Table 7.1.

10.3.2 Theory/Hypotheses

Table 10.6 is the Theory/Hypotheses part of the checklist. This part is essential as it uses the NCA methodology to develop a formal hypothesis. A causal explanation of the necessity relationship is a central part of it. If it is missing, there is no theoretical justification for the hypothesis and the hypothesis should be rejected, regardless of whether the other criteria (effect size, p-value) are satisfied (Chapter 7).

TABLE 10.6 SCoRe checklist – Questions for Theory/Hypotheses.

Item	Priority	Question	Check
7	Must-have	Is the hypothesized necessity relationship explicitly formulated as: X is necessary for Y, or similar?	☐
8	Must-have	Are the four elements of a necessity theory precisely defined: concepts X and Y, direction of the hypothesis, focal unit, theoretical domain?	☐
9	Must-have	Is a causal explanation provided WHY X is necessary for Y using necessity causal logic to justify the necessity hypothesis by addressing three questions? Why does X enable Y? Why does the absence of X lead to the absence of Y? Why is the absence of X not compensable?	☐
10	Must-have	Are temporal aspects considered including specification of the temporal order of X and Y if this is not obvious?	☐
11	Should-have	Is the existing literature about the relationship between X and Y re-considered from a necessity causal perspective?	☐
12	Nice-to-have	If applicable, is an nc-symbol added near the arrow in a diagram that represents a necessity relationship?	☐
13	Nice-to-have	Is it specified if rare exceptions are allowed (typicality perspective)?	☐

7. Formulate the necessity relationship explicitly as a necessary condition hypothesis: X is necessary for Y. For a deductive study, this is done before data analysis; for an explorative study this is done after data analysis when indications for a necessity relationship are found.
   See this book: Section 7.5.

8. To obtain a *formal necessity hypothesis* that is ready for testing, precisely define the four elements of the necessity theory. (1) Define the concepts X and Y according to their meaning in the hypothesis. (2) Specify the hypothesis and its direction, which indicates which corner of the XY-plot is expected to be empty, particularly if the expected corner is not the default upper-left corner. (3) Specify the focal unit of which X and Y of the hypothesis are characteristics. (4) Define the theoretical domain where it is claimed/expected that the hypothesis will hold, given its boundary conditions. Note that the theoretical domain is usually larger than a population from which cases are selected.
   See this book: Section 3.2, Chapter 7, in particular Section 7.5.

9. Explain why X is necessary for Y using a causal explanation (narrative) by answering three questions: (1) Why does X enable Y (X is an enabler). The existing literature may be a source of information; (2) Why does the absence of X lead to

the absence of Y (X is a constraint); (3) Why is the absence of X not compensable (no substitution for X). Making the narrative complete is a creative process, supported by literature and experiences of scholars and practitioners. See this book: Chapter 7, in particular Section 7.5.3.

10. If this is not self-evident, include a plausible explanation of the temporal order (first X then Y) in the description of why the necessity causal relationship exists. If the data support that X is necessary for Y, explain why the alternative reverse explanation, namely that Y is sufficient for X (or in terms of necessity: absence of Y is necessary for absence of X) that would also be consistent with the expected empty space, is not plausible. Also consider other temporal aspects of the hypothesis such as whether X is necessary for the onset or continuation of Y, or whether the hypothesis holds in all time periods.
See this book: Section 2.5, Chapter 7, in particular Section 7.5.3.

11. Review existing literature (that commonly describes probabilistic relationships between your concepts X and Y), to find hints for necessity causality. Avoid just describing, summarizing, or referring to probabilistic findings/ideas. Where possible, include quotes in the literature that hint to necessity logic. Avoid using sufficiency/probabilistic causal logic for making a justification for necessity causal logic. Note that only describing, summarizing, or quoting prior work (often based on probabilistic sufficiency logic) does not constitute a valid theoretical justification of a necessity hypothesis.
See this book: Section 7.4.1, Table 7.1.

12. In a diagram representing an (expected) necessity relationship between two variables using an arrow, add a relevant symbol ($+$ nc $+$, $-$ nc $+$, $-$ nc $-$, or $+$ nc $-$) near the arrow to show the direction of necessity.
See this book: Section 3.5.

13. Explain whether the necessity theory allows for rare exceptions (i.e., a typicality perspective on necessity), or whether any exception leads to a rejection of the theory (i.e., a deterministic perspective on necessity). The typicality perspective is not a solution for poor measurement and should not be used in a probabilistic manner (e.g., ignoring a certain percentage of cases above the ceiling).
See this book: Sections 2.4.3, 9.8, 4.5.4.
Suggested article: Dul (2024a).

10.3.3 Methods – data

Table 10.7 is the Methods – data part of the checklist. Data is input to NCA, and good input is needed for good output. NCA accepts any type for scores of X and Y that are valid (representing the condition and outcome formulated in the hypothesis) reliable and meaningful. The latter refers to the possibility that values are interpretable for theory and practice.

TABLE 10.7 SCoRe checklist – Questions for Methods – data.

Item	Priority	Question	Check
14	Must-have	Is the study design (observational study, longitudinal study, case study, experimental study) adequate?	☐
15	Should-have	For large-n quantitative studies: are adequate sampling approaches used and is a pre-study power analysis reported?	☐
16	Should-have	For small-n qualitative studies: are adequate NCA-specific approaches used for purposive case selection?	☐
17	Must-have	Are adequate measurement approaches used for getting scores of X and Y, and are these measures practically interpretable?	☐
18	Must-have	For simulation studies: are simulated data sampled from bounded distributions of X and Y?	☐
19	Must-have	Are only data transformations that align with NCA employed?	☐

14. Possible study designs for NCA include the large-n observational study (both X and Y are observed), the longitudinal study (X and Y data from different time points), the small-n case study (X and Y are observed), and the experiment (X is manipulated and Y is observed). The general principles for good study design apply. For case selection and experiments, there is an NCA-specific approach.
    See this book: Section 8.2.

15. For large-n quantitative studies the preferred sampling approach from the theoretical domain is random sampling with good coverage of the population of interest. The general principles for good sampling apply. Report if an NCA-specific pre-study power analysis was done to estimate the required sample size for finding necessity if it exists. Note that doing a post-hoc power analysis makes no sense for the present study, but could inform future studies.
    See this book: Sections 5.6, 8.3.1.
    Suggested article: Dul (2024b).

16. For small-n qualitative NCA studies, aim to use purposive selection of cases from the theoretical domain, with cases being selected based on the presence of Y or absence of X. Always clearly state how cases were selected. Reflect on, and report the potential limitations when cases were not selected purposively based on the absence of X or presence of Y.
    See this book: Sections 8.2.3, 8.3.2.
    Suggested article: Dul (2024b).

17. Ensure a proper measurement approach with valid, reliable, and meaningful data (scores of X and Y). Data are input to NCA, not part of NCA. General principles for good data (collection) apply to an NCA study. For interpretation of results, it is important that the scores have a practical meaning in terms of the levels of X and Y. Data are stored in a dataset. The dataset may consist of archival data.
    See this book: Section 8.4.
    Suggested article: Dul et al. (2024).

18. For data generated by simulation, ensure that the variables X and Y are bounded (have minimum and maximum values). Simulated data drawn from, for example, a normal distribution would not be eligible whereas data from a uniform or truncated normal distribution would be appropriate.
See this book: Sections 4.3, 5.4.

19. Explain whether and how the data were transformed, and whether the transformation aligns with NCA. Affine transformations (including linear transformation) produce valid results. NCA's data analysis does not require data transformation and non-transformed data are often more meaningful and better interpretable (e.g., levels of a Likert scale) than transformed data. This particularly applies to necessity-in-degree and interpretations of specific levels of X being necessary for specific levels of Y. Refrain from conducting non-linear data transformations (e.g., log-transformation, logistic transformation) unless the transformed data represent X and Y as defined in the necessity hypothesis (e.g., log-transformed GDP). This recommendation may be violated in studies that combine NCA with QCA when the logistic transformation (S-curve) is applied mechanistically for calibration purposes.
See this book: Section 8.5.2, Appendix D.

10.3.4 Methods – data analysis

The main part of the NCA method is the data analysis which differs from common data analysis approaches. Table 10.8 is the Methods – data analysis part of the checklist. In multimethod studies where the NCA method is combined with other methods, the analyses are independent. The results are considered in combination only during interpretation and discussion. Therefore, the checklist applies to all types of empirical applications of the NCA method.

TABLE 10.8 SCoRe checklist – Questions for Methods – data analysis.

Item	Priority	Question	Check
20	Must-have	Is the selected ceiling line specified and justified?	☐
21	Should-have	Is the bounding box (with empirical or theoretical scope) deliberately selected and specified?	☐
22	Should-have	Are the hypothesis evaluation criteria (thresholds for effect size, p-value, and possibly target outcome) deliberately specified and justified?	☐
23	Should-have	Are the results of a visual inspection of the XY-plot reported?	☐
24	Must-have	Is it explained how potential outliers are analyzed and handled, and are removed cases reported?	☐
25	Should-have	Are all model fit parameters evaluated?	☐
26	Should-have	Are all NCA-related parameters, terms and concepts properly used and described?	☐
27	Nice-to-have	Is the utilized software, including its version number specified?	☐

28	Nice-to-have	Is a short general description of NCA's data analysis approach provided to serve readers who are less familiar with NCA?	☐

20. Specify and justify the selected ceiling line by considering theoretical expectations (border is linear or non-linear), type of data (discrete or continuous), observed border pattern (regular or irregular), balance between accuracy and generalizability (avoiding overfitting), and quality of the data (data with/without measurement error, simulation data).
See this book: Sections 4.3.2, 9.3.2.

21. Specify and justify the selected bounding box (with empirical or theoretical scope). Commonly, the empirical scope is selected to avoid over-estimation of the effect size. The theoretical scope can be considered when X and Y are measured on bounded scales (e.g., Likert scales, percentages).
See this book: Sections 4.3, 9.3.1.

22. Specify and justify the selected threshold values for necessity effect size d and p-value for identifying necessity-in-kind. This threshold depends on the specific context of the study. If no argument is available for a specific value, common benchmark values may be used ($d = 0.10$; $p = 0.05$ with 10,000 permutations). Optionally, select a target outcome Y to evaluate a specific outcome of necessity-in-degree.
See this book: Sections 4.4.1, 5.3, 9.4.1, 9.4.2, 9.4.3.
Suggested article: Dul (2016b).

23. Conduct the analysis with the selected ceiling line and scope to obtain an XY-plot for visual inspection. Evaluate whether the expected corner is empty, if the selected ceiling line is appropriate, if potential outliers are present, and what the data pattern in the rest of the plot is.
See this book: Section 9.5.

24. Conduct an outlier analysis with the selected ceiling line and scope to identify potential ceiling and scope outliers. If outliers are removed, report in detail which cases are removed and why, and compare the results with and without removing the outliers (see robustness checks).
See this book: Section 9.8.
Suggested article: Dul (2016b).

25. Conduct the analysis with the selected ceiling line and scope and evaluate the model fit parameters (complexity, fit, ceiling accuracy, noise, exceptions, support, spread, sharpness), and compare it with benchmark values.
See this book: Sections 4.5, 6.5, 9.9.

26. Describe the NCA-related parameters, terms, and concepts properly, including using the term 'permutation test' (*not* terms like bootstrapping, (Monte Carlo) simulation, robustness check, or t-test) for NCA's statistical test, and using 'multiple NCA' (*not* multivariate NCA) for analyses with several necessity hypotheses.
See this book: Appendix A.

27. Specify which software and which version of the software was used to conduct NCA. The version number for the NCA package in `R` can be obtained with the

command `packageVersion("NCA")`.
See this book: Appendix B.

28. Include a brief description of NCA's main logic and data analysis elements (ceiling line, ceiling zone, scope, effect size) so that the study is easier to understand for readers unfamiliar with NCA. Consider adding references for further reading.
Suggested articles with examples of short descriptions: Dul (2024b); Dul (2025).

10.3.5 Results

Table 10.9 is the Results part of the checklist where the results of the data analysis are presented, and the (non-)support of the hypothesis is evaluated.

TABLE 10.9 SCoRe checklist – Questions for Results.

Item	Priority	Question	Check
29	Must-have	Are XY-tables or XY-plots of all tested/explored relationships shown?	☐
30	Should-have	Is only the ceiling line that is used to draw conclusions about the necessity relationship shown in the XY-plot?	☐
31	Must-have	For large-n quantitative study: are the effect size and its p-value properly reported?	☐
32	Must-have	Is the conclusion about necessity-in-kind based on three criteria: theoretical justification, effect size, and p-value?	☐
33	Should-have	Is the bottleneck table reported and necessity-in-degree evaluated?	☐
34	Nice-to-have	If applicable: is the meaning of NN and NA in the bottleneck table explained?	☐
35	Should-have	Are the type of values used in the bottleneck analysis justified?	☐
36	Must-have	Are the results of the robustness checks summarized?	☐

29. Include the XY-tables or XY-plots of all tested/explored relationships, preferably in the main text. It is essential that readers are able to inspect them visually.
See this book: Section 9.5.

30. In the XY-plot(s), include only the ceiling line that is used to draw conclusions about the necessity relationship. Note that the default output of the NCA software in `R` includes two ceiling lines, but that only one is usually selected for drawing conclusions about the hypothesis (XY-plots with other ceiling lines that were explored before the final selection was made, or that were evaluated during the robustness checks could be provided separately).

31. Report the estimated effect size and its p-value. Preferably, report the effect size in two decimal places. This ensures that the result contains meaningful information about the precision, and that there is no unwarranted level of accuracy. Report the estimated p-value, preferably with three decimal places for the same reasons. Give exact p-values (or if applicable $p < 0.001$) and do not use inequalities or stars (i.e., not $p < 0.05$ or *).

32. Conclude about necessity-in-kind using three criteria: (1) Theoretical justification (the formal hypothesis), (2) Effect size not below the selected threshold value, (3) p-value below the selected threshold value. Note that for support all criteria apply and that missing just one of them is enough for a rejection. This implies, for example, that without theoretical justification the claim for a necessity relationship is rejected.
See this book: Section 9.10.

33. Report the bottleneck table only for necessary conditions that were not rejected. This facilitates the evaluation of necessity-in-degree: which level of X is necessary for which level of Y.
See this book: Section 9.11.

34. Explain the meaning of NN or NA in the bottleneck table if it appears.
See this book: Section 9.11.5.

35. Justify the type of values for X and Y that are displayed in the bottleneck table (e.g., percentage of range, actual values, percentiles). In particular, the actual values of Y and percentile values of X might be informative about the percentage or number of cases unable to achieve a target outcome.
See this book: Section 9.11.

36. Conduct robustness checks and summarize the results: is the main conclusion about necessity (support or rejection of the hypothesis) sensitive to choices by the analyst regarding ceiling line, scope, effect size threshold, p-value threshold, removing/keeping outliers, different choices for the target outcome? Conclude if the results are robust or fragile.
See this book: Section 9.12.

10.3.6 Discussion

Table 10.10 is the Discussion part of the checklist where the results are interpreted and their consequences for theory and practice and for desired future studies are discussed.

TABLE 10.10 SCoRe checklist – Questions for Discussion.

Item	Priority	Question	Check
37	Must-have	Is it explained why all tested necessary conditions were (not) rejected?	☐
38	Must-have	For a multimethod study: is it explained how the results of NCA and of other methods complement each other?	☐
39	Must-have	Referring to the goals of the study (see introduction): is it discussed if the intended contribution is realized and what the key insights are?	☐
40	Should-have	Are the results of the bottleneck analysis explained in terms of necessity-in-degree?	☐
41	Should-have	Are limitations of applying NCA mentioned?	☐
42	Should-have	Are potential future studies with NCA discussed?	☐

37. Summarize the results of using NCA in terms of which necessary conditions were tested/explored, and which were rejected or not rejected. Assuming that (methodological) errors are unlikely, provide theoretical explanations for the results, referring to the necessity theory and hypotheses formulated earlier.
See this book: Section 7.5.3.

38. When NCA is used in a causal-pluralism study by combining it with another method (e.g., regression-based methods or QCA), report how the results of NCA and the other method complement each other using NERT, NEST and BIPMA when applicable.
See this book: Chapter 11.

39. For the primary and possibly secondary goals (see introduction), discuss if the intended contribution is realized and what key insights are gained.
See this book: Section 10.2.1.

40. If necessity-in-kind is accepted, provide an interpretation of necessity-in-degree using the results of the bottleneck analysis. For example, explain what level of X is necessary for a given target level of Y.
See this book: Section 9.11.

41. Mention the study's limitations. Discuss general limitations of NCA (e.g., exclusive focus on necessity; potential sensitivity to outliers), and specific limitations of applying NCA in the current study, and how this may affect the findings.

42. Discuss potential future studies with NCA, for example conducting studies in other parts of the theoretical domain to assess the generalizability of the current findings, or applying NCA in related substantive areas that could benefit from the necessity causal perspective.
Suggested article: Köhler & Cortina (2023).

11

Multimethod studies

11.1 Summary of this chapter

The previous chapters focus on NCA as a standalone methodology and method. This chapter explores how NCA can be applied in combination with other methods. First, a distinction is made between method-triangulation studies that use multiple methods with the same causal perspective, and causal-pluralism studies using multiple methods with distinct causal perspectives (Section 11.2). Next, causal-pluralism multimethod studies that combine NCA with regression-based methods are discussed, showing that NCA can provide insights beyond average effects. With the NCA-Extended Regression Table (NERT) tool, the results can be compared (Section 11.3). This is followed by discussing how NCA can be combined with a specific regression-based method, structural equation modeling (SEM). To compare the results, the Bottleneck Importance Performance Map Analysis (BIPMA) tool is provided (Section 11.4). Both tools acknowledge the different causal perspectives, while facilitating the combined interpretation of results. Subsequently, multimethod studies that combine NCA with Qualitative Comparative Analysis (QCA) are discussed (Section 11.5). Section 11.6 shows how NCA insights on necessity-in-degree can provide additional insights for QCA's necessity and sufficiency analyses. For the combined interpretation of NCA's necessity results and QCA's sufficiency results, the NCA-Extended Solution Table (NEST) tool is provided.

11.2 Method-triangulation versus causal-pluralism

Studies that apply multiple methods are commonly praised for their ability to produce deeper or more robust insights than studies that rely on a single method. For example, a qualitative analysis of interview data can enrich a quantitative analysis of questionnaire data, and vice versa, or regression approaches with different model specifications or parameter estimation techniques may make conclusions more robust.

In hypothesis-testing studies with multiple methods, the methods that are used should be consistent with the causal perspective that is expressed or implied in the hypothesis. For example, if the hypothesis is that X likely has a positive effect on Y, the methods to test this hypothesis should be based on probabilistic sufficiency (i.e., regression-based methods). Similarly, if the hypothesis is that configurations of Xs produce Y, the methods should be based on configurational sufficiency perspective (i.e., QCA).

Theory-method fit ensures a proper link between the hypothesis and the methods that are applied to test it. Table 11.1 shows theory-method (mis)fit depending on causal perspective

DOI: 10.1201/9781003430766-11

and type of method. The selected method for a study should be consistent with the causal perspective that is used in that study. Selecting NCA or QCA for a hypothesis that represents probabilistic sufficiency leads to misfit. The same holds for selecting regression-based methods to test necessity or configurational sufficiency hypotheses.

TABLE 11.1 Theory-method fit.

	Probabilistic sufficiency	Configurational sufficiency	Necessity (in degree)
Regression	Fit	Misfit	Misfit
QCA	Misfit	Fit	Misfit
NCA	Misfit	Misfit	Fit

Terms commonly used for studies with different methods are *multimethod studies*, *multiple method studies* or *mixed method studies*; these terms often refer to studies that use both qualitative and quantitative methods. In this book, however, a multimethod study refers to a hypothesis-testing study that uses different methods with the same or a different causal perspective, and with the same or different data.

Table 11.2 distinguishes three types of multimethod studies. A *method-triangulation* multimethod study uses a single underlying causal perspective, but different methods that fit this perspective. The data can be the same or different. A *causal-pluralism* multimethod study analyses the same data with different causal perspectives and uses different methods that fit these perspectives. If in a multimethod study, both the causal perspectives and the data are different, the study effectively consists of *separate studies* analyzing the same phenomenon, while each study has its own data, method and causal perspective.

TABLE 11.2 Classification of multimethod studies

	Method-triangulation	Causal-pluralism	Separate studies
Different causal perspectives	No	Yes	Yes
Different data	Yes or no	No	Yes

A multimethod study with NCA as one of the methodologies can be conducted as follows:

1. Develop the hypotheses
    a. for method-triangulation studies: same hypothesis for NCA and other method(s) for testing necessity.
    b. for causal-pluralism studies: different hypotheses for NCA and for the other method(s) that are based on another causal perspective.
2. Collect data
    a. for method-triangulation studies: same or different data for NCA and other method(s).
    b. for causal-pluralism studies: same data for NCA and other method(s).
3. Analyze and report the results *separately*, following the guidelines of each method.

4. Combine the findings during interpretation: What are the theoretical and practical implications of the combined results?

11.3 NCA with regression-based methods

Causal analysis from a probabilistic sufficiency perspective has a long tradition, with regression serving as its natural method. The influential works of Fisher, Neyman, Rubin, and Pearl, have emphasized the close link between causality and probability. Journals and reviewers frequently expect regression-based evidence as a marker of methodological rigor and support for causal claims.

Such an alignment makes regression appear to be the only "proper" way of testing causal hypotheses. However, as discussed in Section 2.7, different views on causality can exist, and a necessity causal analysis of the phenomenon of interest is valuable in itself, or can add unique value to the regression analysis.

Regression analysis was invented more than 100 years ago when Francis Galton (Galton, 1886) quantified the pattern in the scores of parental height and child height. Figure 11.1-left shows an original graph.

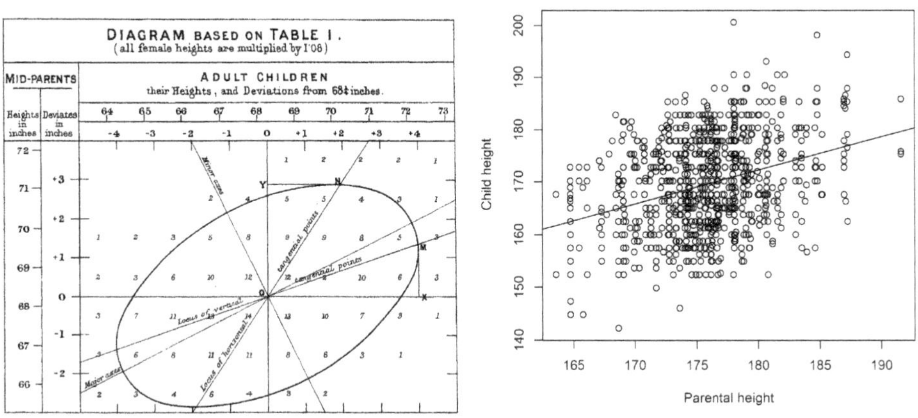

FIGURE 11.1 Francis Galton's (1822–1911) graph with data on Parent height (mid-parents height) and Child height (adult child height). Left: original graph. Right: XY-plot with regression line. The plots use different versions of the original dataset (After Galton, 1886). Data for the right plot were obtained from `HistData` package in `R`.

To describe the observed data pattern, Galton drew lines through the center of the data by eye, describing an average trend between Parental height and Child height and called it the regression line (Figure 11.1-left). He did not explicitly use the method of least squares that had already existed earlier (Gauss, 1809; Legendre, 1805). Karl Pearson formalized Galton's idea mathematically by regression line equations using the method of least squares. Ordinary Least Squares (OLS) is still one of the most common estimation methods in regression analysis.

11.3.1 Principles of regression analysis

Regression analysis is often used to test probabilistic sufficiency hypotheses such as more X likely increases (or decreases) Y or X has a positive (or negative) effect on Y, or similar. The hypothesis is considered to be supported if the regression coefficient is positive (or negative) and statistically significant.

The standard regression equation represents an additive, average-effect model. The simple linear regression model is expressed as:

$$Y = f(X) + \varepsilon \tag{11.1}$$

where the function $f(X)$ represents an additive combination of predictor variables (including, if specified, products of X's or other transformations), with associated coefficients β_i for $i = 1, \dots, k$, where k is the number of terms. The error term ε represents all remaining influences on Y that are not captured by the model. In OLS, the coefficients are chosen so that $f(X)$ is the linear function that best predicts Y in the least-squares sense (minimizing the average squared residual), under the assumption that the predictors are exogenous, meaning that they are uncorrelated with the error term. Measurement error in X is one way this assumption can fail.

Several assumptions are made about the error term. Specifically, it is assumed that ε has an expected value of zero, that it is uncorrelated with (or independent of) the predictors X, and that its variance is constant across all values of X (homoskedasticity). In many applications, it is further assumed that the errors are normally distributed, mainly to justify standard hypothesis tests and confidence intervals. Under this assumption, the conditional distribution of Y given X is normal. Consequently, the model implies an *unbounded* distribution of Y, which can, in principle, take any value between minus and plus infinity.

For the parent–child sample, the regression equation is $y = 57.5 + 0.64x + \varepsilon_x$ (Figure 11.1-right). Since OLS regression assumes that on average $\varepsilon_x = 0$, this equation expresses the average effect. Thus, when x (Parent height) is 175 cm, the estimated average Child height is about 170 cm, but heights between about 140 cm and 200 cm appear to be possible. Given the normality assumption of the error term, very large child heights (e.g., 300 cm) are not observed but are theoretically possible in the population, though unlikely. Because all heights are theoretically possible in such a regression context, an empty space in the upper-left corner (and thus necessity) is theoretically excluded. If an empty space is observed in a sample, the normality assumption implies that this is due to finite sampling, not an indication of necessity.

Most regression models include more than one X. The black box of the error term is opened and additional predictors are added to the regression equation (multiple linear regression, MLR). By including more factors that contribute to Y in the regression equation, Y may be better predicted for given combinations of X's. Adding factors to the equation is not limited to simply adding new predictors. Some variables may be transformed, such as squaring a factor (X_j^2) to represent a non-linear effect of X_j on Y, or taking the product of two factors ($X_1 \cdot X_2$) to represent their interaction. Such transformed or combined factors are added as separate terms in the regression equation. Other regression-based approaches, such as structural equation modeling, also incorporate multiple factors.

The goal of adding more terms to the regression equation is to explain a larger part of the variance (scatter). R^2 represents the proportion of variance explained by the regression model and ranges between 0 and 1. By adding more terms, usually more variance is explained, resulting in higher R^2-values (but also in risks of overfitting).

Another reason to add more factors is to avoid omitted variable bias, for example, due to confounding. A confounder is a relevant variable that correlates with both X and Y, and if excluded from the model may bias the regression coefficient. In regression, it is standard practice to include control variables to reduce omitted variable bias.[1] By adding more *relevant* factors, the prediction of Y becomes more accurate, and the risk of omitted variable bias is reduced.[2]

One characteristic of regression logic is that variables are compensable. It is assumed that the outcome y is obtained by addition or multiplication of the variables of the regression equation (multiple regression), including the error term.[3] As a result, the terms can compensate for each other. For example, when one x_j is low, a high y can still be achieved when other x's are high. Fundamentally, this characteristic therefore implies that any single x_j is not necessary for achieving a certain level of y, as it can be compensated for by other factors.

11.3.2 Comparison NCA and regression-based methods

NCA and regression share several characteristics. Both NCA and regression are variable-based approaches and use linear algebra. Both methods need good (reliable and valid) data without measurement error. For statistical generalization from sample to population, large-n NCA and regression both need to have a sample that is representative for the population; having larger samples usually give more reliable estimations of the population parameters. Large-n NCA and regression analysis both use a p-value to evaluate statistical significance. For generalization of findings, both methods need replications with different samples. Both methods need theoretical reasoning for making causal interpretations.

When data are available that fit both regression analysis and NCA, without knowledge about the underlying causality, it is possible to look at the data from different perspectives (see Section 2.7). One perspective is not inherently superior to another. This also applies when a correlation is observed between two variables. An observed non-zero correlation can be produced by various relationships between X and Y including an average effect relationship and a necessity relationship (see Appendix G). Both perspectives and methods are possible to study the phenomenon. It is the choice of the analyst to use only regression, only NCA, or both, when both underlying causal perspectives are possible.

This means that Galton could also have used his eyes in a different way, namely drawing a line on top of the data, rather than through the center of the data. Then, he would have drawn the border between an area with data, and an area without data describing a possible necessity relationship between Parent height and Child height: the ceiling line (Figure 11.2-right). Instead of estimating the average y for a given x he would have estimated the maximum y for a given x. For example, with a Parent height of 175 cm, the estimated maximum possible Child height is about 195 cm. But Galton did not draw a ceiling line, and the social sciences have adopted the average trend line as the basis for many data analysis approaches. Regression analysis has developed over the years and many variants exist.

[1]Good controls rather than bad controls, see Cinelli et al. (2024).

[2]Omitted variable bias does not occur in NCA (see the "projection method" discussed in Section 4.6).

[3]Also simple regression with one predictor and the error term is an additive model.

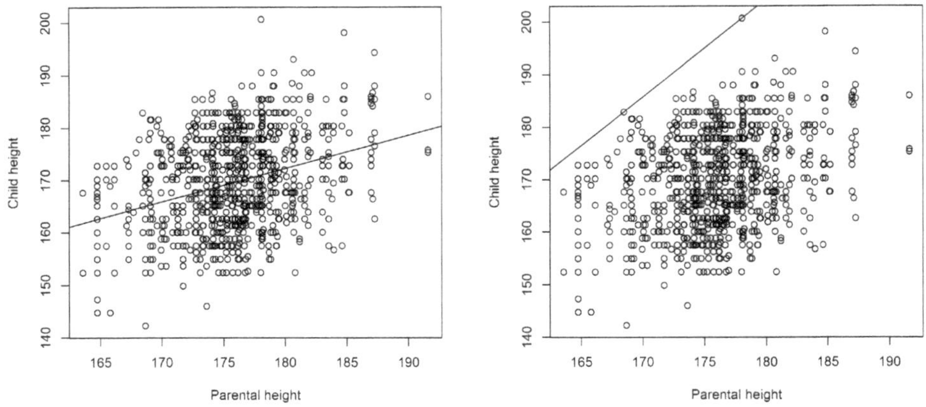

FIGURE 11.2 *XY*-plots of the relationship between Parent height (X) and Child height (Y) (after Galton, 1886). Left: with a regression line. Right: with a ceiling line.

11.3.3 Regression-based statistical concepts do not (always) apply to NCA

In the development of regression analysis, over the years many regression-based statistical concepts relevant for estimating average effects, have been developed and taught to generations of students. These concepts include mediation, moderation, confounding, omitted variable bias, control variables, endogeneity, and spuriousness. However, often these concepts do not apply (directly), or require re-interpretation in the context of NCA.

All differences stem from a fundamental distinction: regression models estimate expected values (average effects), whereas NCA characterizes constraints by estimating an upper boundary (ceiling) in the *XY*-plane and the associated empty space above it. The theoretical ceiling according to Equation (4.15) is a *sharp border* between the area where observations are possible (feasible area) and the area where cases are not possible (ceiling zone) within a *bounding box* (bounded variables). Regression is primarily concerned with how observations are distributed (the full space), while NCA is concerned with combinations of X and Y above a ceiling that do not occur (the empty space).[4] Under the assumptions of proper model specification, sound theory, and adequate data, a regression line may be given an interpretation in terms of probabilistic sufficiency, and a ceiling line may be given an interpretation in terms of necessity.

Table 11.3 shows several regression-based statistical concepts that are defined for average-effect estimation; in NCA they often have no direct analogue or must be reinterpreted in terms of boundary patterns and constraints. The first column lists concepts that are meaningful in regression analysis. The second column refers to correspondence in NCA.

A necessity effect in NCA is the maximum attainable Y given X (boundary), represented by the equations in Section 4.3.2. NCA estimates the overall population ceiling line in the *XY*-plane. When a sample is properly drawn from the population, the observed overall ceiling line can be viewed as an unconditional, marginal, upper envelope across heterogeneous

[4]Information about the ceiling is contained in the upper points (peers), not in the entire distribution below it (Section 4.3.2).

TABLE 11.3 Common regression-based statistical concepts and NCA corresponding concepts.

Regression concept	NCA corresponding concept(s)
Mediation	Necessity chain
Moderation	Subgroup ceiling (conditional ceiling); may indicate additional necessary condition
Confounding	Empty space does not consider third variability; multiple necessary conditions; necessity spuriousness; formal necessity hypothesis
Omitted variable bias	–
Control variable	–
Endogeneity	Reverse causality; measurement error
Spuriousness	Necessity spuriousness; formal necessity hypothesis

cases, including any third variables that operate in the population insofar as they constrain feasible XY-combinations. Consequently, the overall XY-ceiling line already reflects information from all possible third variables (both measured and unmeasured). The ceiling pattern is not sensitive to third variables variability within the feasible area. However, the estimated ceiling can still be affected by data limitations such as limited population support, measurement error, and finite-sample sparsity. In other words, the ceiling pattern is robust to distributional changes within the feasible area.

The average effect, on the other hand, represented by the equality Equation (11.1) and estimated by a regression line, is sensitive to third variable variability and must be corrected for their influence in order to make the line interpretable in terms of an average value of Y for a given X. All common statistical terms in the left column of Table 11.3 are defined in a regression context for estimation of the average effect. Note also that a necessity model represented by the ceiling line is deterministic in the sense that it is not parameterized with a stochastic error term with information about other variables.

In regression, a *mediator* is a third variable Z that helps to explain how X influences Y. Specifically, X has an average effect on Z, and Z has an average effect on Y.

In NCA, a related concept is the *necessity chain* as discussed in Section 4.6.4. This represents a sequence of necessity relations: X is necessary for Z, and Z is necessary for Y. Together, these form a necessity chain ($X \rightarrow Z \rightarrow Y$), meaning that X is necessary for Y via Z. However, the necessity of the chain is covered in the overall necessity ceiling. This overall ceiling line is NCA's main interest as it describes the absolute maximum possible Y for a given X, independently of other variables.

In regression, a *moderator* is a third variable Z that changes the strength or direction of the XY-relationship. Specifically, the average effect of X on Y may become larger or smaller depending on the value of the third variable.

In NCA, the corresponding concept is *subgroup ceiling* (conditional ceiling) which may indicate an additional necessary condition. In an XY-plot, all cases of the sample lie below the XY-ceiling line. Cases with a specific value of a third variable Z also remain below this same ceiling line. Therefore, a third variable cannot raise the ceiling line and cannot reduce the effect size of the overall XY-necessity relationship. However, a subset of cases with a specific value of Z may display a lower ceiling line. This occurs if all these cases lie below the overall XY-ceiling. Such a pattern indicates that Z itself may also be a necessary condition for Y. The lower ceiling, however, does not represent the overall XY-ceiling of the population, but only the ceiling of that particular subgroup (thus a different theoretical domain, see Chapters 3 and 7). The population ceiling remains unchanged. "Moderation" in NCA indicates the presence of a subset constraint (multiple necessity).

In regression, a *confounder* is a third variable Z that influences both X and Y. This dual influence can create or distort the apparent relationship between X and Y: $Z \to X$ and $Z \to Y$. When a causal model that includes the confounder is more credible than a causal model without the confounder, the XY-relationship is considered a spurious association. Unmeasured confounders can bias the regression coefficients.[5]

In NCA, the logic of necessity is fundamentally different, and confounding in the regression sense is not a threat in NCA under a positivity/overlap condition. Then confounding redistributes probability within the feasible area (the area below the overall ceiling line) and cannot make the ceiling zone (the area above the overall ceiling line) populated. This means that a causal necessity interpretation of the empty space does not change with or without a confounding variable. Here, the empty space remains empty under the assumption of a good covered sample that is representative of the population: all combinations of variable values that exist in the population can, in principle, appear in the sample. If this assumption is violated, cases might appear in the empty area when new samples are drawn, because the original sample missed certain combinations. Likewise, if the underlying population changes due to structural changes, cases may emerge in what was previously an empty area. These issues reflect sampling or system changes. In other words, confounding can change densities within the feasible region but cannot create overall ceiling violations if the necessity relationship is true.

In regression, *omitted variable bias* can occur when a relevant variable Z that affects both X and Y is left out of the regression model. When this happens, the estimated effect of X on Y becomes biased because part of the variation attributed to X is actually due to Z: Z causes the estimated average effect. Omitted variable bias is therefore a statistical problem that arises from the covariance structure of the data.

In NCA, however, the idea of 'leaving out a variable' does not apply. The XY-ceiling line already reflects information from other variables. Including or excluding other variable's variability does not affect the overall XY-ceiling line. In NCA, a necessity model by definition is a model of a pair of variables (XY), and the necessity of this pair is not biased by analyzing or not analyzing other pairs of variables (e.g., ZY) for necessity, also not if this 'omitted' third variable would be a mediator, moderator or confounder in the regression analysis (see mediation, moderation, and confounding as discussed above).

In regression, a *control variable* is a third variable Z that is included in the model to remove the influence of alternative explanations for the XY-relationship. This helps to avoid omitted variable bias and to isolate the net average effect of X on Y to find the unique contribution of X to Y.

In NCA, however, the XY necessity effect is the overall effect, which is the necessity effect of interest, and it is independent of other variables. The logic of "controlling for" another variable therefore does not apply. There is nothing to control for (see omitted variable bias as discussed above).

[5]This is one of the main reasons why, in a regression context, experiments are often considered the gold standard research design for causal inference. By using randomized experiments, the effects of confounders are mitigated.

In regression, *endogeneity* occurs when X is correlated with the error term of the regression model. This correlation violates a key assumption of ordinary least squares (OLS) regression, namely that X is exogenous (independent of the error term). When X is endogenous, the estimated average effect of X on Y becomes biased. Endogeneity can arise from several sources such as omitted variable bias, reverse causality, and measurement error.

In NCA, endogeneity in the regression sense is no issue. NCA does not estimate residual errors, and thus cannot suffer from bias caused by correlations between predictors and error terms. An error term represents the scatter of observations below the ceiling line, but NCA does not model this: the NCA model does not have an error term as shown in Equation (4.15). Whereas omitted variable bias does not apply to NCA (see above), the two other underlying sources of endogeneity in regression also affect NCA: reverse causality as discussed in Section 5.7 and measurement error as discussed in Section 5.8.

In regression, *spuriousness* refers to a situation in which an alternative causal explanation is clearly more credible. It often involves confounding, where an apparent average effect between X and Y disappears once a confounder Z is included in the regression model.

In NCA, spuriousness can arise when the underlying necessity theory incorrectly specifies the causal structure (inadequate necessity theory, see Section 5.9) and an alternative theory exists where a causal structure with a third variable Z that is sufficient for X and necessary for Y is more convincing, than the original single XY necessity relationship, making the latter theoretically spurious. In the development of a formal necessity theory (Chapter 7) the original and alternative explanations should be evaluated in a thought experiment (Section 7.5.2) to decide which model is theoretically justified. Thus, as a theoretical idea, spuriousness exists in NCA when an observed empty space in the XY-plot is causally interpreted as necessity, whereas an alternative explanation claims that the empty space is the result of a third variable Z that is *sufficient* for X and *necessary* for Y is stronger. For example, empirical support is available (empty upper-left corner in the XY-plot) that supports the (weak) "theory of the stork" (Höfer et al., 2004) claiming that a high number of stork nesting pairs (X) is necessary for a high human birth rate (Y). The alternative explanation that a third variable Z (Availability of high human-made structures, because people settle in a rural region with storks) is *sufficient* for a high number of stork nesting pairs (X) and *necessary* for a high number of human births (Y) is more plausible, making the original hypothesis spurious (Dul, 2016b). With a formal necessity hypothesis, spurious, unrealistic XY necessity relationships can be avoided (Chapter 7).

11.3.4 Combining NCA and regression-based methods

Given their many shared characteristics but distinct causal perspectives, NCA and regression can be combined in a causal-pluralist multimethod study. In such a study the same data are analyzed with both perspectives, although different assumptions are made about the data. For example, whereas NCA assumes bounded data, regression-based methods do not impose bounds such that fitted values may fall outside NCA's bounding box. While regression-based inference often relies on distributional assumptions (e.g., Central Limit Theorem), NCA does not make an assumption about the distribution of the data. When analyzing data with different causal perspectives, the analyst must be aware of the different assumptions that are inherent to the corresponding methods. Rather than integrating the methods, the methods can be applied separately with their own perspectives and assumptions. Only during the interpretation of the results, additional causal lessons can be learned.

Adding NCA to an existing regression study can be relatively easy from the perspective of availability of data. If the data for the regression study are already available, no additional data are needed for conducting NCA. The scores of X and Y from the regression analysis can be used as input for NCA. However, from the perspective of theorizing for the validity of the results, conducting NCA usually requires considerable additional effort. NCA requires a formal necessity hypothesis (see Section 7.5), which is often not available and must be developed before conducting the analysis. While a probabilistic sufficiency hypothesis may already be available, this does not qualify as a necessity hypothesis. Just changing the phrasing of the hypothesis from "likely", "probably", "on average", or similar, into necessity-based phrasing is certainly not enough.

Adding regression analysis to an existing NCA study may be cumbersome from the perspective of data availability. For testing the necessity hypothesis that X is necessary for Y, data only on X and Y are enough. Such data are often not enough for also conducting a regression analysis for testing the hypothesis that X likely has an effect on Y. Due to the risk of omitted variable bias, regression usually requires additional variables to be measured and included in the regression equation.

Figure 11.3 shows a flowchart for conducting NCA in combination with a regression-based analysis in a causal-pluralism multimethod study. The study starts with theorizing the necessity and probabilistic sufficiency hypotheses independently. Note that the theoretical justification of a necessity hypothesis differs from the theoretical justification of a probabilistic sufficiency relationship (Chapter 7).

The next step is collecting data to obtain valid and reliable scores that represent the X's and Y of the hypotheses (see Chapter 8). Subsequently, two separate data analyses are done: the necessity analysis is conducted with NCA, while the probabilistic sufficiency analysis is done with the selected regression-based analysis (e.g., MLR, SEM). The regression results are often summarized in a regression table that reports the regression coefficients or path coefficients together with their p-values. Additional information is usually included as well, such as standard errors, confidence intervals, and model fit statistics, and a note may add additional information such as sample size. The left side of Table 11.4 shows a minimum example of such a table with artificial data. In this table the NCA results are added to obtain the NCA-Extended Regression Table (NERT). It expands the conventional regression table by adding two columns: the necessity effect size (d) and its p-value, for those predictors that are hypothesized to operate as necessary conditions. A blank line in this table indicates that no theoretical rationale is available, such that the NCA analysis is not conducted for this condition. Additional NCA method information that applies to all tested necessity hypotheses may be added in the note to the table. This applies, for example, to the selected ceiling line for estimating the effect size.

A causal-pluralism multimethod study with NCA and a regression-based method can provide complementary information with the following possible results (Table 11.5).

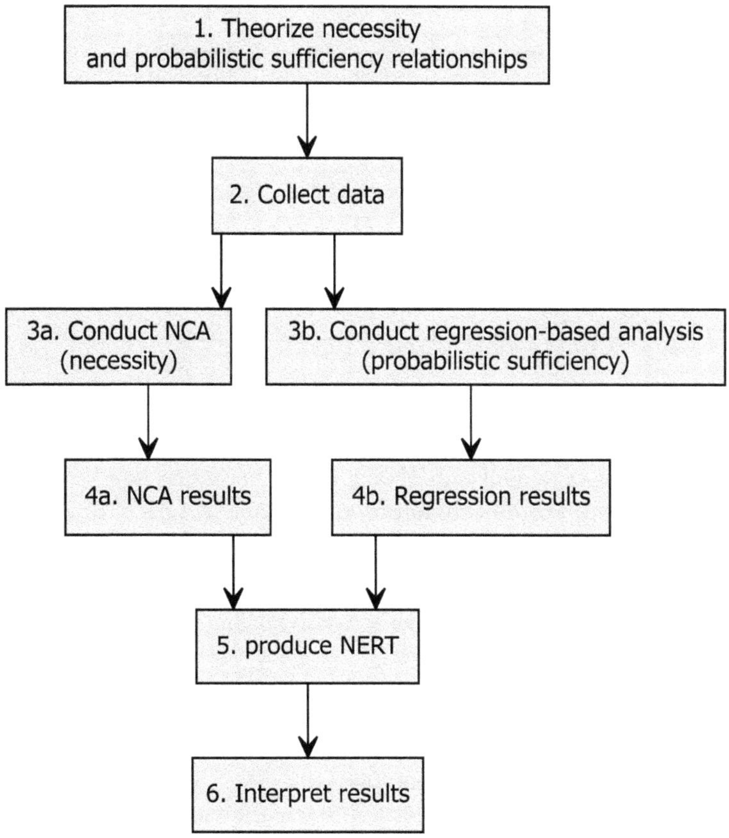

FIGURE 11.3 Flowchart for conducting a causal pluralism multimethod study with NCA and a regression-based method. NERT = NCA-Extended Regression Table.

TABLE 11.4 NERT tool for combining NCA results with regression-based results in a causal-pluralism multimethod study. NERT = NCA-Extended Regression Table. The extension consists of adding two columns to the regression table: the necessity effect size and its *p*-value.

	Regression coefficient	*p*-value	Necessity effect size	*p*-value
Variables/Conditions				
X_1	0.23	0.035	0.20	< 0.001
X_2	0.12	< 0.001	0.00	1.000
X_3	0.43	0.022		
X_4	0.09	0.087	0.03	0.035
X_5	0.18	0.361	0.12	0.028
Model fit (regression)				
$R^2 = 0.28$				

Note:
n = 132. NCA ceiling line: CR-FDH.

TABLE 11.5 Possible outcomes of a causal-pluralism multimethod study with NCA and a regression-based method.

	Probabilistic sufficiency: If X, then probably Y	Necessity: If not X, then not Y
Option 1	Yes	Yes
Option 2	Yes	No
Option 3	No	Yes
Option 4	No	No

- Option 1: X has an average effect on Y and is also necessary for Y. X is not only important but also required for having the outcome (See X_1 in Table 11.4).
- Option 2: X has an average effect on Y but is not necessary for Y. X is important but not essential. Its absence can be compensated by other factors (See X_2 in Table 11.4).
- Option 3: X has no average effect on Y, but is necessary for Y. It may be that X is overlooked in regression analysis as an important factor (possibly due to the existence of many cases with high level of X and low level of Y), but is still an essential factor. (See X_5 in Table 11.4).
- Option 4: X has no average effect on Y and is also not necessary for Y (See X_4 in Table 11.4).

Note that due to the different causal perspectives, one methodological approach cannot logically be a robustness check of the other approach. The results can only be different, not contradictory.

11.4 Combining NCA and SEM in a multimethod study

One of the most common causal-pluralism multimethod applications of combining NCA with a regression-based method is the use of NCA in the context of structural equation modeling (see Appendices C.3 and C.4). Since in many such published applications NCA is not applied according to the standards proposed in this book and elsewhere, this section (as well as Appendix H) provides more details and recommendations on how to conduct such a combined analysis. A SEM-model consists of two parts. The *measurement model* defines how specified indicators relate to the underlying constructs ("latent variables"). It determines how the constructs are operationalized by their indicators and how measurement error is represented. The *structural model* specifies the relationships among the constructs (latent variables). This is the regression-based part of the model.

Different estimation approaches exist for estimating the relationship between the latent variables ("path coefficients"). Conventional covariance-based SEM (CB-SEM, Jöreskog, 1970) estimates model parameters based on the covariance matrix, using techniques such as maximum likelihood. Component-based SEM, for example Partial Least Squares SEM (PLS-SEM, Wold, 1985) estimates latent variable scores as weighted composites of indicators to maximize the explained variance in the dependent variables. A more recent version of PLS is Consistent Partial Least Squares structural equation modeling (PLSc, Dijkstra & Henseler, 2015) that corrects the attenuation bias of traditional PLS estimates when reflective (common-factor) measurement models are used.

If the SEM[6] model is estimated, latent variable scores can be obtained for further analysis, including for use of X- and Y-values as input to NCA. In PLS, these scores are directly estimated as part of the model fitting process as component-based SEM-methods are designed to estimate composites, whereas in CB-SEM the analyst must select (based on the analyst's goals, e.g., unbiasedness, orthogonality, correlation preservation) a factor score estimation method to compute estimated scores for latent factors from observed variables after fitting the model. Thus, depending on the SEM modeling approach that is selected by the analyst, the input scores for NCA and thus the NCA results may differ.

NCA has been applied both in combination with CB-SEM (see Table C.3 in Appendix C.3) and in combination with PLS-SEM (see Table C.4 in Appendix C.4). Using PLS-SEM with NCA is popular, possibly for the following reasons:

1. It is often easier to extract latent variables scores from the estimated PLS models than from estimated CB-SEM models.
2. Leading PLS-SEM proponents introduced NCA as a methodological enrichment of PLS-SEM, and provided specific recommendations (Richter, Schubring, et al., 2020; Richter et al., 2022; Richter, Hauff, Ringle, et al., 2023) and extensions (Hauff et al., 2024; Sarstedt et al., 2024) for combining the methods.
3. A basic version of the NCA software became part of a popular software package originally developed for conducting PLS-SEM (SmartPLS, Section B.4 in the Appendix).

Combining NCA with SEM involves using estimated scores of the latent variables as input for NCA. This allows the analyst to examine whether a relationship between two latent variables is not only related in terms of probabilistic sufficiency (as modeled in SEM) but also in terms of necessity, assuming a theoretical rationale exists. If a formal necessity hypothesis is available or can be derived (see Chapter 7), any pair of latent variables from the structural model can take the role of condition and outcome to study their necessity relationship. When conducting NCA in combination with SEM, all general guidelines for conducting NCA apply (Section 10.3).

The availability of a user-friendly software package for applying NCA within the context of PLS-SEM is an advantage, but it also carries a risk. While the SEM part of the study is often a deductive study for testing a probabilistic sufficiency hypothesis, many published multimethod studies combining NCA and PLS-SEM conduct the NCA part as an exploratory investigation without having a necessity hypothesis. In the absence of a theoretical justification for necessity, such a study cannot claim that necessity has been identified (Section 6.2). This also applies if the theoretical justification is provided after the results are known. An exploratory study does not qualify for a hypothesis testing study (Chapter 7). Having a theoretical rationale before data collection is a prerequisite for making a claim of empirical evidence of necessity (see Section 9.10). After an exploration, a further study is needed to test the hypothesis that was derived from the exploration (Section 7.2).

[6]Whenever the difference between CB-SEM, PLS-SEM and PLSc is not relevant, the general term SEM is used.

It is also not uncommon, in particular in studies that combine CB-SEM with NCA, that the latent variable scores of the SEM-part of the study are *not* used as input to NCA. This happens, for example, when instead of using latent variable scores derived from the SEM study, scores independent from SEM (e.g., means of indicator scores) are used as the latent scores input for NCA. Such a combined NCA-SEM study would qualify as two separate studies according to Table 11.2, as X and Y are different in the two studies.

11.4.1 Steps for conducting NCA with SEM

To outline the steps for an NCA-SEM multimethod study, first some SEM terminology is explained. Since SEM terminology about 'indicators', 'weights', 'latent variables' and 'rescaling' varies across the SEM literature, the terminology described in Table 11.6 is used in this book.

TABLE 11.6 SEM terminology

Terms	Description
General	
Rescaling	Normalization or standardization of original scores.
Normalization (or min-max normalization)	Rescaling of scales with new minimum and maximum values (e.g., 0–1 or 0–100).
Standardization	Rescaling of observed scale values based on their probability distribution. (z-score: mean = 0 and standard deviation = 1)
Indicator	
Indicator (variable)	A manifest/measured variable linked to a latent variable.
Indicator score	The value of an indicator.
Raw indicator score	The value of the indicator measured on its original scale. The scale usually has minimum and maximum values, e.g., Likert scale ranging from 1–5 or 1–7.
Normalized indicator score	The value of an indicator after min-max normalization of the raw indicator score (e.g., 0–1 scale or percentage scale 0–100).
Standardized indicator score	The value of an indicator after standardization of the raw indicator score (z-score).
Original indicator score	The indicator that is used as input for the SEM model, which can be raw or standardized depending on the estimation settings.
Weight and loading	
(Indicator) weight	In PLS, the extent to which an indicator contributes to the latent variable score (as part of a weighted composite).
Standardized indicator weight	Indicator weight using standardized indicators.
Estimated indicator weight	In PLS, the estimated coefficient that multiplies an indicator to construct the latent variable score.
Factor loading	The strength of an indicator–latent variable relationship.
Latent variable	
Latent variable	An unobserved construct linked to multiple indicators.

Latent variable score	The value of a latent variable, calculated as a linear combination (weighted sum) of indicator scores.
Standardized latent variable score	A weighted linear combination of standardized indicator scores (z-scores). In PLS, these scores are computed directly during model estimation, whereas in CB-SEM they can be derived afterward as estimated factor scores using the model's factor loadings.
Unstandardized latent variable score	Score obtained as a weighted linear combination of the original (unstandardized) indicator values. In PLS such scores are computed directly; in CB-SEM these scores can be derived post-estimation as factor scores.
Normalized unstandardized latent variable score	Unstandardized latent variable score after normalization.

Figure 11.4 shows a flowchart for conducting NCA in combination with SEM.

Eight steps are distinguished, and each step is discussed in detail. An example is provided in Appendix H.

11.4.1.1 Step 1: Theorize necessity and probabilistic sufficiency

The first step is a fundamental step for any combined NCA and SEM study. An NCA-SEM study implies that the relationships between the variables are considered from two different causal perspectives: necessity (NCA) and probabilistic sufficiency (SEM), see Section 2.7. For inferring causality, both NCA and SEM require that the study starts with formulating theoretical expectations about the causal relationship (e.g., hypotheses). According to SEM, all relationships between latent variables in the structural model are considered probabilistic sufficiency relationships that are theoretically justified. According to NCA, all, a few, or none of these relationships may be hypothesized as necessity relationships as well, depending on the availability of a theoretical justification in the format of a formal necessity hypothesis (Section 7.5). A theory that includes one or more necessity relationships in addition to probabilistic sufficiency relationships is called an 'embedded necessity theory' (Bokrantz & Dul, 2023).

Latent variables of a structural model can have different roles. It can be a *predictor variable* (independent variable; condition in NCA) that has an effect on a *predicted variable* (dependent variable; outcome in NCA). For example, when in a structural model a latent variable M mediates the relationship between X and Y, then M has two different roles. For the X-M relationship X is the predictor and M is the predicted variable, and for the M-Y relationship M is the predictor and Y the predicted variable. When a relationship is (also) considered a necessity relationship, NCA calls the predictor variable the 'condition' and the predicted variable the 'outcome'.

11.4.1.2 Step 2: Collect indicator data

Conducting SEM assumes that data are available on observed indicators. SEM involves specifying a measurement model, which defines the relationships between indicators and latent variables. In step 2 the indicator data are collected. These data are the input for the measurement model.[7]

[7] Note that it is also possible to hypothesize and test the necessity of an indicator of the latent construct.

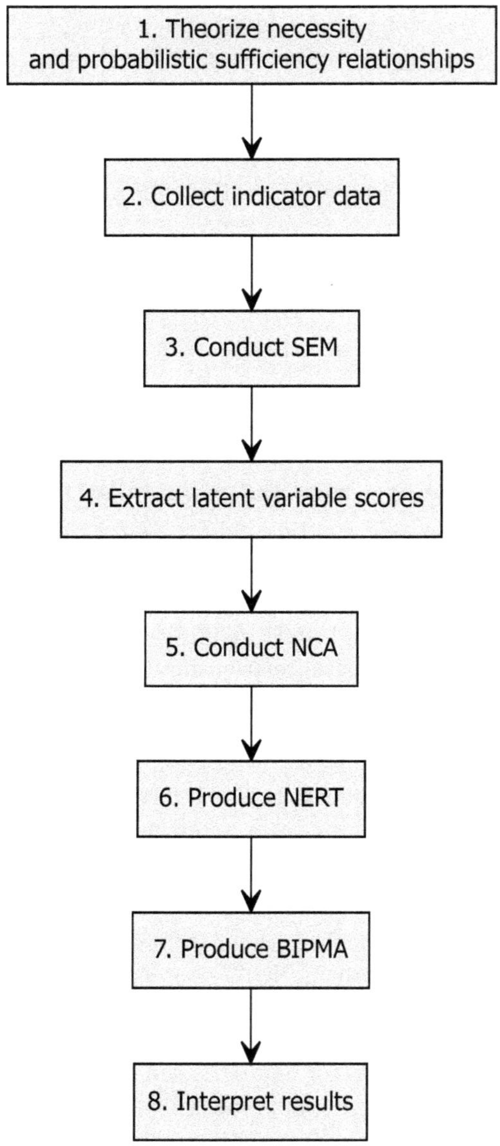

FIGURE 11.4 Flowchart for conducting NCA in combination with structural equation modeling (SEM). NERT = NCA-Extended Regression Table tool. BIPMA = Bottleneck Importance Performance Map Analysis tool.

11.4.1.3 Step 3: Conduct SEM

The structural model specifies the relationships between latent variables. The complete model is then estimated using a chosen algorithm (e.g., PLS or CB-SEM). NCA is not part of the SEM estimation process itself, but it uses latent variable scores obtained from SEM as input for its own analysis. Therefore, it is important that the analyst ensures that the latent variable scores extracted from the SEM model are valid, reliable, and interpretable. Recommendations for good SEM practices are available in the literature and are not discussed here. Both commercial and open-source software tools are available to conduct SEM.

11.4.1.4 Step 4: Extract latent variable scores

The latent variable scores obtained from the SEM model serve as input for NCA. During SEM estimation, latent variable scores are often standardized (z-scores), resulting in scores with a mean of 0 and standard deviation of 1. Although NCA results are invariant to affine transformations (such as standardization or linear rescaling; Section 8.5.2), it is recommended for purposes of interpretation of a necessary condition to first unstandardize standardized latent variable scores. Unstandardized latent variable scores are expressed on a scale that is directly related to the original indicator scale, provided that the same measurement scale was used for all indicators of the latent construct. This allows for more meaningful interpretation of the NCA results, especially when discussing real-world implications and when results are discussed in terms of necessity-in-degree with specified levels of X and Y. Note that not all SEM software packages offer the option to extract unstandardized latent variable scores. Appendix H gives an example showing how to convert standardized scores back to unstandardized scores when needed.

When all unstandardized latent variable scores are derived from indicators measured on the same scale (i.e., with identical minimum and maximum values), min-max-normalization is not required. However, if latent variables have different scales, min-max normalization is recommended. This ensures that the latent variable scores are comparable, which is particularly important in later steps (e.g., Step 7: producing a BIPMA).

A common approach to min-max normalization is to rescale scores to a 0–100 range, allowing latent variable scores to be interpreted as percentages of their original scale range. Alternatively, rescaling to a 0–1 range can be used, which is often more convenient for NCA, as it yields a scope of 1. However, this is optional and not required for the validity of NCA results.

11.4.1.5 Step 5: Conduct NCA

The unstandardized and possibly min-max normalized latent variable scores are input to NCA. In the first part of the analysis, the hypothesized relationships are tested for necessity-in-kind, which is the qualitative statement that X is necessary for Y without specifying levels of X and Y other than absence/presence or low/high. The observed effect size d and the p-value are compared with their threshold values (e.g., $d = 0.10$; $p = 0.05$) that are set by the analyst to decide if a predictor variable is necessary for a predicted variable, as suggested by the hypothesis.

In the second part, identified necessary conditions in kind are selected for a bottleneck analysis for necessity-in-degree, which results in a quantitative statement that level x of X is necessary for level y of Y. A specific format of this table, with the outcome values expressed as actual values or percentiles, and the conditions expressed as percentiles, provides information for each latent variable about the percentage of cases that are bottlenecks.

A bottleneck case is a case that does not have the *required* level of the condition for a *target* level of the outcome.[8]

11.4.1.6 Step 6: Produce NERT

The NCA-Extended Regression Table (NERT) summarizes the main SEM results in terms of direct effects (path coefficients), total effects, and their p-values and the main NCA results in terms of necessity effect sizes and their p-values. Note again, that NCA is not a robustness check of SEM and vice versa. The two approaches provide different insights due to different causal perspectives.

11.4.1.7 Step 7: Produce BIPMA

Whereas NERT presents the results of both NCA and SEM, the Bottleneck Importance Performance Map Analysis (BIPMA) builds on that by providing a tool to assist practitioners to prioritize their actions and resources based on these results. Evidence-based action means that the action is informed by the results of an empirical study, in particular based on hypotheses that were tested and supported (see the NERT). A SEM-hypothesis is a causal claim about average effects: if X then probably Y. The hypothesis is usually considered supported if the criteria for statistical significance are satisfied. An NCA-hypothesis is a causal claim about any case: if not X then not Y. The hypothesis is considered supported if the criteria for statistical significance *and* for practical significance are satisfied (assuming that a formal necessity hypothesis is available). The NERT reports the effect sizes (total effect for SEM; necessity effect size for NCA) and their significance (p-values). Only *supported* hypotheses are considered in BIPMA for evidence-based action.

Taking evidence-based action based only on SEM-results (assuming a supported probabilistic sufficiency hypothesis) means that a predictor variable that has a significant average effect on the predicted variable is a candidate for being changed. The action changes the predictor in order to change the average value of the predicted variable. According to a basic regression assumption, during the change of this predictor all other predictors remain constant (*ceteris paribus*). This ("all else equal") assumption is the idea in regression analysis that a coefficient represents the effect of one predictor while holding the other included predictors constant. This implies that only one predictor can be selected for prioritizing an action based on SEM. Selection of multiple predictors and taking action on them simultaneously violates the *ceteris paribus* requirement. The only evidence-based way to select multiple predictors is through a sequential prioritization, in which a new SEM analysis is conducted after each (hypothetical) change has been made.

A SEM-based action addresses an entire group of cases since SEM results apply to an entire group by definition (average effect). The action is considered most effective if a predictor is "important" and has low "performance." Importance refers to the strength of the probabilistic sufficiency (average effect) relationship of the predictor variable on the predicted variable. Its metric is the total average effect on the outcome including direct and indirect effects. Performance refers to the mean normalized score of the predictor compared to other predictors. A low performance score means that the predictor has room for improvement. In PLS-SEM, the Importance Performance Map Analysis (IPMA) (Ringle & Sarstedt, 2016) is often used for prioritization. In this map Importance is displayed on the horizontal axis and Performance on the vertical axis. The practical meaning of IPMA is that action priority should be given to a predictor variable with a high level of Importance, and a low level of Performance (points in the bottom-right corner assuming positive total effects and average scores).

[8] Chapter 12 discusses the importance of knowing bottleneck cases for practical action.

Taking evidence-based action based only on NCA results (assuming supported necessity hypotheses) implies that a predictor variable is changed such that a selected *target outcome* of the predicted variable is *possible* (though not guaranteed). No assumption is needed about other predictors, including no requirement that these must remain constant. This implies that selection of multiple predictors and taking evidence-based action on them simultaneously is possible. Furthermore, an NCA-based action can address a single case, a small group of cases or an entire group of cases since NCA results apply to all cases. The goal of the action is to ensure that bottleneck cases are resolved: a bottleneck case is a case that does not meet the target condition level that is necessary for the selected target outcome level. An NCA-based action is most effective if only cases are selected that are below the required target condition level (for a detailed discussion see Chapter 12).

In an extension of IPMA, Hauff et al. (2024) introduced the combined Importance Performance Map Analysis (cIPMA) which combines results of NCA and SEM. They add the third dimension that is called here the 'bottleneck' dimension. This dimension takes into account how effective the action is from the perspective of the presence of bottleneck cases according to NCA. This means that for cIPMA a *target outcome* must be selected and that an additional action goal is that the target outcome set by the analyst must become achievable by as many as possible cases. If the target outcome is set at a low level, the goal is to enable a base level of the outcome (e.g., mean or even lower outcome, for example corresponding to a minimum level that is achieved by 50 or 75 percent of the cases). If the target outcome is set at a high level, the goal is to enable high level of the outcome (e.g., corresponding to a minimum level that is achieved by only about 25 or 10 percent of the cases).

NCA's bottleneck table provides the required information for an NCA-based action in the context of cIPMA. This table is produced in Step 5 of a combined NCA-SEM study (see Figure 11.4). After establishing the target outcome, the bottleneck table provides the corresponding target condition for each necessary condition. The version of the table where conditions are expressed as percentiles provides information about the percentage and the number of cases that are bottleneck cases for the conditions. These cases have not achieved the target condition that is necessary for the target outcome.

The cIPMA plot integrates three dimensions: Importance on the horizontal axis, Performance on the vertical axis, and Bottlenecks represented by the size of the points in the Importance-Performance plot. Each point represents a predictor variable with a given combination of Importance and Performance scores. cIPMA suggests that the best choice is the predictor with the largest Importance score and lowest Performance score (according to SEM), and the largest number of bottleneck cases (according to NCA's bottleneck table).

When prioritization is based on the combined results from NCA and SEM, only one predictor variable can be selected for change if this change is (also) based on SEM results (because the *ceteris paribus* assumption of SEM applies). However, it is possible that a bottleneck case is not only a bottleneck for the selected predictor, but also for another predictor. Solving a bottleneck for one predictor (by changing the level of that predictor such that it meets the target condition level) is only effective if the case is not also a bottleneck case for another predictor. A case with multiple bottlenecks (bottlenecks for multiple conditions) remains a bottleneck case after resolving the bottleneck for the selected predictor, because the bottlenecks for the other predictors remain unresolved.

Figure 11.5 shows an example of a Bottleneck Importance Performance Map Analysis (BIPMA).

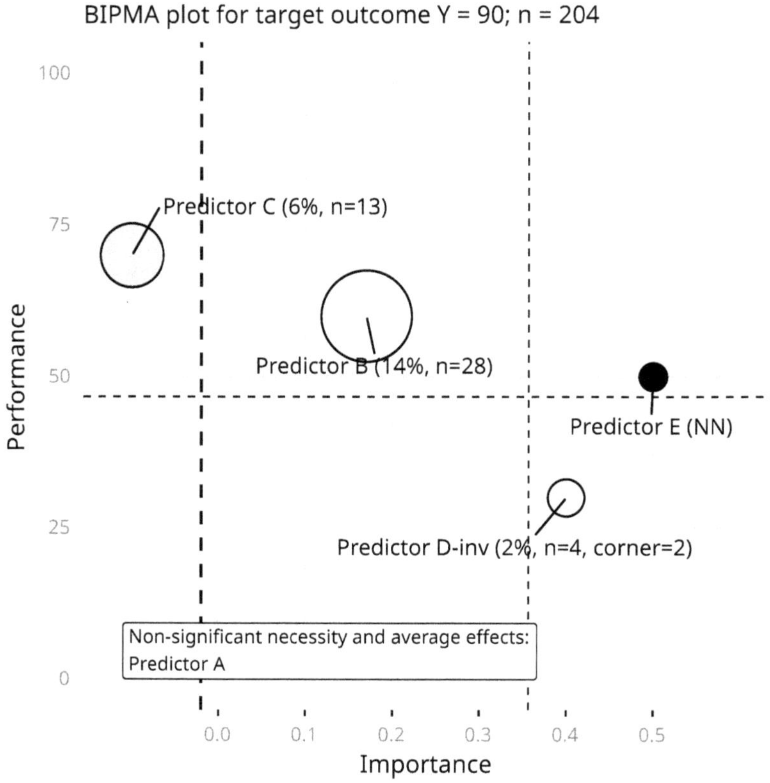

FIGURE 11.5 Example of a Bottleneck Importance Performance Map Analysis (BIPMA) for evidence-based practical advice (prioritizing predictors) based on NCA and SEM results. The size of the dot is an indication of the number of *single* bottleneck cases that cannot achieve the target outcome (here 90% of the maximum of the outcome). White points: predictor has a significant necessity effect according to NCA and a significant total average effect according to SEM. Black points: predictor has a non-significant necessity effect and a significant total average effect. Gray points: predictor has a significant necessity effect and a non-significant total average effect. The horizontal and vertical thin dashed lines correspond to mean Performance and the mean Importance of the predictors with a significant total average effect, respectively.

BIPMA differs from cIPMA regarding the bottleneck cases to be considered during prioritization. BIPMA only considers *single* bottleneck cases. Single bottleneck cases are a bottleneck in only one predictor. These cases are relevant for taking action based on combined results of NCA and SEM. While cIPMA defines the bottleneck dimension for a predictor as the percentage or number of *single and multiple* bottleneck cases, BIPMA defines it as the percentage or number of *single* bottleneck cases, only to avoid ineffective actions.

Three additional adaptations compared to cIPMA are implemented in BIPMA. To avoid the risk of false positive conclusions, as a general rule BIPMA only includes predictors to be considered for prioritization when the hypothesis is supported. Although this is a judgment of the analyst, usually an analyst rejects the hypothesis when the p-value is large (e.g., $p \geq 0.05$), suggesting that the observed effect may be compatible with "no effect."[9] However, cIPMA is asymmetric regarding this statistical significance criterion as it adopts it only for NCA results. cIPMA (and IPMA) allow analysts to consider non-significant SEM results for practical action.[10] BIPMA harmonizes this principle by not only excluding non-significant NCA results for evidence-based practical action, but also non-significant SEM results.[11] Since BIPMA is a tool for practical action and not a tool for reporting results of the two analyses, the BIPMA should always be accompanied with the NERT tool (or similar) to report all observed effect sizes and their significance levels.

The second additional adaptation is that BIPMA deals with different relationship directions between predictor variable and predicted variable. The common situation in SEM is that the predictor variable has a positive (average) effect on the predicted variable. BIPMA also handles predictor variables with hypothesized and observed negative total average effects on the predicted variable. For example, Mwesiumo (in press) added a negative total effect for one of the predictor variables on the negative side of cIPMA's Importance-axis. Since this is inconsistent with the advice (Hauff et al., 2024) that priority should be to points in the bottom-right corner of cIPMA, BIPMA explicitly uses the *absolute* total effect as Importance score, such that importance scores of positive total effects and negative total effects are comparable and priority should be given to points in the bottom-right corner independent of the sign of the total effect. The BIPMA plot identifies an Importance score that is based on a negative total effect by adding '-inv' to the name of the predictor, indicating that this predictor has an inverse effect on the predicted variable. This is shown in Figure 11.5 for predictor D. While an *increase* of the predictor variable with a positive total effect will increase the predicted variable, a *decrease* of the predictor variable with a negative total effect will increase the predicted variable.

A third additional adaptation is that BIPMA also handles different necessity directions. For NCA the default necessity direction is high-high, meaning that a high level of X is necessary for a high level of Y. The other directions are low-high, high-low, or low-low, suggesting that low X is necessary for high Y, high X is necessary for low Y, or low X is necessary for low Y, respectively. This implies that the expected empty corner of the XY-plot is not upper-left, but that another corner of the XY-plot is expected to be empty and should be analyzed. For example, an expected and observed empty corner is the upper-right corner of the XY-plot (low X is necessary for high Y), which implies that for resolving bottleneck cases the predicted variable must decrease, rather than increase. If the corner deviates from corner = 1, the corner is specified in the BIPMA plot in the predictor label. This is shown in Figure 11.5 for predictor D. Predictor D has a negative average effect and an empty upper-right corner in the XY-plot.

BIPMA aims to inform practice as closely as possible in line with the evidence of empirical results both in terms of effect size and in terms of statistical significance. Using points with different shades of gray, BIPMA displays the four possible combinations of statistical (non-)significance of NCA and SEM results (Figure 11.5):

[9] Note that NCA not only rejects a necessity hypothesis if the p-value is large but also if the effect size is small. SEM usually rejects a probabilistic sufficiency hypothesis if the p-value is large (or similarly a zero effect is part of the confidence interval).

[10] The argument is: "While a non-significant effect provides evidence that a total effect is zero in the population, analysts should retain the corresponding construct in the IPMA since this outcome may also represent a valuable finding (e.g., a company invests into the performance of a construct that has no effect), which also can change with different data, for instance, in alternative contexts of the analysis" (Ringle & Sarstedt, 2016, p. 1872). The suggestion to include non-significant SEM results in cIPMA is adopted in Hauff et al. (2024) and Sarstedt et al. (2024).

[11] Harmonizing the approaches by allowing also non-significant NCA results introduces a risk of false positives and ineffective action.

In BIPMA conditions are displayed as points with different sizes and shades of gray:

- The size of the point reflects the number of *single* bottleneck cases for the specific condition. The percentage and number of single bottleneck cases is mentioned near the point.
- White point: the total average effect (Importance) and the necessity effect are both significant. The size of the point reflects the number of *single* bottleneck cases.
- Black point: the total average effect is significant and the necessity effect is non-significant. The black point has a standard size.
- Gray point: the total average effect is non-significant and the necessity effect is significant. The size of the point reflects the number of *single* bottleneck cases. When according to SEM the average effect of the condition is statistically non-significant only the vertical position (Performance) contains relevant information for action. No evidence-based information is available about the total average effect (Importance). For this reason the point is moved to the left outside the horizontal axis, indicating absence of information.[12]
- If the total effect is non-significant (statistically) and the necessity effect is non-significant (statistically or practically) the point is not shown. There is no evidence for practical action. The name of such a non-significant predictor is mentioned at the bottom of the plot.

The goal of BIPMA is to identify predictor variables for action using results from NCA and SEM. Several prioritization approaches and corresponding ways to evaluate the BIPMA plot can be used to combine these criteria. Assuming a positive direction for the average effect and a high-high direction for the necessity effect, these *strategies* can be formulated as follows:

1. Enable a target outcome. This approach only considers NCA results. It is feasible when SEM results do not give a clear preference for predictors to be changed (e.g., no points in the bottom-right corner of BIPMA). Multiple predictor variables can be selected for an action based on the NCA results. This approach implies that the goal of the intervention is to enable a high target score by ensuring that none of the conditions have bottleneck cases. For example, if the target outcome is desirable and X is necessary for high Y, the target can be set at a high level that is achieved by not more than 25% or 10% of the cases (the level corresponding to the best performing cases). The intervention then consists of ensuring that the bottleneck cases are resolved by increasing the values of the conditions with bottleneck cases. The approach is also feasible when the goal is to enable a certain minimum level of the outcome for all cases. The target outcome can for example be set at a minimum level that is achieved by more than 50% or 75% of the cases.
2. Improve the average outcome. This approach only considers the SEM results. It is feasible when NCA results show that none of the conditions are necessary or there are only a small number of bottleneck cases. Only one predictor variable can be selected for action at once given the *ceteris paribus* requirement of regression analysis. For selecting a second predictor, evidence-based action requires a new SEM and BIPMA with a new dataset representing the situation after the intervention on the first predictor.

[12] A non-significant total average effect is statistically compatible with the null hypothesis of no total effect. In reality the true effect could be positive, negative or absent. In this situation only the size of the point (necessity) is informative.

3. Improve the average outcome and enable a target outcome. This approach considers both the NCA and SEM results. Since SEM is part of the consideration, this approach requires that only one predictor is selected for evidence-based action. The combined approach is most effective for a predictor with a large number of bottleneck cases in the bottom-right corner of the BIPMA plot.

Since these stylized situations may not exist in practice, hybrid approaches, for example different approaches for different predictors may be considered.

SEM results are average effect results for the entire group of cases and the goal is to increase the average performance. NCA results allow also custom-made approaches for individual cases or a selection of cases. The case or group of cases gets a specific intervention depending on the case's bottleneck situation. An individualized NCA-based intervention can considerably improve the intervention effectiveness and efficiency. This is discussed in Chapter 12.

11.4.1.8 Step 8: Interpret results

In the final step the results of the two approaches are interpreted using NERT and BIPMA. It is possible to conduct robustness checks to explore the sensitivity of choices made by the analyst on key results of NCA and SEM. Commonly recommended robustness checks for SEM do not apply to NCA. The NCA-related robustness checks in the context of an NCA-SEM deal with the following two NCA key results that may be affected:

- Identification of a predictor from the SEM structural model as a potential necessary condition.
- Prioritization of predictors for intervention.

As discussed in Section 9.12, these key results may be affected by the analyst's choices of:

- Ceiling line.
- Threshold level of the effect size.
- Threshold level of the p-value.
- Scope.
- Outlier handling.
- Target outcome.

A robustness check consists of making an alternative (reasonable) choice rather than the original choice, and studying its influence on the key results. The output of the robustness check is the NCA robustness table. If the influence of changes of the analyst's choices is small, the original results may be considered robust; if not, the original results may be fragile.

11.4.2 Recommendations for combining NCA and SEM

In publications where NCA is applied in combination with SEM, usually SEM is the main analysis and NCA is used as an additional analysis. However, several guidelines for applying NCA are often violated. This applies, for example, to the following recommendations with 'must-have' priority level (see the checklist in Section 10.3):

- Checklist item 4: *When, in the context of an NCA study, a comparison is made between NCA's causal perspective and the causal perspective inferred from conventional regression-based/statistical approaches, use the term probabilistic sufficiency for the latter, and not just sufficiency. The term sufficiency suggests determinism (as applied in QCA's configurational sufficiency logic). This recommendation is often violated in studies that combine NCA with SEM [...].*
  See this book: Sections 2.3, 2.4.
  Suggested articles: Dul (2024a) [...]
  The results of SEM's regression analyses (path coefficients of the structural model) are often interpreted as 'sufficiency', whereas sufficiency (like necessity) refers to deterministic causal logic and not to the probabilistic causal logic. A better description of the presumed causal relationships of the structural model (like any relationship that is analyzed with regression analysis) would be 'probabilistic sufficiency'.

- Checklist item 5. *In multimethod studies, use and refer to NCA as an approach with a specific perspective on causality and data analysis, and do not incorrectly refer to it as a robustness check or add-on to other methods (and vice versa). This recommendation is often violated in studies that combine NCA with SEM.*
  See this book: Sections 2.7, 11.4, 11.5.
  Suggested article: Dul (2024a)
  In studies that combine NCA with SEM, NCA is often suggested as a robustness check of SEM. However, this is incorrect. A good practice is to use NCA as a methodological approach on its own with a different causal perspective, such that the two approaches can give complementary insights.

- Checklist item 7: *Formulate the necessity relationship explicitly as a necessary condition hypothesis: X is necessary for Y. For a deductive study this is done before data analysis; for an explorative study this is done after data analysis when a necessity relationship is found.*
  See this book: Sections 7.5.
  In studies that combine NCA with SEM, well-founded probabilistic sufficiency hypotheses are often formulated (to be tested with the SEM model). However, necessity hypotheses are often not provided. Formulating necessity hypotheses is a requirement since the necessity causal perspective differs from the probabilistic sufficiency causal perspective. This applies also to studies that use NCA for exploration or as an additional analysis. A good practice is to formulate necessity hypotheses prior to conducting the analysis, just as is done with the probabilistic sufficiency hypotheses. However, it is estimated that fewer than 25% of NCA–SEM publications include a necessity hypothesis, and a probabilistic sufficiency hypothesis cannot compensate for this absence.

- Checklist item 9: *Explain why X is necessary for Y using a causal explanation (narrative) by answering three questions: (1) Why does X enable Y (X is an enabler). The existing literature may be a source of information; (2) Why does the absence of X lead to the absence of Y (X is a constraint); (3) Why is the absence of X not compensable (no substitution for X). Making the narrative complete is a creative process, supported by literature and experiences of scholars and practitioners.*
  See this book: Section 7.5.3.
  This often violated recommendation is related to the previous one. Even if a necessity hypothesis is formulated, it should also be explained **why** it is expected that X is necessary for Y (the causal explanation). Less than half of the few NCA - SEM publications with a necessity hypothesis provide a causal explanation why X would be necessary for Y. In a theory-testing study, a necessity hypothesis for NCA should get the same attention and

depth regarding the causal explanation as a probabilistic sufficiency hypothesis for SEM, as for example is done in a study by Battistoni et al. (2023).

- Checklist item 19: *Explain whether and how the data were transformed, and whether the transformation aligns with NCA. Affine transformation (including linear transformation) produces valid results. NCA's data analysis does not require data transformation and non-transformed data are often more meaningful and better interpretable (e.g., levels of a Likert scale) than transformed data. This particularly applies to necessity-in-degree and interpretations of specific levels of X being necessary for specific levels of Y. Refrain from conducting non-linear data transformation (e.g., log-transformation, logistic transformation) unless the transformed data represent X and Y as defined in the necessity hypothesis (e.g., log-transformed GDP). [...].*
See this book: Sections 8.5.2, Appendix D.
In most applications of SEM, latent variable scores are standardized (centering the data around zero and scaling the data to have a standard deviation of one). For proper interpretation of necessity-in-degree (bottleneck table) the use of unstandardized scores may be more meaningful. When the indicators are measured on similar scales like Likert scales, unstandardized scores are more closely related to the original measurement scale than standardized scores, which makes interpretations of levels more meaningful.

- Checklist item 20: *Specify and justify the selected ceiling line by considering theoretical expectations (border is linear or non-linear), type of data (discrete or continuous), observed border pattern (regular or irregular), balance between accuracy and generalizability (avoiding overfitting), and quality of the data (data with/without measurement error, simulation data).*
See this book: Sections 4.3.2, 9.3.2. Often no justification is given for the selection of the ceiling line. The CE-FDH ceiling line is commonly selected without further explanation, possibly because this line was used for good reasons in the original publication that introduced NCA in the SEM context (Richter, Schubring, et al., 2020). However, this reason might not apply in other situations. Similarly, when the two default ceiling lines (CE-FDH and CR-FDH) are used, no explanation is usually given for this selection.

- Checklist item 29: *Include the XY-tables or XY-plots of all tested/explored relationships, preferably in the main text. It is essential that readers are able to inspect them visually.*
See this book: Sections 9.5.
Often, not all XY-plots of tested necessity relationships are presented. In some cases, only the plots of the supported relationships are provided. In others, XY-plots are not provided at all.

- Checklist item 32: *Conclude about necessity-in-kind using three criteria: (1) Theoretical justification (the formal hypothesis), (2) Effect size not below the selected threshold value, (3) p-value below the selected threshold value. Note that for support all criteria apply and that missing just one of them is enough for a rejection. This implies, for example, that without theoretical justification the claim for a necessity relationship is rejected.*
See this book: Sections 9.10.
In many NCA-SEM studies not all three criteria for concluding about necessity (theoretical support, large effect size, small p-value) are taken into consideration. Often the requirement of theoretical support is lacking (see items 7 and 9 above). Usually, only theoretical support for the probabilistic sufficiency relationships of the structural model is provided, but this is not informative about the necessity relationship. Without theoretical support for a necessity relationship, it is invalid to conclude that a necessity relationship was identified, even if the effect size was large enough, and the p-value small enough.

- Checklist item 36: *Conduct robustness checks and summarize the results: is the main conclusion about necessity (support or rejection of the hypothesis) sensitive to choices by the analyst regarding ceiling line, scope, effect size threshold, p-value threshold, removing/keeping outliers, different choices for the target outcome? Conclude if the results are robust or fragile.*
  See this book: Section 9.12.
  In many studies a robustness check is done for the SEM part of the study, but not for the NCA part. Robustness checks for NCA are different from those for SEM, and equally important. For example, outliers for SEM may not be outliers for NCA, and vice versa. NCA's outlier analysis can be done with the NCA software in R (Section H.8 in the Appendix H for an example).

- Checklist item 38: *When NCA is used in a causal-pluralism study by combining it with another method (e.g., regression-based methods or QCA), report how the results of NCA and the other method complement each other using NERT, NEST and BIPMA when applicable.*
  See this book: Chapter 11.
  In many studies, SEM results and NCA results are presented separately without discussing the insights obtained from combining the two methods. The discussion could focus on the consequences of the combined results for theory and practice. The use of NERT (Section 11.3) and BIPMA (Section 11.4.1.7) could be helpful to integrate the results.

The quality of studies that combine NCA with SEM could be enhanced by ensuring that these and other 'must-have' recommendations are fulfilled. Once the must-have requirements are met, further improvements can be achieved by addressing the 'should-have' and 'nice-to-have' suggestions in the checklist (Section 10.3). As the number of combined NCA and SEM studies grows, reviewers may raise the standard for the appropriate application of NCA. Section 10.3 includes recommendations and references to meet the requirements.

Appendix H provides a demonstration of combining NCA and PLS-SEM with R.

11.5 NCA with QCA

Qualitative Comparative Analysis (QCA) has its roots in political science and sociology, and was developed by Charles Ragin (Ragin, 1987, 2000, 2008). QCA has steadily evolved over the years, and currently many types of QCA approaches exist. A common interpretation of QCA as described by Schneider & Wagemann (2012) and Mello (2021) is followed in this book.

11.5.1 Principles of QCA

Set theory is at the core of QCA. Instead of studying relations between variables, set theory focuses on relations between sets. Cases can be either part of a set or not part of a set. The Netherlands, for example, is a case that is 'in the set' of rich countries, and Ethiopia is a case that is 'out of the set' of rich countries. Set membership scores (rather than variable scores) are linked to a case. Regarding the set of rich countries, the Netherlands has a set membership score of 1 and Ethiopia of 0. In the original version of QCA the set membership scores could only be 0 or 1. This version of QCA is called crisp-set QCA (csQCA). Later also fuzzy-set QCA (fsQCA) was developed. Here the membership scores can also have values between 0 and 1. For example, Croatia could be allocated a set membership score of 0.7, indicating that it is 'more in the set' than 'out of the set' of rich countries.

In QCA, relations between sets are studied. Suppose that one set is the set of rich countries (X), and another set is the set of countries with happy people ('happy countries', Y). QCA uses Boolean (binary) algebra and expresses the relationship between condition X and outcome Y: the presence or absence of X is related to the presence or absence of Y. More specifically, the relations are expressed in terms of sufficiency and necessity. For example, the presence of X (being a country that is part of the set of rich countries) could be theoretically stated as *sufficient* for the presence of Y (being a country that is part of the set of happy countries). In this case, the following claims would apply: All rich countries are happy countries; the set of rich countries is a subset of the set of happy countries; no rich country is not a happy country; set X is a subset of set Y. Alternatively, another theory could state that the presence of X (being country that is part of the set of rich countries) is *necessary* for the presence of Y (being a country that is part of the set of happy countries). All happy countries are rich countries; the set of rich countries is a superset of the set of happy countries; no happy country is not a rich country; set X is a superset of set Y.

QCA's main causal interest is sufficiency. QCA assumes that a configuration of single conditions produces the outcome. For example, the condition of being in the set of rich countries (X_1) AND the condition of being in the set of democratic countries (X_2) is sufficient for the outcome of being in the set of happy countries (Y). QCA's Boolean logic statements for this sufficiency relationship is expressed as follows:

$$X_1 * X_2 \rightarrow Y \tag{11.2}$$

where the symbol '$*$' means the logical 'AND', and the symbol '$\rightarrow$' means 'is sufficient for'.

Furthermore, QCA assumes that several alternative configurations may exist that can produce the outcome, known as 'equifinality'. This is expressed in the following example:

$$X_1 * X_2 + X_2 * X_3 * X_4 \rightarrow Y \tag{11.3}$$

where the symbol '+' means the logical 'OR'.

It is also possible that the absence of a condition is part of a configuration. This is shown in the following example:

$$X_1 * X_2 + X_2 * \neg X_3 * X_4 \rightarrow Y \tag{11.4}$$

where the symbol '$\neg$' means 'absence of'.

Single conditions that are necessary/non-redundant in a configuration that is sufficient for the outcome are called INUS conditions (Mackie, 1965). An INUS condition is an 'Insufficient but Non-redundant (i.e., Necessary) part of an Unnecessary but Sufficient condition.' In this expression, the words 'part' and 'condition' are somewhat confusing because 'part' refers to the single condition and 'condition' refers to the configuration that consists of single conditions (parts). Insufficient refers to the fact that a part (single condition) is not itself sufficient for the outcome. Non-redundant refers to the necessity of the part (single condition) for the configurations being sufficient for the outcome. Unnecessary refers to the possibility that other configurations can also be sufficient for the outcome. Sufficient refers to the fact that the configuration is sufficient for the outcome.

Although a single condition may be locally necessary for the configuration to be sufficient to produce the outcome, it is not globally necessary for the outcome because the single condition may be absent in another sufficient configuration. INUS conditions are thus usually not necessary conditions for the outcome (the latter are the conditions that NCA considers). Hence, in above generic logical statements about relations between sets, see Equations (11.2), (11.3), (11.4), X and Y can only be absent or present (Boolean algebra), even though the individual members of the sets can have fuzzy scores. Both csQCA and fsQCA use only binary levels (absent or present) of the condition and the outcome when formulating the solution.

The starting point of QCA's data analysis is to transform variable scores (if available) into set membership scores. Variable scores are called "raw" scores and the transformation process is called 'calibration'. Calibration can be based on the distribution of the data, the measurement scale, or expert knowledge. The goal of calibration is to get scores of 0 or 1 (csQCA) or between 0 and 1 (fsQCA) to represent the extent to which the case belongs to the set (set membership score). In fsQCA, cases with membership score below 0.5 do not count as having the condition present and contribute to the assessment of absence of the condition. Similarly, cases with membership score above 0.5 qualify as having the condition present.

As a qualitative method, QCA's calibration process is preferably guided by knowledge about the studied cases. However, particularly in large-n fsQCA studies a 'mechanistic' (data driven) transformation is often unavoidable due to lack of access to the cases. For this transformation the non-linear logistic function is usually selected. This selection is somewhat arbitrary (but is built into popular QCA software) and moves the variable scores to the extremes (0 and 1) in comparison to just (linear) transformation (e.g., min-max normalization) of the data. With the standard logistic transformation, low values move toward 0 and high values toward 1. When no substantive reason exists for the logistic transformation, and mechanistic transformation is unavoidable, linear transformation from raw scores to set membership scores (normalization) may be preferred to avoid changing the distribution of the data and obscure possible necessary conditions (Dul, 2016b). Moving the X- and Y-scores to the extremes implies that cases in the XY-plot with low to middle values of X move to the left and cases with middle to high values of Y move upwards. As a result, the upper-left corner is filled with more cases. Consequently, potential meaningful empty spaces in the original data (indicating necessity) may not be identifiable. With the normalized transformation, the cases stay where they are; an empty space in a corner of the XY-plot with the original data stays empty. The normalized transformation is an alternative to an arbitrary transformation: it just changes variable scores into set membership scores, without affecting the distribution of the data. A calibration evaluation tool to check the effect of calibration on the necessity effect size is available at https://r.erim.eur.nl/r-apps/qca/.

QCA performs two separate analyses with calibrated data: a necessity analysis for identifying necessary conditions[13], and a sufficiency analysis ('truth table' analysis) for identifying sufficient configurations. Although the focus in QCA is on sufficiency, a necessity analysis precedes the sufficiency analysis because if a necessary condition is not part of the sufficient configuration, the outcome will not be produced:

[13] In QCA both potential single necessary conditions and potential substitutable necessary conditions (OR combinations of conditions) can be identified, although it is often difficult to theoretically justify an OR-combination, see Section 11.5.2.

"If a causal condition passes the analyst's test of necessity, then this condition should be made a component of every causal expression that the analyst examines subsequently in the analysis of sufficiency" (Ragin, 2000, p. 254)

In csQCA, X and Y are expressed as dichotomous set membership scores: present (in the set) or absent (not in the set). In the XY-table, a necessary condition in kind is identified when the upper-left cell is (almost) empty (Figure 11.6). Necessity consistency is the number of cases in the upper-right cell divided by the number of cases in the upper-left plus upper-right cell.

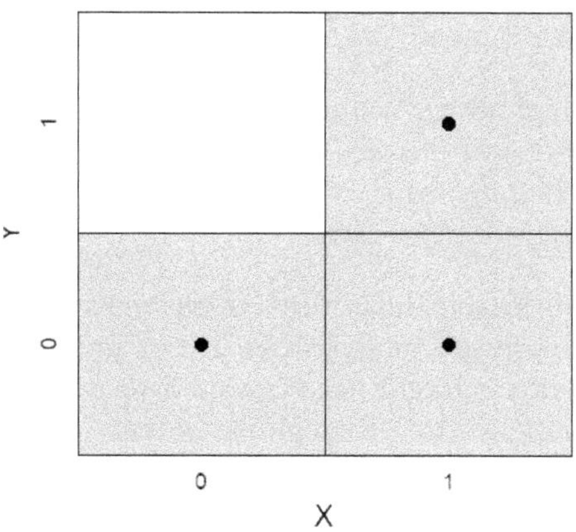

FIGURE 11.6 Necessity analysis by dichotomous NCA and crisp set QCA.

In fsQCA, X and Y are expressed as discrete or continuous set membership scores indicating the extent to which the condition or outcome is in the set. In the XY-plot, a diagonal is drawn and a necessary condition in kind is identified when the area above the diagonal is (almost) empty (Figure 11.7-left). Necessity **in**consistency is the sum of vertical distances of cases above the diagonal, divided by the sum of vertical distances of all cases. Necessity consistency is the complement of this value (1 − inconsistency).

The main analysis of csQCA and fsQCA is the sufficiency analysis. Whereas QCA's necessity analysis uses linear algebra to identify necessity, QCA's sufficiency analysis uses Boolean algebra and truth tables. The truth table is a logical summary of all (2^k) possible combinations of k conditions (X's) that are selected for the analysis. A combination of conditions is referred to as a configuration. In configurations, conditions can only be considered present or absent (with the 0.5 membership score threshold point used in fsQCA; absent: < 0.5; present: > 0.5). The purpose of QCA is to identify which of the possible configurations in the truth table are actually observed in the data, and to establish whether the outcome is present or absent in the observed configurations. This leads to QCA's *solution table* consisting of configurations that are considered to be able to produce the outcome ($Y = 1$). A second solution table consists of configurations that are considered to not produce the outcome ($Y = 0$).

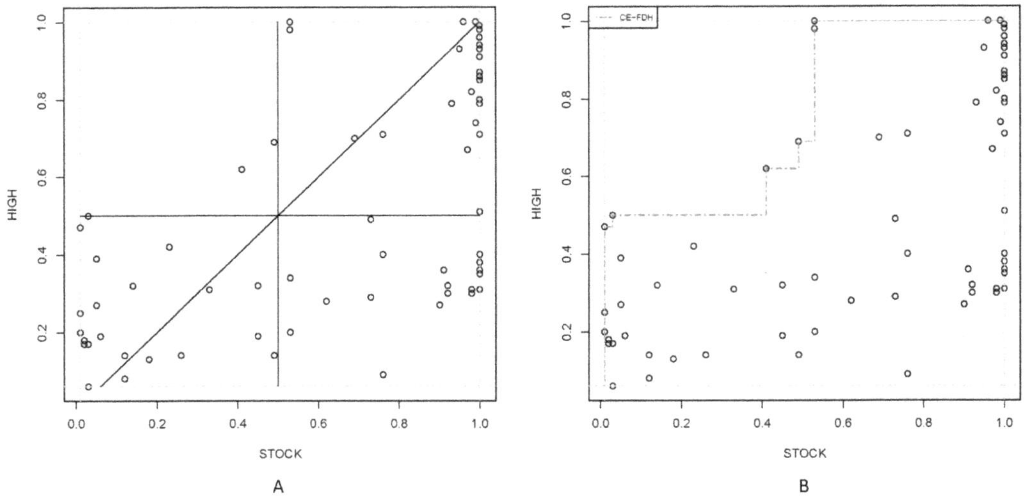

FIGURE 11.7 Comparison of necessity analysis with fsQCA (Left) and with NCA (Right). The ceiling in QCA is predefined (diagonal) and in NCA estimated (with CE-FDH). Data from Rohlfing & Schneider (2013); see also Vis & Dul (2018).

Over the years, different approaches have been suggested to integrate an observed necessary condition from QCA's necessity analysis (for $Y = 1$ and $Y = 0$, respectively) into the sufficiency solution. For example, Ragin has suggested to *add* the necessary condition if it is absent in an observed configuration, to *remove* the necessary condition from all configurations, and to *ignore* necessity.[14] It appears that the necessity analysis of QCA is often ignored and if this analysis is done, different approaches are used for integration of the necessity analysis with the sufficiency analysis.

11.5.2 Comparison NCA and QCA

NCA and QCA share several characteristics. For a causal interpretation of data, both NCA and QCA rely on conditional logic (necessity and sufficiency), and theoretical argumentation for causality. Although NCA primarily uses quantitative variable-based approaches, it can also be applied (like QCA) with calibrated set membership scores. Both methods can be used in large-n or small-n settings. In a large-n setting, both methods preferably use probability sampling and statistical generalization, although this is not common in QCA (Braumoeller, 2015).

Despite these similarities there are also major differences (Dul, 2016a; Vis & Dul, 2018). NCA focuses on necessity, whereas QCA focuses on sufficiency. NCA is mainly a deductive approach requiring a formal necessity hypothesis whereas QCA is mainly an inductive approach. Although QCA carefully selects conditions, a hypothesis how these are related to the

[14] Ragin (2000, p. 254): "If several conditions pass the test of necessity, they are all *inserted* into the causal expressions that are subsequently tested for sufficiency." Ragin (2009, p. 110): "It is often useful to check for necessary conditions before conducting the fuzzy truth table procedure. Any condition that passes the test and that 'makes sense" as a necessary condition can be *dropped* from the truth table procedure, which, after all, is essentially an analysis of sufficiency" Mendel & Korjani (2010, p. 83.) quote Ragin's view on this approach as follows: "That being said (*ignore* necessity), there is still a lot of interest in necessary conditions in the social sciences" (emphasis added).

outcome is usually not provided. The difference is evident when analyzing OR-combinations of necessity. An OR-combination (X_1 or X_2 is necessary for Y) is often an indication that a higher order construct is necessary for the outcome. A green apple is not necessary for an apple pie, nor is a red apple, but apple is necessary for an apple pie. In NCA's procedures for formulating a formal necessary condition hypothesis (Chapter 7), a relevant higher order construct, rather than the related lower order construct, would likely be recognized and become part of the analysis. An inductive QCA approach may not consider OR-combinations before the analysis, but may identify them afterwards. However, many identified necessity OR-combinations have no substantive meaning as by logic, any necessary condition that is combined with another condition is a necessary OR-combination, even if the combination does not have a substantive meaning. If apple is necessary for an apple pie, then the combination [apple OR blue sky] is also necessary by logic, but this does not mean that blue sky is necessary or relevant. The combination is logically true but substantively meaningless.

Since NCA is an approach for analyzing necessity, it can be compared with QCA's necessity analysis only. When NCA is applied with dichotomous set membership scores as input for NCA's necessity analysis, the conclusions are usually the same as with csQCA's necessity analysis. Without cases in the upper-left cell, QCA's consistency = 1, which supports necessity. In this situation NCA's effect size is 1, and assuming theoretical support and statistical significance, NCA supports necessity as well. However, NCA differs completely from fsQCA's necessity analysis. Whereas in fsQCA the 'ceiling line' is predefined as the diagonal in the XY-plot (Figure 11.7-left), in NCA the ceiling line is usually estimated from data[15], which is usually above the diagonal. When fsQCA rejects necessity when too many cases are above the diagonal (using necessity consistency as model fit measure), NCA's necessity analysis may conclude that necessity does exist, though not for all levels of X and Y. Whether or not necessity-in-kind should be rejected depends in NCA on the effect size, the availability of theoretical support for necessity and the p-value (Section 9.10), but not directly on its model fit measures (Section 4.5). Additionally, when necessity-in-kind is observed, NCA can analyze necessity-in-degree to evaluate which level of X is necessary for which level of Y (Section 9.11).

When fsQCA supports necessity, some cases may be in the 'empty' zone above the diagonal as *deviant cases*. According to NCA the same cases may be close to but below the ceiling line and may be considered *best cases*: cases that are able to achieve a high outcome with a low level of the conditions (assuming that X is an effort and Y is a desired outcome). In NCA, cases above the ceiling are considered not feasible and classified as noise (close above to the ceiling) or outliers/exceptions (further away from the ceiling) as discussed in Section 4.5.

11.5.3 Combining NCA and QCA

There are two ways to combine NCA and QCA in a multimethod study. First, since both NCA and QCA employ a necessity analysis based on necessity causal logic, these two approaches can be used jointly in a *method-triangulation multimethod study*. Second, since QCA also employs a sufficiency analysis based on configurational sufficiency causal logic, NCA can be combined with QCA's sufficiency analysis in a *causal-pluralism multimethod study*. Both types of multimethod studies are discussed below in more detail.

---

[15] It is also possible to predefine a ceiling line, see Section 9.3.2.

11.6 Combining NCA and QCA in a multimethod study

11.6.1 Steps for conducting NCA with QCA's necessity analysis

Figure 11.8 shows a flowchart for conducting a necessity analysis with NCA and QCA in a method-triangulation multimethod study.

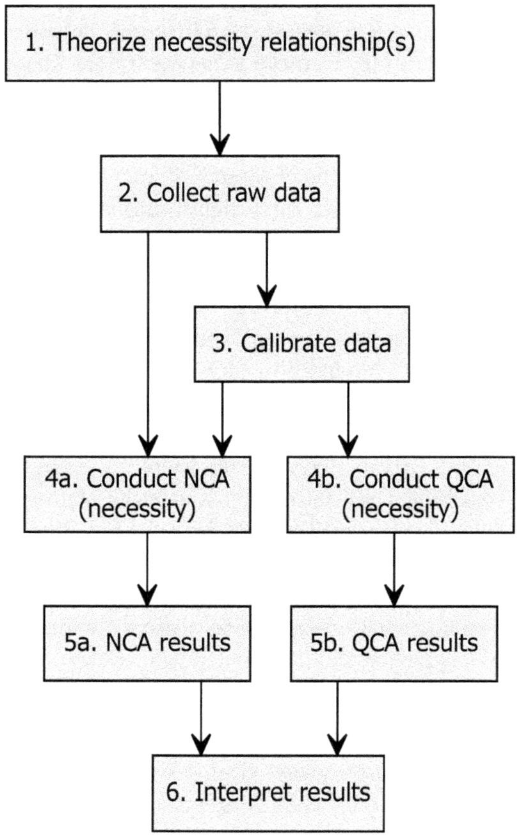

FIGURE 11.8 Flowchart for conducting a casual-triangulation multimethod study with NCA and QCA's necessity analysis.

The study starts with theorizing the necessity relationship(s). In NCA this means the formulation of a formal necessity hypothesis (Chapter 7) in which not only X and Y are selected, but also the causal relationship is theoretically justified. In QCA the relationship between the conditions and outcome is usually not discussed extensively before the analysis is done and usually no necessity hypothesis is formulated. This reflects that QCA is a more inductive approach.

The next step is collecting (raw) data. In QCA the data need to be calibrated to transform the data into meaningful set membership scores. Qualitative data may be calibrated based on case knowledge of the user and expert judgment (the direct method), and quantitative data can be calibrated by mechanistically applying a transformation function (the indirect method). NCA accepts both calibrated data and raw data (when the raw data are valid and reliable), but QCA accepts only calibrated data.

Therefore, two types of NCA-QCA method-triangulation for analyzing necessity can be done:

1. NCA with raw data and QCA with calibrated data.
2. NCA and QCA both with calibrated data.

After the NCA and QCA analyses are conducted, in the last step the results are interpreted in combination.

11.6.2 Example NCA-QCA method-triangulation multimethod study

This section presents an example of integrating NCA in QCA according to a method-triangulation multimethod necessity study. The example is based on a study by Lipset (1959), discussed in Ragin (2009). The outcome variable is SURV: survival of democracy during the inter-war period. Assuming that a theoretical rationale for necessity exists, five conditions were analyzed for necessity using both QCA and NCA:

- DEV: Level of development: expressed as GDP per capita (USD) in the raw data.
- URB Level of urbanization: percent of the population in towns with 20,000 or more inhabitants in the raw data.
- LIT Level of literacy: percent of the literate population (raw data).
- IND Level of industrialization: percent of the industrial labor force (raw data).
- STB Government stability: "political-institutional (in)stability" expressed in the raw data by the number of cabinets which governed in the period under study.

The raw data were calibrated into fuzzy set membership scores (the calibration process is not discussed here). The analyses can be done as follows:

```
library(QCA)
library(NCA)

# Data
conditions <- c("DEV", "URB", "LIT", "IND", "STB") #names of the conditions
outcome <- "SURV" #name of the outcome
data(LR) #Lipset raw data
data(LF) #Lipset calibrated data

#QCA-calibrated (fuzzy)
pofind(LF, outcome = "SURV") #single necessary conditions

##
##              inclN   RoN   covN
## -----------------------------
## 1   ~DEV  0.285  0.587  0.274
## 2    DEV  0.831  0.811  0.775
## 3   ~URB  0.568  0.452  0.402
## 4    URB  0.539  0.899  0.771
## 5   ~LIT  0.096  0.764  0.168
```

```
## 6     LIT  0.991  0.509  0.643
## 7    ~IND  0.417  0.576  0.367
## 8     IND  0.669  0.786  0.684
## 9    ~STB  0.218  0.687  0.269
## 10    STB  0.920  0.680  0.707
## ------------------------------

#NCA-calibrated (fuzzy)
set.seed(123) # for reproducible results
model_calibrated<- nca_analysis(LF,
                                1:5,6,
                                ceilings = "ce_fdh",
                                test.rep = 10000)

## Setting up parallelization, this might take a few seconds...
##      Do test for  :  ce_fdh - DEVDone test for:  ce_fdh - DEV
## Do test for  :  ce_fdh - URBDone test for:  ce_fdh - URB
## Do test for  :  ce_fdh - LITDone test for:  ce_fdh - LIT
## Do test for  :  ce_fdh - INDDone test for:  ce_fdh - IND
## Do test for  :  ce_fdh - STBDone test for:  ce_fdh - STB

model_calibrated

##
## --------------------------------------------------------------------------------
## Effect size(s):
##     ce_fdh p
## DEV 0.37   0.000
## URB 0.01   0.361
## LIT 0.82   0.002
## IND 0.02   0.366
## STB 0.46   0.004
## --------------------------------------------------------------------------------

# NCA-raw data
set.seed(123) # for reproducible results
model_raw <- nca_analysis(LR,
                          1:5,6,
                          ceilings = "ce_fdh",
                          corner = c(1,1,1,1,2),
                          test.rep = 10000)

## Setting up parallelization, this might take a few seconds...
##       Do test for  :  ce_fdh - DEVDone test for:
##       ce_fdh - DEV
## Do test for  :  ce_fdh - URBDone test for:  ce_fdh - URB
## Do test for  :  ce_fdh - LITDone test for:  ce_fdh - LIT
## Do test for  :  ce_fdh - INDDone test for:  ce_fdh - IND
## Do test for  :  ce_fdh - STBDone test for:  ce_fdh - STB
```

```
model_raw

##
## --------------------------------------------------------------------------
## Effect size(s):
##      ce_fdh p
## DEV 0.27   0.000
## URB 0.09   0.337
## LIT 0.74   0.001
## IND 0.11   0.097
## STB 0.53   0.008
## --------------------------------------------------------------------------
```

Table 11.7 shows the results. Using the common necessity consistency threshold of 0.9 two necessary conditions are identified by QCA: LIT and STB. NCA's necessity analysis with the raw and with calibrated data show similar results. Also, NCA shows that LIT and STB are necessary conditions. However, NCA identifies a third necessary condition: DEV.

TABLE 11.7 Comparison of NCA's and QCA's necessity analyses in a method-triangulation multimethod study.

	QCA Consistency (calibrated)	Necessity effect size (calibrated)	p-value, (calibrated)	Necessity effect size (raw)	p-value, (raw)
DEV	0.831	0.37	< 0.001	0.27	< 0.001
URB	0.539	0.01	0.361	0.09	0.336
LIT	0.991	0.82	0.002	0.74	0.001
IND	0.669	0.02	0.366	0.11	0.097
STB	0.920	0.46	0.004	0.53	0.008

Generally, NCA identifies more necessary conditions than fsQCA. This particularly happens when raw data, rather than (mechanistically) calibrated data, are used as input to NCA. For example, Torres & Godinho (2022) conducted a method-triangulation NCA-QCA necessity study. They examined which elements of digital entrepreneurial ecosystems are necessary to produce outcomes. In their study the authors used only calibrated data and concluded that "[...] fsQCA should be complemented with NCA to better understand the necessary conditions. The results show that the former identifies considerably less necessary conditions in data sets than NCA[...]" (Torres & Godinho, 2022, p. 41).

11.6.3 Steps for conducting NCA with QCA's sufficiency analysis

Another type of NCA-QCA multimethod study is the causal-pluralism multimethod study in which NCA's necessity causal perspective is combined with QCA's configurational sufficiency causal perspective. For a causal-pluralism study, it is necessary that the two methods use the same data. Hence, NCA must be applied with set membership scores, otherwise comparing and combining necessity results with sufficiency results becomes problematic.[16]

[16] According to the definitions proposed in Section 11.2 a QCA standard study that combines QCA's necessity analysis with QCA's sufficiency analysis is a causal-pluralism multimethod study. However, most

Figure 11.9 shows a flowchart for conducting a necessity analysis of NCA with a sufficiency analysis of QCA in a causal-pluralism multimethod study.

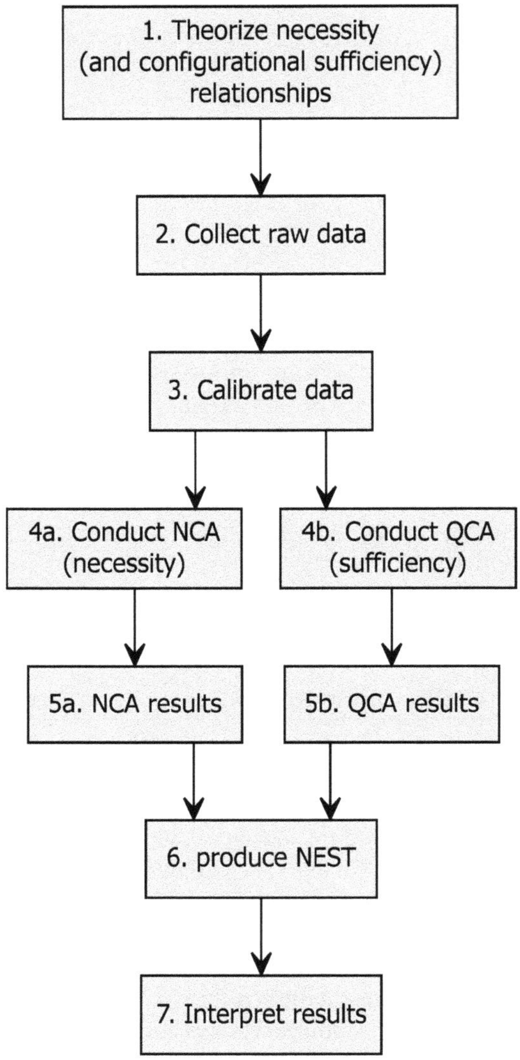

FIGURE 11.9 Flowchart for conducting a causal-pluralism multimethod study with NCA and QCA's sufficiency analysis.

The study starts with theorizing the necessity relationships for NCA: the formulation of a formal necessity hypothesis (Chapter 7). For the QCA sufficiency analysis the conditions that are part of the necessity hypothesis, and possibly other non-necessary contributing conditions are selected. Often, QCA is used inductively and does not formulate sufficiency hypotheses (claims about configurations that can produce the outcome) although this is also possible as, for example, suggested by Mello (2021). Next, the raw data are collected and subsequently calibrated. The set membership scores are input to both NCA and QCA. Then NCA and QCA are conducted according to their own standards.

QCA studies focus on the sufficiency analysis, and often largely ignore the necessity analysis, possibly because necessity is often not identified by QCA.

After the two analyses are done, the integration of the results may be challenging. As indicated in the above quote by Ragin, a necessary condition must become a part of the sufficient configuration, otherwise the configuration will not produce the outcome. Ragin's views on how to realize that goal has changed over the years, and new views have been added. Currently, at least five different views exist in the QCA community on how to integrate the identified necessary conditions (with QCA) with the identified sufficient configurations. These views may also apply to integrating a necessary condition identified with NCA.

In the first view, only sufficient configurations that include the necessary conditions are considered. This ensures that all selected configurations have the necessary condition. In the second view, the truth table analysis to find the sufficient configurations is done without the necessary conditions, and afterwards the necessary conditions are added to the configuration. This also ensures that all configurations have the necessary conditions. In the third view, configurations that do not include the necessary condition are excluded from the truth table before this table is further analyzed to find the sufficient configurations. This 'ESA' approach (Schneider & Wagemann, 2012) also ensures that all configurations have the necessary conditions. In the fourth view, sufficient configurations are analyzed without worrying about necessary conditions. Afterwards, the necessary conditions are discussed separately. In the fifth view, a separate necessity analysis is not done, or necessity is ignored.

All views have been employed in QCA; no consensus exists yet. An additional complexity of integrating NCA's necessity with QCA's sufficient configurations is that NCA produces necessary conditions in degree, rather than QCA's necessary conditions in kind. In QCA's solution table, the conditions that are part of the sufficient configurations can only be absent or present.

One way to proceed with integrating NCA's necessity approach with QCA's sufficiency approach is to combine the last part of the second view (adding necessary conditions afterwards), with the fourth view (discussing necessary conditions separately):

1. After NCA has identified necessity-in-kind in step 4a and QCA has identified the solution table in step 4b, NCA's bottleneck analysis for necessity-in-degree is conducted in step 5a, NCA's necessity-in-degree analysis with the bottleneck table is conducted as follows. Since in QCA's solution table an outcome is considered to be in the set if the set membership score is $Y > 0.5$, NCA's necessity-in-degree analysis consists of identifying the required level of the condition membership score X that is necessary for an outcome membership score Y of at least 0.5. In other words, the target outcome in the bottleneck table is $Y = 0.5$ membership score.
2. Next, the resulting required membership score of the condition according to NCA is added to all QCA's sufficient configurations in the solution table, if this condition is not already part of the configuration. Specifically, the NCA-Extended Solution Table (NEST) is produced. This extension specifies the required level of the necessary condition identified by NCA, such that the configuration can produce the outcome (for an example see below).
3. Finally, the result of the NEST and the full results of NCA's necessity analysis and QCA's sufficiency analysis are interpreted in combination. In particular, it could be discussed that specific levels of necessary membership scores found by NCA must be present in each sufficient configuration found by QCA. If that membership in degree is not in place, the configuration will not produce the outcome.

11.6.4 Example NCA-QCA causal-pluralism multimethod study

This section presents an example of integrating NCA in QCA according to a causal-pluralism multimethod necessity-sufficiency study. The example is based on a study by Emmenegger (2011) about job security regulation (JSR) in Western European countries. In particular, the effect of six conditions is investigated. These six conditions are: S = state-society relationships, C = non-market coordination, L = strength labor movement, R = denomination, P = strength religious parties and V = veto points. The study conducts a necessity analysis and a sufficiency analysis, both with fsQCA. The study can be replicated as follows:

```
library(QCA)

# Data
data("d.Emm")
conditions <- c("S","C","L","R","P","V") #names of the conditions
outcome <- "JSR" #name of the outcome

##### QCA #####
pofind(d.Emm, outcome = "JSR") #single necessary conditions

##
##          inclN   RoN   covN
## ----------------------------
##  1   ~S  0.438  0.639  0.453
##  2    S  0.667  0.783  0.714
##  3   ~C  0.543  0.615  0.509
##  4    C  0.620  0.833  0.743
##  5   ~L  0.571  0.677  0.571
##  6    L  0.714  0.843  0.793
##  7   ~R  0.388  0.661  0.431
##  8    R  0.791  0.812  0.791
##  9   ~P  0.528  0.612  0.498
## 10    P  0.672  0.863  0.800
## 11   ~V  0.648  0.692  0.627
## 12    V  0.500  0.738  0.577
## ----------------------------

#published solution: no necessary conditions with consistency > 0.9
ttEmm <- truthTable(d.Emm, condition =  conditions,
                    outcome = outcome,
                    incl.cut = 0.90,
                    show.cases = TRUE)
# conservative/complex solution
minimize(ttEmm)

##
## M1: S*~C*L*R*P + S*C*R*P*V + S*~C*~L*R*~P*~V + ~S*C*L*~R*P*~V -> JSR

#M1: S*~C*L*R*P + S*C*R*P*V + S*~C*~L*R*~P*~V + ~S*C*L*~R*P*~V -> JSR
```

```
# parsimonious solution
minimize(ttEmm, include =  "?")

##
## M1: S*R + (L*P) -> JSR
## M2: S*R + (P*~V) -> JSR

#M1: S*R + (L*P) -> JSR
#M2: S*R + (P*~V) -> JSR

# intermediate solution
minimize(ttEmm, include = "?", dir.exp = c(1,1,1,1,1,0), details=T)

##
## From C1P1, C1P2:
##
## M1:    S*R*~V + S*C*R*P + S*L*R*P + C*L*P*~V -> JSR
##
##              inclS   PRI   covS   covU   cases
## ---------------------------------------------------
## 1     S*R*~V  0.990  0.983  0.402  0.152  FR,PT; IT
## 2    S*C*R*P  0.965  0.921  0.277  0.041  BE,DE; AT
## 3    S*L*R*P  1.000  1.000  0.354  0.027  IT; ES; AT
## 4  C*L*P*~V  0.964  0.872  0.297  0.138  NO
## ---------------------------------------------------
##          M1  0.965  0.941  0.685

#M1:    S*R*~V + S*C*R*P + S*L*R*P + C*L*P*~V -> JSR

#published solution
#M1:    S*R*~V + S*C*R*P + S*L*R*P + C*L*P*~V -> JSR
```

The necessity analysis of fsQCA shows that none of the six conditions were necessary for job security regulation (necessity consistency of each condition is < 0.9). The following four sufficient configurations for the outcome JSR were identified:

S1. S*R*¬V (presence of S AND presence of R AND absence of V)

S2. S*C*R*P (presence of S AND presence of C AND presence of R AND presence of P)

S3. S*L*R*P (presence of S AND presence of L AND presence of R AND presence of P)

S4. C*L*P*¬V (presence of C AND presence of L AND presence of P AND absence of V: ¬V)

This combination of solutions can be expressed by the following logical expression:

$$S^*R^*\neg V + S^*C^*R^*P + S^*L^*R^*P + C^*L^*P^*\neg V \rightarrow JSR$$

For demonstration purposes, an NCA necessity analysis is added to this study. For the sake of this illustration, it is assumed that a formal necessity hypothesis exists for each condition (Chapter 7). Producing the *XY*-plots can be done as follows:

```
library(NCA)
model <- nca_analysis(d.Emm,
                      conditions,
                      outcome,
                      corner = c(1,1,1,1,1,2),
                      ceilings = "ce_fdh")
nca_output(model, summaries = FALSE)
```

Figure 11.10 shows the XY-plots for the six conditions using the CE-FDH ceiling line.

NCA's necessity-in-kind analysis is conducted as follows:

```
library(NCA)
set.seed(123) # for reproducible results
model <- nca_analysis(d.Emm,
                      conditions,
                      outcome, corner = c(1,1,1,1,1,2),
                      test.rep = 10000, #This may take a while
                      ceilings = "ce_fdh"
                      )
# necessity-in-kind
print(model)

## Setting up parallelization, this might take a few seconds...
##     Do test for  :  ce_fdh - SDone test for:  ce_fdh - S
## Do test for  :  ce_fdh - CDone test for:  ce_fdh - C
## Do test for  :  ce_fdh - LDone test for:  ce_fdh - L
## Do test for  :  ce_fdh - RDone test for:  ce_fdh - R
## Do test for  :  ce_fdh - PDone test for:  ce_fdh - P
## Do test for  :  ce_fdh - VDone test for:  ce_fdh - V

##
## --------------------------------------------------------------------------

##   ce_fdh p
## S 0.14   0.132
## C 0.00   1.000
## L 0.29   0.003
## R 0.31   0.001
## P 0.17   0.009
## V 0.05   0.533
## --------------------------------------------------------------------------
```

The results indicate that three out of six conditions are necessary for Job security regulations (JSR) given the effect size and p-value thresholds of 0.10 and 0.05, respectively. The necessary conditions are: L = strength of labor movement, R = denomination, and P = strength of religious parties.

NCA's analysis of necessity-in-degree for these three conditions, adapted to an outcome of Y = 0.5, can be conducted as follows:

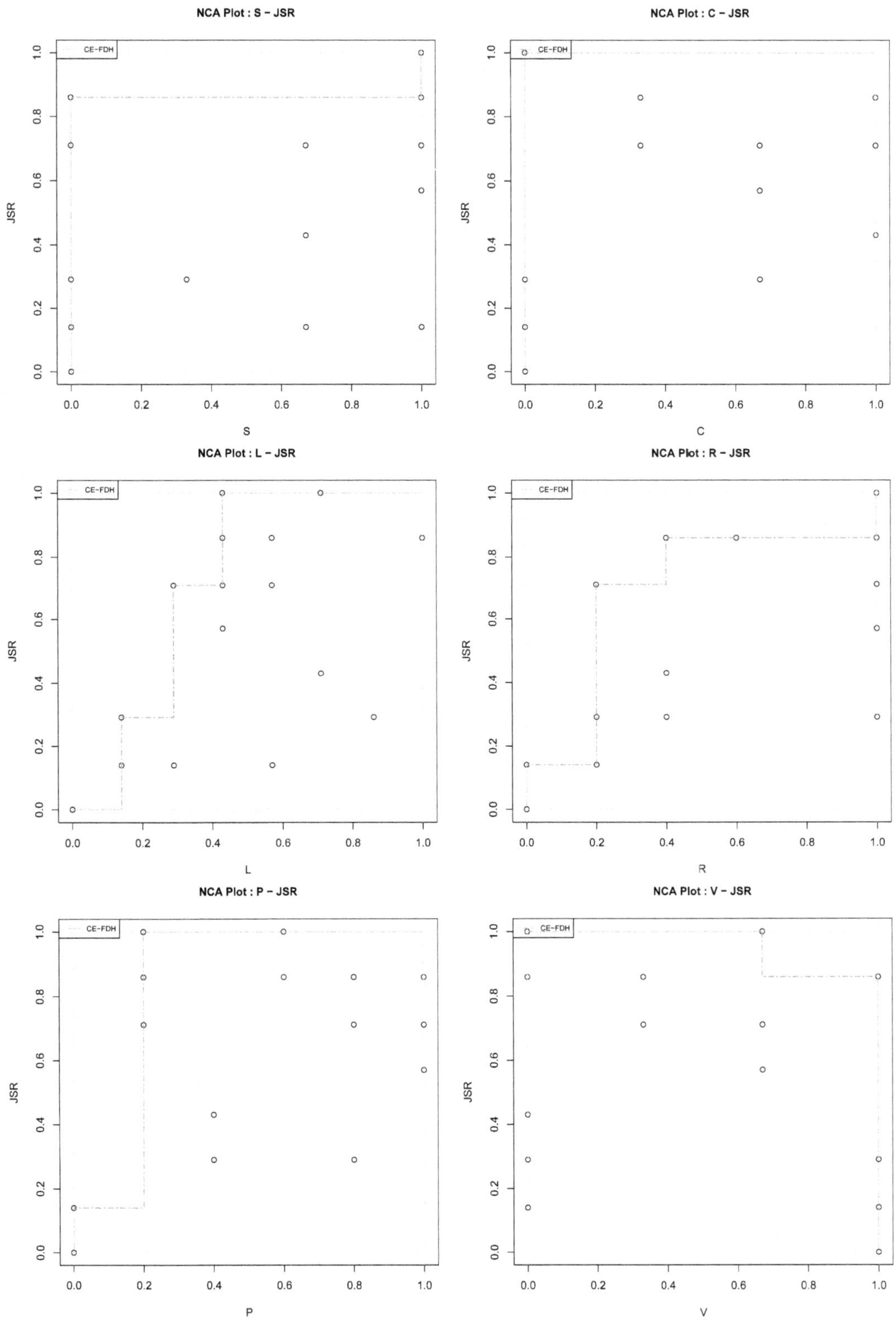

FIGURE 11.10 Example of a necessity analysis with NCA for a causal-pluralism multimethod study with fsQCA. (Data from Emmenegger, 2011).

```
library(NCA)
model_bottl <- nca_analysis(d.Emm,
                            conditions,
                            outcome,
                            corner = c(1,1,1,1,1,2),
                            ceilings = "ce_fdh",
                            bottleneck.x =  'actual',
                            bottleneck.y =  'actual',
                            steps = seq(from = 0, to = 1, by = 0.1) #include 0.5
                            )

# necessity-in-degree (reduced bottleneck table)
nca_output(model_bottl, summaries = FALSE, plots= FALSE,
           bottlenecks = TRUE,
           selection = c("L", "R", "P")
           )
```

```
##
## --------------------------------------------------------------------------

## --------------------------------------------------------------------------
## Y     1     2     3
## 0.0   NN    NN    NN
## 0.1   0.140 NN    NN
## 0.2   0.140 0.200 0.200
## 0.3   0.290 0.200 0.200
## 0.4   0.290 0.200 0.200
## 0.5   0.290 0.200 0.200
## 0.6   0.290 0.200 0.200
## 0.7   0.290 0.200 0.200
## 0.8   0.430 0.400 0.200
## 0.9   0.430 1.000 0.200
## 1.0   0.430 1.000 0.200
```

From the bottleneck table it can be observed that the following conditions are necessary for JSR > 0.5:

- L > 0.29 is necessary for JSR > 0.5 (presence of JSR)
- R > 0.20 is necessary for JSR > 0.5 (presence of JSR)
- P > 0.20 is necessary for JSR > 0.5 (presence of JSR)

Although in QCA's binary logic these small necessary membership scores of L, R, P (all < 0.5) would be framed as 'absence' of the condition, in NCA these membership scores are considered small, yet must be present for having the outcome. Thus, according to NCA the low level of membership scores must be present, otherwise the sufficient configurations identified by QCA will not produce the outcome.

The presence of a condition and the outcome means that the membership score is > 0.5. The absence of a condition means that the membership score is < 0.5. A common way to summarize the results is the QCA solution table. The Fiss-style solution table (Fiss, 2011) is shown in Table 11.8.

TABLE 11.8 QCA Solution Table (without NCA information).

	S1	S2	S3	S4
S	●	●	●	
C		•		•
L			●	●
R	●	●	●	
P		•	●	●
V	⊗			⊗
Consistency	0.990	0.965	1.000	0.964
Raw coverage	0.402	0.277	0.354	0.297
Unique coverage	0.152	0.041	0.027	0.138
Solution consistency: 0.965				
Solution coverage: 0.685				

S1, ... = Sufficient configurations
●/• = Presence of core/contributory condition
⊗/⊗ = Absence of core/contributory condition

The NCA necessity results can be combined with the QCA sufficiency results using the *NEST tool*. The NEST tool is the NCA-Extended Solution Table. It combines QCA results with NCA's necessity-in-degree results by adding a column and new symbols to the conventional QCA solution table (Table 11.9).

TABLE 11.9 NEST tool for combining NCA results with fsQCA results in a causal-pluralism multimethod study. NEST = NCA-Extended Solution Table. The extension consists of adding a necessity column to the fsQCA solution table: the necessary membership score for making the outcome possible (necessity-in-degree).

	S1	S2	S3	S4	NiD
S	●	●	●		
C		•		•	
L	◢	◢	●	●	≥ 0.29
R	●	●	●	◢	≥ 0.20
P	◢	•	●	●	≥ 0.20
V	⊗			⊗	
Consistency	0.990	0.965	1.000	0.964	
Raw coverage	0.402	0.277	0.354	0.297	
Unique coverage	0.152	0.041	0.027	0.138	
Solution consistency: 0.965					
Solution coverage: 0.685					

S1, ... = Sufficient configurations
NiD = Necessity in Degree: Required membership score according to NCA
●/• = Presence of core/contributory condition
⊗/⊗ = Absence of core/contributory condition
◢/◿ = 'Don't care' according to QCA; high/low level required according to NCA

The extra column NiD shows the required membership score of a condition such that the condition allows 0.5 membership score of the outcome. If this level of the condition is not achieved, the outcome will not occur. This required membership score of a necessary condition applies to each configuration. The *minimum* required level of the condition is shown as ($\geq$) when the hypothesized necessity relationship is high-high (presence/high level if X is necessary for presence/high level of Y). In this situation NCA analyzes corner 1 of the XY-plot, which is the upper-left corner that is expected to be empty. Similarly, the maximum allowable level ($\leq$) is shown when the hypothesized necessity relationship is low-high and corner 2 (upper-right) is the expected empty corner.[17]

In the conventional QCA solution table, blank cells in a configuration indicate that the condition has the *'don't care'* status. It is considered a non-essential part of the configuration to produce the outcome. However, it may be that according to NCA's necessity-in-degree result such a condition may be necessary with a certain level of membership score. When the condition is necessary according to NCA and 'don't care' according to QCA, the blank cell in the conventional QCA solution table is filled with a triangle symbol ◢ or ◿ . This ensures that the minimum or maximum required necessity membership score (according to NCA) is fulfilled. Only when this level is achieved, the solution can produce the outcome.[18]

The NCA results that the presence of L $\geq$ 0.29 is necessary for JSR $>$ 0.5 is already achieved in the QCA sufficient configurations S3 and S4, but not in configurations S1 and S2. According to QCA, in these latter configurations L is a 'don't care' condition. However, the necessity requirement of L $\geq$ 0.29 still applies. Therefore, the symbol (◢) is added to the configuration indicating that a 'don't care' condition is necessary as well. Similarly, R $\geq$ 0.20 is added to configuration 4, and P $\geq$ 0.20 is added to configuration 1. Without adding these requirements to the configurations, they cannot produce the outcome. Only configuration 3 includes all three necessary conditions according to NCA, without a need for adding them. If the NCA results would be ignored, configurations 1, 2, and 4 may not produce the outcome. Additionally, the NCA results can show what levels of the condition would be necessary for a higher level of the outcome than a membership score $>$ 0.5. This can be observed in Figure 11.10. For example, for a membership score of JSR of 0.9, it is necessary to have membership scores of L $>$ 0.43, R $=$ 1, P $>$ 0.2.

In summary, the following suggestions can be made for an NCA-QCA causal-pluralism multimethod study:

- Use membership scores for both analyses;
- Conduct these analyses separately;
- Integrate NCA's necessity-in-degree results into QCA's sufficiency solution using the NEST tool as in Table 11.9;

---

[17]For high-low and low-low relationships, NCA analyzes corners 3 and 4, respectively, and the NEST tool shows the corresponding inequality-signs: ($\geq$) and ($\leq$).

[18]The symbol ◢ for necessity-in-degree differs from the square symbol ■ proposed by Greckhamer (2016), and the diamond symbol ♦ used by Holtrop et al. (2024). Both symbols indicate the presence of a necessary condition *in kind* according to QCA's necessity analysis. The NEST table uses black dot symbols to indicate presence of the condition as originally and commonly used (e.g., Fiss, 2011; Rubinson, 2019). To indicate absence of a condition, often a circle, a circled minus, or a crossed circle are in the QCA solution table. In the NEST table crossed circles ⊗/⊗ are selected as such symbols are originally proposed by Fiss (2011) and often used (e.g., Mello, 2021). Note that the symbols may slightly differ in publications depending on selected fonts. Rubinson (2025) provides dedicated, interactive software for producing publication-quality "Fiss configuration charts", including NEST tables that incorporate the ◢ symbol. A template for producing the NEST in R is available via the book-page https://jandul.github.io/NCA/.

- Additionally, discuss the full results of NCA for deeper understanding of both necessity and configurational sufficiency.

11.6.5 Recommendations for combining NCA and QCA

In publications where NCA is applied in combination with QCA, QCA is usually the main analysis with the focus on sufficiency. NCA could replace QCA's necessity analysis. The two main advantages of NCA are that NCA can detect necessity-in-degree (which QCA cannot do) and that NCA can detect necessary conditions that QCA might miss. When NCA is combined with QCA, several guidelines for applying NCA are often violated. This applies, for example, to the following recommendations with 'must-have' priority level (see the SCoRe checklist in Section 10.3):

- Checklist item 4: *[...] When NCA is compared with QCA's necessity analysis, use the term necessity analysis of QCA for the latter (and not NCA) and explain the differences between the necessity analysis of QCA (only necessity-in-kind) and that of NCA (also necessity-in-degree).*
  See this book: Sections 2.3, 2.4. Suggested articles: Dul (2024a); Dul (2016a); Vis & Dul (2018); Dul (2022).
  When NCA is combined with QCA the following misinterpretations have occurred: The necessity analysis of QCA is confused with the necessity analysis of NCA. This has happened, for example, in QCA studies that do not use NCA. Some studies call QCA's necessity analysis an 'NCA'. However, the two types of necessity analyses are different. An INUS condition of a configuration is confused with the necessary condition for the outcome. An INUS condition is an *Insufficient but Necessary part of an Unnecessary but Sufficient configuration.* INUS conditions are the necessary elements of a configuration to make that configuration sufficient. NCA captures necessary conditions for the outcome, not the INUS conditions for the configuration, see also Dul, Vis, et al. (2021). Although an identified necessary condition for the outcome *must* be part of each sufficient configuration to be able to produce the outcome, the opposite is not true: a factor that is observed to be part of a particular sufficient *configuration* (INUS condition) is not automatically a global necessary condition for the *outcome*. An INUS condition that is present in all observed sufficient configurations is not automatically a necessary condition for the outcome. A separate necessity analysis is always needed to investigate if such an INUS condition is indeed necessary for the outcome.

- Checklist item 7: *Formulate the necessity relationship explicitly as a necessary condition hypothesis: X is necessary for Y. For a deductive study, this is done before data analysis; for an explorative study this is done after data analysis when indications for a necessity relationship are found.*
  See this book: Section 7.5.
  In studies that combine NCA with QCA, often no formal necessity hypothesis is formulated. NCA is primarily a deductive approach, and having a theoretical justification for necessity is one of the fundamental requirements to identify necessity. QCA is often an inductive approach and in small-n QCA case knowledge, and in large-n QCA general expert knowledge possibly illustrated with cases is often used to justify necessity. This case and general expert knowledge could be used for formulating a formal necessity hypothesis according to NCA before (deductive study) or after (inductive study) data collection.

- Checklist item 9: *Explain why X is necessary for Y using a causal explanation (narrative) by answering three questions: (1) Why does X enable Y (X is an enabler). The existing literature may be a source of information; (2) Why does the absence of X lead to the*

absence of Y (X is a constraint); (3) Why is the absence of X not compensable (no substitution for X). Making the narrative complete is a creative process, supported by literature and experiences of scholars and practitioners.
See this book: Chapter 7, in particular Section 7.5.3.
As part of the formal necessity hypothesis it should also be explained **why** it is expected that X is necessary for Y.

- Checklist item 19: *Explain whether and how the data were transformed, and whether the transformation aligns with NCA. Affine transformation (including linear transformation) produces valid results. NCA's data analysis does not require data transformation and non-transformed data are often more meaningful and better interpretable (e.g., levels of a Likert scale) than transformed data. This particularly applies to necessity-in-degree and interpretations of specific levels of X being necessary for specific levels of Y. Refrain from conducting non-linear data transformation (e.g., log-transformation, logistic transformation) unless the transformed data represent X and Y as defined in the necessity hypothesis (e.g., log-transformed GDP). [...].*
See this book: Sections 8.5.2, Appendix D.
In QCA applications that transform raw variable scores into set membership scores, often calibration is done mechanistically without using case or expert knowledge, though this is not preferred. The most commonly used transformation is the logistic transformation (S-curve). Such a non-linear transformation changes the distribution of the data such that points near NCA's ceiling line change. Unless there is theoretical motivation for it, non-linear transformation violates NCA's requirement of affine transformations (D). With logistic transformation necessity often diminishes or even disappears due to the calibration. The effect of calibration on NCA's necessity should then be discussed.
- Checklist item 38: *When NCA is used in a causal-pluralism study by combining it with another method (e.g., [...] QCA), report how the results of NCA and the other method complement each other using [...] NEST [...] when applicable.*
See this book: Chapter 11.
In many NCA-QCA multimethod studies, QCA results and NCA results are presented separately without discussing the insights obtained from combining the two different methods. The discussion could focus on the consequences of the combined results for theory and practice. The use of NEST (see Table 11.9) could be helpful for this.

The quality of studies that combine NCA with QCA could be enhanced by ensuring that these and other 'must-have' recommendations are fulfilled. Once the must-have requirements are met, further improvements can be achieved by addressing the 'should-have' and 'nice-to-have' suggestions in the checklist (Section 10.3). As the number of combined NCA and QCA studies grows, reviewers may raise the standard for the appropriate application of NCA. Section 10.3 includes recommendations and references to meet the requirements.

12

NCA in practice

12.1 Summary of this chapter

This chapter may be of interest to practitioners and scholars who are engaged in the practical application of NCA. First, in Section 12.2, the value of NCA for practice is explained by distinguishing between must-have factors for desired outcomes (e.g., performance), and stop factors for undesired outcomes (e.g., risks, disorders) on which practitioners can act through intervention. It discusses the added value and challenges of shifting attention from the common average effect thinking to necessity thinking. This section also mentions application areas by referring to academic work that suggests such an application. The next section (Section 12.3) discusses NCA's contribution to change and innovation in more detail. In an NCA-based intervention, levels of actionable factors are changed to make a specific practical outcome possible. This starts with the formulation of an intervention strategy, in which goals, barriers/drivers, and target groups are specified, culminating in necessity hypotheses. After that, the intervention is designed, based on collected data that are analyzed using NCA. The next two sections (Sections 12.4 and 12.5) discuss how interventions can make a desired outcome possible, and can prevent undesired outcomes, respectively. Concepts like intervention effectiveness, efficiency and waste are introduced. Section 12.6 discusses interventions with multiple necessary conditions. In the final section (Section 12.7), the application of an NCA-based intervention is illustrated with an example.

12.2 The practical value of necessity logic

Practice refers to the real-world context in which practitioners operate and implement interventions. For example:

- Teachers, school leaders, and others create effective learning environments in education.
- Employees, managers, and entrepreneurs create, produce, and deliver products and services in business and civil services.
- Physicians, nurses, and related professionals diagnose and treat health problems in health care.
- Ecologists, policymakers, and consumers prevent environmental degradation through collaborative actions in environmental management.

In any domain, interventions can benefit from knowledge about necessary conditions. By acting on these conditions, interventions can either make a *desired outcome possible* (e.g., student learning, business success) or make an *undesired outcome impossible* (e.g., disease,

DOI: 10.1201/9781003430766-12

environmental harm). Ignoring necessary conditions can make interventions ineffective or inefficient.

Applying necessity logic often tells a previously undiscovered story for many practitioners. NCA offers a straightforward and practical way to analyze complex problems that may initially seem "too simple to be true." It identifies *must-have* factors that are non-compensable requirements for a desired outcome, and *stop* factors that can block an undesired outcome. Achieving a desired outcome requires all must-have factors to be present, while the absence of a single factor guarantees failure. Preventing an undesired outcome is possible by removing just a single stop factor.

Figure 12.1-left shows what happens if necessity applies.

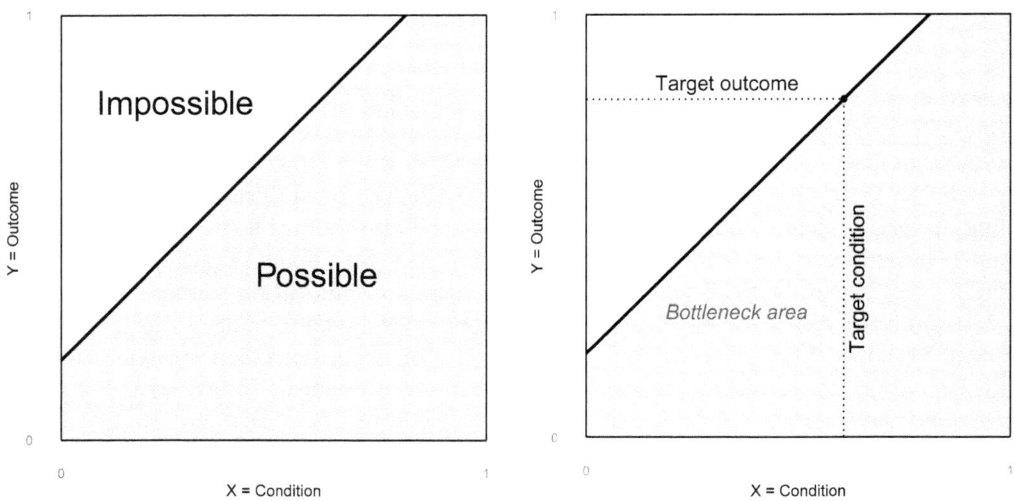

FIGURE 12.1 XY-plot with the ceiling line dividing the area where cases are impossible from the area where cases are possible (Left). The target condition is the threshold level of X that is necessary for the target outcome: a desired or undesired level of Y. Cases that stay in the bottleneck area cannot reach the target outcome (Right).

12.2.1 Must-have factors for desired outcomes

Necessary conditions for a desired outcome (e.g., performance) are must-have factors: without them, the desired level of the outcome cannot be reached. In the XY-plot in Figure 12.1-left, the ceiling marks a boundary that cases cannot cross. A certain desired target level of the outcome (Figure 12.1-right) is only possible if the condition value is at or above the target condition level. The target condition level acts as a binding constraint for the target outcome. It does not determine the outcome by itself, but it sets a minimum requirement that must be satisfied before the target becomes attainable. For a case to have any chance of attaining the target outcome, it must fall within the feasible area on the right side of the target condition.

Cases with condition values below the target condition are *bottleneck cases*. They lie in the *bottleneck area*, which is the region with X-values below the threshold needed for the target outcome. No matter how strongly other contributing conditions are improved, these cases cannot reach the desired outcome level as long as their condition level remains too low.

An intervention can therefore create opportunity by removing this constraint. Increasing the condition level to at least the threshold moves a case out of the bottleneck area and into the non-bottleneck area to the right of it. Therefore, X functions as a must-have factor for Y: it is the entry ticket to the set of cases for which the target is achievable. From this perspective, the goal of an intervention is to raise X for bottleneck cases so that they meet or exceed the threshold and can, in principle, attain the desired outcome.

12.2.2 Stop factors for undesired outcomes

Necessary conditions also matter when the target outcome is undesirable (e.g., risk, disease). In that setting, the same threshold logic flips interpretation: rather than asking what is required to enable a good outcome, the question is what is required for a bad outcome to be possible. A condition that is necessary for an undesired outcome becomes a stop factor: keeping it below its threshold blocks the undesired outcome from occurring.

In the XY-plot, cases in the bottleneck area (with X below the threshold condition level) cannot reach the undesired target level. In that sense, they are protected: the necessary condition is missing or not high enough, so the harmful outcome is not attainable. By contrast, cases with X at or above the threshold lie in the feasible region for the undesired outcome; they are the cases "at risk", because the necessary condition is present and the bad outcome is, at least in principle, possible.

Intervening in this context therefore has a different aim. Instead of lifting cases over a threshold, the goal is to move at-risk cases below it. Reducing X to a value under the threshold removes the necessary condition for the undesired outcome, closing off the possibility of reaching that harmful target. Thus, when the outcome is undesirable, effective prevention focuses on maintaining the condition level below its critical threshold: if the stop factor is kept low enough, the undesired outcome is blocked by necessity. In other words, a single condition can block an undesired outcome if the value of that condition is below the threshold.

12.2.3 Shifting attention from average to individual interventions

NCA's necessity logic is inherently threshold-based. It shifts attention away from interventions based on average effects for a group of cases (Figure 12.2) to interventions based on constraints for individual cases.

Regression-based approaches estimate how much a factor contributes on average to an outcome (nearly always assuming compensatory effects among predictors). Under an average-effect intervention logic, to achieve a desired outcome, the goal generally is to increase contributing factors *as much as possible*, and for an undesired outcome, the goal generally is to decrease them. In contrast, necessity logic identifies which factors are indispensable: those that must be present for a desired or undesired outcome to be possible at all. Necessity-based intervention logic focuses on ensuring that a condition reaches at least the target condition X (for a desired outcome) or remains below it (for an undesired outcome), which applies to every single case. NCA's necessity logic provides such a framework for designing interventions that are effective, efficient, and targeted. It can focus on the relevant cases where the intervention will work: cases that must exit or enter the bottleneck area. This approach complements, and in some situations challenges, the dominant average-effect paradigm that underpins most evidence-based practices (e.g., Dul, 2025). In this way, NCA does not replace but rather enriches causal reasoning in evidence-based practice: it highlights what must (not) exist rather than what merely increases (decreases) an outcome on average.

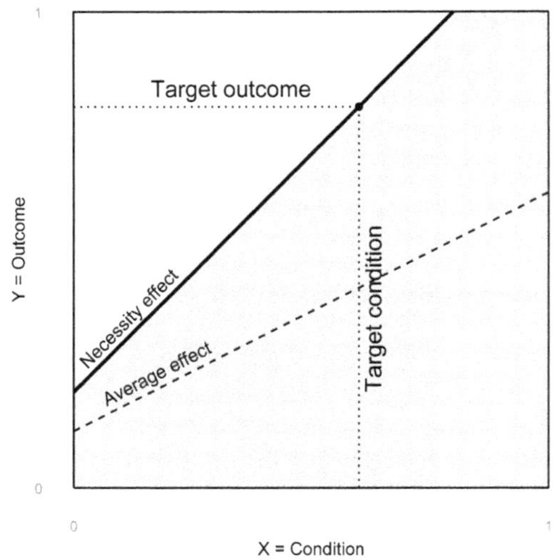

FIGURE 12.2 Shifting intervention perspective from average effect logic to necessity logic.

However, just as it can be difficult to shift the perception from the first woman that is noticed in the picture in Figure 2.5 to the second woman hidden in it, shifting from an average-effect mindset to a necessity-logic mindset is equally challenging (and maybe even more!). Research in the psychology of causality (e.g., Mandel & Lehman, 1998) shows that people naturally overestimate sufficiency ("this condition produces the outcome") and underestimate necessity ("this condition must be present, otherwise the effect cannot occur"). Sufficiency-based causal reasoning is also dominant in learning and working environments. Statistical education is dominated by probabilistic sufficiency models, and many scientific methods are designed to detect average effects. As a result, large parts of society operate on the assumption that optimizing averages for the whole group or for subgroups is the only way to understand and improve systems.

The dominance of average-effect logic is visible, for example, in governmental policies that allocate support based on mean income of (sub) groups, in medical protocols built around average treatment effects from clinical trials, and in organizational performance programs aimed at improving average satisfaction scores. Because actions based on overall averages often fail to address the needs of individual cases, systems frequently try to compensate by creating increasingly fine-grained subgroups and then acting on the average within each subgroup. This approach adds complexity without solving the underlying problem: decisions remain driven by averages, not by the specific conditions that determine what is possible for each individual case.

Recognizing necessity logic requires a deliberate shift from asking "What may help?" to asking "What is indispensable?" This change in perspective is subtle but difficult when the standard is average-effect thinking. With a necessity perspective, it becomes possible to act on the specific cases that require attention, rather than designing interventions for broad groups based on averages.

12.2.4 Examples of application areas

Scholars have applied NCA in areas that are relevant for practice. Their studies show that necessity exists and provides relevant information beyond what is currently known. The overview of Table 12.1 shows that NCA can be applied in virtually any area of practice where success must be realized, or risks must be prevented. Most studies have found potentially important necessary conditions on which the practitioner could act. It illustrates that in any specific practical setting necessary conditions can be identified for action.

TABLE 12.1 Examples of NCA studies that test a necessity hypothesis X is necessary for Y in several practice fields.

	X = condition	Y = outcome	Focal unit
Economics			
1	National adoption of support policies and low-intensity agricultural practices	National environmental sustainability performance in the agricultural sector	Agricultural firm
2	Human capital, institutional business facilities, and market opportunities	Open innovation	Country
Education			
3	Individual characteristics and study-related behaviors (e.g., conscientiousness, attendance, study time)	Academic success (grades/GPA)	Student
4	SAT mathematic scores	GPA in mathematics major	Student
Entrepreneurship			
5	Entrepreneurial role stress	Burnout	Self-employed individual
6	Entrepreneurial ecosystem domains (e.g., political environment, financing)	Deep-tech entrepreneurship	Country
Environmental Sciences			
7	Access to and experience of urban nature	Well-being	Urban resident
8	Planning, goal setting, and R&D capabilities	Eco-innovation	Agricultural firm
Human Resource Management			
9	High performance work practices	Job satisfaction & managerial effectiveness	Organization
10	Technostress related to workplace social media use	Work performance	Organization
International Business/Management			
11	Organizational context (e.g., top management support, technical competencies.)	Stage of adoption of artificial intelligence	Organization
12	Environmental, social, and governance performance scores	R&D expenditure (innovation)	Multinational life sciences firm
Marketing			
13	Personality traits	Impulsive buying behavior	Consumer
Medicine and Health			
14	Metacognition	Motivation	Individual with schizophrenia spectrum disorder

15	Baseline depressive symptoms, self-criticism, rumination and stressful events	Major depressive episodes across two years	Adolescent
Operations and Supply Chain management			
16	Critical success factors for lean practices (e.g., leadership, employee involvement, continuous improvement)	Implementation of lean practices in manufacturing SMEs	Manufacturing firm
17	Formal contracts and interpersonal trust in buyer–supplier relationships	Innovation in buyer–supplier relationships	Buyer-supplier relationship
Psychology/Organizational Behavior			
18	Satisfaction of basic psychological needs for autonomy, competence and relatedness	Work engagement	Employee
19	Levels of fear of missing out and social networking addiction	Psychological well-being of young adults	Young adult
Public Administration			
20	Delay of first response, political decentralization, elderly populations, and urbanization	Early COVID-19 Mortality	Country
21	Institutional trust and individualism	COVID-19 first-response stringency	Country
Strategy			
22	Managers' strategic myopia	Firm performance	Manager
23	Market PL-driven orientation	Market-driving orientation	Company
Technology			
24	Cable features (void fraction and critical current)	Change in current sharing temperature	Twisted Cable-in-Conduct Conductor
25	Evolutionary conservation	Dynamic cooperativity	Amino acid (protein system)
Tourism and Hospitality			
26	Positive affect and carefreeness	Eudaimonic experience	Tourist
27	Organizational capabilities (e.g., transparency, flexibility)	Resilience	Hotel
Transportation			
28	Favorable meteorological conditions (temperature, precipitation, humidity, wind)	Bike rental demand in Washington, DC	Hour (across a two-year period)

Note. 1. Lankoski & Lankoski (2023); 2. Galindo-Martin et al. (2025); 3. Tynan et al. (2020); 4. Zhou et al. (2024); 5. Manchiraju et al. (2024); 6. Dionisio et al. (2023); 7. Allard-Poesi & Massu (2023); 8. Chaparro-Banegas et al. (2024); 9. Garg et al. (2019); 10. Wang & Beh (2025); 11. Solaimani & Swaak (2023); 12. Subramanian et al. (2024); 13. Shahjehan & Qureshi (2019); 14. Luther et al. (2017); 15. Marchetti et al. (2025); 16. Knol et al. (2018); 17. Valk et al. (2016); 18. Ding & Kuvaas (2023); 19. Sirisety et al. (2025); 20. Yan et al. (2023); 21. Liu et al. (2025); 22. Czakon et al. (2023); 23. Vu & Tolstoy (2025); 24. Kwon (2022); 25. Chong & Ham (2023); 26. Lee & Jeong (2020); 27. Ofori (2024); 28. Kumar (2021).

12.3 NCA's contribution to change and innovation

Using NCA in practice means that an action must be taken on identified necessary conditions. NCA does not have its own change or innovation approach, but can add to existing ones. It provides a new perspective by focusing on necessity logic and by selecting and prioritizing (non-)bottleneck cases to be acted on. In this way NCA can add new value to any existing change process be it a classical Plan-Do-Check-Act for continuous improvement, more recent methods for agile product development and management like Scrum, or any other approach that is used in a specific practice.

Each phase of the change process from planning, execution and evaluation can benefit from NCA. NCA can help to decide about the strategic direction of the intervention (focus on must-have and stop factors, and not on nice-to-have factors and average effects). NCA's analytic tools can help to operationalize these strategies into concrete effective and efficient interventions by identifying which conditions and cases must be acted on. Specifically, NCA clarifies what potential must-have factors must be in place, and what potential stop factors must be removed or reduced, for interventions to succeed.

The development of an NCA-based intervention consists of two main parts. The first part is the intervention *strategy*, which establishes the team of stakeholders involved in the intervention, sets the goal and target group for the intervention, suggests the drivers and barriers to achieve that goal, and formulates the necessary condition hypotheses for prioritized drivers and barriers to be analyzed further. In this phase, NCA's strengths as a broader methodological/theoretical approach based on necessity logic are leveraged.

The second part is the intervention *design* based on a study of a group of cases that represent the target group. For this group, existing or new data are collected for the outcome Y and the selected barriers and drivers (X's) expressed in the hypotheses. These data are analyzed using the standard procedures described in Chapter 9. Based on the results, an effective and efficient intervention can be designed that focuses on bottleneck conditions and bottleneck cases.

The following sections 12.3.1 and 12.3.2 outline the details of these two parts.

12.3.1 Part 1: Intervention strategy

12.3.1.1 Establishing the team

After a decision has been made to include NCA in the change process, the NCA-based intervention approach can be integrated as follows. It is assumed that a broad team consisting of stakeholders that affect or are affected by the intervention, is established to allow a successful change project. Having a broad team is not only the common advice for successful change, it is also particularly important for an NCA-based intervention as different stakeholders may have different views on the goals, target outcome and target group of the intervention. Having a diversity of stakeholders helps to avoid that important drivers and barriers (potential necessary conditions) are overlooked (see below). In a business environment, for example, a diverse team could consist of representatives from strategy, marketing, operations, customer services and business analytics, etc., who all could have a different stake in the intervention project and have different views about important necessary conditions.

It can be helpful to assign one team member to focus specifically on the necessity perspective, ideally as a rotating role so everyone develops this skill. During change processes and team discussions, it is easy to lose sight of necessity logic or to slip back into probabilistic sufficiency thinking. A designated team member helps keep the group attentive to necessity-oriented reasoning and supports the collective learning process.

12.3.1.2 Establishing the goal and target group

The *intervention goal* must be clearly defined and preferably shared among all those involved in the design and implementation of the intervention. The goal specifies the intended result of the intervention, and specifies the desired or undesired outcome. Whenever possible, a specific *target outcome* is to be expressed quantitatively. This could be a specific level of the outcome that should be achieved (desired outcome) or prevented (undesired outcome). It is also possible to express this level in terms of the percentage of cases that should be able to achieve the outcome or be prevented from having the outcome. Such a quantification of the goal helps to guide the second part of intervention development: the intervention design. The *target group* is the broader collection of cases for which the intervention goal is set. It may be that for reasons of effectiveness and efficiency, only part of the cases from the target group (the *intervention group*) will actually receive the intervention (see below).

12.3.1.3 Establishing drivers/barriers

Based on the team's knowledge, a list of potential drivers and barriers to achieve the goal is developed through a creative process, for example brainstorming. The goal is to have a long list of potential drivers and barriers for the outcome. If there are multiple outcomes, each outcome has its own list of drivers and barriers. Drivers and barriers are identified based on the team members' expert knowledge and experience, and possibly academic knowledge (see Section 7.4). The reason for this important step is that, in contrast to interventions based on for example big-data or AI-based approaches, an NCA-based intervention is selective and transparent regarding the conditions and cases to be acted on. It only focuses on crucial conditions and relevant cases in order to be effective, efficient, and avoid waste of resources.

12.3.1.4 Formulating necessity hypotheses

The goal of this phase is to reduce the number of drivers and barriers to the essential ones: the potential necessary conditions. Often the number of drivers and barriers that are developed in the previous phase is large. However, many drivers and barriers are "just" important from the perspective of probability and average effects, but not from the perspective of necessity. Through a careful evaluation of the drivers/barriers, it is possible to identify the potential necessary conditions. This evaluation requires a team discussion, and results in a selection of drivers and barriers that the team considers necessary conditions for achieving the goal. If the set of necessary conditions is still large, prioritization may be needed, in particular when an NCA-based intervention is new to the team.

The intervention development continues with the selected potential necessary conditions. For each necessary condition a preliminary necessity hypothesis is formulated and further developed toward a formal necessity hypothesis, as outlined in Chapter 7. For example, this phase specifies variables X (measurable drivers/barriers) and Y (measurable goal), or the selection of a group of cases where the hypothesis is expected to apply (entire target group, selection of target group, or beyond the target group). This information is important for data collection (in Part 2 of the intervention) and the design of an effective intervention. The team should critically evaluate whether the potential necessary condition is *causal*: if the single condition is removed, will the target outcome not be possible? An intervention only works if the condition is causal (see Chapter 2). Since causality cannot be observed, the hypothesis should make causality plausible. If not, the necessary condition, is not a candidate for an intervention based on necessity causal logic.[1]

The result of the strategic phase is one or more causal necessary condition hypotheses formulated as X is necessary for Y where X is the set of expected *necessary* drivers or barriers, and Y is the desired or undesired target outcome (possibly with quantification of the level of the target outcome).

12.3.2 Part 2: Intervention design

An evidence-based intervention is based on an established relationship supported by empirical data. For an NCA-based intervention this means that the team's formulated formal necessary conditions hypotheses is tested with archival or new data obtained about X and Y from the target group, or from a group that is representative for the target group. The aim of this analysis is to find the condition thresholds that are assumed to apply to the entire target group. This is the basis for the intervention in the target group.

Specifically, each hypothesis that is selected in the previous phase is tested to evaluate if the hypothesis is supported and the condition is indeed necessary. If necessity holds, an XY-plot similar to Figure 12.1-right is generated to observe where the ceiling exists and which cases are inside or outside of the bottleneck area. Once the target conditions for the target outcome are estimated, the intervention focuses on removing cases from the bottleneck area (if the outcome is desirable) by increasing the value of the necessary conditions, or moving cases into the bottleneck area (if the outcome is undesirable) by decreasing the value of the necessary condition. Different intervention designs are possible depending on whether it focuses on all or a selection of cases, and on all or a selection of necessary conditions.

It may be concluded that only a specific group of cases from the target group should receive the intervention to benefit from it (*intervention group*). An effective and efficient intervention focuses only on cases in the target group that have condition levels below the threshold (if the target outcome is desired) or have condition levels above the threshold (if the target outcome is undesired).

12.3.2.1 Data collection

A sample of the target group or another group of cases that is representative for the target group is selected for testing the hypothesis. For each case in the sample, scores for X_j (j is the specific index of a necessary barrier/driver) and Y must be available. For proper results, the data must be reliable (same score for repeated measurements under the same circumstances) and valid (scores must correspond to the meaning of the variables in the hypothesis) (Chapter 8).

12.3.2.2 Data analysis

After data collection the data are analyzed according to NCA's standards (Chapter 9). Each hypothesis is rejected or supported (= not rejected). If supported, the condition is considered for the NCA-based intervention. If rejected, the condition may be "just" a contributing factor but is not essential.

[1]The requirement of causality also applies to average effect interventions.

12.3.2.3 Displaying crucial results

For each supported hypothesis, the cases are shown in an XY-plot. In the example of Figure 12.3-left, five cases (A-E) are shown.

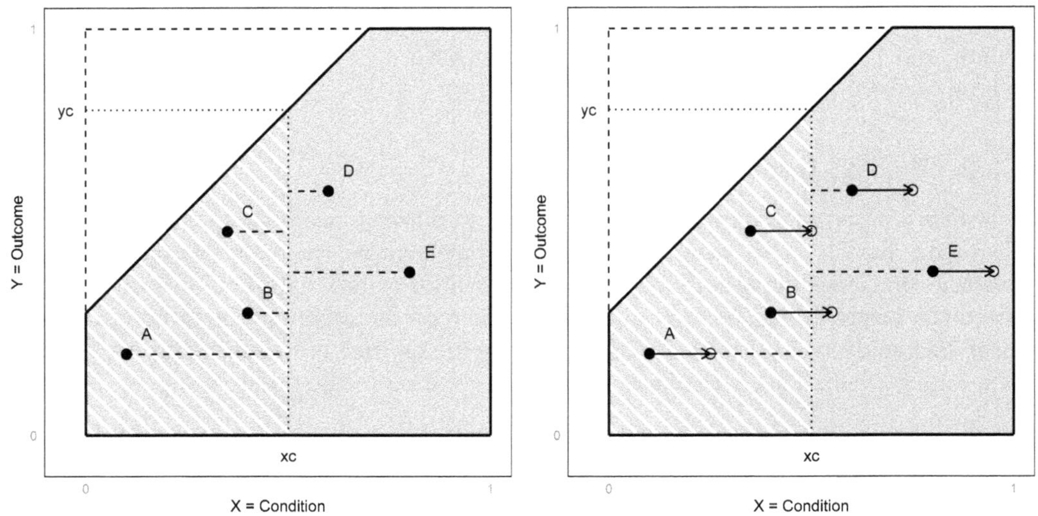

FIGURE 12.3 XY-plot with ceiling line, target outcome y_c, target condition x_c, and five cases. Left: dashed lines are the bottleneck distances; Right: A generic intervention on all cases.

In this plot, the condition X is displayed on the horizontal axis, the outcome Y on the vertical axis, and the five cases with specific X- and Y-values are displayed as dots. The plot also shows the estimated ceiling line, separating the white ceiling zone (empty space) where cases are impossible from the gray feasible area below the ceiling line where cases are possible. In the plot, the target level $Y = y_c$ and the corresponding target condition level x_c are displayed as well. These points indicate that level x_c of X is necessary for target outcome y_c of Y. The threshold value x_c defines the bottleneck area. Cases in this area are the bottleneck cases that cannot reach the target outcome y_c unless their condition X is increased to $x \geq x_c$.

In Figure 12.3-left, cases A, B, and C are bottleneck cases ($x < x_c$) and cases D and E are *non-bottleneck cases.* For the latter, $y = y_c$ is achievable from the perspective of this X. The *bottleneck distance* is the horizontal distance from a case to the target value x_c, indicated by a dashed line. If the outcome is desirable, cases A, B and C cannot achieve the desired outcome y_c. These cases must overcome the bottleneck distance to make y_c possible. For case A more *intervention effort* is needed to overcome the bottleneck distance (larger bottleneck distance) than for cases B and C.

12.4 Interventions for making desired outcomes possible

When the outcome is desirable, cases inside the bottleneck area (bottleneck cases) should be removed from it by increasing the condition value. For example, when it was found that students' study time X is necessary for exam grade Y, and a certain target pass grade y_c

can only be achieved with at least x_c study time, the intervention consists of motivating students with study time $x < x_c$ to increase their study time to $x \geq x_c$.

Figure 12.3-right shows an example of in intervention to make a *desired* target outcome possible. The intervention consists of a generic condition *increase* of 0.15 (15% of the range of 1.00) and is applied to all five cases.

12.4.1 Intervention effort

An *intervention effort* is an externally applied change of the value of the necessary condition. The effort is applied to a case from the target group. A case to which the intervention is applied is called an *intervention case*. For practical or strategic reasons, it may be that not all cases from the target group are intervention cases. The *total intervention effort* is the sum of efforts applied to the intervention cases. The example of Figure 12.3-right shows that the generic intervention is applied to all five cases, such that all these cases are intervention cases. The total intervention effort is $5 \times 0.15 = 0.75$.

An *effective case* is an intervention case that was in the bottleneck area before the intervention and left that area due to the intervention effort. An *ineffective case* is an intervention case that was in the bottleneck area before the intervention, but did not leave the area despite the intervention effort (the effort was too small). An *irrelevant case* is an intervention case that was not in the bottleneck area before and also not after the intervention. An *ignored case* is a bottleneck case to which an intervention is not applied.

So, an effective case is a *bottleneck case* that becomes a *non-bottleneck* case after intervention. An ineffective or ignored case stays in the bottleneck area, and an irrelevant case stays in the non-bottleneck area.

Figure 12.3-right shows that bottleneck cases B and C are effective cases as they left the bottleneck area. Case A is an ineffective case and cases D and E are irrelevant cases.

12.4.2 Intervention effectiveness, waste and efficiency

The goal of an NCA-based intervention is to move the bottleneck cases out of the bottleneck area by raising their condition value to at least the target condition level. *Intervention effectiveness* is the share of treated bottleneck cases (that received the intervention) that leave the bottleneck area after the intervention (effective cases). These are cases that, after intervention, have a condition value at or above the target condition level and are therefore no longer constrained by that condition in achieving the target outcome. Cases that were already outside the bottleneck area before the intervention are irrelevant cases with respect to this goal.

$$\text{Intervention effectiveness} = \frac{\text{Number of effective cases}}{\text{Number of treated bottleneck cases}} \times 100\% \tag{12.1}$$

In Figure 12.3-right, all five cases received the intervention. Two of the three bottleneck cases leave the bottleneck area; moving case A out of the bottleneck area is not successful. The effectiveness is therefore $(2/3) \times 100\% = 66.7\%$. After the intervention, four cases are non-bottleneck cases and are able, in principle, to achieve the desired outcome.

Intervention waste is the amount of intervention effort that does not contribute to moving cases out of the bottleneck area. It has three components: (1) *undershoot*, effort on bottleneck cases that remain bottlenecks; (2) overshoot, effort on effective cases beyond the minimum required effort to reach the target condition; and (3) irrelevant effort on cases that were not bottlenecks before the intervention. In Figure 12.3-right, case A is ineffective because its bottleneck distance ($x_c - x$) exceeds the intervention effort, producing 0.15 waste due to undershoot. Case B is an effective case but shows 0.05 waste due to overshoot. Cases D and E are irrelevant cases, producing 0.30 waste. Total waste is therefore 0.50. Case C is the only effective case without waste.

Intervention efficiency is the share of total intervention effort that is effective (i.e., not wasted) in removing cases from the bottleneck area. It is the complement of waste and is expressed in percentages:

$$\text{Intervention efficiency} = \frac{\text{Total intervention effort} - \text{Total waste}}{\text{Total intervention effort}} \times 100\% \qquad (12.2)$$

In Figure 12.3-right, total intervention effort is 0.75 and total waste is 0.50, so intervention efficiency is (0.75 − 0.50) / 0.75 × 100% = 33.3%.

Evidence-based action following NCA focuses on reducing bottleneck distances by increasing cases' condition values to the target condition level or above ($x \geq x_c$), thereby removing the bottleneck and enabling attainment of the target outcome y_c.

12.4.3 Ideal intervention

It is clear that with an effectiveness of 66.7% and an efficiency of 33.3% the intervention is not ideal. The reason is twofold:

- The intervention is applied to all cases, rather than to the cases in the bottleneck area only.
- The intervention effort is the same for all cases.

An ideal NCA-based intervention strives for 100% effectiveness and 100% efficiency (and 0% waste) so that the intervention goals can be achieved with minimum effort. This can be achieved with a selective intervention that only focuses on cases in the bottleneck area and provides case-specific intervention efforts corresponding to the distance of the case to the condition threshold required for the target outcome.

For the example of Figure 12.3-right, this can be achieved by an intervention design with the following efforts:

Case A: 0.40

Case B: 0.10

Case C: 0.15

Case D: 0

Case E: 0

An ideal intervention targets bottleneck cases that is just enough to be 100% effective and 100% efficient. Some overshoot might be desirable as "safety margin." Although effectiveness would remain 100%, efficiency would slightly decrease as the price for more resilience.

The practical advice for an NCA-based intervention is:

1. Apply the intervention only to cases in the bottleneck area: the cases that are constrained by the specific condition: Act only where something is blocking improvement, and do not intervene where no real bottleneck exists.
2. Match the intervention effort to the case's need. Each case in the bottleneck area has a bottleneck distance indicating how far it is from its target. Rather than a one-size-fits-all approach, the tailored approach overcomes the case's bottleneck distance: how far it is from the target. The intervention effort reflects that distance: the larger the distance the stronger the effort.
3. For quick results start with the easiest cases first: those that are closest to the target condition.

The potential of an NCA-based intervention compared to an average effect-based intervention can be illustrated as follows. Suppose an organization with 500 employees sets new performance goals (target Y). To be able to achieve that goal, employees must have a higher skill level (target X). An average effect analysis finds that higher skills increases performance on average, and an NCA analysis finds that skill is necessary for performance. Based on the average effect results the whole group of employees receives an intervention consisting of a skill training. The training results in each person's skill increase of 0.15 units. This situation corresponds to Figure 12.3-right where each point stands for a group of 100 employees with the starting same skill level. Suppose further that the training costs 1,000$ per unit skill increase (intervention effort). Consequently, a person's 0.15 skill increase costs $0.15 \times 1{,}000 = 150$\$. The total intervention effort is therefore $500 \times 0.15 = 75$, such that the total intervention costs is $75 \times 1000 = 75{,}000$\$ (or 500 persons $\times$ 150\$ $= 75{,}000$\$).

After the intervention, 400 employees have the required skills to make the target outcome possible (employees of groups B, C, D, E). For group A the intervention effort was not strong enough to become non-bottleneck cases. Groups D and E were already non-bottleneck cases before the intervention. Thus, 200 from the 300 employees that were in the bottleneck area and received the intervention left that bottleneck area. Therefore, the *effectiveness* of the intervention is $200/300 \times 100\% = 66.7\%$.

Part of the total intervention effort of 75 is wasted. All effort for group A is wasted due to undershoot ($100 \times 0.15 = 15$ waste). A part of the effort for group B is wasted due to overshoot ($100 \times 0.05 = 5$ waste), and the entire effort for groups D and E is irrelevant and is thus wasted ($200 \times 0.15 = 30$ waste). The *efficiency* of the intervention is therefore $(75 - 50) / 75 \times 100\% = 33.3\%$).

Suppose now that the intervention is based on the NCA study. The NCA-based intervention focuses only on bottleneck cases and takes bottleneck distances into account. Suppose further that a person's skill improvement is maximized to 0.15 (or maximum training cost per person of 150$). Therefore, the NCA-based intervention does not select group A for intervention because this group is too far from minimum required skill level for achieving the target outcome, given the maximum budget per person. This intervention also does not select groups D and E as these groups have already enough skills.

The NCA-based intervention gives an intervention effort of 0.10 to group B and 0.15 to group C. This means that the total intervention effort is $(100 \times 0.10) + (100 \times 0.15) = 25$. The total intervention costs $= 25 \times 1000 = 25{,}000$\$. While after intervention the same number of employees (400) have the required skills for making the new target outcome possible, the intervention costs are considerably lower. The intervention effectiveness and efficiency are both 100% because all 200 cases in the bottleneck area that received the intervention were effective (100% effectiveness) and the intervention does not produce waste (100% efficiency). The results are summarized in 12.2.

TABLE 12.2 Example comparing average effect intervention with NCA-based intervention

	Average effect intervention	NCA-based intervention
Intervention costs	75,000$	25,000$
Intervention effectiveness	66.7%	100%
Intervention efficiency	33.3%	100%

12.5 Interventions for preventing undesired outcomes

In the previous section it is assumed that the outcome is desirable. If the outcome is undesirable, the target outcome is the maximum allowable level of undesired outcome. When the outcome is undesirable, cases outside the bottleneck area (non-bottleneck cases) should enter the bottleneck area by decreasing the condition value. For example, it was found that adolescent rumination (overthinking) may be necessary for depression (Marchetti, Koster, et al., 2025). Adolescents are not at risk for a clinically relevant depression level y_c when the rumination level is below x_c. Here, the intervention consists of decreasing adolescents' rumination to below x_c so that their level of depression y_c does not occur.

Figure 12.4-right shows an example of a generic intervention to reduce the condition value by 0.15.

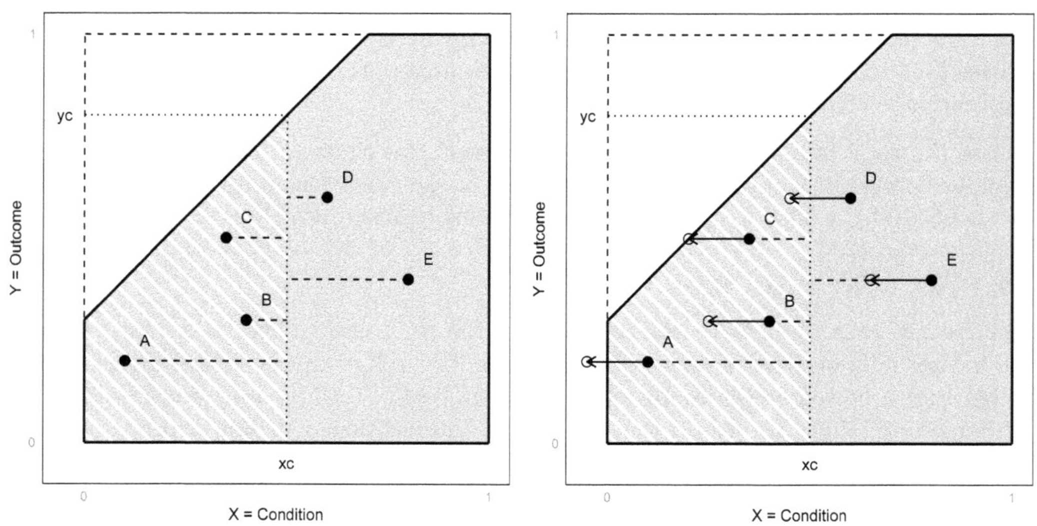

FIGURE 12.4 XY-plot with ceiling line with undesired target outcome y_c, threshold condition x_c, and five cases. Left: dashed lines are the bottleneck distances; Right: A generic intervention on all cases.

Cases A, B, and C are in the bottleneck area. This means that they are protected against level y_c of the undesired outcome and their intervention efforts are a waste. However, cases D and E are at risk. To protect these cases against the undesired level y_c, an intervention effort is needed to reduce their x-levels toward a safe level of $x < x_c$. To achieve this goal, more effort is needed for case E compared to case D. However, the generic effort on case D produces some waste, whereas the intervention effort for case E is not enough to bring this case to the bottleneck area.

An ideal NCA-based intervention (100% effective and 100% efficient) for preventing the undesired target outcome is possible when providing the following intervention efforts:

Case A: 0

Case B: 0

Case C: 0

Case D: −0.10

Case E: −0.30

Evidence-based action, based on the results of NCA, implies designing an intervention that moves cases into or out of the bottleneck area. The goal is to increase or decrease the condition value of cases in order to avoid or ensure that cases are in the bottleneck area, depending on whether the outcome is desirable or undesirable.

12.6 Interventions with multiple necessary conditions

The situation of multiple necessary conditions may be more complex. Each condition has its own bottleneck area and it may be that a case is present in more than one bottleneck area. For a desired outcome (having the possibility to achieve the target outcome), a case must leave *all* bottleneck areas. If the intervention is focused on a specific condition, but the case is also constrained by another condition that is not intervened on, it will not be effective. For a desired outcome, information must be available for all multiple necessary conditions and all conditions must be considered when designing an intervention.

For an ideal intervention (100% effective and 100% efficient) with multiple necessary conditions, two approaches can be used (Table 12.3): The *case-focused* intervention strategy focuses on all conditions in one case at the same time and then moves to the next case (parallel on conditions, sequential on cases). It addresses bottleneck conditions together. The intervention can start with the easiest cases. These are the cases with the lowest (weighted) *total* bottleneck distance for all conditions.[2].

The *condition-focused* intervention strategy focuses on one condition at a time, across all cases (parallel on cases, sequential on conditions). The intervention begins with the condition that has the greatest impact. This is the condition with the largest number of *single* bottleneck cases. Then the intervention moves to the next condition in order of how many single-bottleneck cases it affects, etc. The BIPMA approach discussed in Section 11.4.1 is an example of a condition-focused intervention.

[2]Assuming min-max normalized condition scales such that bottleneck distances and effort are comparable between conditions

TABLE 12.3 Intervention strategies for desired outcomes and multiple conditions

	Work on ...	Prioritize ...
Case-focused	All conditions in one case, one case after another	Cases from low to high bottleneck distance
Condition-focused	One condition after another across all cases	Conditions from high to low number of single bottleneck cases

For an undesired outcome (having no possibility to achieve the target outcome), a case must be in *any* bottleneck area. If the intervention focuses on a specific condition, a case that enters the bottleneck area is effective, regardless of whether the case is outside the bottleneck area of another condition.

TABLE 12.4 Intervention strategies for undesired outcomes and multiple conditions

	Work on ...	Prioritize ...
Case-focused	Selected condition in one case, one case after another	Cases from low to high bottleneck distance
Condition-focused	One condition across all cases	Preferred condition with lowest number of bottleneck cases

In practice it may not be possible to act on all bottleneck cases, to apply case-specific intervention efforts, or to act on all bottleneck conditions. This will negatively affect the effectiveness and efficiency of the intervention.

When developing interventions based on bottlenecks, the following general advice can be given to practice:

- Be selective: only act in bottleneck areas.
- Be proportional: match effort to bottleneck distance.
- Be strategic: choose the right sequencing of the intervention.

12.7 Illustrative example

The illustrative example describes a stylized and partly constructed situation (with existing data).[3] A company wants to improve the frequency of use of a recently launched technological product for daily use. Data from 174 customers show that the product is not used as frequently as expected (Figure 12.5).

[3]The data are the same data as those used in Appendix H.

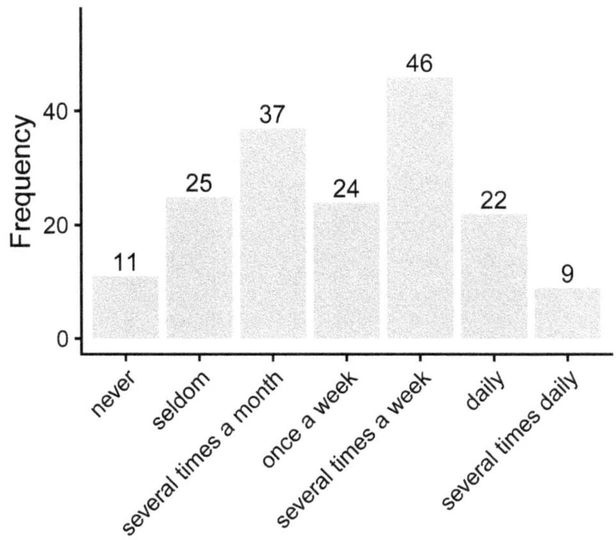

FIGURE 12.5 Example: Frequency of product use by 174 customers.

The company wants to improve this product use with an NCA-based intervention approach. The target outcome is daily use of the product.

12.7.1 Example Part 1: Intervention strategy

In part 1, the intervention strategy is defined. The company establishes a team with representatives from marketing, new product development, operations and customer services. The team follows the change strategy that is common in the organization. One of the team members takes the role of the necessity-perspective-advocate and ensures that this perspective is applied and safeguarded during the process. This role is rotated among team members for learning purposes. The team establishes the goal of daily product use. Via brainstorming sessions the team identifies causal factors related to limited product use, in particular factors that the company can influence. The team identifies the following drivers and barriers that relate to the company's communication with the customer:

- The use of benefit-focused messaging that emphasizes the daily value of using the product, both functionally and in terms of experience.
- Tips on how to use the product.
- The availability of quick and effective customer support, strengthening trust and satisfaction.

Furthermore, the team identifies drivers and barriers related to the product design:

- An acceptable level of physical and cognitive effort to operate the product.
- The compatibility of the product with users' daily routines/devices.
- The extent to which displays are understandable and readable by a diversity of users.

By selecting and combining these drivers and barriers the team identifies four potential critical success factors for daily product use that can be influenced by marketing or re-design of the product. For each of them the hypotheses are formulated:

1. A high level of customer perceived usefulness (X_1) of the product is necessary for daily use of the product (Y);
2. A high compatibility (X_2) of the product is necessary for daily use of the product (Y).
3. A high level of customer perceived ease of use (X_3) of the product is necessary for daily use of the product (Y);
4. A high emotional value (X_4) of the product is necessary for daily use of the product (Y);

12.7.2 Example Part 2: Intervention design

In part 2, the intervention is designed. First, data are collected for the four expected necessary conditions (X) and the frequency of product use (Y). The data are analyzed with NCA. The results indicate that the four conditions are necessary (see Appendix H.5). This means that the selected four conditions are relevant to be acted upon during the intervention. The XY-plots are displayed in Figure 12.6.

Each plot has 174 dots (customers). The stepwise linear ceiling line separates the empty impossible area from the possible area with cases. The horizontal dotted line in each plot at product use level 6 corresponds to daily product use. This is the desired target outcome y_c. The vertical dotted line is the corresponding threshold value of the condition (x_c). The black dots in each plot are cases in the bottleneck area: the area to the left of the threshold condition and below the ceiling line. Black dots are customers that have not achieved the level of the condition that is necessary for the desired target outcome. These customers will not use the product daily because Perceived usefulness is too low (30 = 17% of the customers), Compatibility is too low (19 = 11%), Perceived ease of use is too low (5 = 3%), and/or Emotional value is too low (10 = 6%).

The number and percentage of cases between brackets are obtained from the bottleneck distances table of Table I.1 in Appendix I. This table shows all 38 bottleneck cases (22%), their distances to the threshold conditions, the number of conditions where they are constrained, and their total bottleneck distance to be overcome by the intervention. This means that 22% of the customers have too low values of one or more conditions. The other 78% are able to achieve the target outcome from the perspective of their values of the four necessary conditions. They may or may not have achieved the target outcome as this depends on other factors that are not measured (the necessary condition is not sufficient). The company's intervention should focus only on the 22% cases that are in the bottleneck area. Acting on other cases that already meet the threshold condition is ineffective in an NCA-based intervention. These cases already have the possibility to achieve the outcome. If the company acts only on bottleneck cases, the intervention effort should correspond to the bottleneck distance to avoid ineffectiveness and inefficiency. For 100% effectiveness and 100% efficiency, all cases' bottlenecks must be handled that way.

The example illustrates how NCA can identify multiple necessary conditions for a practical goal (daily product use) and determine their threshold levels; how bottleneck areas and bottleneck cases (22% of customers) can be made visible for each necessary condition; how an intervention can be focused only on those customers in the bottleneck area, rather than

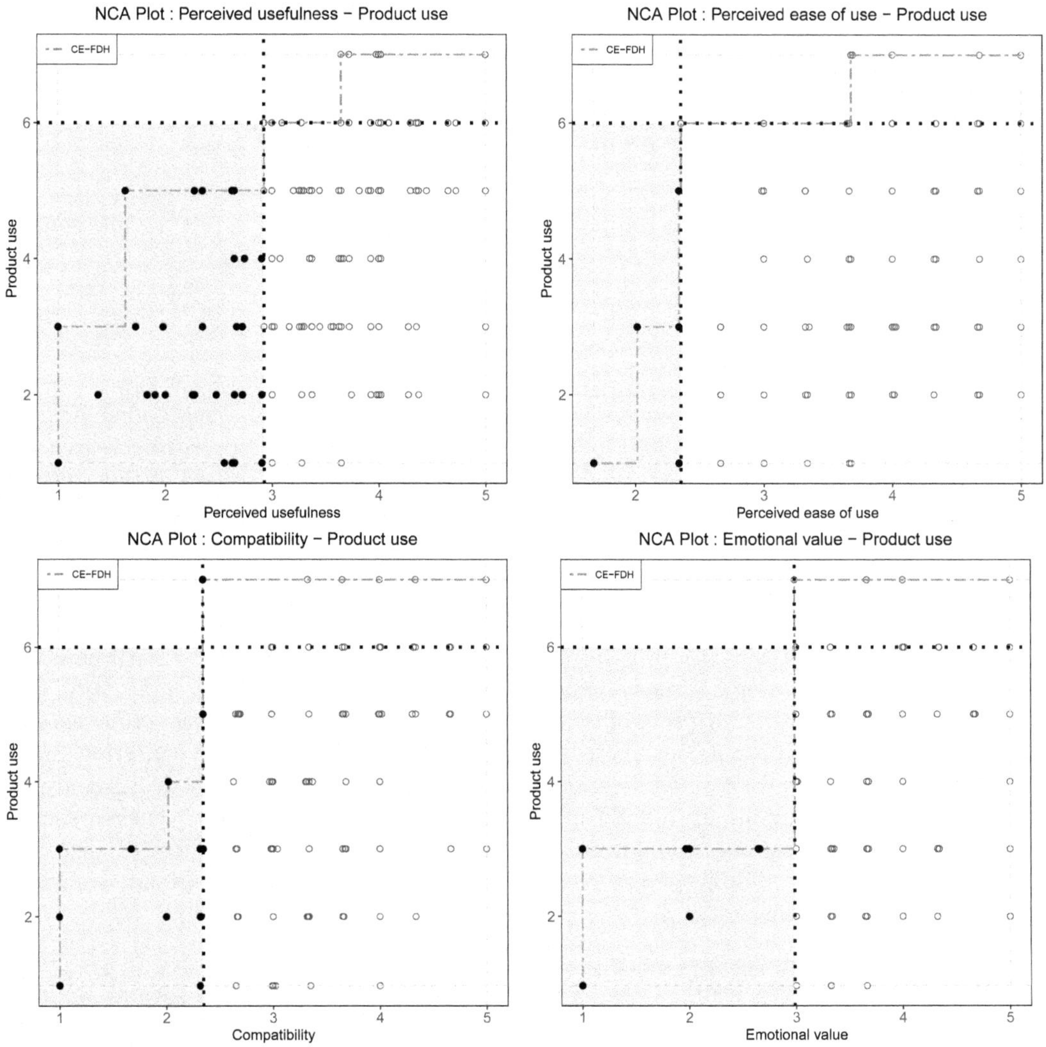

FIGURE 12.6 Example: bottleneck areas (gray) for the four necessary conditions.

on the entire target group; how this selective, threshold-based focus supports more effective and efficient interventions than approaches guided by average effects alone.

Summary and personal reflection

Summary of the book

This book has introduced Necessary Condition Analysis (NCA) as a methodological approach that adds a distinct necessity perspective to the toolbox of empirical research and practical decision-making. NCA starts from a simple idea: some factors are not just helpful contributors to an outcome, they are indispensable. If a necessary condition is not present (at a required level), a desired outcome cannot occur, and no other factor can compensate for this absence. Conversely, for undesirable outcomes, certain stop factors can be kept below a threshold to make the outcome impossible.

Part I of the book set out the principles of NCA. It began in Chapter 2 with causality, clarifying how necessity logic (if not X, then not Y) differs from sufficiency logic (if X, then Y) and from configurational, probabilistic, or typicality-based perspectives. NCA extends classical binary necessity to a continuous view in which different levels of a condition enable different levels of an outcome, while still maintaining a clear zone in which cases are impossible, above a ceiling line in the XY-plot. The theory chapter (Chapter 3) then showed how necessity relationships can be formulated as parsimonious necessity theories, discussed directions of necessity, double necessity, and clarified what "necessary but not sufficient" means in a causal framework.

The mathematical foundations of NCA (Chapter 4) characterize necessity in geometric terms: bounded variables, ceiling lines, degrees of freedom, effect size, and several model fit measures such as ceiling accuracy, sharpness, and spread. Together, these measures describe how well the ceiling line separates the "impossible" empty corner from the feasible area, and how sharply the necessity pattern appears in the data.

The statistics chapter (Chapter 5) then explained how NCA uses permutation tests, simulations, and power analyses to test whether an observed necessity effect could have arisen by chance. It also showed how regression-based concepts such as mediation, moderation, omitted variable bias, and confounding are different or do not apply from the perspective of necessity logic and NCA.

The credibility chapter (Chapter 6) synthesized these foundations into a set of criteria for deciding whether a necessary condition claim is believable: theoretical support (a well-developed necessity hypothesis), practical relevance (a substantively meaningful effect size), statistical significance, robustness checks, and model fit. Together they provide a structured way to evaluate whether an observed ceiling pattern truly supports a necessity claim rather than reflecting a statistical artifact or a trivial result.

Part II translated these principles into concrete steps for conducting NCA studies. First, the hypothesis chapter (Chapter 7) described how to develop formal necessity hypotheses by drawing on explicit and tacit knowledge, building a preliminary necessity theory, performing thought experiments, and adding a causal explanation.

The data chapter (Chapter 8) discussed study designs (observational, longitudinal, case-based, experimental), sampling, case selection, and measurement for quantitative, qualitative, and set-membership data, as well as the use of archival data and data transformations.

The data analysis chapter (Chapter 9) presented NCA's core analytic approach, which involves specifying the model bounding box and the ceiling line, visually inspecting the XY-plot, estimating the ceiling line, computing effect size and p-values, identifying and handling outliers, and assessing model fit. It then distinguished between necessity-in-kind (whether a condition is necessary at all) and necessity-in-degree (how much of a condition is required for a specific target outcome), operationalized via bottleneck tables. These bottleneck tables show which levels of conditions constrain particular target levels of the outcome, thus directly connecting NCA results to potential action. Robustness checks (varying ceiling functions, scopes, thresholds, outlier decisions, and target outcomes) help to assess the stability of necessity claims.

The reporting chapter (Chapter 10) explained how to write up NCA studies for different purposes (empirical, theoretical, methodological introduction, and practical publications) and introduced the SCoRe checklist for high-quality reporting, covering theory, methods, data, analysis, results, and discussion.

The multimethod chapter (Chapter 11) positioned NCA within a broader landscape of methods and explained how NCA can be combined with regression-based approaches (MLR, SEM, PLS) and with QCA. Rather than treating methods as competing, the book advocates a causal-pluralist view where necessity, sufficiency, and probabilistic perspectives complement one another: they are first analyzed separately and then interpreted collectively using tools like NERT, NEST and BIPMA. Each method answers a different causal question and combining them allows a richer understanding of the same phenomenon.

The final chapter (Chapter 12) translated NCA into practice. Here, necessary conditions become levers for change. NCA helps to identify must-have factors for desirable outcomes and stop factors for preventing risks. Its threshold logic and bottleneck areas make it possible to design interventions that target those cases that cannot reach a target outcome unless a specific condition is improved, or that will remain protected from an undesired outcome as long as a risk factor stays below its threshold. The chapter introduced concepts such as intervention effort, effectiveness, waste and efficiency, and illustrated how NCA-based interventions can be more targeted and cost-effective than interventions guided by average-effect logic.

Taken together, the book shows that NCA is not a replacement for existing methods, but a complementary approach that adds a missing causal lens: necessity. It invites analysts to move beyond the question "What works on average?" and to also ask "What is indispensable?" For researchers, NCA encourages the development of necessity theories and provides tools to test them rigorously. For practitioners, NCA offers a way to detect and remove bottlenecks in real systems, whether in education, business, healthcare, public policy, or any other field. With its conceptual framework, analytic tools, reporting standards, software support, and practical guidance, the book aims to enable thoughtful, high-quality use of NCA in both academic and applied work, and to stimulate further development of necessity-based thinking in the years to come.

Reflections on a decade of NCA

The publication of this book coincides with the tenth anniversary of Necessary Condition Analysis. The first formal exposition of NCA appeared in the journal *Organizational Research Methods* in 2016, but the idea itself required another decade of thinking, experimenting, and refining before it was ready to enter the scientific world. My own path toward NCA was far from linear. Trained first in mechanical engineering, then in biomedical engineering, I worked in applied research and consultancy in the field of human factors and ergonomics, and then moved into academia in business and management in the social sciences. I did not begin my career in the fields where NCA would ultimately take root.

I held managerial and administrative positions. These experiences gave me a deep appreciation for practical decision-making: understanding constraints, identifying what is truly essential, and distinguishing what merely helps from what must be present. With my strong connections to engineering and medical sciences, I was accustomed to environments where replication, empirical validation, and evidence-based practice were indispensable. Predictions had to be precise and actionable.

Entering the social science field late and as an outsider was, in some ways, a disadvantage. I lacked the traditional background and networks. But it was also an advantage: I came without disciplinary blinders, free to be curious, to ask naive questions, and to challenge established assumptions. When I moved into the social sciences, I found a landscape overwhelmed by theories and concepts, often complex, difficult to interpret, and hard to translate into action. I saw a surprising tolerance for drawing strong conclusions from single-shot studies. The lack of replication, limited availability of high-quality data, and the growing complexity of models left me wondering how decisions could be made with confidence.

At the same time, it became clear to me that this situation was not due to negligence. The social sciences are young compared to engineering and medicine, and the scale of funding available for research is simply not comparable. Despite these constraints, the social sciences have achieved great advancements. Real-life problems involve people, decisions, organizations, and institutions, and these problems cannot be understood and solved without the insights of the social sciences. Addressing real-world challenges requires the combined strengths of sciences working together.

In this context, I sensed that something essential (literally, something necessary) was missing. The deterministic logic of necessity, which can perfectly predict the absence of an outcome and is foundational in how real systems function, offered a way out. Although the idea emerged during my encounter with the social sciences, it applies equally in engineering and medicine. Indeed, it is now taking hold in those fields as well. I often wondered how it was possible that the necessity perspective of NCA had not been developed earlier.

Before 2016, I also took time to think carefully about dissemination. I reflected on how new ideas spread and where resistance may lie: among gatekeepers, reviewers, editors, and entrenched methodological traditions. I knew that for NCA to have a chance, it should not be perceived as a competitor to established methods but as a complementary lens that adds something genuinely new. I also knew that it would need openness, transparency, and accessibility. From the start, my principle was that NCA should be freely available: free software, free materials, free support, and freely shared ideas. Dissemination meant planting seeds among doctoral students, researchers, and practitioners, and letting those seeds grow where they found fertile ground.

The years following the first publication were a period of extremes. On one side were strong negative reactions, sometimes surprisingly intense. There were science trolls (yes, they exist) who attacked not only the method but also its intentions. There were editors who feared that using NCA was "naive", reviewers who issued strong opinions without knowing the method, and peers who advised others not to engage with the method. Some speculated that "if applied in medicine, people will die." Others insisted that NCA should first prove itself in *the* top-tier econometrics journal before it could be taken seriously. And there were cynical remarks about miraculous empty spaces or trivial data transformations. These moments were not easy, but in areas where NCA has been accepted, they are largely behind us now. As Schopenhauer observed, new ideas first encounter ridicule, then opposition, and finally acceptance as self-evident.

Yet there were equally strong and uplifting experiences. Brave editors such as James LeBreton were willing to give the method a fair chance. Scholars like Herman Aguinis immediately saw the potential of necessity thinking, and many colleagues contributed enthusiastically (as recognized in the Acknowledgments). Even some early critics who once dismissed the method later became among the strongest supporters once they understood what NCA actually offers. A few even speculated wildly that the invention merited a Nobel Prize or could be a panacea for all methodological challenges (it is not!).

Alongside these positive developments, I also encountered weaker applications: rushed implementations, superficial readings, and a kind of "science populism" when people present NCA with great certainty while missing its underlying logic, sometimes even giving incorrect answers. Others tried to monetize the method, which only strengthened my commitment to keep NCA free and openly available.

Even with this turbulence, the growth of NCA has been remarkable. More scholars have used the method than I ever imagined, and many have helped push it forward. I am proud of what has been achieved: the development of concepts, terminology, mathematical foundations, statistical procedures, and tools that previously did not exist. Much had to be invented from scratch like names for ceiling lines, new concepts like the bottleneck table, or new measures of fit, and many other components that now form the method's core.

Why has NCA not yet been more widely adopted in practice, despite its potential? One reason is academic incentives: researchers are encouraged to move quickly to the next publication, leaving limited time for sustained engagement with practical application. Many who use NCA strive for both academic rigor and practical relevance, but the system rarely rewards that combination. Nevertheless, I remain optimistic. With growing experience, more applications in practice, and continued methodological development, NCA is becoming increasingly mature, and its practical value is gradually being recognized.

Looking forward, I see three important directions:

- Broader uptake in academia, including in fields such as the medical sciences, where NCA can help prevent undesired outcomes by identifying and eliminating indispensable single conditions, as well as in disciplines that still focus primarily on explaining why outcomes occur on average.

- Further deepening of the method. Once a method is established and accepted, the following decades allow for refinement, extension, and specialization.

- More real-world applications demonstrating how NCA can guide action, decision-making, and practical interventions. Practitioners who currently rely solely on average-effect logic can benefit from necessity logic as well.

At the same time, there are no shortcuts in NCA. Applying NCA requires methodological competence, respecting the fundamentals of NCA: understanding its deterministic necessity logic (with possible exceptions), its mathematical and statistical characteristics, and its assumptions, without slipping into probabilistic or conventional statistical thinking. Founders and early methodologists of other methodological approaches such as SEM, QCA, and grounded theory have warned in their own traditions that users should refrain from mechanical application, neglected assumptions, and overclaiming results. I have no illusion that NCA is immune to these tendencies, or that this book can fully prevent them. NCA, too, will be (and from the very beginning already has been) misinterpreted, routinized, overclaimed, and at times misused.

I therefore call upon NCA users to respect its principles: build strong theoretical support with explicit necessity hypotheses; avoid tweaking analyses to "find" necessity (since non-necessity can be equally informative); and report approaches, results and conclusions transparently. These principles are operationalized in the SCoRe checklist for analysts, authors, readers, peers, editors, reviewers and others to assess and enhance the quality of NCA applications.

Finally, on a personal note, with thankfulness, many people have joined me along the way, contributing to the development and growth of NCA. But only one person has witnessed every effort, doubt, frustration, and joy, and has supported me unconditionally throughout the entire journey. For that, Renata, I am deeply grateful.

Part III

Additional materials

A

Nomenclature and glossary

A.1 Nomenclature

α	Threshold value for statistical significance
a	Slope of a linear function
b	Intercept of a linear function
C	Size of the ceiling zone; A point on the ceiling line (x_c, y_c)
ca	Model fit metric ceiling accuracy
cp	Model fit metric complexity
d	Effect size
df	Degrees of freedom of a necessity model
ex	Model fit metric exceptions
f	General symbol for a mathematical function
F	Feasible area
ft	Model fit metric fit
h	Horizontal segment of a ceiling line
H	Hypothesis
i	Index for an observation or of a segment of a ceiling line
j	Index for a variable
J	Number of necessary conditions
k	Number of outliers in a set of outliers
$lowP$	Number of cases in the lowP-zone
$lowS$	Number of cases in the lowS-zone
max	Maximum value of a given range
$medP$	Number of cases in the medP-zone
$medS$	Number of cases in the medS-zone
min	Minimum value of a given range
n	Sample size
N	Necessary condition (in NEST)
nc	Necessary condition; necessary cause (near arrow in causal graph)
ns	Model fit metric noise
p	p-value for a statistical significance test
P	Number of peers that define the ceiling line
S	Scope: the size of the bounding box
sd	Standard deviation

DOI: 10.1201/9781003430766-A

sh	Model fit metric sharpness
sp	Model fit metric spread
su	Model fit metric support
T	Time indicator
v	Vertical segment of a ceiling line
X, Y, Z	Variable names (capitals)
x, y, z	Variable values (lower case)

A.2 Glossary

This glossary defines important terms used throughout this book. Defined terms are printed in **bold**. When a definition contains another defined term, it is printed in *italic*. When a defined term appears more than once within a single definition, only its first occurrence is highlighted.

 # above. An *NCA parameter* indicating the number of *cases* above the *ceiling line*.

Absence/low value of condition/outcome. A value of a *condition* or *outcome* that is close to its minimum. Also see *Presence/high value of condition/outcome*, *Direction (of a necessary condition)*.

Absolute inefficiency. The area of the *bounding box* where the *necessary condition* does not constrain the *outcome* and the *outcome* is not constrained by the necessary condition. Also see *Relative inefficiency*, *Condition inefficiency*, *Outcome inefficiency*.

Academia. The world of higher education and research, like universities, focusing on theory, discovery, and teaching. Also see *Practice*.

Accuracy. See *Ceiling accuracy*, *p-value accuracy*.

Adjacent corner. A corner in the *bounding box* next to the *corner* of interest. Also see *Opposite corner*, *Expected empty corner*.

Alternative hypothesis. A statistical concept describing a competing claim to the *null hypothesis*. Also see *Formal necessity hypothesis*.

Analyst. A *scholar* or *practitioner* who uses NCA or other methodological approaches.

Binary logic. Two-valued logic where statements can only be true or false. Also see *Conditional logic*, *Causal logic*, *Necessity logic*, *Sufficiency logic*.

BIPMA. *Bottleneck Importance Performance Map Analysis*. An *NCA* tool for giving practical advice based on *SEM* results and NCA results, based on *single bottleneck cases*. Also see *NERT*, *IPMA*, *cIPMA*.

Bivariate analysis. An analysis of two variables. Also see *Multiple bivariate analyses*, *XY-plot*, *XY-table*, *Multivariate analysis*.

Boolean logic. See *Binary logic*.

Bottleneck case. A *case* without the necessary level(s) of the *condition(s)*, such that it is unable to achieve the *target outcome*. Also see *Non-bottleneck case*, *Ignored case*, *Irrelevant case*, *Single bottleneck case*, *Multiple bottleneck case*.

Bottleneck distance. The gap between a *case*'s *condition* value (x) and *target condition* value (x_c).

Bottleneck Importance Performance Map Analysis. See *BIPMA*.

Bottleneck table. A tabular representation of the *ceiling line* showing which values of the *condition(s)* is/are necessary for a particular value of the *outcome*.

Bounding box. The rectangle defined by observed or theoretical limits of the *condition* and the *outcome*. Also see *Scope*, *Empirical scope*, *Theoretical scope*, *Tight bounding box*.

c-accuracy. See *Ceiling accuracy*.

C-LP. Ceiling - Linear Programming. A linear *ceiling line* based on minimization of heights of the CE-FDH peers by linear programming. Also see *CE-FDH*, *CR-FDH*, *CR-VRS*, *QR*.

Case. An instance of a *focal unit*.

Case selection. The selection of one or a small number of *cases* from a set of cases for inclusion in a *small-n study*. Also see *Sampling*.

Case study. A *study design* in which one or a small number of *cases* is selected for a *small-n* study. Also see *Observational study*, *Longitudinal study*, *Experimental study*.

Causal explanation. A story that connects *causes* to effects.

Causal interpretation. The causal explanation of a relationship judged as credible based on extra-data considerations such as domain knowledge, simplicity, stability, and the plausibility of underlying assumptions. Often data allow multiple interpretations.

Causal logic. The logic in which statements are causal relationships. Also see *Binary logic*, *Conditional logic*, *Necessity logic*, *Sufficiency logic*.

Causal narrative. See *Causal explanation*.

Causal perspective. The perspective selected by the *analyst* to theorize or infer causality from data. Also see *Necessity perspective*, *Typicality perspective*, *Sufficiency perspective*, *Probabilistic sufficiency perspective*.

Causal-pluralism study. A *multimethod study* that uses multiple *causal perspectives* and corresponding different methods with the same data. Also see *Method-triangulation study*, *Separate studies*, *Theory-method fit*.

Causal relationship. A relationship between two variable characteristics X and Y of a *focal unit* in which a value of X (or its change) permits, or results in a value of Y (or in its change).

Causal underdetermination. The idea that data are compatible with multiple distinct causal perspectives and explanations, so the causal structure cannot be uniquely identified from the data alone.

Cause. A variable characteristic X of a focal unit of which the value (or its change) permits, or results in a value (or its change) of another variable characteristic Y. Also see *Necessity cause*, *Sufficiency cause*.

CB-SEM. Covariance-Based SEM. A *SEM* approach that estimates model parameters based on the covariance matrix. Also see *PLS-SEM*.

CE-FDH. Ceiling Envelopment - Free Disposal Hull. A stepwise linear *ceiling line* based on the free disposal hull. Also see *Ceiling line*, *CR-FDH*, *CE-VRS*.

CE-VRS. Ceiling Envelopment - Variable Returns to Scale. A piecewise linear *ceiling line* consisting of oblique line segments such that envelope is concave when the *ceiling zone* is in *corner 1*. Also see *Ceiling line*, *CE-FDH*, *CR-FDH*, *CR-VRS*.

Ceiling accuracy. A *model fit* metric expressing the extent to which *cases* are on or below the *ceiling line*.

Ceiling Envelopment - Free Disposal Hull. See *CE-FDH*.

Ceiling Envelopment - Variable Returns to Scale. See *CE-VRS*.

Ceiling line. The borderline within the *bounding box* between the area where *cases* are impossible (*ceiling zone*) and the area where cases are possible (*feasible area*), such that a point (x, y) is on the ceiling if and only if, for any other point (x', y') in the bounding box, it holds that if $x' < x$ and $y' > y$, then (x', y') is in the ceiling zone and if $x' \geq x$ and $y' \leq y$, then (x', y') is in the feasible area. Also see *CE-FDH*, *CR-FDH*, *C-LP*, *CE-VRS*, *CR-VRS*, *QR*.

Ceiling outlier. An *outlier* in the *ceiling zone*. Also see *Scope outlier*.

Ceiling Regression - Free Disposal Hull. See *CR-FDH*.

Ceiling zone. The area in a *corner* of the *bounding box* where points cannot exist, except for *exceptions* and *noise*. Also see *Feasible area*.

Chain (of) necessity. See *Necessity chain*.

Chain necessity effect. The necessity effect of the first element in a *necessity chain* on the final *outcome* (the extent to which the initial condition constrains the maximum attainable level of the chain's endpoint through the intermediate necessary links).

Chained necessity. See *Chain necessity effect*.

cIPMA. Combined Importance Performance Map Analysis. An extended version of *IPMA* that includes *NCA* results by considering a condition's number of bottleneck cases. Also see *BIPMA*.

Combined Importance Performance Map Analysis. See *cIPMA*.

Complete hypothesis. A *necessity hypothesis* with a plausible *causal explanation* about why the *condition* is necessary for the *outcome*. Also see *Formal necessity hypothesis*, *Trivial hypothesis*, *Testable hypothesis*, *Plausible hypothesis*.

Completeness. See *Complete hypothesis*.

Complexity. A *model fit* metric expressing the number of parameters that are needed for describing the necessity model (*degrees of freedom* $- 4$). Also see *Model specification*.

Concept. The varying characteristic of a *focal unit* of a *theory*. Also see *Condition*, *Outcome*.

Conceptual model. A visual representation of a *hypothesis* of a *necessity theory* in which the *condition* and *outcome* are presented by rectangles and the relationship between them by an arrow. The arrow originates in the condition and points to the outcome and represents the causal direction with the letters *nc* near it to express the *direction* of the necessary condition. Also see *Formal necessity hypothesis*.

Condition. A varying characteristic X of a *focal unit* of which the value (or its change) permits, or results in a value (or its change) of another varying characteristic Y (which is called the *outcome*). Also see *Necessary condition*, *Sufficient condition*, *Outcome*, *Predictor (variable)*.

Condition inefficiency. The area of the *bounding box* where the *condition* does not constrain the *outcome*. Also see *Absolute inefficiency*, *Relative inefficiency*, *Outcome inefficiency*.

Conditional logic. If-then statements that can only be true or false. Also see *Binary logic*, *Causal logic*, *Necessity logic*, *Sufficiency logic*.

Confounder. A term used in regression-based analysis indicating the theoretical role of a *concept* as a common cause of two other concepts and that may account for part of the observed relationship between them. Also see *Mediator*, *Moderator*.

Contingency table. See *XY-table*.

Continuous necessity. A *necessity relationship* in which the *condition* and the *outcome* can have infinite numbers of levels (values). Also see *Dichotomous necessity*, *Discrete necessity*.

Contrast test. A *permutation test* in *NCA* in which for each *case* the labels X_1 and X_2 are switched with 50/50 probability. Also see *Null test*, *Independent test*, *Paired test*.

Control variable. A *variable* that is added in a regression-based data analyses for improving the prediction of the *outcome* and avoiding biased estimation of regression coefficients. Also see *Predictor (variable)*, *Predicted variable*.

Convenience sample. A *sample* in which the instances are selected for convenience of the *analyst*. Also see *Random sample*, *Purposive case selection*.

Corner 1. Upper-left corner of the *XY-plot* or *bounding box*.

Corner 2. Upper-right corner of the *XY-plot* or *bounding box*.

Corner 3. Lower-left corner of the *XY-plot* or *bounding box*.

Corner 4. Lower-right corner of the *XY-plot* or *bounding box*.

Corner of interest. The corner of the *bounding box* implied by the *necessity hypothesis*. Also see *Formal necessity hypothesis*, *Corner 1*, *Corner 2*, *Corner 3*, *Corner 4*.

Counterexample. A case that is inconsistent with the *necessity hypothesis*. Also see *Exception*, *Outlier*, *Noise*.

Covariance-Based SEM. See *CB-SEM*

CR-FDH. Ceiling Regression Free Disposal Hull. A linear *ceiling line* based on a trend line through the *CE-FDH* peers. Also see *C-LP*, *QR*, *CR-VRS*.

Credibility. The trustworthiness of conclusions about necessity identified from data. Also see *Statistical credibility*, *Empirical credibility*.

Crisp-set Qualitative Comparative Analysis (csQCA). See *csQCA*.

CR-VRS. Ceiling Regression Variable Returns on Scale. A linear *ceiling line* based on a trend line through the *CE-VRS* peers. Also see *CR-FDH*, *C-LP*, *QR*.

csQCA. Crisp-set Qualitative Comparative Analysis. A variant of *QCA* in which all conditions and outcomes are expressed as binary (0/1) *set membership scores*, indicating full non-membership (0) or full membership (1) in a set. Also see *fsQCA*.

d. See *Effect size*.

Data. Recordings of evidence generated in the process of *data collection*. Also see *Measurement*, *Qualitative data*, *Quantitative data*, *Longitudinal data*.

Data analysis. The interpretation of *scores* obtained in a *study* in order to generate the result of the study. Also see *Qualitative data analysis*, *Quantitative data analysis.*

Data collection. The process of identifying and selecting one or more objects of measurement, extracting evidence of the value of the relevant variable properties or characteristics from these objects, and recording this evidence. Also see *Data*, *Measurement*, *Score*, *Dataset.*

Data Generation Process. See DGP.

Dataset. A collection of *scores* obtained from *data collection.*

Degrees of freedom. The minimum number of parameters that are needed for describing a necessity model. See also *Model specification*, *Complexity.*

Deterministic necessity. Necessity from the *deterministic perspective.* Also see *Typicality necessity.*

Deterministic perspective. A *causal perspective* taken by the *analyst* that a *cause* always influences an effect without *exceptions.* Also see *Probabilistic perspective*, *Typicality perspective.*

DGP. Data Generation Process. The hypothetical mechanism that produces the data that are observed.

Dichotomous necessity. A *necessity relationship* in which the condition or the outcome has only two levels (values). Also see *Discrete necessity*, *Continuous necessity.*

Direct effect. A term used in regression-based analyses indicating the effect of X on Y without going through an intermediate variable (*mediator*). Also see *Indirect effect*, *Total effect*, *Moderator.*

Direction (of a necessary condition). An indication whether the *condition* and *outcome* in the *necessity hypothesis* are absent or present. Also see *nc*, *Formal necessity hypothesis.*

Discrete necessity. A *necessity relationship* in which the condition or the outcome has a finite number of levels (values). Also see *Dichotomous necessity*, *Continuous necessity.*

Domain. See *Theoretical domain.*

Effect size. The magnitude of the constraint that a *necessary condition* poses on the *outcome* expressed as the size of the *ceiling zone* relative to the size of the *bounding box* (*scope*). Also see *p-value.*

Effect size threshold. The value of the *effect size* selected by the *analyst* for evaluating the *necessity hypothesis.* Also see *Formal necessity hypothesis*, *Statistical significance threshold.*

Effective case. An *intervention case* that is in the *bottleneck area* (*bottleneck case*) before the *intervention* and removed from it after the intervention. Also see *Ineffective case*, *Irrelevant case*, *Ignored case.*

Embedded necessity theory. A *necessity theory* with one or more *necessity relationships* and one or more other relationships. Also see *Pure necessity theory.*

Emerging theory. A *theory* that is not (yet) broadly accepted in *academia* or *practice.* Also see *Established theory.*

Empirical credibility. The trustworthiness of the *necessity hypothesis*, the *study design*, the *data*, and the *data analysis* used to draw conclusions about necessity. Also see *Formal necessity hypothesis*, *Statistical credibility*, *Effect size*, *p-value*, *Robustness*, *Model fit.*

Empirical data. *Data* from the real world. Also see *Simulation data*

Empirical scope. The *bounding box* (or its area) defined by empirically observed minimum and maximum values of the *condition* and the *outcome.* Also see *Theoretical scope*, *Scope.*

Empty area. See *Ceiling zone.*

Empty corner. See *Ceiling zone.*

Empty space. See *Ceiling zone.*

Established theory. A *theory* that is broadly accepted in *academia* or *practice.* Also see *Emerging theory.*

Evidence-based intervention. An *intervention* that is based on a *hypothesis* tested with *empirical data.*

Exception. A *case* in the *lowP-zone* of the *ceiling zone* that is not the result of error and is accepted by the *analyst* as a rare counterexample of necessity. Also see *Outlier*, *Noise*, *Typicality perspective.*

Exceptions. A *model fit* metric expressing the number of *cases* in the *lowP-zone* of the *ceiling zone* that the *analyst* considers an *exception.*

Expected empty corner. The area in the *bounding box* that is predicted to be empty according to the *necessity hypothesis.* Also see *Formal necessity hypothesis*, *Ceiling zone.*

Experimental study. A *study design* in which the *condition* is manipulated and the *outcome* is observed. Also see *Observational study*, *Longitudinal study*, *Case study.*

Expert knowledge. *Tacit knowledge* of a *scholar* or *practitioner.* Also see *Explicit knowledge.*

Explicit knowledge. Documented knowledge in the literature from *academia* and *practice* to be used for formulating a *formal necessity hypothesis.* Also see *Tacit knowledge*, *Expert knowledge.*

Extension. A horizontal or vertical line segment added to the ceiling line that connects *peers* to envelop the *feasible area.*

Falsification. The view that theories and hypotheses cannot be proven true, but can only be proven false.

Feasible area. The area within the *bounding box* below the *ceiling line* for *corner 1* or *corner 2*, or above the ceiling line ("floor line") for *corner 3* or *corner 4*. Also see *Ceiling zone.*

Fiss chart. A solution table of *QCA* with specific notation for presence or absence of a *condition*, and for core or peripheral conditions. Also see *Solution table*, *NEST.*

Fit. A *model fit* metric expressing the *effect size* of a selected *ceiling line* as percentage of the effect size of the *CE-FDH* ceiling line.

Floor line. A *ceiling line* in *corner* 3 or 4.

Focal unit. The unit of a *theory* or *hypothesis.* Examples are 'employee', 'team', 'company', 'country'. Also see *Theory*, *Theoretical domain.*

Formal necessity hypothesis. A *necessity relationship* that is *theory-grounded*, *testable*, *non-trivial*, *plausible* and *complete.* Also see *Preliminary necessity hypothesis.*

Formal necessity theory. A *necessity theory* in which the *necessity hypotheses* are *formal necessity hypotheses*. Also see *Preliminary necessity theory*.

Fragility. The absence of *robustness*.

fsQCA. Fuzzy-set Qualitative Comparative Analysis. A variant of *QCA* in which conditions and outcomes are expressed as fuzzy *set membership scores* ranging from 0 to 1. Also see *csQCA*.

Fuzzy-set Qualitative Comparative Analysis. See *fsQCA*.

High-high necessity. A *necessity relationship* where the *presence/high value* of the *condition* is necessary for the *presence/high value* of the *outcome* ('+ nc +'). Also see *nc*, *Low-high necessity*, *High-low necessity*, *Low-low necessity*, *direction*.

High-low necessity. A *necessity relationship* where the *presence/high value* of the *condition* is necessary for the *absence/low value* of the *outcome* ('+ nc -'). Also see *nc*, *High-high necessity*, *Low-high necessity*, *Low-low necessity*, *direction*.

Hypothesis. A statement about the relationship between *concepts* or *variables*. Also see *Hypothesis testing*, *Necessity hypothesis*, *Formal necessity hypothesis*.

Hypothesis testing. The procedure of making a decision about the credibility of a *hypothesis*. Also see *Statistical credibility*, *Empirical credibility*, *Null hypothesis test*, *Thought experiment*.

Ignored case. A *non-intervention case* that is in the *bottleneck area* (*bottleneck case*) before and after the *intervention*. Also see *Irrelevant case*, *Effective case*, *Ineffective case*.

Importance Performance Map Analysis. See *IPMA*.

Independent test. A *permutation test* in which from a pool of two groups of *cases* are randomly reassigned to two groups of the same sizes. Also see *Null test*, *Contrast test*, *Paired test*.

Indirect effect. A term used in regression-based analyses indicating the effect of X on Y through a *mediator*. Also see *Direct effect*, *Total effect*, *Moderator*.

Ineffective case. An *intervention case* that is in the *bottleneck area* (*bottleneck case*) before and after the *intervention*. Also see *Effective case*, *Irrelevant case*, *Ignored case*.

Inefficiency. A set of *NCA parameters* related to the extent to which the *necessary condition* covers the range of the *condition* and the *outcome*. Also see *Absolute inefficiency*, *Relative inefficiency*, *Condition inefficiency*, *Outcome inefficiency*.

Infeasible area. See *Ceiling zone*.

Influential case. A *case* that has a large influence on the necessity *effect size* when removed. Also see *Outlier*.

Informant. A person who is the object of measurement for a *variable* and who is knowledgeable about that variable and informs the *analyst* about it. Also see *Subject*.

Instance of a focal unit. A single *case*.

Intervention. An external action applied to a *case* in order to increase or decrease the *condition* value with the intention to remove the *case* from the *bottleneck area* (for a desired *target outcome*) or move it into the bottleneck area (for an undesired target outcome).

Intervention case. A *case* to which an *intervention* is applied. Also see *Non-intervention case*, *Effective case*, *Ineffective case*, *Irrelevant case.*

Intervention design. The approach to collect and analyze data and displaying crucial results using the *bottleneck area.*

Intervention effectiveness. The number of *treated bottleneck cases* that have left the *bottleneck area* (*effective cases*) as a percentage of *intervention cases.*

Intervention efficiency. The amount of *intervention effort* on *intervention cases* that was effective in removing cases from the *bottleneck area.* It is the complement of *intervention waste* and is expressed in percentages.

Intervention effort. The magnitude of change in the *condition* value produced by the *intervention.*

Intervention goal. The intended result of an *intervention* specified as desired or undesired *outcome* and possibly by a specific *target outcome.*

Intervention group. A group consisting of *intervention cases.*

Intervention strategy. The approach to establish the team for the *intervention*, set the goals and the *target groups*, evaluate drivers and barriers, and formulate hypotheses. See also *Intervention design.*

Intervention waste. The amount of *intervention effort* that does not contribute to moving *cases* out of or into the *bottleneck area.* Also see *Undershoot waste*, *Overshoot waste.*

Irrelevant case. An *intervention case* that is not in the *bottleneck area* (*non-bottleneck case*) before (and after) the intervention. Also see *Ignored case*, *Effective case*, *Ineffective case.*

IPMA Importance Performance Map Analysis. A tool for giving practical advice based on *SEM* results. Also see *cIPMA*, *BIPMA.*

Iso-purity line. See *Iso-P line.*

Iso-P line. A line representing points with the same *purity* value. Also see *Iso-S line*, *Model fit*, *NCA ribbon.*

Iso-solidity line. See *Iso-S line.*

Iso-S line. A line representing points with the same *solidity* value. Also see *Iso-P line*, *Model fit*, *NCA ribbon.*

Large-n study. A study with a large number of *cases.* n stands for the number of cases. Also see *Small-n study.*

Likert scale. A *rating scale* in the format of a limited number of points (e.g., 5 or 7) that can represent a person's response. Also see *Informant*, *Subject.*

Logic. See *Binary logic*, *Causal logic*, *Conditional logic*, *Necessity logic*, *Sufficiency logic.*

Longitudinal data. *Scores* of *condition* and *outcome* that are measured at multiple time points. Also see *Quantitative data*, *Qualitative data*, *Set membership data.*

Longitudinal study. A *study design* with *data collection* at multiple time points. Also see *Observational study*, *Case study*, *Experimental study.*

Low-high necessity. A *necessity relationship* where the *absence/low value* of the *condition* is necessary for the *presence/high value* of the *outcome* ('- nc +'). Also see *nc*, *High-high necessity*, *High-low necessity*, *Low-low necessity*, *direction*.

Low-low necessity. A *necessity relationship* where the *absence/low value* of the *condition* is necessary for the *absence/low value* of the *outcome* ('- nc -'). Also see *nc*, *High-high necessity*, *Low-high necessity*, *High-low necessity*, *direction*.

Low-purity zone. See *lowP-zone*.

Low-solidity zone. See *lowS-zone*.

lowP-zone. The area of the *ceiling zone* far from the *ceiling line*. Also see *lowS-zone*, *medP-zone*.

lowS-zone. The part of the *feasible area* far from the *ceiling line*. Also see *lowP-zone*, *medS-zone*.

Measurement. The process in which *scores* are generated for data analysis. Also see *Data*, *Measurement validity*, *Measurement reliability*.

Measurement reliability. The degree of precision of a *score* when the measurement is repeated. Also see *Measurement*, *Measurement validity*.

Measurement validity. The extent to which procedures of *data collection* and of scoring can be considered to meaningfully capture the ideas contained in the concept of which the value is measured. Also see *Measurement*, *Measurement reliability*.

Mediator. A term used in regression-based analyses indicating the theoretical intermediate role of a *concept* between two other concepts. Also see *Moderator*, *Confounder*, *Necessity chain*.

Medium-purity zone. See *medP-zone*.

Medium-solidity zone. See *medS-zone*.

medP-zone. The area of the *ceiling zone* close to the *ceiling line*. Also see *lowP-zone*, *Iso-P line*.

medS-zone. The part of the *feasible area* close to the *ceiling line*. Also see *lowS-zone*, *Iso-S line*.

Membership score. See *Set membership data*.

Method-triangulation study. A *multimethod study* that uses a single *causal perspective* and corresponding different methods with the same or different data. Also see *Causal-pluralism study*, *Separate-studies*, *Theory-method fit*.

Min-max normalization. A linear transformation that maps values from an original range [min,max] to a chosen range (e.g., 0–1, or 0–100). Also see *Standardization*.

Model fit. The extent to which a model captures the pattern in the data. Also see *Complexity*, *Fit*, *Ceiling accuracy*, *Noise*, *Exceptions*, *Support*, *Spread*, *Sharpness*, *Purity*, *Solidity*.

Model specification. The process of selecting the functional form of the *ceiling line* and the *bounding box* for chosen *condition(s)* and *outcome*.

Moderator. A term used in regression-based analyses indicating the theoretical role of a *concept* as qualifier of the relationship between two other concepts. Also see *Mediator*, *Confounder*, *Subgroup ceiling line*.

Monte Carlo simulation. A computational method that uses repeated random sampling to approximate the behavior of a system or the value of a quantity. Also see *Power*.

Multimethod study. A hypothesis-testing *study* that uses different methods for testing one or more *hypotheses* with the same or a different *causal perspective* and with the same or different *data*. Also see *Method-triangulation study*, *Causal-pluralism study*, *Separate studies*, *Theory-method fit*.

Multiple bivariate analyses. A series of *bivariate analyses*. Also see *XY-plot*, *XY-table*, *Multivariate analysis*.

Multiple bottleneck case. A *case* that is a *bottleneck case* for multiple *conditions*.

Multiple NCA. The application of *NCA* with the same *outcome* and different *conditions*, where each condition-outcome pair is analyzed with NCA.

Multivariate analysis. An analysis that involves multiple variables simultaneously. Also see *bivariate analysis*, *Multiple bivariate analyses*.

Natural distribution. The distribution describing how a system behaves under its unaltered, observable conditions, without forced manipulation.

nc. An abbreviation of 'necessary condition' or 'necessity cause' placed near the arrow in a *conceptual model* and preceded and followed by '+' or '-' to indicate the *direction* of the necessary condition. Also see *high-high necessity*, *low-high necessity*, *high-low necessity*, *low-low necessity*.

NCA. Necessary Condition Analysis. A methodological approach developed by Jan Dul that uses a necessity causal logic (methodology) for identifying necessary conditions from data (method). Also see *NESS*, *QCA*, *SCC*.

NCA approach. See *NCA*.

NCA method. The part of *NCA* concerned with data analysis and empirical testing. Also see *NCA methodology*.

NCA methodology. The part of *NCA* concerned with causal logic and theory. Also see *NCA method*.

NCA parameter. A quantity to evaluate a *necessary condition*. Also see *Ceiling zone*, *Effect size*, *# above*, *Model fit*, *Inefficiency*.

NCA Ribbon. The zone in the *bounding box* consisting of the *medP-zone* and *medS-zone*. Also see *Purity*, *Solidity*.

NCA statistical test. One of NCA's *permutation tests* to estimate the p-value. Also see *Null test*, *Contrast test*, *Independent test*, *Paired test*.

NCA-Extended Solution Table. See *NEST*.

Necessary condition. A *cause* that must exist in order for the *outcome* to exist. Also see *Sufficient condition*.

Necessary Condition Analysis. See *NCA*.

Necessary condition hypothesis. See *Necessity hypothesis*, *Formal necessity hypothesis*.

Necessary condition in degree. See *NiD*.

Necessary condition in kind. See *NiK*.

Necessary Element of a Sufficient Set. See *NESS*.

Necessity cause. See *Necessary condition*.

Necessity chain. A sequence of linked necessary conditions where each condition is required for the next outcome in the sequence to be possible. Also see *Chain necessity effect*.

Necessity corner. See *Expected empty corner*.

Necessity hypothesis. A *hypothesis* about a *necessity relationship*. Also see *Preliminary necessity hypothesis*, *Formal necessity hypothesis*.

Necessity-in-degree. See *NiD*.

Necessity-in-kind. See *NiK*.

Necessity logic. A *causal logic* describing a *necessity relationship*. Also see *Binary logic*, *Conditional logic*, *Sufficiency logic*.

Necessity model. The *ceiling line* and the *bounding box* representing *necessity-in-degree*. Also see *Model specification*.

Necessity perspective. The causal perspective on necessity selected by the *analyst*. In *NCA* the *Deterministic perspective* or the *Typicality perspective* can be selected. Also see *Sufficiency perspective*, *Probabilistic sufficiency perspective*.

Necessity relationship. A *causal relationship* in which the *cause* always, probably or typically enables or constrains the *outcome*. Also see *Necessity hypothesis*, *Formal necessity hypothesis*, *Deterministic necessity*, *Typicality necessity*.

Necessity theory. A theory consisting of at least one *necessity hypothesis*, with specified *focal unit*, *condition(s)* and *outcome(s)*, and a defined *theoretical domain*. Also see *Preliminary necessity theory*, *Formal necessity theory*.

NESS. Necessary Element of a Sufficient Set. A tool developed by Richard W. Wright that argues that an event is a legal cause of an outcome if and only if the event was a necessary element of a set of conditions that together were sufficient for the outcome to occur. Also see *NCA*, *QCA*, *SCC*.

NEST. NCA-Extended Solution Table. A *solution table* to with an extra column representing the *condition's* minimum required necessity level according to *NCA*. Also see *Fiss chart*.

NHST. Null Hypothesis Statistical Test. A procedure for deciding whether *sample* data provide enough evidence to reject a *null hypothesis*. Also see *Null test*, *Contrast test*, *Independent test*, *Paired test*, *Alternative hypothesis*, *Formal necessity hypothesis*.

NiD. Necessity-in-degree. A *necessary condition* that is quantitatively formulated as level x of X is necessary for level y of Y. Also see *NiK*.

NiK. Necessity-in-kind. A *necessary condition* that is qualitatively formulated as X is necessary for Y. Also see *NiD*.

Noise. A *model fit* metric expressing the number of *cases* in the *medP-zone* of the *ceiling zone* as a percentage of all cases in the *bounding box*. Also see, *Counterexample*, *Exception*, *Outlier*.

Non-bottleneck case. A *case* with satisfied *target condition* level. Also see *Bottleneck case*, *Necessity-in-degree*, *Bottleneck table*.

Non-intervention case. A *case* to which an *intervention* is not applied. Also see *Intervention case*, *Ignored case.*

Non-trivial hypothesis. A *hypothesis* where both the absence of the *condition* and the presence of the *outcome* are possible; otherwise it is trivial. Also see *Formal necessity hypothesis.*

Normal distribution. A continuous probability distribution in which values cluster around a mean in a symmetric, bell-shaped pattern, defined by its mean and standard deviation. Also see *Uniform distribution*, *Truncated normal distribution*, *Skewed distribution.*

Normalization. See *Min-max normalization.*

Null hypothesis. The statistical concept describing the claim of no (nil) effect or relationship in the population. Also see *Null test*, *Contrast test*, *Independent test*, *Paired test*, *Alternative hypothesis.*

Null hypothesis test. See *NHST.*

Null test. A *permutation test* in which Y is randomly shuffled across *cases* while keeping X fixed. Also see *Contrast test*, *Independent test*, *Paired test.*

Observational study. A *study design* in which *variables* are observed in the real life context without manipulation by the *analyst.* Also see *Longitudinal study*, *Case study*, *Experimental study.*

Omitted variable bias. The estimation error that is made in a regression-based analysis when a *variable* is omitted from the regression model specification.

Opposite corner. A corner in the *bounding box* diagonally across the *corner of interest.* Also see *Adjacent corner.*

Outcome. The varying characteristic Y of a *focal unit* of which the value (or its change) is the result of, or is permitted by a value (or its change) of another varying characteristic X (which is called the *condition*). Also see *Predicted variable.*

Outcome inefficiency. The area of the *bounding box* where the *outcome* is not constrained by the *condition.* Also see *Absolute inefficiency*, *Relative inefficiency*, *Condition inefficiency.*

Outlier. A point (*case*) in the *bounding box* that is considered to be 'far away' from the other points (cases) and has a large influence on the *effect size* if removed. Also see *Ceiling outlier*, *Scope outlier*, *Counterexample*, *Exception*, *Noise*, *LowP-zone*, *MedP-zone.*

Overall ceiling line. A *ceiling line* that applies to the total group of cases. Also see *Subgroup ceiling line.*

Overshoot waste. *Intervention waste* caused by too much *intervention effort* beyond what is required for reaching the *target condition.* Also see *Undershoot waste.*

p-value. The probability of obtaining a result that is greater than or equal to the observed result when the *null hypothesis* is true. Also see *Permutation test*, *NCA statistical test.*

p-value accuracy. The estimated difference between the exact *p-value* and the estimated p-value Also see *Permutation test*, *NCA statistical test*

p-value threshold. See *Statistical significance threshold.*

Paired test. A *permutation test* in which for each *case* it is randomly decided whether to switch the labels. Also see *Null test*, *Contrast test*, *Independent test.*

Partial Least Squares SEM. See *PLS-SEM*.

Peer. A point in the *XY*-plane that is used to define a *ceiling line*. Also see *CE-FDH*, *CE-VRS*, *CR-FDH*, *CR-VRS*, *C-LP*.

Permutation test. A statistical test that builds an empirical null distribution by repeatedly randomly permuting the observed *X-Y* data to break any systematic relationship, recomputing *NCA*'s *effect size* each time, and then comparing the original effect size to this distribution to obtain the *p-value*. Also see *Null test*, *Contrast test*, *Independent test*, *Paired test*.

Plausible hypothesis. A *hypothesis* for which virtually no *cases* are reasonably expected in the *ceiling zone*. Also see *Formal necessity hypothesis*, *Trivial hypothesis*, *Testable hypothesis*, *Complete hypothesis*.

Plausibility. See *Plausible hypothesis*.

PLS-SEM. Partial Least Squares SEM. A component-based SEM approach that estimates model parameters to maximize explained variance in the predicted variables. Also see *CB-SEM*.

Population. The set of instances of a *focal unit* defined by one or a small number of criteria and that is defined in the *theoretical domain*. Also see *Sample*.

Potential necessary condition. A *condition* that is suggested to be necessary based existing knowledge and is a candidate for formulating a *necessity hypothesis*. Also see *Preliminary necessity hypothesis*, *Formal necessity hypothesis*, *Expert knowledge*, *Tacit knowledge*, *Explicit knowledge*.

Potential outlier. A *case* that has a large influence on the *effect size* if removed. Also see *Outlier*.

Power. The probability that a statistical test correctly rejects the null hypothesis when a specific alternative hypothesis is true. Also see *p-value*, *Statistical credibility*.

Practice. The real-world application of knowledge, like business, policy, or services, focusing on action, outcomes, and impact. Also see *Academia*, *Practitioner*.

Practitioner. An investigator, a data analyst, a data scientist, etc. active in *practice*. Also see *Analyst*, *Scholar*.

Predicted variable. Name used for *outcome* in the context of a regression-based analysis. Also see *Predictor (variable)*.

Predictor (variable). Name used for a *condition* in the context of a regression analysis. Also see *Predicted variable*.

Preliminary necessity hypothesis. A "best guess" *necessity hypothesis* based on existing sources of knowledge about *potential necessary conditions* and to be developed toward a *formal necessity hypothesis*.

Preliminary necessity theory. A *necessity theory* in which the *necessity hypotheses* are not yet *formal necessity hypotheses*. Also see *Formal necessity theory*.

Presence/high value of condition/outcome. A value of a *condition* or *outcome* that is close to its maximum. Also see *Absence/low value of condition/outcome*, *Direction (of a necessary condition)*.

Probabilistic sufficiency perspective. The probabilistic perspective selected by the analyst that a cause probably produces an effect. Also see *Necessity perspective Deterministic perspective*, *Typicality perspective*, *Sufficiency perspective.*

Proposition. A statement about the relationship between *concepts* of a *theory.* Also see *Hypothesis.*

Pure necessity theory. A *necessity theory* with only necessity relationships. Also see *Embedded necessity theory.*

Purity. A measure of closeness of a point in the *ceiling zone* to the *ceiling.* Also see *Solidity*, *Model fit*, *NCA ribbon.*

Purposive case selection. A selection of *cases* from the *theoretical domain* in which instances are selected purposefully (e.g., instances with a certain value of the *outcome*). Also see *Convenience sample*, *Random sample.*

Qualitative Comparative Analysis. See *QCA.*

QCA. A comparative method developed by Charles Ragin that analyzes combinations of conditions (configurations) that are sufficient for the outcome using set theory. Also see *fsQCA*, *csQCA*, *NCA*, *NESS*, *SCC.*

QR. Quantile Regression. A linear (pseudo) *ceiling line* obtained through a regression method that estimates the relationship between X and a chosen conditional high quantile of Y. Also see *Ceiling line, CR-FDH, CR-VRS, C-LP.*

Qualitative data. *Scores* expressing in words or letters the extent to which a *case* has a property or characteristic. Also see *Quantitative data*, *Longitudinal data*, *Set membership data.*

Qualitative data analysis. Identifying and evaluating a pattern in qualitative data.

Quantitative data. *Scores* expressing in numbers the extent to which a *case* has a property or characteristic. Also see *Qualitative data*, *Longitudinal data*, *Set membership data.*

Quantitative data analysis. Generating and evaluating a pattern in *quantitative data.* Also see *Qualitative data analysis.*

R. A programming language and environment for statistical computing and graphics, used to import, manage, analyze, and visualize data.

Random sample. A *sample* in which instances have the same probability of being selected from the *population* into the sample. Also see *Convenience sample*, *Purposive case selection.*

Rating scale. A method in which a person assigns a value to an object. Also see *Informant*, *Subject.*

Relative inefficiency. The total area of the *bounding box* where the *necessary condition* does not constrain the *outcome* and the outcome is not constrained by the necessary condition, expressed as percentage of the *scope.* Also see *Absolute inefficiency*, *Condition inefficiency*, *Outcome inefficiency.*

Replication. Conducting a test of a *hypothesis* in another instance, or in another group or *population* of instances of the *focal unit.*

Ribbon. See *NCA ribbon.*

Robustness. The extent to which the conclusions of a *study* remain essentially unchanged when the *analyst* makes other plausible choices (e.g., model specification, evaluation criteria). Also see *Fragility*.

Robustness check. An analysis where the original analysis is re-run using other plausible choices by the *analyst* to conclude if the results are *robust* or *fragile*.

Sample. A set of *instances* selected from a *population* of the *theoretical domain*. Also see *Convenience sample*, *Random sample*, *Purposive case selection*.

Sampling. The process of drawing a *sample*. Also see *Sampling frame*.

Sampling frame. A list of all instances of a *population*. Also see *Random sample*.

SatP. Saturated purity zone. The *feasible area* including the ceiling line where *Purity* = 1.

SatS. Saturated solidity zone. The *feasible area* including the ceiling line where *Solidity* = 1.

Saturated purity zone. See *SatP*.

Saturated solidity zone. See *SatS*.

Scatter plot. See *XY-plot*

SCC. Sufficient-Component Cause model. A conceptual framework in epidemiology (also called causal pie) developed by Kenneth J. Rothman, that explains how an *outcome* occurs when a complete set of component causes collectively forms a sufficient cause. Also see *NCA*, *NESS*, *QCA*.

Scholar. A researcher active in *academia*. Also see *Practitioner*, *Analyst*.

Scope. The area of the *bounding box*. Also see *Empirical scope*, *Theoretical scope*.

Scope outlier. An *outlier* on the bounds of the *bounding box*.

Score. A value assigned to a *condition* or *outcome* based on *data* or *expert knowledge*. Also see *Quantitative data*, *Longitudinal data*. *Set membership data*.

SCoRe checklist. The Strengthening (theoretical rigor), Conducting (data & analysis quality), and Reporting (transparency) checklist. A tool for authors, editors and reviewers to evaluate the quality of an *NCA study* and its reporting.

SEM. Structural Equation Modeling. A model consisting of a measurement model specifying how indicators relate to the underlying constructs ("latent variables") and a structural model specifying the (average) relationships among the constructs. (latent variables). Also see *PLS-SEM*, *CB-SEM*.

Sensitivity. See *TPR*.

Separate studies. *Studies* that use multiple *causal perspectives* and corresponding different methods with different *data*. Also see *Method-triangulation study*, *Causal-pluralism study*, *Theory-method fit*.

Set membership data. *Scores* expressing the extent to which a *case* belongs to a set. Also see *Qualitative data*, *Quantitative data*, *Longitudinal data*

Sharpness. A *model fit* metric expressing the difference in density of *cases* in the *lowS-zone* and cases in the *lowP-zone*.

Significance. See *Statistical significance*, *Substantive significance*.

Simulated data. *Data* that are fabricated, for example, for a *Monte Carlo simulation.* Also see *Empirical data.*

Single bottleneck case. A *bottleneck case* for only one *condition.* Also see *BIPMA*.

Skewed distribution. A distribution with a longer tail on one side (right-skewed or left-skewed). Also see *Truncated normal distribution*, *Uniform distribution*, *Normal distribution.*

Small-n study. A *study* with one or a small number of *cases.* The letter n stands for the number of cases. Also see *Large-n study.*

Solidity. A measure of closeness of a point in the *feasible area* to the *ceiling.* Also see *Purity*, *Model fit*, *NCA ribbon.*

Solution table. A table presenting the results of a *QCA* sufficiency analysis as the configurations (combinations of *conditions*) that consistently led to the *outcome* (pass the frequency and consistency thresholds set by the analyst) together with their consistency and coverage information. Also see *QCA*, *Fiss chart.*

Specificity. See *TNR.*

Spread. A *model fit* metric expressing the evenness of the X positions of the *cases* in the *medS-zone* and on the *ceiling line.*

Spurious relationship. An observed relationship that seems causal but is more plausibly explained by another model.

Standardization. A linear transformation that centers and scales values using the sample mean μ and standard deviation σ, producing values in standard deviation units. Also see *Normalization.*

Statistic. A number computed from a *sample* that summarizes the evidence against the *null hypothesis.* Also see *Effect size.*

Statistical credibility. The trustworthiness of conclusions about necessity identified from *data* when necessity is absent or present in the *population.* Also see *Empirical credibility*, *TPR*, *TNR.*

Statistical generalization. The statement that the *study* results that are obtained in a *sample* of a *population* also apply to the population from which the sample is drawn.

Statistical significance. The meaningfulness of the *effect size* from a statistical perspective. Also see *p-value*, *Substantive significance.*

Statistical significance threshold. The *p-value* (α) selected by the *analyst* for evaluating the *formal necessity hypothesis.* Also see *Effect size threshold.*

Structural Equation Modeling. See *SEM.*

Study. An academic research activity or a project in practice.

Study design. A category of procedures for selecting or generating one or more instances of a *focal unit* as well as for analyzing the *data* that are observed or generated in the selected or generated instance or instances. Also see *Observational study*, *Longitudinal study*, *Experimental study*, *Case study.*

Subgroup ceiling line. A *ceiling line* that applies to a subgroup of the total group of *cases.* Also see *Overall ceiling line.*

Subject. A person who is the object of measurement and an instance of the *focal unit* of the *theory*. Also see *Informant*.

Substantive significance. The meaningfulness of the effect size *from a practical perspective*. Also see *Statistical significance*.

Sufficiency cause. See *Sufficient condition*.

Sufficiency logic. A causal logic describing a sufficiency relationship. Also see *Sufficiency relationship*, *Necessity logic*.

Sufficiency perspective. The causal perspective on sufficiency selected by the *analyst*. Also see *Necessity perspective*, *Probabilistic sufficiency perspective*.

Sufficient condition. A *cause* that always results in an outcome. Also see *Necessary condition*.

Sufficient-Component Cause model. See *SCC*.

Support. A *model fit* metric expressing the number of *cases* in the *medS-zone* of the *feasible area* including the cases on the ceiling line as a percentage of all cases in the *bounding box*.

Tacit knowledge. Undocumented knowledge in *academia* and *practice* to be used for formulating a *formal necessity hypothesis*. Also see *Explicit knowledge*, *Expert knowledge*.

Target condition. The level of a *condition* that is necessary for the *target outcome*. Also see *Necessity-in-degree*.

Target group. The group of *cases* for which an *intervention* goal is set.

Target outcome. A desired or undesired level of the *outcome* set by the *analyst* that is to be achieved or to be avoided. Also see *Target condition*, *Necessity-in-degree*.

Test. Determining whether a *hypothesis* is rejected or not rejected (supported) in an instance or in a group or *population* of instances selected from the *theoretical domain*.

Test statistic. See *Statistic*.

Testable hypothesis. A *hypothesis* with measurable *variables* (variables that can have *scores*). Also see *Formal necessity hypothesis*, *Trivial hypothesis*, *Plausible hypothesis*, *Complete hypothesis*.

Testability. *See Testable hypothesis*.

Theoretical domain. The universe of instances of a *focal* unit of a *theory* or *hypothesis* where the theory or hypothesis is supposed to hold. Also see *Necessity theory*, *Formal necessity hypothesis*.

Theoretical justification. The availability of a *formal necessity hypothesis* when evaluating a *necessity relationship*.

Theoretical scope. The *bounding box* (or its area) defined by specified minimum and maximum values of the *condition* and the *outcome*. Also see *Empirical scope*, *Scope*.

Theorizing. Development of a *formal (necessity) hypothesis*.

Theory. A (set of) *hypotheses* (*propositions*) regarding the relationships between the varying characteristics (*concepts*) of a *focal unit*, and the description why the relations exist (*causal explanation*) in a *theoretical domain*. Also see *Formal necessity hypothesis*.

Theory-grounded hypothesis: A *hypothesis* that is embedded in a *necessity theory.* Also see *Formal necessity hypothesis.*

Theory-in-use. A more or less consistent set of beliefs in *practice* about the world. Also see *Theory*, *Formal necessity hypothesis.*

Theory-method fit. The situation that the *causal perspective* is aligned with the data analysis method.

Thought experiment. The mental evaluation to check the validity of the *necessity theory* and its *hypothesis.* Also see *Hypothesis testing*, *Formal necessity hypothesis.*

Tight bounding box. The pair of *bounding box* and *ceiling line* when the *ceiling* zone holds the single *corner of interest.*

TNR. True Negative Rate (*specificity*). The ability of an approach to correctly identify that necessity does not exist in the *population.* Also see *TPR*, *Credibility*, *Statistical credibility*, *Empirical credibility*, *Formal hypothesis. Effect size*, *p-value*, *Robustness*, *Model fit.*

Total effect. A term used in regression-based analyses indicating the combined effect of the *direct effect* and the *indirect effect* of X on Y. Also see *Direct effect*, *Indirect effect.*

Total intervention effort. The sum of *intervention efforts* applied to the *intervention cases.*

TPR. True Positive Rate (*sensitivity*). The ability of an approach to correctly identify that necessity exists in the *population.* Also see *TNR*, *Credibility*, *Statistical credibility*, *Empirical credibility*, *Formal hypothesis. Effect size*, *p-value*, *Robustness*, *Model fit.*

Treated bottleneck cases. *Bottleneck cases* that received the *intervention.*

Trivial hypothesis. A hypothesis in which the absence of the condition or the presence of the outcome are not possible. Also see *Formal necessity hypothesis*, *Testable hypothesis*, *Plausible hypothesis*, *Complete hypothesis.*

Triviality. See *Trivial hypothesis.*

True Negative Rate. See *TNR.*

True Positive Rate. See *TPR.*

Truncated normal distribution. A *normal distribution* that is restricted to a range. Also see *Uniform distribution*, *Skewed distribution*

Typicality necessity. Necessity from the *typicality perspective.* Also see *Deterministic necessity.*

Typicality perspective. A lens taken by the *analyst* that a *cause* always influences an effect, but that there can be rare *exceptions.* Also see *Necessity perspective*, *Sufficiency perspective*, *Probabilistic sufficiency perspective.*

Undershoot waste. *Intervention waste* caused by insufficient *intervention effort* such that *cases* do not move from or into the *bottleneck area.* Also see *Overshoot waste.*

Uniform distribution. A distribution in which all values within a specified range are equally likely. Also see *Truncated normal distribution*, *Normal distribution*, *Skewed distribution.*

Unit box. A special case of a *bounding box* in which the *condition* and the *outcome* are both limited to the interval [0,1]. Also see *Min-max normalization.*

Variable. The operationalized *concept* (varying characteristic of a *focal unit* of a *hypothesis*).

Visual inspection. The procedure by which patterns are discovered or compared by looking at the *scores* or a graphical representation of the *scores*.

XY-plot. A graphical representation of the relationship between *condition* and *outcome* in a coordinate system with *cases* shown as points. Also see *XY-table*.

XY-table. A matrix representation of the relationship between *condition* and *outcome* with the number of *cases* shown in the cells. Also see *XY-plot*.

B

Software

B.1 NCA with R

The main NCA software is a free package for the R platform called NCA (Dul & Buijs, 2026), which supports conducting a quantitative analysis of the data. The package is built on several other R packages:

gplots (Warnes et al., 2024),

quantreg (Koenker, 2024),

KernSmooth (Wand, 2023),

lpSolve (Berkelaar et al., 2023),

ggplot2 (Wickham et al., 2024),

doParallel (Corporation & Weston, 2022),

foreach (Analytics & Weston, 2022a),

iterators (Analytics & Weston, 2022b),

plotly (Sievert et al., 2024),

truncnorm (Mersmann et al., 2023),

RSQLite (Müller et al., 2026).

The first version of this software was released just before the online publication of NCA's core paper (Dul, 2016b) on July 15, 2015. Since then, the software has been updated several times to correct bugs, integrate new NCA developments, and make improvements based on feedback from users. The following versions have been released through 17 June 2026:

NCA_1.0 (2015-07-02)

NCA_1.1 (2015-10-10)

NCA_2.0 (2016-05-18)

NCA_3.0 (2018-08-01)

NCA_3.0.1 (2018-08-21)

NCA_3.0.2 (2019-11-22)

NCA_3.0.3 (2020-06-11)

NCA_3.1.0 (2021-03-02)

NCA_3.1.1 (2021-05-03)

DOI: 10.1201/9781003430766-B

NCA_3.2.0 (2022-04-05)

NCA_3.2.1 (2022-09-15)

NCA_3.3.0 (2023-02-06)

NCA_3.3.1 (2023-02-10)

NCA_3.3.2 (2023-06-27)

NCA_3.3.3 (2023-09-05)

NCA_4.0.0 (2024-02-16)

NCA_4.0.1 (2024-02-23)

NCA_4.0.2 (2024-11-09)

NCA_4.0.3 (2025-10-14)

NCA_4.0.4 (2025-11-14)

NCA_4.0.5 (2025-12-21)

NCA_5.0.0 (2026-03-20)

NCA_5.0.1 (2026-04-28)

NCA_5.0.2 (2026-06-17)

When the first digit of the version number increases, major changes are implemented. Version 1 was the first limited version, version 2 was the first comprehensive version, version 3 introduced the statistical test for NCA, and version 4 introduced new functions. The second digit refers to larger changes and the third digit minor changes. Until 31 December 2025, the software was downloaded 69,983 times from the CRAN website (the Comprehensive R Archive Network).

During the publication of this book, version 5.0.0 of the software is released. This version is an extensive update with new main functions and NCA-specific helper functions (name starts with `nca_` ...) and helper functions that are general helper functions (name starts with `nca_util_` ...). Version 5.0.2 is used in this book.

To facilitate the analysis, some main functions of the NCA software are:

`nca_analysis`: conducts the core analysis (estimating ceiling line, effect size, *p*-value, etc.).

`nca_output`: displays the results.

`nca_outlier`: evaluates potential outliers.

`nca_difference`: conducts statistical difference test for comparing two effect sizes.

`nca_power`: conducts a power analysis.

`nca_random`: generates a random dataset representing population necessity.

B.2 Additional R-code

For some specialized applications, the book uses additional helper functions that are not (yet) integrated in the NCA package. For example, functions are available with templates for tables and figures that can be used to compare NCA results with results of other methods (Chapter 11). This includes NERT for combining NCA results with regression results (Table 11.4), NEST for combining NCA results with QCA results (Section 11.6.4), and BIPMA for giving practical advice based on combined NCA and SEM results (Section 11.4.1.7).

Additional R code can be downloaded via the book-page: https://jandul.github.io/NCA/.

B.3 Demonstration of `NCA` with `R`

This demonstration assumes little or no knowledge about `R` and `RStudio` and helps the user run the main `NCA` functions mentioned above.[1]

Example

The demonstration uses an example of necessary conditions for innovation in buyer-supplier co-operation (Van der Valk et al., 2016). The dataset of this early example of an NCA application is part of the NCA software.[2] This demonstration follows the four stages of the NCA approach:

1. Formulate a formal necessity hypothesis (Chapter 7).
2. Collect data (Chapter 8).
3. Analyze data (Chapters 9, 11).
4. Report results (Chapter 10).

B.3.1 Stage 1: Formulate the formal necessity hypothesis

It is assumed that the hypotheses of the example were developed according to the requirements of a formal necessity hypothesis (Chapter 7). Only a limited explanation is provided here.

H1: Contractual detail (X_1) is necessary for innovation performance (Y); high-high direction.

H2: Goodwill trust (X_2) is necessary for innovation performance (Y); high-high direction.

H3: Competence trust (X_3) is necessary for innovation performance (Y); high-high direction.

- Focal unit = Buyer-supplier relationship.
- Concept X_1 = Contractual detail: the degree of specificity and completeness in a contract.

[1] Many helpful sources to get started with `R` and `RStudio` are publicly available online.

[2] At the time that this example was published, not all software functions were available yet.

- Concept X_2 = Goodwill trust: the confidence that the supplier intends to fulfill its role and will act fairly.
- Concept X_3 = Competence trust: the confidence in the supplier's ability/capability to fulfill agreed obligations.
- Concept Y = Innovation: the extent to which the supplier continuously improves the maintenance process and increases asset utilization.
- Theoretical domain = buyer-supplier relations in manufacturing.
- Causal explanation H1 = Detailed contracts create a minimum set of formal agreements that make high innovation performance possible. Without these agreements, the work cannot be reliably coordinated, protected, and adapted. Other factors cannot compensate for the absence of detailed contracts.
- Causal explanation H2 = Goodwill trust creates the minimum relational confidence that makes high innovation performance possible. Without goodwill trust, partners will not share sensitive information, voice concerns early, or collaborate openly; instead they protect themselves, which reduces learning, slows joint problem solving, and blocks the experimentation needed for innovation. Other factors cannot compensate for the absence of goodwill trust.
- Causal explanation H3 = Competence trust creates the minimum capability confidence that makes high innovation performance possible. Without competence trust, coordination becomes inefficient because work must be checked, reworked, or duplicated; errors increase, response times slow, and adaptation suffers, making it impossible to execute complex, uncertain innovation tasks reliably. Other factors cannot compensate for the absence of competence trust.

B.3.2 Stage 2. Collect data

- Study design = observational study.
- Sampling = convenience sampling of n = 48 buyer-supplier relations.
- Measurement = questionnaires for buyer informants.
- Dataset = `nca.example2` (part of the `NCA` software for `R`).

Install and load the NCA package and data

After `R` and `RStudio` are installed and `RStudio` is opened, the script window can be used to type and run instructions. The first instruction is to install (download) the `NCA` package with the `install.packages` function in `R`. Installing the `NCA` package must be done once. Afterwards, for each new session, the `NCA` package must be loaded (activated) using the `library` function. `R` runs the script line by line. Text after the #-symbol is ignored and may be used for deactivating a line or for comments. To limit the computation time for this demonstration, the `test.rep` argument of the `nca_analysis` function for the number of resamples for NCA's statistical tests is reduced.

```
# Install and load the NCA package
# install.packages("NCA")  # remove "#" in front if package is not installed
library(NCA)  # load the package
```

```
# Limit computation time
test.rep <- 10000 # set lower for less computation time;
                  # for final analysis select 10000
```

After loading the NCA package, the example data in the package are loaded and afterwards renamed for convenience. The number of rows (cases) and the first rows of the dataset are shown as well.

```
# Load the data
data("nca.example2")  # load the data
data <- nca.example2  # rename the data for convenience
nrow(data)            # number of rows (cases)

## [1] 48

head(data)            # display the first rows of the data

##   Innovation Contractual detail Goodwill trust Competence trust
## 1       3.57               3.24           2.71              4.0
## 2       3.57               2.71           2.43              3.0
## 3       1.29               2.29           4.00              4.0
## 4       2.14               4.14           3.71              4.0
## 5       1.00               2.43           3.29              3.5
## 6       3.43               1.86           3.86              4.5
```

The dataset has 48 cases. Each row represents a case. The four columns correspond to the four variables of the study, starting with outcome Innovation, and followed by the three conditions. The variables are measured on Likert scales with scale values ranging from 1 to 5, except for Contractual detail where the Likert scale values ranges from 1 to 7.

B.3.3 Stage 3. Analyze data

Model specification

- Selected ceiling line = CE-FDH, because data are discrete (Likert scales).
- Selected bounding box = empirical scope to avoid possible risk of overestimating the effect size.
- Selected effect size threshold = 0.10 (often used when no specific threshold is available).
- Selected p-value threshold = 0.05 (often used in the social sciences).

Visual inspection

First, three XY-plots are produced for a visual inspection of the data, one for each condition. Although XY-plots can be done with the NCA package (see below), here it is first done with standard R functions:

```
# Create XY-plots for visual inspection
# Using base R functions (not part of the NCA package)
plot(data$`Contractual detail`, data$Innovation,
     xlab = "Contractual detail",
     ylab = "Innovation")
```

```
plot(data$`Goodwill trust`, data$Innovation,
     xlab = "Goodwill trust",
     ylab = "Innovation")
plot(data$`Competence trust`, data$Innovation,
     xlab = "Competence trust",
     ylab = "Innovation")
```

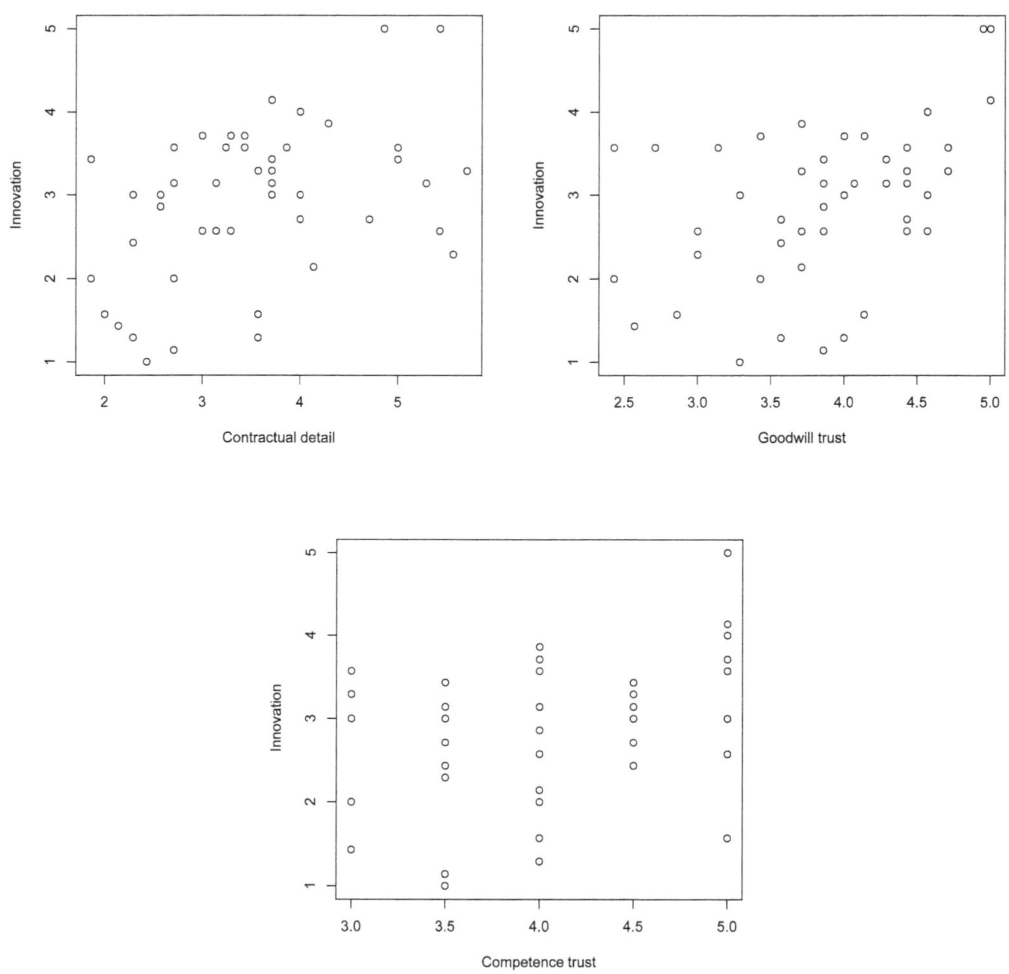

FIGURE B.1 XY-plots for visual inspection (Data from Van der Valk et al., 2016).

Since the directions of the three hypotheses are high-high (presence/high X is necessary for presence/high Y), it is expected that in each XY-plot the upper-left corner is empty. The visual inspection confirms this.

Furthermore, it is observed that for each plot the borderlines in the upper-left corner between the area with cases and the area without cases are somewhat irregular. This suggests the use of the CE-FDH stepwise linear ceiling line that can follow the border more precisely than a linear ceiling line, although the other default ceiling line (CR-FDH) could have been selected as well.

Another observation is that no clear outliers are present, although the two points the upper-right corner may be candidates as both ceiling and scope outliers. Note that in the plot of Competence trust only one point is visible in the upper-right corner, although here two points have the same XY-combination (duplicates). The default advice to not remove potential outliers unless these are errors is followed. However, uncertainty about outliers warrants an outlier check.

The plots also show that all possible values of the outcome (1–5) are represented in the dataset. This does not apply for the conditions where in particular low values of the conditions are missing. If these values are theoretically possible, this suggests that also a theoretical scope corresponding to the Likert scale could be selected, which warrants a robustness check of it (Sections 4.3 and 9.3.1).

Based on the visual inspection of the XY-plots, it is decided to first conduct the primary analysis with the selected CE-FDH ceiling line, to keep all cases, and to use the empirical scope. Afterwards robustness checks are conducted to evaluate the sensitivity of the choices for the main conclusion. Particularly, the effects of the use of CR-FDH ceiling line, and of a theoretical scope corresponding to the Likert scale values is evaluated. Also, a detailed outlier analysis is done. This possibly results in a different way of handling the outliers than is done in the original primary analysis (removing outliers), which may result in a different conclusion about the hypotheses.

Estimate effect size and p-value: necessity-in-kind

The main analysis consists of estimating the three effect sizes and corresponding p-values for making a decision about the necessity-in-kind for each hypothesis. The analysis is done with the function `nca_analysis`. The selected name of the analysis is model1.

```
# Conduct NCA
set.seed(123) # for reproducible results
model1 <- nca_analysis(data,
                       x = c('Contractual detail',
                             'Goodwill trust',
                             'Competence trust'),
                       y = 'Innovation',
                       ceilings = 'ce_fdh',
                       corner = 1,          # default upper-left corner
                       scope = NULL,        # default empirical scope
                       test.rep = test.rep)
```

```
## Setting up parallelization, this might take a few seconds...
##         Do test for  :  ce_fdh - Contractual detailDone test for:
##            ce_fdh - Contractual detail
## Do test for  :  ce_fdh - Goodwill trustDone test for:
##          ce_fdh - Goodwill trust
## Do test for  :  ce_fdh - Competence trustDone test for:
##            ce_fdh - Competence trust
```

The `nca_analysis` instruction first specifies the dataset, then the names of the conditions, followed by the name of the outcome. Instead of names, also column numbers can be selected: `x = c(2,3,4); y = 1`. The `ceilings` argument selects the ceiling line. Given the high-high hypotheses, the expected corner for all conditions is the upper-left corner (`corner = 1`). The `test.rep` argument activates the p-value estimation by selecting the number of permutations to create the null distribution with which the observed effect size is compared. The selection of the number of permutations is a trade-off between p-value accuracy and computation time. For the final analysis 10,000 permutations are recommended.

After running `nca_analysis`, the progress of the computation is printed. When the computations are done, no output is shown yet. The main summary of the output can be displayed by calling `print(model1)` or just `model1`. This prints the effect sizes and p-value for each condition.

```
# Print main results
print(model1)

##
## --------------------------------------------------------------------------------

## Effect size(s):

##                      ce_fdh p
## Contractual detail 0.24   0.008
## Goodwill trust     0.31   0.002
## Competence trust   0.32   0.003
## --------------------------------------------------------------------------------
```

The results show support for the three hypotheses because the effect sizes and the p-values satisfy the threshold values that were selected.

By using the `nca_output` function, more output can be displayed. The XY-plots with the ceiling lines can be shown as follows.

```
# Display XY-plots with ceiling lines
nca_output(model1, summaries = FALSE)
```

Evaluate model fit

With the argument `summaries = TRUE`, further details of the results are printed for each condition separately.

```
# Print detailed output
nca_output(model1, plots = FALSE, summaries = TRUE)

##
## --------------------------------------------------------------------------------

## NCA Parameters : Contractual detail - Innovation

## --------------------------------------------------------------------------------
##
## Number of observations 48
## Scope                  15.40
## Xmin                    1.86
## Xmax                    5.71
## Ymin                    1.00
## Ymax                    5.00
##
##                      ce_fdh
```

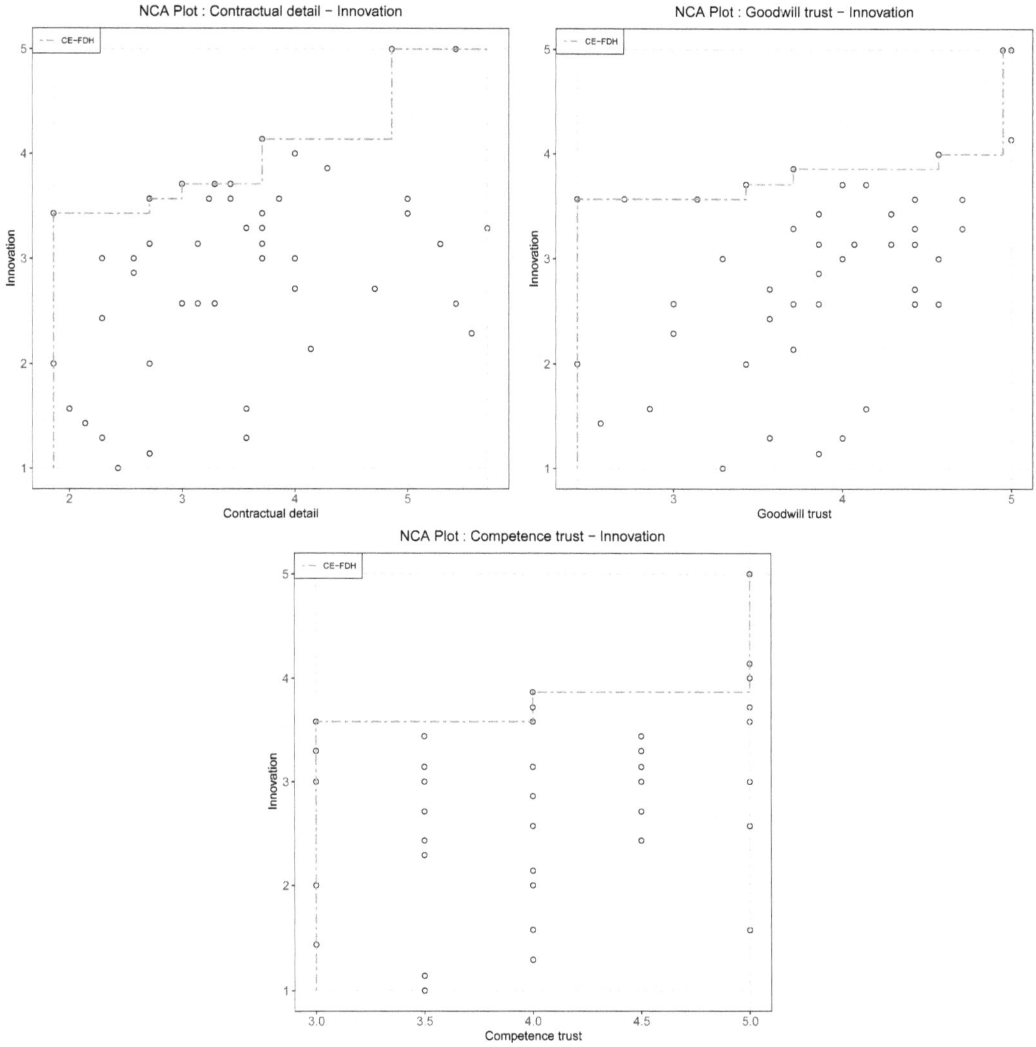

FIGURE B.2 XY-plots with CE-FDH ceiling line.

```
## Ceiling zone       3.654
## Effect size        0.237
## # above            0
## Slope
## Intercept
## p-value            0.008
## p-accuracy         0.002
##
## Complexity         5
## Fit              100  %
## Ceiling accuracy 100  %
## Noise              0  %   (0)
## Exceptions         0  %   (0)
## Support          33.3%  (16)
## Spread             0.49
```

```
## Sharpness          1.00
##
## Abs. ineff.       10.690
## Rel. ineff.       69.416
## Condition ineff.  22.078
## Outcome ineff.    60.750
##
##
## --------------------------------------------------------------------------
## NCA Parameters : Goodwill trust - Innovation
## --------------------------------------------------------------------------
##
## Number of observations 48
## Scope                      10.28
## Xmin                        2.43
## Xmax                        5.00
## Ymin                        1.00
## Ymax                        5.00
##
##                   ce_fdh
## Ceiling zone       3.152
## Effect size        0.307
## # above            0
## Slope
## Intercept
## p-value            0.001
## p-accuracy         0.000
##
## Complexity         5
## Fit              100  %
## Ceiling accuracy 100  %
## Noise              0  %   (0)
## Exceptions         0  %   (0)
## Support          33.3%  (16)
## Spread             0.53
## Sharpness          1.00
##
## Abs. ineff.        6.676
## Rel. ineff.       64.946
## Condition ineff.   1.946
## Outcome ineff.    64.250
##
##
## --------------------------------------------------------------------------
## NCA Parameters : Competence trust - Innovation
## --------------------------------------------------------------------------
##
## Number of observations 48
## Scope                      8
```

```
## Xmin                   3
## Xmax                   5
## Ymin                   1
## Ymax                   5
##
##                   ce_fdh
## Ceiling zone       2.570
## Effect size        0.321
## # above            0
## Slope
## Intercept
## p-value            0.002
## p-accuracy         0.000
##
## Complexity         3
## Fit              100  %
## Ceiling accuracy 100  %
## Noise              0  %   (0)
## Exceptions         0  %   (0)
## Support           37.5%  (18)
## Spread             0.28
## Sharpness          1.00
##
## Abs. ineff.        5.140
## Rel. ineff.       64.250
## Condition ineff.   0.000
## Outcome ineff.    64.250
```

From the output, only model fit is discussed and model fit metrics are extracted by using the `nca_extract` function.

```
# Extract model fit metrics
xs = c('Contractual detail', 'Goodwill trust', 'Competence trust')
ceiling <- "ce_fdh"
metrics <- c("Complexity", "Fit", "Ceiling accuracy", "Noise", "Exceptions",
             "Support", "Spread", "Sharpness")
tab1 <- sapply(xs, function(x) {
  sapply(metrics, function(p) {
    nca_extract(model1, x = x, ceiling = ceiling, param = p)
  })
})

# Print model fit metrics
tab1_df <- data.frame(Metric = rownames(tab1), tab1,
                      row.names = NULL, check.names = FALSE)
tab1_df
```

```
##             Metric Contractual detail Goodwill trust Competence trust
## 1       Complexity               5.00           5.00             3.00
## 2              Fit             100.00         100.00           100.00
## 3 Ceiling accuracy             100.00         100.00           100.00
## 4            Noise               0.00           0.00             0.00
```

```
## 5      Exceptions          0.00         0.00          0.00
## 6         Support         33.30        33.30         37.50
## 7          Spread          0.49         0.53          0.28
## 8       Sharpness          1.00         1.00          1.00
```

By definition, CE-FDH does not allow observations above the ceiling such that model fit in terms of Fit, Ceiling accuracy, Noise, Exceptions, and sharpness is 'best'. However, this comes at the cost of model Complexity (and thus a greater risk of overfitting). The points on or just below the ceiling line show acceptable level of Support (> 5%, see Table 6.3), while the Spread metric indicates that these points are weakly dispersed along the ceiling line (compared to a suggested minimum value of 0.5).

Robustness checks

To evaluate if the support for the hypotheses is robust, three robustness checks are done. First, the sensitivity of the results for selecting another plausible ceiling line is evaluated. The CR-FDH ceiling line is an attractive alternative because it models the ceiling line as a linear ceiling line with low Complexity of 1. The analysis is repeated with this line.

```
# Perform robustness check with other ceiling line
set.seed(123) # for reproducible results
model2 <- nca_analysis(data,
                       x = c('Contractual detail',
                             'Goodwill trust',
                             'Competence trust'),
                       y = 'Innovation',
                       ceilings = "cr_fdh",
                       corner = 1,          # default upper-left corner
                       scope = NULL,        # default empirical scope
                       test.rep = test.rep)
print(model2)
nca_output(model2, summaries = FALSE)

# Extract model fit metrics
xs = c('Contractual detail', 'Goodwill trust', 'Competence trust')
ceiling <- "cr_fdh"
metrics <- c("Complexity", "Fit", "Ceiling accuracy", "Noise", "Exceptions",
            "Support", "Spread", "Sharpness")

tab2 <- sapply(xs, function(x) {
  sapply(metrics, function(p) {
    nca_extract(model2, x = x, ceiling = ceiling, param = p)
  })
})

# Print model fit metrics
tab2_df <- data.frame(Metric = rownames(tab2), tab2,
                      row.names = NULL, check.names = FALSE)
tab2_df

## Setting up parallelization, this might take a few seconds...
##         Do test for  :  cr_fdh - Contractual detailDone test for:
```

```
##          cr_fdh - Contractual detail
## Do test for :  cr_fdh - Goodwill trustDone test for:
##          cr_fdh - Goodwill trust
## Do test for :  cr_fdh - Competence trustDone test for:
##          cr_fdh - Competence trust

##
## --------------------------------------------------------------------------

## Effect size(s):

##                    cr_fdh p
## Contractual detail 0.19   0.008
## Goodwill trust     0.26   0.003
## Competence trust   0.21   0.010
## --------------------------------------------------------------------------

##             Metric Contractual detail Goodwill trust Competence trust
## 1       Complexity            1.00000        1.00000          1.00000
## 2              Fit           79.17335       83.22455         64.66219
## 3 Ceiling accuracy           95.83333       91.66667         93.75000
## 4            Noise            4.20000        4.20000          6.20000
## 5       Exceptions            0.00000        4.20000          0.00000
## 6          Support           22.90000       16.70000         16.70000
## 7           Spread            0.49000        0.56000          0.38000
## 8        Sharpness            0.82000        0.78000          0.57000
```

The results show that the effect sizes are reduced, but that they are still above the selected threshold level. Also the p-value remains below the selected threshold level, such that the conclusion about the hypothesis does not change with changing ceiling line from CE-FDH to CR-FDH.

The model fit measures confirm that the CR-FDH line is attractive regarding complexity. Since the CR-FDH line is a trend line through the peers (upper-left edges of CE-FDH line), cases exist above this line (ceiling accuracy $< 100\%$). Although most of these cases remain close to the ceiling according to the Noise metric, for Goodwill trust two cases are relatively far from the ceiling (see Exceptions metric). These cases are potential outliers and were also identified by visual inspection. The Support and Spread metrics are comparable to those for CE-FDH.

The second robustness check tests the sensitivity of the results for selecting another scope. The analysis is repeated with a theoretical scope according to the minimum and maximum values of the Likert scales, rather than with the empirical scope based on observed minima and maxima.

```
# Robustness check with different scopes corresponding to Likert scales
set.seed(123)
model3 <- nca_analysis(data,
                       x = c('Contractual detail',
                             'Goodwill trust',
                             'Competence trust'),
                       y = 'Innovation',
                       ceilings = "ce_fdh",
                       corner = 1,          # default upper-left corner
```

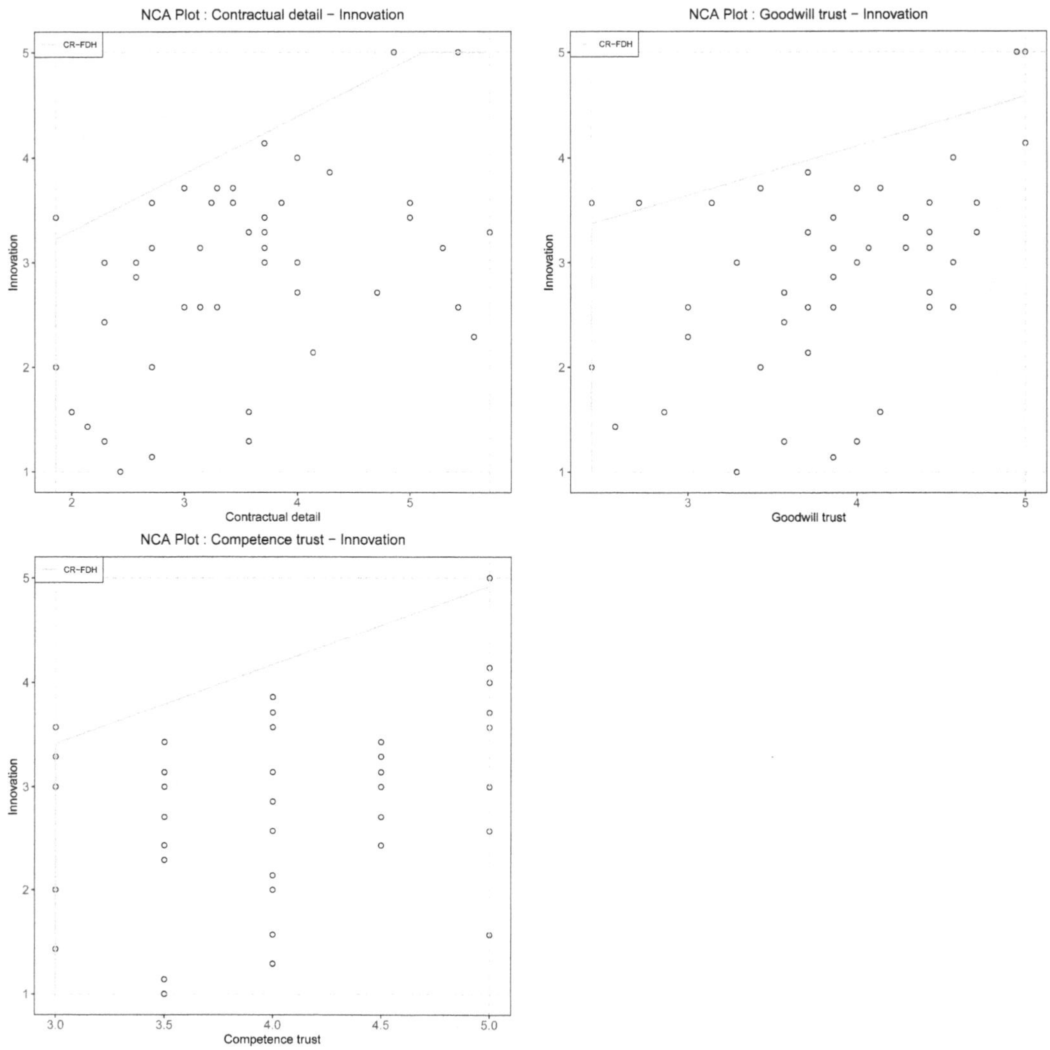

FIGURE B.3 *XY*-plots with CR-FDH ceiling line.

```
                        scope = list (c(1,7,1,5), c(1,5,1,5), c(1,5,1,5)),
                        test.rep = test.rep)
print(model3)
nca_output(model3, summaries = FALSE)

# Extract model fit metrics
xs = c('Contractual detail', 'Goodwill trust', 'Competence trust')
ceiling <- "ce_fdh"
metrics <- c("Complexity", "Fit", "Ceiling accuracy", "Noise", "Exceptions",
             "Support", "Spread", "Sharpness")
tab3 <- sapply(xs, function(x) {
  sapply(metrics, function(p) {
    nca_extract(model3, x = x, ceiling = ceiling, param = p)
  })
```

```
})

# Print model fit metrics
tab3_df <- data.frame(Metric = rownames(tab3), tab3,
                      row.names = NULL, check.names = FALSE)
tab3_df

## Setting up parallelization, this might take a few seconds...
##       Do test for  :  ce_fdh - Contractual detailDone test for:
##         ce_fdh - Contractual detail
## Do test for  :  ce_fdh - Goodwill trustDone test for:
##          ce_fdh - Goodwill trust
## Do test for  :  ce_fdh - Competence trustDone test for:
##          ce_fdh - Competence trust

##
## --------------------------------------------------------------------------

## Effect size(s):

##                    ce_fdh p
## Contractual detail 0.30   0.008
## Goodwill trust     0.55   0.001
## Competence trust   0.66   0.003
## --------------------------------------------------------------------------

##             Metric Contractual detail Goodwill trust Competence trust
## 1       Complexity               5.00           5.00             3.00
## 2              Fit             100.00         100.00           100.00
## 3 Ceiling accuracy             100.00         100.00           100.00
## 4            Noise               0.00           0.00             0.00
## 5       Exceptions               0.00           0.00             0.00
## 6          Support              33.30          33.30            37.50
## 7           Spread               0.49           0.53             0.28
## 8        Sharpness               1.00           1.00             1.00
```

As expected, the empty spaces have increased such that the effect sizes are considerably larger than the original effect sizes and thus again above the selected threshold level. Also, the p-values remain below the selected threshold level, such that the conclusion about the hypothesis does not change with changing from empirical scope to theoretical scope.

Third, the sensitivity of the results for a different handling of outliers is evaluated. First, potential outliers are identified, then a decision is made to accept them as outlier or not, and finally to decide to include them in the analysis or not. If the results of this analysis are different than the original choice, the analysis is repeated with the new choice.

An outlier is defined as a case that has a large influence on the effect size when it is removed. The first outlier analysis is done by removing one potential outlier at a time. The `nca_outlier` function has similar arguments as the `nca_analysis` function. One additional argument is `plotly`, which can be selected to obtain an interactive plot to evaluate the cases.

```
# Single outlier analysis
nca_outliers(data, 2, 1, plotly = TRUE, ceiling = 'ce_fdh')  # Contractual detail
```

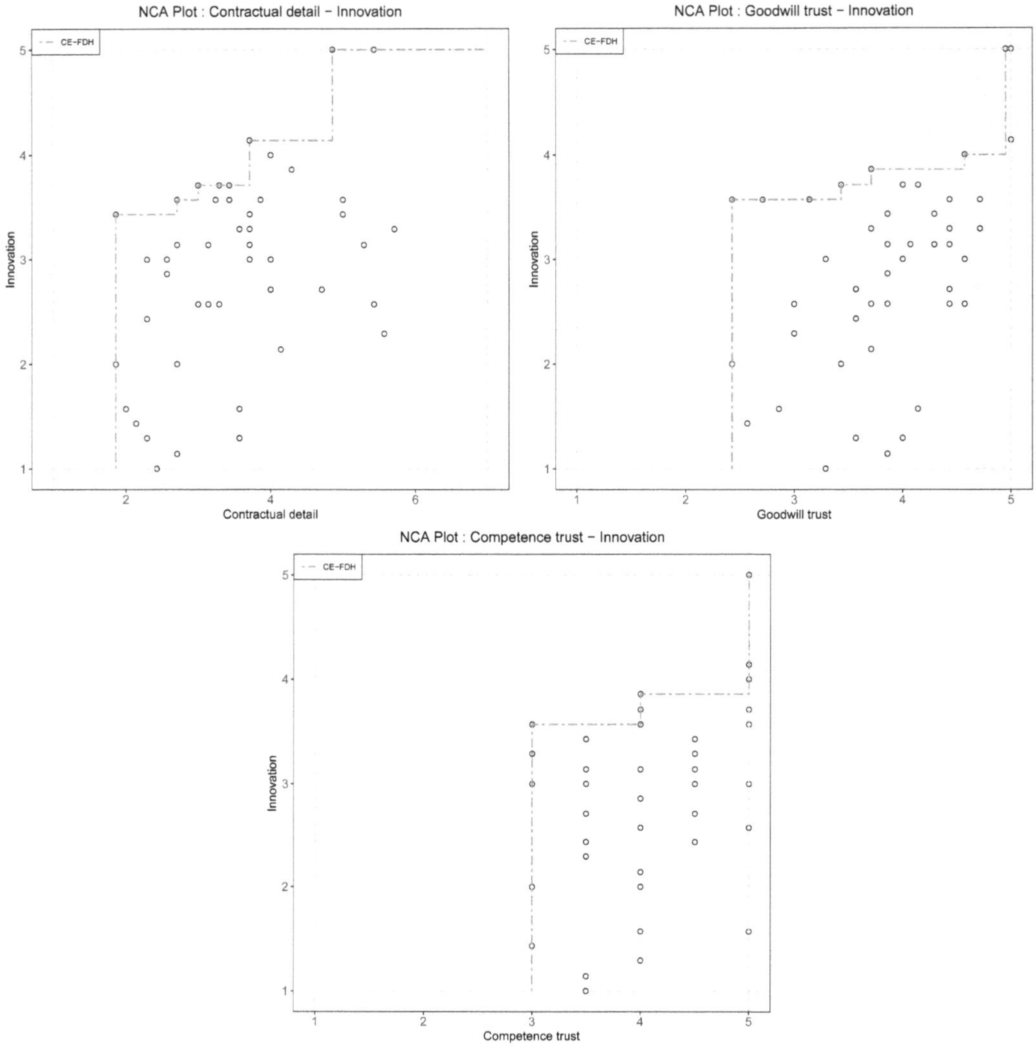

FIGURE B.4 XY-plots with CE-FDH ceiling line and theoretical scope.

```
##    outliers eff.or eff.nw dif.abs dif.rel ceiling scope
## 1        6   0.24   0.29    0.05    21.8       X
## 2       45   0.24   0.27    0.03    13.4       X
## 3        9   0.24   0.25    0.02     6.7       X
## 4       18   0.24   0.25    0.01     3.8             X
## 5        5   0.24   0.25    0.01     3.6             X
## 6        2   0.24   0.24    0.00     1.1       X
## 7       14   0.24   0.24    0.00     1.1       X
```

```
nca_outliers(data, 3, 1, plotly = TRUE, ceiling = 'ce_fdh')  # Goodwill trust
```

```
##    outliers eff.or eff.nw dif.abs dif.rel ceiling scope
## 1        2   0.31   0.35    0.04    13.9       X
## 2       44   0.31   0.32    0.01     4.1       X
## 3        5   0.31   0.32    0.01     3.6             X
```

```
## 4        42   0.31   0.31    0.01      1.7        X
## 5        45   0.31   0.31    0.00      1.6        X
## 6        14   0.31   0.31    0.00      1.2        X
```

```
nca_outliers(data, 4, 1, plotly = TRUE, ceiling = 'ce_fdh')  # Competence trust
```

```
##    outliers eff.or eff.nw dif.abs dif.rel ceiling scope
## 1         2   0.32   0.35    0.03      8.2       X
## 2        44   0.32   0.34    0.02      5.8       X
## 3         5   0.32   0.33    0.01      3.6             X
```

The printed results show that there are no cases that could be classified as outliers, because the change in effect size is relatively moderate when the case is removed. The maximum change is an effect size increase of 21.8% when case 6 (the upper-left case for Contractual detail) is removed.[3]

The outlier analysis continues with a removal of two conditions at the same time, because two case closely together or in the upper-right corner are a possible outlier pair.

```
# Double outliers analysis (two potential outliers)
nca_outliers(data, 2, 1, plotly = TRUE, k = 2, ceiling = 'ce_fdh')  # Contractual detail
```

```
##     outliers eff.or eff.nw dif.abs dif.rel ceiling scope
## 1   45 - 8     0.24   0.09   -0.15   -62.6       X     X
## 2   6 - 45     0.24   0.32    0.08    35.2       X
## 3   6 - 9      0.24   0.30    0.07    28.5       X
## 4   6 - 18     0.24   0.30    0.06    26.4       X     X
## 5   6 - 5      0.24   0.30    0.06    26.2       X     X
## 6   6 - 10     0.24   0.30    0.06    26.1       X
## 7   6 - 2      0.24   0.30    0.06    25.2       X
## 8   6 - 14     0.24   0.29    0.05    22.9       X
## 9   45 - 9     0.24   0.29    0.05    22.3       X
## 10  6          0.24   0.29    0.05    21.8       X
## 11  6 - 8      0.24   0.29    0.05    21.8       X
## 12  6 - 38     0.24   0.29    0.05    21.8       X
## 13  6 - 31     0.24   0.29    0.05    21.8       X
## 14  6 - 1      0.24   0.29    0.05    21.8       X
## 15  6 - 28     0.24   0.29    0.05    21.8       X
## 16  6 - 42     0.24   0.29    0.05    21.8       X
## 17  6 - 24     0.24   0.28    0.04    18.0       X     X
## 18  45 - 18    0.24   0.28    0.04    17.7       X     X
## 19  45 - 5     0.24   0.28    0.04    17.5       X     X
## 20  45 - 2     0.24   0.27    0.03    14.5       X
## 21  45 - 14    0.24   0.27    0.03    14.5       X
## 22  45         0.24   0.27    0.03    13.4       X
## 23  45 - 24    0.24   0.27    0.03    13.4       X
## 24  45 - 38    0.24   0.27    0.03    13.4       X
## 25  45 - 10    0.24   0.27    0.03    13.4       X

## # Not showing 59 possible outliers
```

[3]Cases names can be identified in the interactive plotly XY-plot, see Figure 9.2.

```
nca_outliers(data, 3, 1, plotly = TRUE, k = 2, ceiling = 'ce_fdh')  # Goodwill trust
```

```
##    outliers eff.or eff.nw dif.abs dif.rel ceiling scope
## 1   45 - 8    0.31   0.12   -0.18   -59.9       X     X
## 2   2 - 1     0.31   0.41    0.10    32.8       X
## 3   2 - 5     0.31   0.36    0.06    18.1       X     X
## 4   2 - 44    0.31   0.36    0.06    18.0       X
## 5   2 - 42    0.31   0.35    0.05    15.6       X
## 6   2 - 45    0.31   0.35    0.05    15.5       X
## 7   2 - 14    0.31   0.35    0.05    15.2       X
## 8   2         0.31   0.35    0.04    13.9       X
## 9   2 - 8     0.31   0.35    0.04    13.9       X
## 10  2 - 9     0.31   0.35    0.04    13.9       X
## 11  2 - 36    0.31   0.35    0.04    13.9       X
## 12  2 - 28    0.31   0.35    0.04    13.9       X
## 13  2 - 41    0.31   0.33    0.03     9.1       X     X
## 14  44 - 5    0.31   0.33    0.02     7.9       X     X
## 15  44 - 42   0.31   0.33    0.02     7.6       X
## 16  44 - 14   0.31   0.33    0.02     6.6       X
## 17  44 - 45   0.31   0.32    0.02     5.7       X
## 18  5 - 42    0.31   0.32    0.02     5.4       X     X
## 19  5 - 45    0.31   0.32    0.02     5.3       X     X
## 20  5 - 14    0.31   0.32    0.02     4.9       X     X
## 21  44        0.31   0.32    0.01     4.1       X
## 22  44 - 41   0.31   0.32    0.01     4.1       X
## 23  44 - 8    0.31   0.32    0.01     4.1       X
## 24  44 - 9    0.31   0.32    0.01     4.1       X
## 25  44 - 36   0.31   0.32    0.01     4.1       X

## # Not showing 33 possible outliers
```

```
nca_outliers(data, 4, 1, plotly = TRUE, k = 2, ceiling = 'ce_fdh')  # Competence trust
```

```
##    outliers eff.or eff.nw dif.abs dif.rel ceiling scope
## 1   8 - 45    0.32   0.14   -0.19   -57.9       X     X
## 2   2 - 44    0.32   0.37    0.04    14.0       X
## 3   2 - 40    0.32   0.37    0.04    13.8       X
## 4   2 - 5     0.32   0.36    0.04    12.1       X     X
## 5   2 - 30    0.32   0.36    0.03    10.9       X
## 6   44 - 5    0.32   0.35    0.03     9.7       X     X
## 7   2         0.32   0.35    0.03     8.2       X
## 8   2 - 8     0.32   0.35    0.03     8.2       X
## 9   2 - 45    0.32   0.35    0.03     8.2       X
## 10  2 - 14    0.32   0.35    0.03     8.2       X
## 11  2 - 16    0.32   0.35    0.03     8.2       X
## 12  2 - 9     0.32   0.35    0.03     8.2       X
## 13  44        0.32   0.34    0.02     5.8       X
## 14  44 - 8    0.32   0.34    0.02     5.8       X
## 15  44 - 45   0.32   0.34    0.02     5.8       X
## 16  44 - 40   0.32   0.34    0.02     5.8       X
## 17  44 - 30   0.32   0.34    0.02     5.8       X
## 18  44 - 14   0.32   0.34    0.02     5.8       X
```

```
## 19  44 - 16   0.32   0.34    0.02     5.8      X
## 20  44 - 9    0.32   0.34    0.02     5.8      X
## 21  5         0.32   0.33    0.01     3.6            X
## 22  5 - 8     0.32   0.33    0.01     3.6      X     X
## 23  5 - 45    0.32   0.33    0.01     3.6      X     X
## 24  5 - 40    0.32   0.33    0.01     3.6            X
## 25  5 - 30    0.32   0.33    0.01     3.6            X

## # Not showing 3 possible outliers
```

It turns out that the pair 8-45 considerably reduces the effect size for all conditions (> 50%) when this pair is removed. This pair indeed consists of the two cases in the upper-right corner identified by visual inspection. Given the large influence that removing this pair has on the effect size, the robustness check consists of removing these cases from the dataset and repeating the analysis.

```
# Remove two potential outliers and perform NCA again
data1 <- data[-c(8, 45), ]
set.seed(123)
model4 <- nca_analysis(data1,
                        c('Contractual detail',
                          'Goodwill trust',
                          'Competence trust'),
                        'Innovation',
                        ceilings = 'ce_fdh',
                        corner = 1,
                        test.rep = test.rep)

## Setting up parallelization, this might take a few seconds...
##            Do test for  :  ce_fdh - Contractual detailDone test for:
##             ce_fdh - Contractual detail
## Do test for  :  ce_fdh - Goodwill trustDone test for:
##              ce_fdh - Goodwill trust
## Do test for  :  ce_fdh - Competence trustDone test for:
##              ce_fdh - Competence trust

print(model4)

##
## --------------------------------------------------------------------------

## Effect size(s):

##                    ce_fdh p
## Contractual detail 0.09   0.111
## Goodwill trust     0.12   0.143
## Competence trust   0.14   0.053
## --------------------------------------------------------------------------

nca_output(model4, summaries = FALSE)
```

The results show that after removal of the two outliers, the effect sizes decrease considerably. The effect size for Contractual detail is below the threshold level. Also all p-value are above the threshold level of 0.05. Consequently, without the outliers, none of the hypotheses is supported. This implies that the original results can be considered fragile and do not justify making strong conclusions.

Bottleneck analysis: necessity-in-degree

Despite the fragility of the original results, for this demonstration the results of model 1 are used for conducting a bottleneck analysis.

To show the bottleneck table, the `nca_output` function includes the argument `bottlenecks = TRUE`. To suppress further printed and plotted output, the argument `summaries = FALSE` and `plots = FALSE` are added.

```
# Default bottleneck analysis with percentage range for X and Y
nca_output(model1, bottlenecks = TRUE, summaries = FALSE, plots = FALSE)
```

```
##
## --------------------------------------------------------------------------

## Bottleneck CE-FDH (cutoff = 0)

## Y Innovation           (percentage.range)

## 1 Contractual detail (percentage.range)

## 2 Goodwill trust     (percentage.range)

## 3 Competence trust   (percentage.range)

## --------------------------------------------------------------------------
## Y        1    2    3
## 0        NN   NN   NN
## 10       NN   NN   NN
## 20       NN   NN   NN
## 30       NN   NN   NN
## 40       NN   NN   NN
## 50       NN   NN   NN
## 60       NN   NN   NN
## 70       48.1 49.8 50.0
## 80       77.9 98.1 100.0
## 90       77.9 98.1 100.0
## 100      77.9 98.1 100.0

##
```

The first column of the bottleneck table is a set of values of the outcome Y, and the next columns show the corresponding required levels of the conditions. The table can be read row by row to find the levels of the conditions that are required for a given level of the outcome. NN means that the condition is not necessary for the corresponding level of the outcome. By default, bottleneck X and Y-values are expressed as percentages of the range. This means that 100 corresponds to the maximum value, 0 to the minimum value, 50% to the middle value, etc.

The output shows that up to a level of 60% Innovation performance, none of the conditions is necessary for Innovation performance. However, at 70% Innovation performance, each condition must have a level of about 50%. For a higher level of Innovation performance, a level of about 80% of Contractual detail and nearly maximum levels of Goodwill trust and Competence trust are required.

As discussed in Section 9.11, the values can also be expressed as actual values, percentiles, and percentage of maximum, which allows useful interpretations of the bottleneck table.

An alternative bottleneck table is one with levels of outcome and the conditions expressed as actual values. To obtain this table, the `nca_analysis` function must run with arguments that specify the desired type of value of X and Y for the bottleneck table (`bottleneck.x` and `bottleneck.y`). Furthermore, values of Y in the bottleneck table can be specified using the `steps` argument, for example, to specify the exact values of the anchors of the Likert scale of Innovation performance. To reduce computation time, the statistical analysis is not repeated (`test.rep = FALSE` or the argument is not included).

```
# Bottleneck analysis with actual values for X and Y
model5 <- nca_analysis(data,
                       c('Contractual detail',
                         'Goodwill trust',
                         'Competence trust'),
                       'Innovation',
                       ceilings = 'ce_fdh',
                       bottleneck.x = 'actual',
                       bottleneck.y = 'actual',
                       steps = seq(1, 5, by = 1)) # values of Likert scale
nca_output(model5, bottlenecks = TRUE, summaries = FALSE, plots = FALSE)
```

```
##
## --------------------------------------------------------------------------

## Bottleneck CE-FDH (cutoff = 0)

## Y Innovation         (actual)

## 1 Contractual detail (actual)

## 2 Goodwill trust     (actual)

## 3 Competence trust   (actual)

## --------------------------------------------------------------------------
## Y     1     2     3
## 1     NN    NN    NN
## 2     NN    NN    NN
## 3     NN    NN    NN
## 4     3.710 4.570 5.000
## 5     4.860 4.950 5.000

##
```

The output shows that Likert scale level 4 for Innovation performance is only possible if the Likert scale levels of Contractual detail, Goodwill trust, and Competence trust are at least 3.7, 4.5 and 5.0, respectively.

When actual values are used for the outcome and percentile values for the conditions, the formatting of the bottleneck table is somewhat different. The values of the conditions can now be interpreted as the percentage of cases that are unable to achieve the outcome. In brackets are the corresponding number of cases.

```
# Bottleneck analysis with actual values for Y and percentile values for X
model6 <- nca_analysis(data,
                       c('Contractual detail',
                         'Goodwill trust',
                         'Competence trust'),
                       'Innovation',
                       ceilings = 'ce_fdh',
                       bottleneck.x = 'percentile',
                       bottleneck.y = 'actual',
                       steps = seq(1, 5, by = 1))
nca_output(model6, bottlenecks = TRUE, summaries = FALSE, plots = FALSE)
```

```
##
## ---------------------------------------------------------------------------
## Bottleneck CE-FDH (cutoff = 0)
## Y Innovation         (actual)
## 1 Contractual detail (percentile)
## 2 Goodwill trust     (percentile)
## 3 Competence trust   (percentile)
## ---------------------------------------------------------------------------
## Y     1          2          3
## 1     NN (0)     NN (0)     NN (0)
## 2     NN (0)     NN (0)     NN (0)
## 3     NN (0)     NN (0)     NN (0)
## 4     58.3 (28) 83.3 (40) 79.2 (38)
## 5     83.3 (40) 93.8 (45) 79.2 (38)
##
```

This output shows that 58.3 percent of the cases (28 cases) did not achieve the required level of Contractual detail to make an Innovation performance of level 4 possible.

Advanced functions

Advanced functions in `NCA` for `R` that are demonstrated here are `nca_difference`, `nca_power`, and `nca_random` for statistical difference tests, power calculation, and random data generation, respectively.

Statistical difference tests

For certain studies it may be desirable to compare two effect sizes, for example, the difference in effect size between two time points, between two groups, or between two conditions. Assuming that the null hypothesis of no difference is not an unrealistic assumption, namely that both samples are drawn from the same population, NCA's statistical difference test could be helpful in evaluating if the observed difference is compatible with the null of no difference (Section 5.3). For comparability of two effect sizes, a common scope should be selected. Depending on the question one tries to answer, this can be the pooled empirical scope (based on observed minimum and maximum scale values), a theoretical scope (e.g., based on theoretical minimum and maximum scale values), or a min-max normalized scope (e.g., [0,1] × [0,1]) that makes the scales dimensionless and numerically equal.

NCA's contrast test compares effect sizes of two comparable conditions (in one sample).

```
# Contrast test: effect size difference between two conditions
# Common scope: normalized (for making the scales dimensionless)

# Min-max normalize data
scale <- c (0,1) # unit box
min_max <- c(min(data$Innovation), max(data$Innovation),
             min(data$`Contractual detail`), max(data$`Contractual detail`),
             min(data$`Goodwill trust`), max(data$`Goodwill trust`),
             min(data$`Competence trust`), max(data$`Competence trust`)
             )
data_n <- nca_util_normalize(data = data, scale = scale, min_max = min_max)

# Conduct test
set.seed(123)
difference_contrast <- nca_difference(data1 = data_n,
                                      x = c("Goodwill trust", "Competence trust"),
                                      y = "Innovation",
                                      ceilings = "ce_fdh",
                                      test.rep = 1000,
                                      common_scope = c(0,1,0,1),
                                      test.type = "contrast")

## Doing ce_fdh

print(difference_contrast)

##
## --------------------------------------------------------------------------

## Effect size (difference):

##                   ce_fdh     p
## Goodwill trust      0.31
## Competence trust    0.32
## Difference          0.01     0.128
## --------------------------------------------------------------------------
```

During computation, a message (Doing ...) is shown. The output shows a minor and non-significant difference between the two effect sizes. NCA's independence test compares effect sizes of two independent samples (e.g., two subgroups). To simulate this situation it is assumed that the first set of cases in the dataset may have a different characteristic than the second set of cases and it is tested if the effect sizes of the two groups are different.

```
# Independence test: effect size difference between two datasets
data1 <- data[1:24, ]
data2 <- data[25:48, ]

# Common scope: pooled scope or common theoretical scope
# Conduct test with pooled empirical scope
# Common scope:
```

```
min_x <- min(min(data1$`Contractual detail`),min(data2$`Contractual detail`))
max_x <- max(max(data1$`Contractual detail`),max(data2$`Contractual detail`))
min_y <- min(min(data1$`Innovation`),min(data2$`Innovation`))
max_y <- max(max(data1$`Innovation`),max(data2$`Innovation`))
common_scope <- c(min_x,max_x,min_y,max_y)

# Conduct test
set.seed(123)
difference_independent <- nca_difference(data1 = data1,
                                         data2 = data2,
                                         x = c("Contractual detail"),
                                         y = "Innovation",
                                         ceilings = "ce_fdh",
                                         common_scope = common_scope,
                                         test.rep = 1000,
                                         test.type = "independent")

## Doing ce_fdh

print(difference_independent)

##
## --------------------------------------------------------------------------
## Effect size (difference):
##                        ce_fdh     p
## Contractual detail.1   0.27
## Contractual detail.2   0.37
## Difference             0.10   0.331
## --------------------------------------------------------------------------
```

The results show no statistically significant difference between the groups.

NCA's paired test compares effect sizes of two measurements (e.g., two time points, same group). This situation is simulated assuming that the data were obtained a T1 and by constructing a new sample (same sample size) where these cases get new scores for X and Y for T2.

```
# Paired test: effect size differences between 2 measurements
data1 <- subset(nca.example2, select = c(1,2)) # select X and Y for T1

# Simulate changes of X and Y at T2
set.seed(123)
data2 <- data1
data2$`Contractual detail` <- pmin(5, pmax(1, data2$`Contractual detail`
                                           + sample(-1:1, nrow(data2), TRUE)
                                           )
                                   )
data2$`Innovation` <- pmax(1, ifelse(data2$`Innovation`
                                     >= 4, data2$`Innovation`
                                     - 1, data2$`Innovation`)
                           )
```

```
# Conduct test with pooled empirical scope
# Common scope:
min_x  <- min(min(data1$`Contractual detail`),min(data2$`Contractual detail`))
max_x  <- max(max(data1$`Contractual detail`),max(data2$`Contractual detail`))
min_y  <- min(min(data1$`Innovation`),min(data2$`Innovation`))
max_y  <- max(max(data1$`Innovation`),max(data2$`Innovation`))
common_scope  <- c(min_x,max_x,min_y,max_y)

# Conduct test
set.seed(123)
difference_paired <- nca_difference(data1=data1,
                                    data2=data2,
                                    x = "Contractual detail",
                                    y = "Innovation",
                                    ceiling = "ce_fdh",
                                    common_scope = common_scope,
                                    test.rep = 1000,
                                    test.type = "paired")
```

```
## Doing ce_fdh
```

```
print(difference_paired)
```

```
##
## --------------------------------------------------------------------------------
## Effect size (difference):
##                         ce_fdh       p
## Contractual detail.1      0.34
## Contractual detail.2      0.33
## Difference               -0.01       0.887
## --------------------------------------------------------------------------------
```

The output shows that the effect size is slightly different, and that the difference is not statistically significant.

Power analysis

For a planned study, a power analysis may be useful to estimate the probability that NCA's statistical test rejects the null when necessity is true. Assumptions must be made for the (planned) sample size, the expected effect size, the slope of the true linear ceiling line, and the true distribution of data under the ceiling. Power also depends on the selected threshold p-value and ceiling-line estimation technique. For example, consider a planned study with sample size of 50, and an expected effect size of 0.2, with the default values of slope (1), distribution (uniform), threshold p-value (0.05) and ceiling technique (CE-FDH).

```
# Power analysis
set.seed(123)
power <- nca_power(n = 50, effect = 0.2) # This may take some time
print(power)
```

```
## [1] "\r
## [2] "    n  ES slope ceiling distribution.x distribution.y power"
## [3] "1 50 0.2     1  ce_fdh         uniform         uniform  0.98"
```

The results show that power is 0.98 in the defined situation. This means that with high probability (though not guaranteed) the defined necessity effect will be detected in a random sample of n = 50 that is drawn from the population.

Random sample

For simulation studies (e.g., Monte Carlo simulations) the `nca_random` function produces a random sample from a population in which the necessity is represented by a linear ceiling line with given intercept and slope, and where the data under the ceiling and within the scope (unit box) have a uniform or truncated normal distribution. For example, a random sample of size 100 from a population where a necessary condition has a ceiling with an intercept of 0.2 and a slope of 1 with the default values of distribution under the ceiling (uniform) and upper left empty corner (1), the first six cases of the sample are:

```
# Get random sample with necessity
set.seed(123)
random <- nca_random (n = 100, intercepts = 0.2, slopes = 1)
head(random)
```

```
##          X         Y
## 1 0.7883051  0.2875775
## 2 0.8830174  0.4089769
## 3 0.8924190  0.5281055
## 4 0.4566147  0.5514350
## 5 0.5726334  0.6775706
## 6 0.8998250  0.1029247
```

Save results

When selected, the results from `summaries`, `plots`, and `bottlenecks` can be saved in pdf format using the argument `pdf = TRUE` in the `nca_output` function. The results are saved in the working directory.

```
# Save main results
nca_output(model6, summaries = TRUE, plots = TRUE, bottlenecks = TRUE, pdf = TRUE)
```

A `plotly` plot can be saved from the Viewer window of `RStudio` by selecting 'Download plot as png', which saves the plot in the download folder of the computer. Saving is also possible via the Export tab → Save as Image.

Help

The `NCA` package for `R` includes a manual to support users. The following instructions open the help.

```
# Access help documentation for the NCA package and specific functions
help(package = "NCA")
help("nca_analysis")
```

```
help("nca_output")
help("nca_extract")
help("nca_outliers")
help("nca_difference")
help("nca_random")
help("nca_power")
```

B.3.4 Stage 4. Report results

The results of a study that applies NCA can be reported according to guidelines and recommendations presented in the SCoRe checklist (Section 10.3). This includes presenting NCA-specific elements such as:

- Details of the formal necessity hypothesis and how it was developed.
- The selected necessity model (primary ceiling line, bounding box).
- The evaluation criteria for (non)-rejection of the hypothesis (effect size and p-value thresholds, and possibly the target outcome).
- The result for each necessity hypothesis (including effect size, p-value, XY-plot and model fit).
- The bottleneck tables for non-rejected hypotheses.
- The details and results of the robustness checks.

For a pre-registered study the first three elements could be part of the study description.

B.4 Other software packages with NCA

Based on the `NCA` `R` package, a `Stata` module is available for conducting NCA (Spinelli et al., 2023). NCA is also part of the SmartPLS software for conducting NCA in combination with PLS-SEM (Magno et al., 2023; SmartPLS Development Team, 2023). These software packages enable conducting a basic NCA with platforms other than `R`, but lack more advanced functions and recent developments.

B.4.1 Demonstration of nca with **`Stata`**

It is assumed that the analyst has access to the `Stata` software. For this demonstration the same dataset is used as in the demonstration of `NCA` with `R` (see Section B.3). In `Stata` the name of the NCA module is `nca`.

Prepare the analysis

After opening `Stata`, the `nca` package can be installed by entering **`ssc install nca, all`** in the Command window. Installing the `nca` package must be done once. If `nca` is already installed and a new version is available, the command is **`ssc install nca, all replace`**.

After installing the package, a folder (working directory) can be defined for the NCA-project using the command: **`cd "C:/path/to/your/nca-project"`**, where 'path' is the specific path to the folder.

The package is now ready for use. The command `help nca` in the command window opens a new window with help for the different **nca** functions.

Load the data

Enter `use ncaexample.dta` to load the built-in dataset of `nca`. If there is a message 'file ncaexample.dta not found' the file location can be found with this command: `findfile ncaexample.dta` and the command can be run as follows: `use "[path]/ncaexample.dta", clear`, where [path] is obtained from the output of `findfile`. The dataset is now activated. Information about the active dataset appears in the Variables and Properties windows.

Estimate effect size and the p-value

To conduct NCA enter `nca_analysis individualism risktaking innovationperformance, testrep(10000)`, in the Command window. This command first shows the names of the two conditions, followed by the name of the outcome. The next argument `testrep(10000)` indicates that the p-value is estimated and that 10,000 permutations are used for it.

Create output

After the computations are done, the textual output is displayed in the Results window and the graphical output (XY-plots) in a pop-up window called 'Graph'.

Perform the bottleneck analysis

The command for the bottleneck analysis while suppressing the standard **nca** textual and graphical output is `nca_analysis individualism risktaking innovationperformance, nograph nosummaries bottlenecks`.

The bottleneck table is displayed in the Results window with default values for conditions and outcome in terms of percentage of the range (0% is the lowest value; 100% is the highest value). The values can also be expressed as, for example, actual values and percentiles. The command for actual values for the outcome and percentiles for the conditions is: `nca_analysis individualism risktaking innovationperformance, nograph nosummaries xbottlenecks(percentile) ybottlenecks(actual) bottlenecks`.

Advanced functions

The command for the power analysis with a given sample size and effect size and default values for the other inputs is: `nca_power, n(30) effect(0.2)`. For generating random data under a given ceiling line, the `nca_random` function can be used, for example: `nca_random, n(100) intercepts(0.2) slopes(1)`.

C

Publications

This appendix classifies the 583 English-language scholarly articles published or in press in journals listed in Clarivate's Journal Citation Reports. Articles with problematic applications of NCA (see Dul, 2022) and articles in journals that are considered 'predatory journals' (https://predatoryjournals.org/) are excluded from the list. Being listed does not automatically imply that NCA is properly described and applied (see Section 10.3 for a checklist with guidelines and recommendations).

Figure C.1 shows the growth of the cumulative number of scholarly articles that apply NCA. About two-thirds of the articles appear in the highest journal rank category by impact factor (Q1) with Q1 = first quartile, Q2 = second quartile, etc.; Q? indicates that no quality rank is available yet.

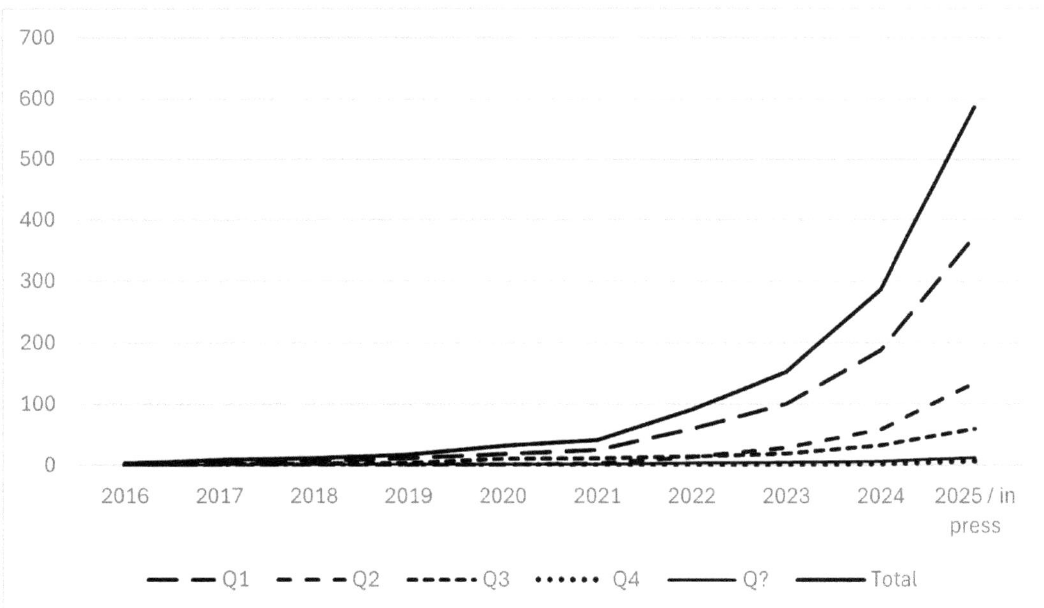

FIGURE C.1 Cumulative number of articles that apply NCA published in Web of Science (Clarivate) listed journals. (Situation of 31 December 2025).

DOI: 10.1201/9781003430766-C

C.1 Journal articles: NCA alone

Table C.1 shows articles that use NCA as the sole or primary research method.

TABLE C.1 Journal articles with application of NCA alone. (Situation of 31 December 2025).

Year	Author(s)	Journal	Journal quality
in press	Hulme	American Political Science Review	Q1
in press	Liu, Yan, Chen, Wu & Zhang	Policy Studies Journal	Q1
in press	Marchetti, Colpizzi, De Caro, Pedretti, Iannattone, Di Blas & Bottesi	Eating and Weight Disorders – Studies on Anorexia, Bulimia and Obesity	Q2
in press	Olafsen & Marescaux	Motivation and Emotion	Q2
in press	Oliveira, Van der Heijden, Finsel, Axelrad, Derous, Gu, Marzec, Mykletun, Vignoli, Pajic, Wöhrmann & Deller	Human Resource Management	Q1
in press	Torres & Godinho	Competitiveness Review: An International Business Journal	Q2
in press	Wang & Beh	Technology Analysis & Strategic Management	Q2
in press	Zaerpour, Gharehgozli & Li	Journal of Manufacturing Technology Management	Q1
2026	Becker, Richter, Ringle & Sarstedt	Journal of Business Research	Q1
2025	Colino, Maiche & Antino	High Ability Studies	Q3
2025	Colpizzi, Pössel & Marchetti	Journal of Youth and Adolescence	Q1
2025	Conde	Journal of Business & Industrial Marketing	Q2
2025	Huang & Wu	Technology Analysis & Strategic Management	Q2
2025	Liang, Li & Yang	Humanities and Social Sciences Communications	Q1
2025	Marchetti, Koster & Hankin	Journal of the American Academy of Child & Adolescent Psychiatry	Q1
2025	Marchetti, Pedretti, Iannattone, Colpizzi, Farina, Di Blas, Ghisi & Bottesi	Journal of Anxiety Disorders	Q1
2025	Morgado Oliveira & Gomes	Total Quality Management & Business Excellence	Q1
2025	Nikookar, Ali, Stevenson & Shokouhyar	Journal of Supply Chain Management	Q1
2025	Ofori, Agyapong & Buer	Journal of Manufacturing Technology Management	Q1
2025	Okeke & Onyemere	Discover Sustainability	Q2

2025	Schelvis, van Veen, Tamminga, Oude Hengel, Nieuwenhuijsen, Boot & Dul	BMJ Open	Q2
2025	Sirisety, Nekkanti, Datti, Bikkina, Patrakov & Baturina	European Journal of Mental Health	Q4
2025	Teston, Zawadzki, Machado & Cazella	BAR – Brazilian Administration Review	Q?
2024	Chaparro-Banegas, Sánchez-Garcia, Calafat-Marzal & Roig-Tierno	Business Strategy and the Environment	Q1
2024	Hermans, Borda, Newburry, Cordova Chea, Finchelstein, Gonzalez-Perez, Montoya-Bayardo, Velasco & Velez-Ocampo	European Business Review	Q2
2024	Ofori	Heliyon	Q1
2024	Subramanian, Jain & Bhattacharyya	Benchmarking: An International Journal	Q1
2023	Abner, Alam & Cho	Public Administration Review	Q1
2023	Acquah	RAUSP Management Journal	Q3
2023	Allard-Poesi & Massu	Landscape and Urban Planning	Q1
2023	Czakon, Klimas & Kawa	Industrial Marketing Management	Q1
2023	Ding & Kuvaas	Journal of Managerial Psychology	Q2
2023	Dionisio, Inacio Junior, Morini & Carvalho	RAUSP Management Journal	Q3
2023	Lankoski & Lankoski	Ecological Economics	Q1
2023	Lee & Lu	Journal of Hospitality & Tourism Research	Q1
2023	Linder, Moulick & Lechner	Entrepreneurship Theory and Practice	Q1
2023	Manchiraju, Akbari & Seydavi	Current Psychology	Q2
2023	Solaimani & Swaak	Journal of Engineering and Technology Management	Q2
2023	Yan, Liu, Chen, Zhang & Wu	Public Administration	Q1
2022	Costa, Daher, Neves & Velez	European Journal of Work and Organizational Psychology	Q1
2022	Glück & Scherpf	Psychology and Aging	Q1
2022	Greco, Guilera, Maldonado-Murciano, Gómez-Benito & Barrios	International Journal of Environmental Research and Public Health	Q?
2022	Kwon	Cryogenics	Q3
2022	Lee & Borgonovi	Comparative Education Review	Q2
2021	Kumar	Case Studies on Transport Policy	Q3
2021	Kwon	Superconductor Science and Technology	Q2
2020	Chaurasia, Kaul, Yadav & Shukla	Journal of Knowledge Management	Q1
2020	Lee & Jeong	Current Issues in Tourism	Q1

2020	Messner	Computational Management Science	Q3
2020	Tynan, Credé & Harms	Learning and Individual Differences	Q1
2019	Shahjehan & Qureshi	Economic research-Ekonomska istraživanja	Q?
2018	Knol, Slomp, Schouteten & Lauche	International Journal of Production Research	Q1
2017	Arenius, Engel & Klyver	Journal of Business Venturing Insights	Q?
2017	Karwowski, Kaufman, Lebuda, Szumski & Firkowska-Mankiewicz	Intelligence	Q1
2017	Luther, Bonfils, Firmin, Buck, Choi, Dimaggio, Popolo, Minor & Lysaker	The Journal of nervous and mental disease	Q3
2016	Golini, Deflorin & Scherrer	International Journal of Operations & Production Management	Q1
2016	Van Der Valk, Sumo, Dul & Schroeder	Journal of Purchasing and Supply Management	Q1

C.2 Journal articles: NCA and MLR

Table C.2 shows articles that apply NCA in combination with multiple linear regression (MLR).

TABLE C.2 Journal articles with application of NCA and MLR (Situation of 31 Decemb 2025).

Year	Author(s)	Journal	Journal quality
in press	Liao, Wu & Chen	Psychological Reports	Q2
in press	Costa, Gomes & Linardi	Innovation	Q3
in press	de Andrés-Sánchez, Almahameed, Arias-Oliva & Pelegrín-Borondo	International Journal of Human–Computer Interaction	Q1
in press	Sekar, Veerasamy, Michellin Kiruba & Ranganathan	Journal of Human Behavior in the Social Environment	Q2
in press	Yang, Bai, Li, Feng, Zhang & Jiang	Journal of Asian Architecture and Building Engineering	Q3
2025	Charles & Kanani	Cogent Business & Management	Q2
2025	Gavín-Chocano, Luque de la Rosa & García-Martínez	Psychology, Society & Education	Q3
2024	Lu, Yang & Yu	International Journal of Emerging Markets	Q1

2024	Dinh, Mühlbacher & Torchia	Journal of Brand Management	Q2
2024	Bachmann, Rose, Maul & Hölzle	Journal of Business Research	Q1
2024	Muhlbacher, Einarsdóttir & Castello	Industrial Marketing Management	Q1
2024	Ciacci, Balzano & Marzi	Journal of Management and Organization	Q2
2024	Kellberg, Keller, Nordine, Moser & Lewalter	International Journal of Science Education Part B-Communication and Public Engagement	Q2
2024	Indihar Štemberger, Bosilj Vuksic, Morelli & Jaklič	Business Process Management Journal	Q1
2024	Hernaus, Dragičević & Hauff	European Journal of Work and Organizational Psychology	Q1
2024	Yiling, He, Wang, Deng & Zhang	Engineering, Construction and Architectural Management	Q1
2024	Dechaume, Mercier, Feybesse, Lubart, Chouvelon, Kermarrec & Tordjman	Learning and Individual Differences	Q1
2024	Xu, Siddik, Sobhani & Rahman	Humanities and Social Sciences Communications	Q1
2023	Jankowska & Karwowski	European Journal of Psychological Assessment	Q2
2023	Zielińska, Lebuda & Karwowski	Personality and Individual Differences	Q1
2023	Servajean-Hilst & Suurmond	International Journal of Innovation Management	Q3
2022	Klimas, Czakon & Fredrich	European Management Journal	Q1
2022	Korman, Van Quaquebeke & Tröster	Journal of Business and Psychology	Q1
2022	Widge, Zhang, Gosai, Papadimitrou, Wilson-Braun, Tsintou, Palanivelu, Noecker, McIntyre, O'Donnell, McLaughlin, Greenberg, Makris, Dougherty & Rathi	Neuropsychopharmacology	Q1
2022	Jain, Bhaskar & Jain	Research in Transportation Business and Management	Q2
2022	Jaiswal & Zane	Thunderbird International Business Review	Q3
2021	Stek & Schiele	Journal of Purchasing and Supply Management	Q1
2020	Garg	International Journal of Educational Management	Q3
2020	Garg	Global Business Review	Q3
2020	Garg & Sarkar	Journal of Organizational Effectiveness: People and Performance	Q3

2020	Ben Jabeur	Environmental Modeling & Assessment	Q3
2020	Karwowski, Jankowska, Brzeski, Czerwonka, Gajda, Lebuda & Beghetto	Creativity Research Journal	Q2
2020	Kobarg, Stumpf-Wollersheim, Schlägel & Welpe	Industry and Innovation	Q1
2020	Richter, Schlaegel, Bakel & Engle	European Journal of International Management	Q3
2019	Tho	Baltic Journal of Management	Q3
2018	Tho	Journal of Management Development	Q3
2017	De Vries, De Koster, Rijsdijk & Roy	Transportation research part E: logistics and transportation review	Q1
2017	Garg	Journal of Human Values	Q3
2016	Karwowski, Dul, Gralewski, Jauk, Jankowska, Gajda, Chruszczewski & Benedek	Intelligence	Q1

C.3 Journal articles: NCA and CB-SEM

Table C.3 shows articles that apply NCA in combination with covariance-based Structural Equation Modeling (SEM).

TABLE C.3 Journal articles with application of NCA and CB-SEM (Situation of 31 December 2025).

Year	Author(s)	Journal	Journal quality
2026	Pinthong, Inprasertkul, Rattanopas, Fakfare, Han & Chua	International Journal of Hospitality Management	Q1
2026	Xie	International Journal of Information Management	Q1
2025	Ahsan & Tahir	Measuring Business Excellence	Q3
2025	Faiella, Zielińska, Karwowski & Corazza	The Journal of Creative Behavior	Q2
2025	Fu, Hu, Zhang & Li	British Food Journal	Q1
2025	Helinski, Westmattelmann & Schewe	Journal of Cleaner Production	Q1
2025	Niu, Liu, Yao, Xiao & Wu	Transport Policy	Q1
2025	Tetteh, Owusu Kwateng, Obiri-Yeboah, Amoako & Nyamekye	Journal of Manufacturing Technology Management	Q1
2025	Tho, Phong, Quan & Trang	Baltic Journal of Management	Q3

2025	Wiśniewska	Journal of Cleaner Production	Q1
2025	Wiśniewska, Liczmańska-Kopcewicz & Żemigała	Sustainable Futures	Q1
2024	Ahsan, Gyensare, Adomako & Zheng	International Small Business Journal	Q2
2024	Blom, Voorn & Borst	Public Management Review	Q1
2024	Fakfare, Manosuthi, Lee, Han, Jarumaneerat & Kim	Asia Pacific Journal of Tourism Research	Q1
2024	Fakfare, Manosuthi, Lee, Lee & Han	Journal of Hospitality & Tourism research	Q1
2024	Gutiérrez-Broncano, Rubio-Andrés, Linuesa Langreo & Sastre-Castillo	Journal of International Marketing	Q1
2024	Ibrahimi, Berisha, Krasniqi & Ramosaj	Journal of Organizational Effectiveness: People and Performance	Q3
2024	Kakkar, Ghulyani, Kuril & Chaudhuri	International Journal of Educational Management	Q3
2024	Su, Lai & Zhang	Leisure studies	Q2
2023	Gorgievski, Bakker, Petrou & Gawke	International Public Management Journal	Q2
2023	Thu, Kadirov, Fam & Tho	Journal of Macromarketing	Q2
2023	Zhu, Miao, Wang & Li	Electronic Commerce Research and Applications	Q1
2022	Frommeyer, Wagner, Hossiep & Schewe	Journal of Business Research	Q1
2022	Lee, Park & Jeong	Journal of Hospitality and Tourism Management	Q1
2021	Lee & Jeong	Journal of Hospitality and Tourism Management	Q1
2020	Franke & Foerstl	Journal of Business Logistics	Q1
2019	Jada & Mukhopadhyay	Personnel Review	Q1
2019	Tho	Education and Training	Q1

C.4 Journal articles: NCA and PLS-SEM

Table C.4 shows articles that apply NCA in combination with Partial Least Squares Structural Equation Modeling (PLS-SEM).

TABLE C.4 Journal articles with application of NCA and PLS-SEM (Situation of 31 December 2025).

Year	Author(s)	Journal	Journal quality
in press	Abbass, Khan & Tao	International Journal of Human–Computer Interaction	Q1
in press	Asheq, Hasan & Coughlan	Journal of Travel Research	Q1
in press	Bawa, Jain & Goel	Interactive Technology and Smart Education	Q1
in press	Casadiego-Alzate, Arroyo-Cañada, Castañeda-Quirama, Sánchez-Torres & Luzardo Briceño	Journal of Modelling in Management	Q3
in press	Choudhary & Ahuja	Journal of Global Marketing	Q2
in press	Dang, Quach & Roberts	International Journal of Human–Computer Interaction	Q1
in press	Das, Kondasani, Deb & Batabyal	Tourism and Hospitality Research	Q2
in press	De Sousa, Bizarrias, Raposo & da Cruz	Journal of Services Marketing	Q2
in press	Díez-de-Castro, Vázquez-Sánchez & Agustí-Pérez	European Journal of Management and Business Economics	Q1
in press	Garg, Wadhwa, Gomes & Sahasrabudhe	Education and Information Technologies	Q1
in press	Haverila, Twyford, McLaughlin & Arora	Management Decision	Q1
in press	Hussain, Wong, Stephenson & Chong	Current Issues in Tourism	Q1
in press	Kashive & Raina	International Journal of Organizational Analysis	Q2
in press	Kayser & Gradtke	Journal of Decision Systems	Q2
in press	Koay & Ngang	British Food Journal	Q1
in press	Li, Wang & Wang	Energy & Environment	Q2
in press	Loh, Shyu, Lee, Huang, Tan & Ooi	Journal of Computer Information Systems	Q2
in press	Low, Tang & Yap	Management of Environmental Quality	Q2
in press	Low, Wut & Pok	Educational Gerontology	Q2

in press	Ly	Journal of the Knowledge Economy	Q1
in press	Mkedder, Das, Bakır, Lachachi & Aldhabyani	International Journal of Emerging Markets	Q1
in press	Murtaza, Hussain, Salman, Mustafa & Han	European Journal of Management and Business Economics	Q1
in press	Mwesiumo	Studies in Higher Education	Q1
in press	P. & Mekonnen	International Journal of Ethics and Systems	Q2
in press	Pérez-Tapia, Cabrera-Sánchez & Hernández Correa	Journal of Vacation Marketing	Q1
in press	Poku, Nuertey, Agbemude, Owusu & Buabeng	Benchmarking: An International Journal	Q2
in press	Rath & Jain	Benchmarking: An International Journal	Q2
in press	Samagaio, Francisco, Felício & Verga Matos	Management Decision	Q1
in press	Sánchez-Medina, Pellejero, C-Sánchez & Cerviño-Cortínez	Current Issues in Tourism	Q1
in press	Silalahi	International Journal of Human–Computer Interaction	Q1
in press	Tetteh & Owusu Kwateng	Journal of Enterprise Information Management	Q1
in press	Tetteh, Owusu Kwateng, Owusu-Antwi & Opoku-Mensah	Business Process Management Journal	Q1
in press	Tingli, Aw, Huam & Tan	Journal of Hospitality and Tourism Technology	Q1
in press	Twyford, Hosseini, Haverila & Haverila	Social Responsibility Journal	Q2
in press	Velikova, Bouzdine-Chameeva, Ben Nasr, Sanz-Ibáñez, Anton Clavé & Rabassa-Figueras	Current Issues in Tourism	Q1
in press	Wang, Aw, Tan, Sthapit & Li	Current Issues in Tourism	Q1
in press	Wu, Aw, Tan & Liu	Journal of Marketing Analytics	Q2
in press	Wu, Aw, Tan & Ooi	Internet Research	Q1
in press	Xu, Yang, Chen, Chen & Nan	International Journal of Human–Computer Interaction	Q1
in press	Zeng, Murayama, Nishikizawa & Suwanteep	Sustainable Development	Q1
in press	Zhang, Lim, Cham, Huang & Tapsir	Asia Pacific Business Review	Q3
in press	Zhao, Ru & Jia	International Journal of Human–Computer Interaction	Q1
in press	Zhong & Luo	International Journal of Human–Computer Interaction	Q1
2026	Arpaci, Al-Sharafi & Mahmoud	Transportation Research Part F: Traffic Psychology and Behaviour	Q1
2026	Baah, Agyabeng-Mensah, Afum, Acquah & Agyeman	Industrial Marketing Management	Q1

2026	Bao, Xie & Yao	Journal of Retailing and Consumer Services	Q1
2026	Chen	The International Journal of Management Education	Q1
2026	Choi, Yang, Ahmad & Lee	Travel Behaviour and Society	Q1
2026	Darwati, Yusron, Purnomo, Suryadi, Rusmin, Trisilawati, Yatno, Pitono, Syakir, Fidiyawati & Hartati	Scientia agropecuaria	Q3
2026	Hasselwander, Sunio, Lah & Mogaji	Journal of Retailing and Consumer Services	Q1
2026	Mobeen, Kabir, Schneider, Ahmed, Ahmad & Scheffran	Heliyon	Q1
2026	Nguyen & Le	The International Journal of Management Education	Q1
2026	Nguyen, Nguyen, Le, Pervan, Vu & Nguyen	Journal of Retailing and Consumer Services	Q1
2026	Pérez-Portabella, Arias-Oliva, Padilla-Castillo & de Andrés-Sánchez	Computers and Education Open	Q1
2026	Rahman, Carlson, Chowdhury, Gudergan, Wetzels, Ringle & Grewal	Journal of Business Research	Q1
2026	Sen, Roy, Laradi, Das, Ray, Parida & Das	Journal of Cleaner Production	Q1
2026	Wang & Guo	Journal of Retailing and Consumer Services	Q1
2025	Agustí-Pérez, Vázquez-Sánchez & Orta-Perez	International Journal of Quality & Reliability Management	Q2
2025	Al Mamun, Reza, Yang & Gao	Mitigation and Adaptation Strategies for Global Change	Q3
2025	Alajhar, Bajaba & Basahal	The International Journal of Human Resource Management	Q1
2025	Aziz, Li & Khan	Technology in Society	Q1
2025	Baah, Saleem, Greenland, Tenakwah & Chakrabarty	Environmental Development	Q1
2025	Bajaba & Basaad	Stress and Health	Q2
2025	Bazan	Sage Open	Q1
2025	Bhatt, Pandey, Patel, Ghodake, Jariwala & Thomas	Journal of Nonprofit & Public Sector Marketing	Q3
2025	Bowden, Eshaghi, Afshardoost & Rahman	Journal of Consumer Marketing	Q2
2025	Chaity, Islam, Masud, Islam, Rahman & Kabir	Business Strategy and the Environment	Q1
2025	Chu, Ai, Wu & Zhang	Transport Policy	Q1
2025	Contreras Pinochet, Pardim, Mangini & de Souza	Case Studies on Transport Policy	Q2
2025	de Andrés-Sánchez, Arias-Oliva, Souto-Romero & Llorens-Marín	Sage Open	Q1

2025	Do, Tho & Le	International Journal of Emerging Markets	Q1
2025	Duan & Xiang	Transport Policy	Q1
2025	Duan, Si & Xia	Transportation Research Part F: Traffic Psychology and Behaviour	Q1
2025	Duarte, Meneses, Silva & Tharakan	International Journal of Gastronomy and Food Science	Q2
2025	Erdmann & Toro-Dupouy	European Journal of Innovation Management	Q1
2025	Galindo-Martin, Mendez-Picazo & Perez-Pujol	International Journal of Information Management	Q1
2025	Gangakhedkar & Khan	Energy Efficiency	Q3
2025	Garg, Ahmad & Madsen	Cleaner and Responsible Consumption	Q1
2025	Gomes, Tontini & Picolo	International Journal of Operations & Production Management	Q1
2025	Gordon, Balta-Ozkan, Haq & Nabavi	Sustainable Energy & Fuels	Q2
2025	Harun, Salema & Matto	Future Business Journal	Q3
2025	Hassan, Mai, Wahab, Amin, Hassan & Oláh	Computers in Human Behavior Reports	Q1
2025	Hasselwander	Transportation Research Part F: Traffic Psychology and Behaviour	Q1
2025	Ji & Li	Teaching and Teacher Education	Q1
2025	Koay & Low	British Food Journal	Q1
2025	Liao, Li, Al Mamun, Tseng & Helmi Ali	International Journal of Logistics Research and Applications	Q1
2025	Lima, Bahli, Arbulu, Hamdi and Saikouk	Journal of Business & Industrial Marketing	Q2
2025	Liu, Zhu & Wu	Electronic Commerce Research and Applications	Q1
2025	Lourenco, Madeira & Santos	ReMark – Revista Brasileira de Marketing	Q4
2025	Low, Cham & Ramayah	Quality & Quantity	Q?
2025	Low, Rahim & Wut	Sustainable Futures	Q1
2025	Malibari, Bajaba, Bajaba & Basahal	Journal of Innovation & Knowledge	Q1
2025	Manzi-Puertas, Agirre-Aramburu & López-Pérez	Cogent Business & Management	Q2
2025	Mazhar & Zilahy	British Food Journal	Q1
2025	Milovan, Dobre & Moisescu	Journal of Business Research	Q1
2025	Mohammed, Al-Swidi, Ferraris & Jabeen	International Journal of Consumer Studies	Q1
2025	Moon, Lee & Lee	Acta Psychologica	Q2
2025	Munir, Abbas, Qamar, Farooq & Cohen	Sustainable Futures	Q1
2025	Mwesiumo, Similä & Langseth	Journal of Public Procurement	Q2

2025	Niranjan, Garg, Gligor & Hawkins	Journal of Business & Industrial Marketing	Q2
2025	Norena-Chavez, Chavez-Limaylla & Romani-Torres	Cogent Business & Management	Q2
2025	Osei & Ofori	Journal of Business Logistics	Q1
2025	Ouazzani, Calderón-García & Tubillejas-Andrés	Empirical Studies of the Arts	Q3
2025	P. & Mekonnen	Accounting Research Journal	Q2
2025	Pangarso, Yunani, Umbara, Hendriyanto, Sisilia & Panutur	Sustainable Futures	Q1
2025	Pillai, Sivathanu & Rana	Journal of Business Research	Q1
2025	Quansah, Lang & Mühlbacher	Journal of Business Research	Q1
2025	Rafi, Roy, Rehman & Hasni	Decision Support Systems	Q1
2025	Rahimi & Daneshvar Ghorbani	Computers in Human Behavior Reports	Q1
2025	Rahimi & Sevilla-Pavón	Smart Learning Environments	Q1
2025	Rahman & Sadik	Discover Sustainability	Q2
2025	Riggs, Felipe, Roldán, & Real	Journal of Marketing Analytics	Q2
2025	Rocha, Ed-Dafali & Sharma	Journal of Small Business and Enterprise Development	Q1
2025	Romero-Charneco, Casado-Molina, Alarcón-Urbistondo & Cabrera Sánchez	Journal of Hospitality and Tourism Technology	Q2
2025	Rulandari & Silalahi	Transforming Government: People, Process and Policy	Q2
2025	Santana, Martens, Bizarrias & Martens	Brazilian Administration Review	Q?
2025	Santos-Jaén, Pemartín, Gimeno-Arias & León-Gómez	Journal of Business & Industrial Marketing	Q2
2025	Sathatip, Agmapisarn, Suttipun & Fakfare	British Food Journal	Q1
2025	Shan, Deng, Deng & Song	IEEE Transactions on Engineering Management	Q1
2025	Sharma & Paço	Journal of Information, Communication and Ethics in Society	Q2
2025	Shen, Zhang, Wong, Chen & Ren	Humanities and Social Sciences Communications	Q1
2025	Siddik, Yong, Du & Goodell	IEEE Transactions on Engineering Management	Q1
2025	Singh Yadav & Kumar Kar	Journal of Cleaner Production	Q1
2025	Sohaib, Alshemeili & Bhatti	Cleaner and Responsible Consumption	Q1
2025	Suganthi & Mohamed Jasim	Journal of Internet Commerce	Q2
2025	Tan, Leong & Richter	International Journal of Human–Computer Interaction	Q1

2025	Tho, Trang & Thu	Journal of Knowledge Management	Q1
2025	Troiville, Moisescu & Radomir	Journal of Retailing and Consumer Services	Q1
2025	Vázquez-Sánchez, Tuesta-Tapia & Amador-Hidalgo	Quality Management Journal	Q2
2025	Venciute, Lukosiute, Correia, Meneses & Kuslys	European Business Review	Q2
2025	Verma & Kapoor	British Food Journal	Q1
2025	Vu & Tolstoy	Journal of Business Research	Q1
2025	Wijesinghe, Mapitiyage, Wickramarathne, Wickramage, Wisenthige & Aluthwala	PLOS ONE	Q2
2025	Wu, Aw & Phau	Journal of Internet Commerce	Q2
2025	Yoon & Wang	International Journal of Tourism Research	Q1
2025	Zhao & Furuoka	Humanities and Social Sciences Communications	Q1
2024	Aldhamiri, Carlson, Vilches-Montero, Rahman & Gudergan	Journal of retailing and Consumer Services	Q1
2024	Aw, Thomas, Patel, Bhatt & Cham	International Journal of Bank Marketing	Q1
2024	Baah, Saleem, Greenland, Tenakwah & Chakrabarty	Journal of Environmental Management	Q1
2024	Baber, Fanea-Ivanovici, Pana & Manole	Online information review	Q1
2024	Brazo, Velicia-Martín, Palos-Sanchez & Roldán	European Business Review	Q2
2024	Cassia & Magno	Journal of Business Research	Q1
2024	Choksi & Bhatt	International Journal of Human–Computer Interaction	Q1
2024	Chong, Lim, Cheah & Tee	Current Issues in Tourism	Q1
2024	Corchuelo Martínez-Azúa, Dias & Sama-Berrocal	International Journal of Innovation Science	Q2
2024	Damberg, Liu & Ringle	Journal of Marketing Analytics	Q2
2024	Dávila, Andreeva & Sætre	European Management Review	Q2
2024	De Andrés-Sánchez & Gené-Albesa	Journal of Organizational Computing and Electronic Commerce	Q3
2024	Dos Santos Leite Gonella, Filho, Ganga & Lizarelli	Business Strategy and the Environment	Q1
2024	Dutta, Thomas, Shiva, Papa & Cuomo	Journal of Knowledge Management	Q1
2024	Eccarius & Chen	Technological Forecasting and Social Change	Q1
2024	Escadas, Jalali, Septianto & Farhangmehr	Business Ethics, the Environment & Responsibility	Q2

2024	Gao, Luo, Wang, Zhang & Deng	Journal of Construction Engineering and Management	Q1
2024	Gavín-Chocano, García-Martínez & De La Rosa	Current Psychology	Q2
2024	Gollagari, Birega & Mishra	African Journal of Economic and Management Studies	Q3
2024	Gordon, Balta-Ozkan, Haq & Nabavi	International Journal of Hydrogen Energy	Q1
2024	Gyensare, Miri, Zahoor & Alajaty	Journal of Islamic Marketing	Q2
2024	Horvatinović, Mikić & Dabić	International Journal of Entrepreneurial Behavior and Research	Q1
2024	Hui, Al Mamun, Masukujjaman, Makhbul & Ali	Heliyon	Q1
2024	Islam, Rahman, Taher, Quaosar & Uddin	Future Business Journal	Q2
2024	Li, Sampene, Khan & Wiredu	Operations Management Research	Q1
2024	Lim, Munikrishnan, Leong, Hiew, Leong & Yang	Information and Computer Security	Q3
2024	Liying & Mengying	Sage Open	Q1
2024	Ma & Bennett	Journal of Higher Education	Q1
2024	Magno & Cassia	Journal of retailing and Consumer Services	Q1
2024	Mansoor, Rahman & Bowden	Journal of Consumer Behaviour	Q2
2024	Oliveira & Gomes	Business Strategy and the Environment	Q1
2024	Oppong & Bannor	Journal of Food Products Marketing	Q2
2024	Osakwe, Říha, Elgammal & Ramayah	International Journal of Retail & Distribution Management	Q1
2024	Pandey, Thomas, Bhatt, Patel & Malkar	Journal of Marketing Theory and Practice	Q2
2024	Pinochet, Souza, Viana & Rodríguez-Abitia	REGE-Revista de Gestao	Q3
2024	Quaicoe, Acquah & Gatsi	Journal of Cleaner Production	Q1
2024	Rahimi & Sevilla-Pavón	Computers and Education: Artificial Intelligence	Q?
2024	Rico, Calvo-Mora, Medina-Molina & Alves	Total Quality Management & Business Excellence	Q2
2024	Rishi, Shiva & Israney	Asia Pacific Journal of Marketing and Logistics	Q2
2024	Rosli, Binti Mohamad, Johar, Hashim & Lim	Heliyon	Q1
2024	Safdar, Moazzam, Ahmed, Khan, Manzoor & Raziq	Benchmarking: An International Journal	Q1
2024	Sanyal, Jain & Choudhury	International Social Science Journal	Q?

2024	Sarstedt, Richter, Hauff & Ringle	Journal of Marketing Analytics	Q2
2024	Sepe, Luongo, Di Gioia & Della Corte	European Journal of Innovation Management	Q1
2024	Sharma, Kar & Gupta	Government Information Quarterly	Q1
2024	Similä & Mwesiumo	Journal of Cleaner Production	Q1
2024	Sun, Ooi, Wei-Han Tan & Lee	Supply Chain Management: An International Journal	Q1
2024	Suryavanshi, Thomas, Trivedi, Patel & Bhatt	Journal of Philanthropy and Marketing	Q3
2024	Talib, Papastathopoulo & Ahmad	Digital Policy Regulation and Governance	Q2
2024	Thien, Chuah & Low	European Journal of Education	Q1
2024	Tiwari, Kaurav & Koay	Journal of Marketing Analytics	Q2
2024	Tuesta-Tapia, Vázquez-Sánchez & Amador-Hidalgo	International Journal of Quality & Reliability Management	Q2
2024	Ullibarriarana-Garate, Agirre-Aramburu, Miguel & Sánchez-Urien	Journal of Infrastructure, Policy and Development	Q4
2024	Veiga, Ferreira, Zhang & Liu	Journal of Computer Information Systems	Q2
2024	Wang & Wang	Annals of Public and Cooperative Economics	Q2
2024	Yadav, Parida, Choudhary & Kumar	Journal of Environmental Management	Q1
2024	Zhou, Kumar & Furuoka	Humanities and Social Sciences Communications	Q1
2023	Acquah, Issau, Dei Mensah & Vanderpuye	Heliyon	Q1
2023	Anand, Arya, Suresh & Sharma	Tourism Planning & Development	Q2
2023	Battistoni, Gitto, Murgia & Campisi	International Journal of Production Economics	Q1
2023	Delladio, Caputo, Magrini & Pellegrini	Management Decision	Q2
2023	Harun, Hogset & Mwesiumo	Sustainable Development	Q1
2023	He & Ismail	Humanities and Social Sciences Communications	Q1
2023	Kaushik, Jain & Choudhary	Case Studies on Transport Policy	Q3
2023	Kumar, Saha & Anand	International Journal of Organizational Analysis	Q3
2023	Lim, Leong, Lim, Lim, Lim & Heng	Young Consumers	Q2
2023	Liu, Wang, He, Ma & Fu	Journal of Cleaner Production	Q1
2023	Mwesiumo, Halpern, Bråthen, Budd & Suau-Sanchez	Transportation Research Part A: Policy and Practice	Q1
2023	Mwesiumo, Harun & Hogset	Industrial Marketing Management	Q1

2023	Ngoc Su, Quy Nguyen-Phuoc, Thi Kim Tran, Van Nguyen, Trong Luu & Pham	Travel Behaviour and Society	Q1
2023	Si, Li, Duan, Cheng & Bao	Sustainable Production and Consumption	Q1
2023	Singh, Jain, Sharma & Nigam	International Journal of Project Management	Q1
2023	Tumaku, Ren, Boakye, Ofori & Abubakari	International Journal of Quality and Service Sciences	Q2
2022	Abdalla, Mwesiumo, Öztüren & Kılıç	International Journal of Tourism Research	Q1
2022	Adam & Dzang Alhassan	African Journal of Science, Technology, Innovation and Development	Q2
2022	Afrin & Prybutok	Health Promotion International	Q2
2022	Bolívar, Roldán, Castro-Abancéns & Casanueva	Management International Review	Q2
2022	Cheung, Leung, Aw & Koay	Journal of Retailing and Consumer Services	Q1
2022	Frezatti, Mucci & Bido	BAR-Brazilian Administration Review	Q?
2022	Hayat, Salameh, Malik & Yaacob	Technology in Society	Q1
2022	Koay, Cheah & Chang	British Food Journal	Q1
2022	Kopplin & Rausch	Review of Managerial Science	Q1
2022	Kopplin, Gantert & Maier	Review of Managerial Science	Q1
2022	Liu, Yu & Damberg	Journal of Sports Marketing and Sponsorship	Q2
2022	Loh, Lee & Leong	Expert Systems With Applications	Q1
2022	Lyu, Chin, Zhang & Liu	Journal of Consumer Behaviour	Q2
2022	Pangarso, Winarno, Aulia & Ritonga	Leadership & Organization Development Journal	Q2
2022	Prakash, Jeyakumar Nathan, Kini & Victor	PLOS ONE	Q1
2022	Saraf, Saha & Anand	International Journal of Public Sector Management	Q2
2022	Sharma, Dwivedi, Mariani & Islam	Technological Forecasting and Social Change	Q1
2022	Sukhov, Olsson & Friman	Transportation Research Part A: Policy and Practice	Q1
2022	Yu, Cheah & Liu	International Journal of Contemporary Hospitality Management	Q1
2022	Zahrai, Veer, Ballantine, De Vries & Prayag	Journal of Consumer Affairs	Q2
2021	Della Corte, Aria, Del Gaudio, Barney, Cobanoglu & Sepe	International Journal of Hospitality Management	Q1

2021	Richter, Martin, Hansen, Taras & Alon	Journal of Business Research	Q1

NCA + PLS-SEM + QCA applications are listed under 'NCA + combination of other methods'.

C.5 Journal articles: NCA and QCA

Table C.5 shows articles that combine NCA with Qualitative Comparative Analysis (QCA).

TABLE C.5 Journal articles with applications of NCA and QCA (Situation of 31 December 2025).

Year	Author(s)	Journal	Journal quality
in press	Chen & Huang	Technology Analysis & Strategic Management	Q2
in press	Chen, Zhao, Wang & Shi	Environment, Development and Sustainability	Q2
in press	Fakfare, Manosuthi, Lee, Han & Jin	Current Issues in Tourism	Q1
in press	Fakfare, Manosuthi, Lee, Jin, Han & Kim	Journal of Vacation Marketing	Q1
in press	Farmaki, Pappas, Stergiou, Apostolakis & Camilleri	Tourism Review	Q1
in press	Gao, Wang, Sethanan & Tseng	Journal of Industrial and Production Engineering	Q1
in press	Gong, Wu, Tan, He & Lin	Electronic Commerce Research	Q2
in press	Hu, Jie & Zhu	Technological and Economic Development of Economy	Q1
in press	Huang, Bu & Li	Small Business Economics	Q1
in press	Huang, Wu, Li & Zhang	Aslib Journal of Information Management	Q1
in press	Lai, Zhang & Su	Environment, Development and Sustainability	Q2
in press	Ledesma-Chaves, Gil-Cordero & Maldonado-López	Journal of Enterprise Information Management	Q1
in press	Li, Huang & Yao	Environment, Development and Sustainability	Q2
in press	Li, Ma, Hao, Xu & Liu	Current Issues in Tourism	Q1
in press	Li, Yan & Ma	Journal of Asian Architecture and Building Engineering	Q3
in press	Li, Zhu & Kong	The Electronic Library	Q2
in press	Manickavasagam, Ayyalusamy, Bhaiyat, Jha, Jain & Poddar	International Journal of Finance & Economics	Q2
in press	Piriyakul, Kunathikornkit & Piriyakul	Journal of Vacation Marketing	Q1

in press	Sun & Zhou	Review of Policy Research	Q1
in press	Wang, Feng, Wang & Shen	Polish Journal of Environmental Studies	Q4
in press	Wang, Tang & Zhao	Environmental Progress & Sustainable Energy	Q3
in press	Yang	Environmental Research Communications	Q3
in press	Yin, Ye, Yu, Li & Liu	Education and Information Technologies	Q1
in press	Yuan, Tao, Chen, Zhang, Chen & Lai	Engineering, Construction and Architectural Management	Q1
in press	Yuan, Zhuang & Liu	Engineering, Construction and Architectural Management	Q1
in press	Zhang	Current Issues in Tourism	Q1
in press	Zhang	Journal of Sustainable Tourism	Q1
in press	Zhang, Dong, Wang, Yao & Huang	Engineering, Construction and Architectural Management	Q1
2026	An, Lin & Benitez	Decision Support Systems	Q1
2026	Choi	Journal of Destination Marketing & Management	Q1
2026	Ding, Wang & Tan	Land Use Policy	Q1
2026	Huang, Ding & Long	Journal of Cleaner Production	Q1
2026	Huang, Huang, Rong & Xie	Journal of Innovation & Knowledge	Q1
2026	Kang & Shang	Technology in Society	Q1
2026	Li, Hu, Tong, Xie, Tong & Hu	Sustainable Cities and Society	Q1
2026	Porlán-Ariza, Villarejo-Ramos, Ledesma-Chaves & Gil-Cordero	Thinking Skills and Creativity	Q1
2026	Zhang, Cui & Chai	Journal of Innovation & Knowledge	Q1
2025	Acquah, Naude & Oguji	Technological Forecasting and Social Change	Q1
2025	Arcidiacono, Podrecca & Schupp	Industrial Management & Data Systems	Q2
2025	Cangialosi & Odoardi	Small Group Research	Q3
2025	Cao & Zhang	Sage Open	Q1
2025	Chai & Wu	Scientometrics	Q1
2025	Chen, Wu & Wang	International Journal of Innovation Management	Q4
2025	Du, Kim, Fan & Su	Journal of Business Research	Q1
2025	Guevara, Medina-Molina, Pérez-Macías & Fabra Florit	Quality & Quantity	Q?
2025	Guo, Wang & Li	Scientific Reports	Q1
2025	He, Chen & Gu	Cities	Q1
2025	He, Cheng & Wang	PLOS ONE	Q2
2025	Huang & Yang	Industrial Management & Data Systems	Q2
2025	Huang, Jiang & Hou	Sage Open	Q1

2025	Huang, Sun, Yuan & Bu	Corporate Social Responsibility and Environmental Management	Q1
2025	Huang, Zhang, Xu, Yuan, Zhang & Huang	Humanities and Social Sciences Communications	Q1
2025	Jing, Qianqiang, Yantai & Qi	Technological Forecasting and Social Change	Q1
2025	Li, Che, Yao & Cui	PLOS ONE	Q2
2025	Li, Gong & Song	Scientific Reports	Q1
2025	Li, Ma, Zheng, Ma, Zuo & Goodsite	Scientific Reports	Q1
2025	Li, Yanrui, Fan, Yuqing & Xu	China: An International Journal	Q2
2025	Meddeb & St-Jean	Mentoring & Tutoring: Partnership in Learning	Q2
2025	Mladenović, Monashev, Jirásek & Bruni	Computers in Human Behavior	Q1
2025	Mohsin, Rashed, Gerschberger, Plasch, Ahmed, Dritan & Islam	The International Journal of Logistics Management	Q1
2025	NA, Corbett, Nishant, NA, Cherki El Idrissi, NA, Chung & NA	Communications of the Association for Information Systems	Q3
2025	Nguyen, Ngo, Phan, Tao & Bui	Journal of Health Organization and Management	Q2
2025	Nie, Qiu, Wu, Yang, Li, Feng & Wang	Environmental Research Communications	Q3
2025	Peng & Hu	The American Journal of Economics and Sociology	Q3
2025	Qin, Luo, Xu, Wang & Škare	Technological and Economic Development of Economy	Q1
2025	Rashed, Uddin, Islam, Mohsin, Rahman, Kumar & Rana	Sustainable Futures	Q1
2025	Ren & Liu	Social Indicators Research	Q1
2025	Ren, Hu & Zhang	European Journal of Innovation Management	Q1
2025	Shan, Wang & Luo	Acta Psychologica	Q2
2025	Shen, Yang & Zhu	PLOS ONE	Q2
2025	Shen, Zhang, Zheng & Ruan	SAGE Open	Q1
2025	Tang, Zhao, Zhou, Li, Qian, Hou & Liu	Environmental Geochemistry and Health	Q1
2025	Tao, Wang, He, Zhang & Liu	Gender in Management	Q1
2025	Tian, Wang, Li, Chen, Kuang & Mao	Scientific Reports	Q1
2025	Wan, Ke & Liang	Forest Policy and Economics	Q1
2025	Wang & Wang	Learning and Instruction	Q1
2025	Wang & Zhao	Managerial and Decision Economics	Q3
2025	Wang, Wang, Foley & Lin	The American Journal of Economics and Sociology	Q3
2025	Wang, Xu, Wang, Wang, Kang, Qiu, Xue, Fang & Zhang	Sustainable Cities and Society	Q1

2025	Wang, Zhao, Wang, Song, Feng, Yu, & Li	BMC Nursing	Q1
2025	Wei & Yang	Economic Analysis and Policy	Q1
2025	Xia, Cheng, Jiang, Anand & He	Journal of Knowledge Management	Q1
2025	Xiao, Dong & Li	Journal of Cleaner Production	Q1
2025	Xie, Zainudin & Mahdzan	Future Business Journal	Q3
2025	Yan & Wang	Chinese Management Studies	Q3
2025	Ye, Ruan, Xia & Gu	Humanities and Social Sciences Communications	Q1
2025	Yin	Applied Spatial Analysis and Policy	Q2
2025	Zhai, Zeng & Wang	Regional Studies in Marine Science	Q1
2025	Zhang	Tourism Management	Q1
2025	Zhang & Hao	Technological Forecasting and Social Change	Q1
2025	Zhao, Yang, Chen, Yu, Chen & Yang	PLOS ONE	Q2
2025	Zheng, Lin, Yan & Guo	International Review of Economics & Finance	Q1
2025	Zhou, Wu, Chen & Li	Science, Technology and Society	Q3
2025	Zitao, Zixiao & Huaxia	Sustainable Futures	Q1
2024	Acquah	Heliyon	Q1
2024	Calispa-Aguilar	Entrepreneurial Business and Economics Review	Q1
2024	Cezar	Journal of Organizational Computing and Electronic Commerce	Q3
2024	Cezar	Telematics and Informatics	Q1
2024	Chen & Chen	Journal of Innovation & Knowledge	Q1
2024	Do Prado & De Moraes	Journal of Technology Transfer	Q1
2024	Dong, Jiang, Dai & Chi	Grey Systems: Theory and Application	Q1
2024	Dou, Hou, Yan, Jin & Li	Engineering, Construction and Architectural Management	Q1
2024	Gabay-Mariani, Bastian, Caputo & Pappas	International Journal of Entrepreneurial Behavior and Research	Q1
2024	Gong, Wang & Li	PLOS ONE	Q1
2024	Huang, Shuaib, Rahman, Rahman & Hossain	Natural Resources Forum	Q2
2024	Huang, Zhang, Xu & Bu	The International Journal of Management Education	Q1
2024	Jiang & Wang	Current Issues in Tourism	Q1
2024	Kazemzadeh, Fuinhas, Salehnia, Silva & Radulescu	Energy	Q1

2024	Li, Du, Sun & Xie	Chinese Management Studies	Q3
2024	Li, Fan, Kumar & Ananthram	Journal of Management Studies	Q1
2024	Li, Niu, Liu & Wu	Engineering, Construction and Architectural Management	Q1
2024	Lin, Yao & Yu	Journal of Urban Planning and Development	Q3
2024	Medina-Molina & Rey-Moreno	Cuadernos de Gestión / Management Letters	Q3
2024	Min, Zhu & Peng	Heliyon	Q1
2024	Shao, Jiang, Liou, Su, Yuan & Dan	Managerial and Decision Economics	Q2
2024	Sun, Xia, Huang & Tong	Science of the Total Environment	Q1
2024	Vargas-Zeledon & Lee	Journal of Innovation & Knowledge	Q1
2024	Vasist & Krishnan	Information Systems Frontiers	Q1
2024	Wang & Cheng	Scientific Reports	Q1
2024	Wang, Sun & Tiong	International Journal of Managing Projects in Business	Q3
2024	Wattanacharoensil, Fakfare, Manosuthi, Lee, Chi & Heesup	Tourism Management	Q1
2024	Yan, Işık, Ongan, Pinzon & Tillaguango	Sustainable Development	Q1
2024	Yu, Fan & Zhang	PLOS ONE	Q1
2024	Zang & Guo	Journal of Chinese Governance	Q1
2024	Zhang	Journal of Sustainable Tourism	Q1
2024	Zhang	Current Issues in Tourism	Q1
2024	Zhang	Journal of Environmental Planning and Management	Q1
2024	Zhang & Zhang	Economic Change and Restructuring	Q1
2024	Zhang, Chen & Wu	Journal of Environmental Planning and Management	Q1
2024	Zhang, Li, Li & Lu	Internet Research	Q1
2024	Zhang, Teirlinck, Su & Yu	International Journal of Production Economics	Q1
2024	Zhang, Yang, Zhang, Sun & He	Polish Journal of Environmental Studies	Q4
2024	Zhao, Ye & Cai	Emerging Markets Finance and Trade	Q1
2024	Zhou, Chen, Jin & Yuan	European Journal of Information Systems	Q1
2023	Bhattacharyya, Balaji & Jiang	Journal of Business Research	Q1
2023	Bouncken, Fredrich & Kraus	Global Strategy Journal	Q1
2023	Bu, Li & Huang	The International Journal of Management Education	Q1
2023	Carmona, Stef, Ben Jabeur & Ben Zaied	Journal of Environmental Management	Q1
2023	Guo, Fu & Sun	Scientific Reports	Q1

2023	Huang, Bu & Long	Journal of Innovation & Knowledge	Q1
2023	Huang, Li & Li	Technological Forecasting and Social Change	Q1
2023	Huang, Li, Bu & Zhao	Journal of Cleaner Production	Q1
2023	Huang, Li, Chen & Wang	Journal of Innovation & Knowledge	Q1
2023	Jain, Chakraborty & Choudhary	International Journal of Consumer Studies	Q1
2023	Kazemzadeh, Fuinhas, Salehnia, Koengkan & Silva	Environmental Science and Pollution Research	Q?
2023	Kopplin	Australasian Journal of Information Systems	Q2
2023	Li & Larivière	Psychology & Marketing	Q1
2023	Liang, Wang, Foley & Ma	Applied Economics	Q2
2023	Liu, Wen, Peng & Cao	Journal of Cleaner Production	Q1
2023	Luo, Yang & Wang	PLOS ONE	Q1
2023	Sun, Fu & Yang	Journal of Innovation & Knowledge	Q1
2023	Tan & Wei	Journal of the Knowledge Economy	Q1
2023	Wagner, Raymond & Paré	JMIR Medical Education	Q1
2023	Wang, Li, Gao, Zeng, Wang, Li & Li	Land Use Policy	Q1
2023	Wang, Zhang, Luo & Abedin	Journal of Environmental Management	Q1
2023	Wang, Zhang, Zeng, Meng & Lin	Corporate Social Responsibility and Environmental Management	Q1
2022	Deng, Zhuang, Li & Yang	Industrial Management & Data Systems	Q1
2022	Ding	Journal of Innovation & Knowledge	Q1
2022	Eggers, Risselada, Niemand & Robledo	Journal of Business Research	Q1
2022	Fredrich, Bouncken & Tiberius	Journal of Business Research	Q1
2022	Gantert, Fredrich, Bouncken & Kraus	Journal of Business Research	Q1
2022	Gu, Wu, Wu, He, Tao, Ye, Wu, Hao & Qiu	BMC public health	Q1
2022	Han & Zhang	Managerial and Decision Economics	Q3
2022	Huang, Tung, Wang, Xu, Lam & Zhang	Heliyon	Q1
2022	Huang, Zhang, Wang, Li & Li	Journal of Innovation & Knowledge	Q1
2022	Luo, Wang, Liu, Zhang & Fang	Ecological Indicators	Q1
2022	Peng, Zhao, Elahi & Wan	Technological Forecasting and Social Change	Q1

2022	Torres & Godinho	Small Business Economics	Q1
2022	Zheng, Shi & Jiang	PLOS ONE	Q1
2021	Delgosha, Saheb & Hajiheydari	Information Systems Frontiers	Q1
2021	Kopplin & Rösch	Journal of Retailing and Consumer Services	Q1
2021	Mazumder & Garg	International Journal of Information Management	Q1
2020	Bouncken, Fredrich, Sinkovics & Sinkovics	Journal of Business Research	Q1
2020	Bouncken, Ratzmann, Barwinski & Kraus	Long Range Planning	Q1
2019	Fredrich, Bouncken & Kraus	Journal of Business Research	Q1

NCA + QCA + PLS-SEM applications are listed under 'NCA + combination of other methods'.

C.6 Journal articles: NCA and other methods

Table C.6 shows articles that apply NCA in combination with other methods.

TABLE C.6 Journal articles with application of NCA and other method (Situation of 31 December 2025).

Year	Author(s)	Journal	Journal quality
in press	El Manzani & El Idrissi	Management Review Quarterly	Q1
in press	İnce & Orhan	Kybernetes	Q2
in press	İnce, Şahin & Durmaz	Journal of Hospitality and Tourism Insights	Q1
in press	Ismael, Ibrahim, Abdulmuhsin, Khan, Rehman & Alkhwaldi	Journal of Health Organization and Management	Q2
2025	Calafat-Marzal, Piñeiro, Cervera & García-Álvarez-Coque	New Medit	Q2
2025	Gómez-Gómez, Bustos-Gutierrez, Díaz Tautiva & Quezada-Baier	Gender Issues	Q2
2025	İnce, Şahin & Akbiyik	DECISION	Q4
2025	Mohsin, Rashed, Gerschberger, Ahmed, Plasch, Rahman & Noor	Sustainable Futures	Q1
2025	Novales, Mocker, Van Heck & Dul	Decision Support Systems	Q1
2025	Sathatip, Senachai, Leruksa & Fakfare	Journal of Hospitality, Leisure, Sport & Tourism Education	Q1
2025	Yu, Wang, Jiang & Lu	Resources Policy	Q?
2024	Agmapisarn, Seetha & Fakfare	Sustainable Development	Q1

2024	Inacio Junior, Dionisio & Gimenez	Journal of Entrepreneurship in Emerging Economies	Q2
2024	Lévy-Garboua, Gazel, Berlin, Dul & Lubart	Learning and Individual Differences	Q1
2024	Medina-Molina, Pérez-Macías & Rey-Tienda	International Review on Public and Nonprofit Marketing	Q3
2024	Roxas	Journal of Asian Business and Economic Studies	Q1
2024	Wattanacharoensil, Stettler, Zimmermann & Wagenseil	Tourism Recreation Research	Q1
2024	Yang, Lv, Xu & Chen	Pacific-Basin Finance Journal	Q1
2023	Bokrantz & Skoogh	International Journal of Production Economics	Q1
2023	Cheng & Jiang	Management Decision	Q2
2023	Chong & Ham	Journal of Computational Chemistry	Q2
2023	Peltokorpi & Sekiguchi	European Journal of Work and Organizational Psychology	Q1
2022	Bi, Li, Hua, Zhu & Ye	Transport Policy	Q1
2022	Bokhorst, Knol, Slomp & Bortolotti	International Journal of Production Economics	Q1
2022	Glück, Weststrate & Scherpf	Journal of Happiness Studies	Q1
2022	Yang & Hurmelinna-Laukkanen	Technovation	Q1
2021	Batey, Hughes, Crick & Toader	Ergonomics	Q2
2019	Knol, Slomp, Schouteten & Lauche	International Journal of Operations & Production Management	Q1
2018	Kerwin, Hittle, Stredney, De Boeck & Wiet	Journal of surgical education	Q1

C.7 Journal articles: NCA and a combination of other methods

Table C.7 shows articles that apply NCA and a combination of other methods.

TABLE C.7 Journal articles with application of NCA and combinations of other methods (Situation of 31 December 2025).

Year	Author(s)	Journal	Journal quality
in press	Keimasi, Giwa, Chitsazan & NasirpourOsgoei	Assessment & Evaluation in Higher Education	Q1
in press	Lisboa, Dias, Abrantes, Ferreira & Sousa	Journal of Marketing Analytics	Q2
in press	Quan, Wu & Zhang	Asia Pacific Journal of Education	Q2
2025	Cao, Hassan & Sampene	Journal of Environmental Management	Q1

2025	Fakfare, Manosuthi, Lee, Han & Jin	International Journal of Hospitality Management	Q1
2025	Liébana-Cabanillas, Abbasi, Higueras-Castillo & Wagner	Technological Forecasting and Social Change	Q1
2025	Pourazad, Stocchi & Simmonds	Information Technology & Tourism	Q1
2025	Rahman, Tafsirun, Faisal-E-Alam, Ferreira, Loures & Castanho	PLOS ONE	Q2
2025	Tripathi, Husain & Rana	Journal of Hospitality Marketing & Management	Q1
2025	Wang	Economics of Governance	Q3
2025	Wang & Beh	Technological Forecasting and Social Change	Q1
2025	Yuan, Shah, Tajeddini, Gamage & Acevedo-Duque	International Journal of Human–Computer Interaction	Q1
2025	Zhang & Cheng	Journal of Retailing and Consumer Services	Q1
2025	Zhou, Xing, Xu, Tian, Niu & Lin	Journal of Environmental Management	Q1
2025	Zhu & Song	Computers & Industrial Engineering	Q1
2024	Foroughi, Huy, Iranmanesh, Ghobakhloo, Rejeb & Nikbin	The Service Industries Journal	Q1
2024	Jain, Chakraborty & Choudhary	Journal of Business and Industrial Marketing	Q2
2024	Khan, Ali, Xiaobao & Zhiying	IEEE Transactions on Engineering Management	Q1
2024	Liu & Meng	Journal of Hospitality and Tourism Management	Q1
2024	Sun, Wang & Zhang	Journal of Environmental Management	Q1
2024	Wang, Rahman, Siddik, Wen & Sobhani	Scientific Reports	Q1
2024	Zielińska, Lebuda, Gop & Karwowski	Contemporary Educational Psychology	Q1
2023	Jiang & Huang	Womens Studies International Forum	Q2
2023	Ruiz-Equihua, Romero, Loureiro & Ali	International Journal of Contemporary Hospitality Management	Q1

D

Affine transformations

This appendix shows that affine transformations preserve the NCA effect size d.

The following definitions are used:

- Let $\Omega \subseteq \mathbb{R}^2$ be the data in the XY-plane.
- Let S be the *scope* defined by:

$$S = [X_{\text{min}}, X_{\text{max}}] \times [Y_{\text{min}}, Y_{\text{max}}]$$

- Let $C \subseteq S$ be the *ceiling zone* or *empty area* (above the ceiling line).

The *effect size* in NCA is defined as:

$$d = \frac{\text{Area}(C)}{\text{Area}(S)}$$

Let $f : \mathbb{R}^2 \to \mathbb{R}^2$ be a general *differentiable and invertible* transformation:

$$f(x, y) = (u, v) = (f_1(x), f_2(y))$$

The *Jacobian determinant* $J_f(x, y)$ indicates how an infinitesimal area around point (x, y) is transformed:

$$J_f(x, y) = \begin{vmatrix} \frac{df_1}{dx} & 0 \\ 0 & \frac{df_2}{dy} \end{vmatrix} = f_1'(x) \cdot f_2'(y)$$

The area of a transformed region R becomes:

$$\text{Area}(f(R)) = \iint_R |J_f(x, y)| \, dx \, dy$$

So the *transformed effect size* becomes:

$$d' = \frac{\iint_C |J_f(x, y)| \, dx \, dy}{\iint_S |J_f(x, y)| \, dx \, dy}$$

For the affine case:

If $f_1(x) = ax + b$ and $f_2(y) = cy + e$, then:

DOI: 10.1201/9781003430766-D

$$f_1'(x) = a, \quad f_2'(y) = c \quad \Rightarrow \quad J_f(x, y) = ac = \text{constant}$$

So:

$$d' = \frac{ac \cdot \text{Area}(C)}{ac \cdot \text{Area}(S)} = d$$

Thus, affine transformations preserves the effect size.

E

Simulations for TPR and TNR

This appendix provides additional simulation results for the True Positive Rate (TPR; sensitivity) and the True Negative Rate (TNR; specificity), as discussed in Section 6.3.

E.1 True Positive Rate (TPR)

With the same simulation settings as in Figure 6.3, Figure E.1 shows the simulation results for an effect size $d = 0.20$ instead of $d = 0.10$. All TPR curves shift to the left. Consequently, with a larger true effect size, the required sample size is smaller to achieve the same TPR.

With the same simulation settings as in Figure 6.3, Figures E.2 and E.3 show the simulation results for ceiling slopes of 0.60 and 1.67, respectively, instead of 1. This comparison allows evaluation of the (non)-evenness of the TPR curves for scenarios B and C as a function of the slope of the ceiling line. Only the slope of the ceiling line in the necessity corner of interest is changed. Because the slopes of the lines in the other empty corners remain the same (-1), this slope change introduces asymmetry. Due to the intersection of the lines within the bounding box (Figure 6.2), only an effect size of 0.12 could be evaluated. The results indicate that the slope change has little or no effect on the TPR curves.

The results show that the evenness of the curves for scenarios B and C remains. This may not be the case if the sizes of empty corners 2 and 3 differ, or if the X and Y distributions in the feasible area differ.

E.2 True Negative Rate (TNR)

With the same simulation settings as in Figure 6.4, Figure E.4 shows TNR (p- and d-based) when the distribution in the feasible space is based on a normal distribution instead of a uniform distribution.

The high TNR for empty corners 2 and 3 remains, whereas the curves for corner 4 shift to the right. This means that with a normal-based distribution under the ceiling, a larger sample size is required compared to a uniform distribution to obtain the same TNR.

DOI: 10.1201/9781003430766-E

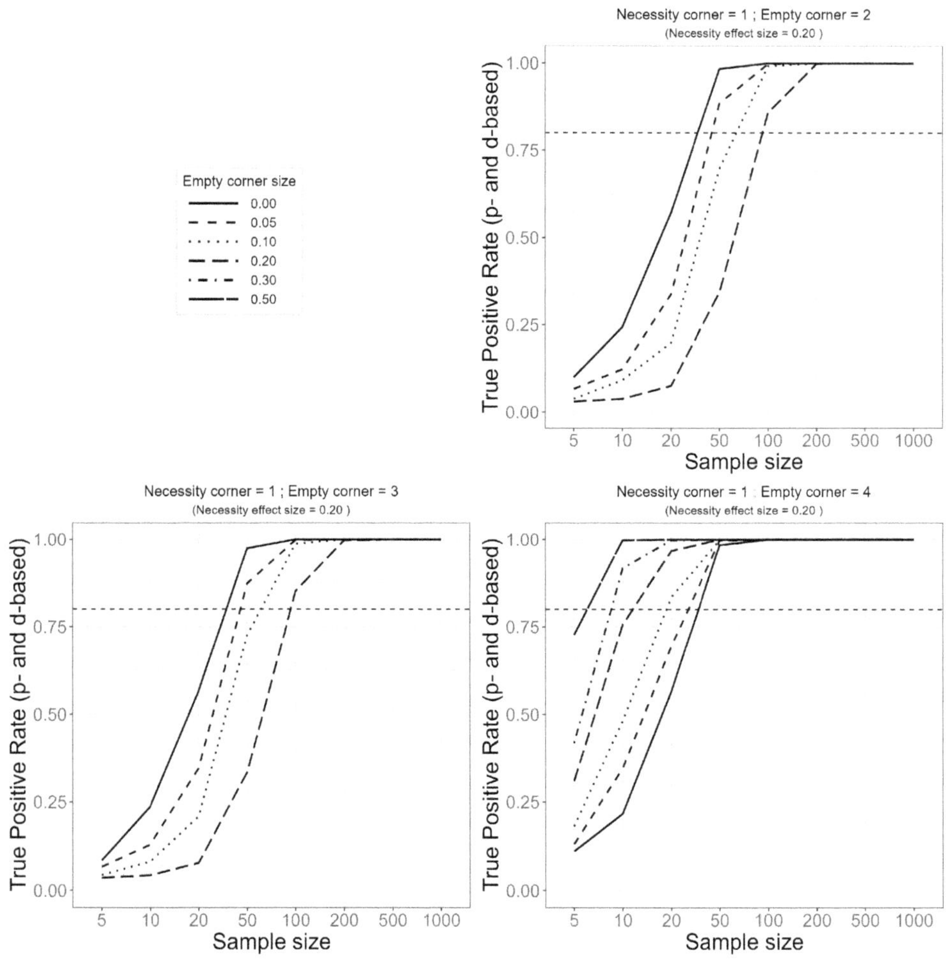

FIGURE E.1 True Positive Rate (TPR, sensitivity) of the NCA approach for identifying necessity in the population. True necessity effect size in corner 1 = 0.20. Slopes of ceiling/floor lines are 1 (corners 1 and 4) or −1 (corners 2 and 3). Top-Right: other empty corner = 2. Bottom-Left: other empty corner = 3. Bottom-Right: other empty corner = 4.

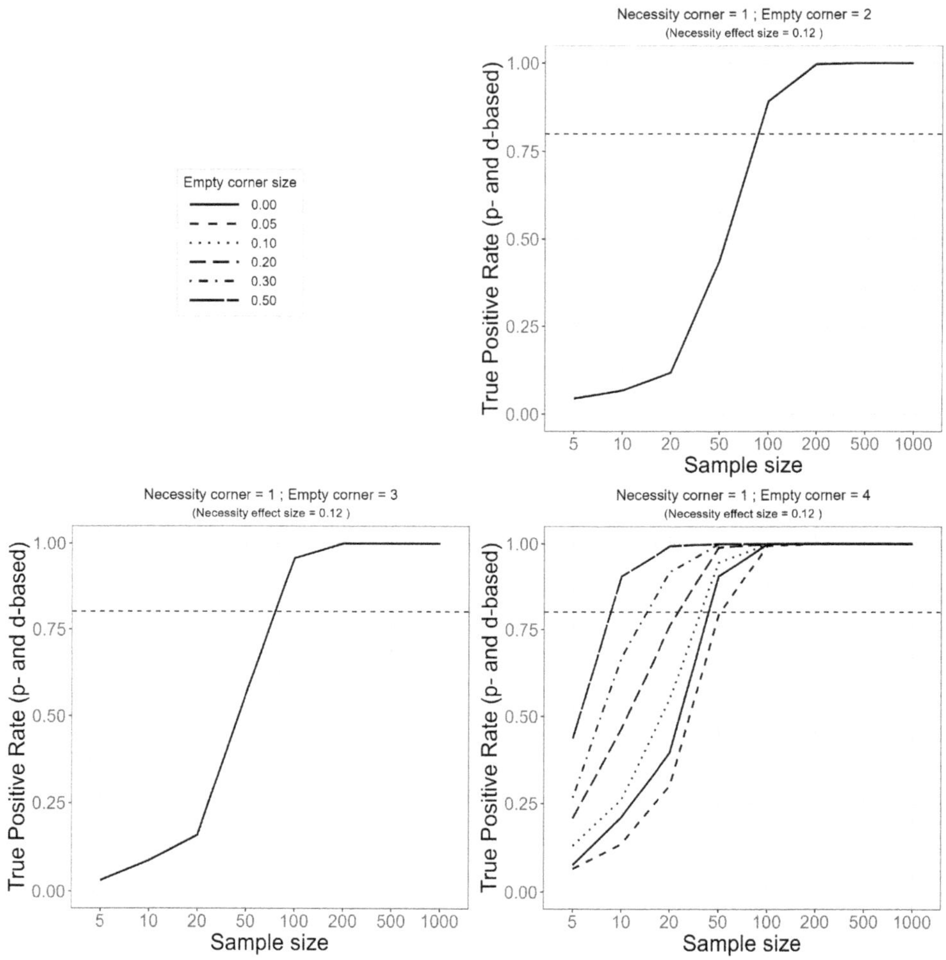

FIGURE E.2 True Positive Rate (TPR, sensitivity) of the NCA approach for identifying necessity in the population. True necessity effect size in corner 1 = 0.12. Slope of necessity ceiling line (corner 1) = 0.60. Slopes of ceiling/floor lines of other empty corners are 1 (4) or −1 (corners 2 and 3). Top-Right: other empty corner = 2. Bottom-Left: other empty corner = 3. Bottom-Right: other empty corner = 4.

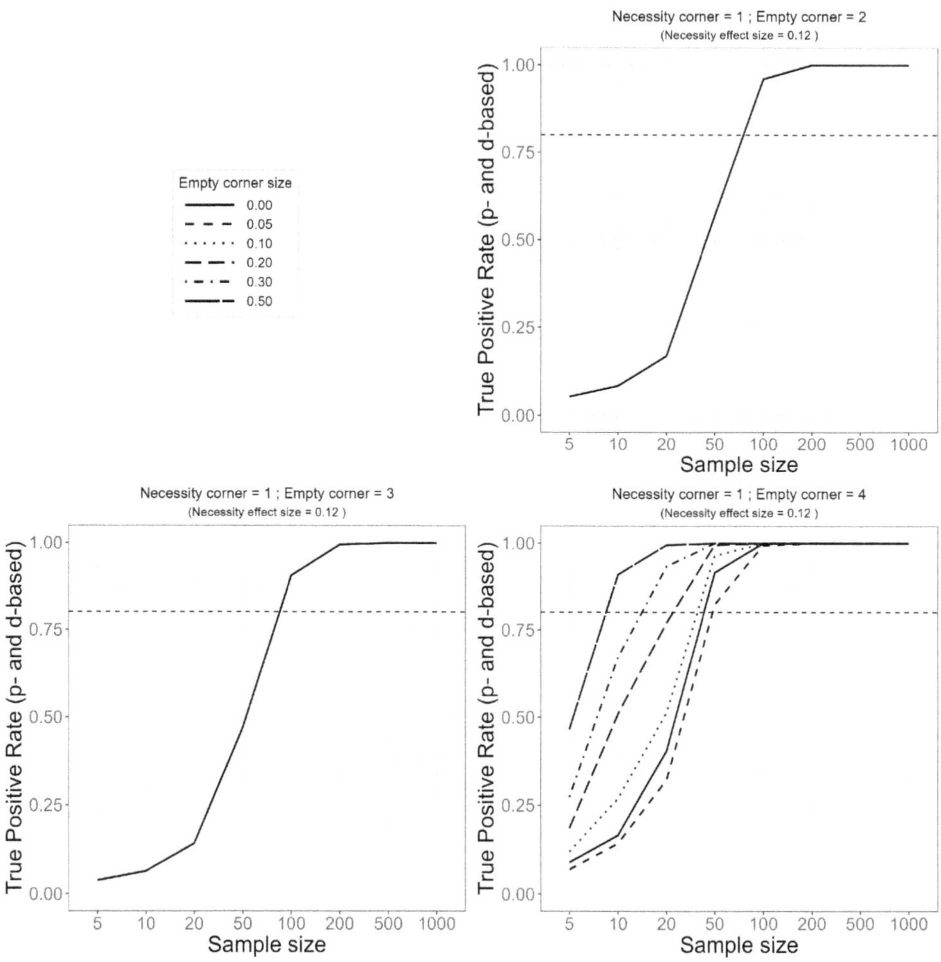

FIGURE E.3 True Positive Rate (TPR, sensitivity) of the NCA approach for identifying necessity in the population. True necessity effect size in corner 1 = 0.12. Slope of necessity ceiling line (corner 1) = 1.67. Slopes of ceiling/floor lines of other empty corners are 1 (4) or −1 (corners 2 and 3). Top-Right: other empty corner = 2. Bottom-Left: other empty corner = 3. Bottom-Right: other empty corner = 4.

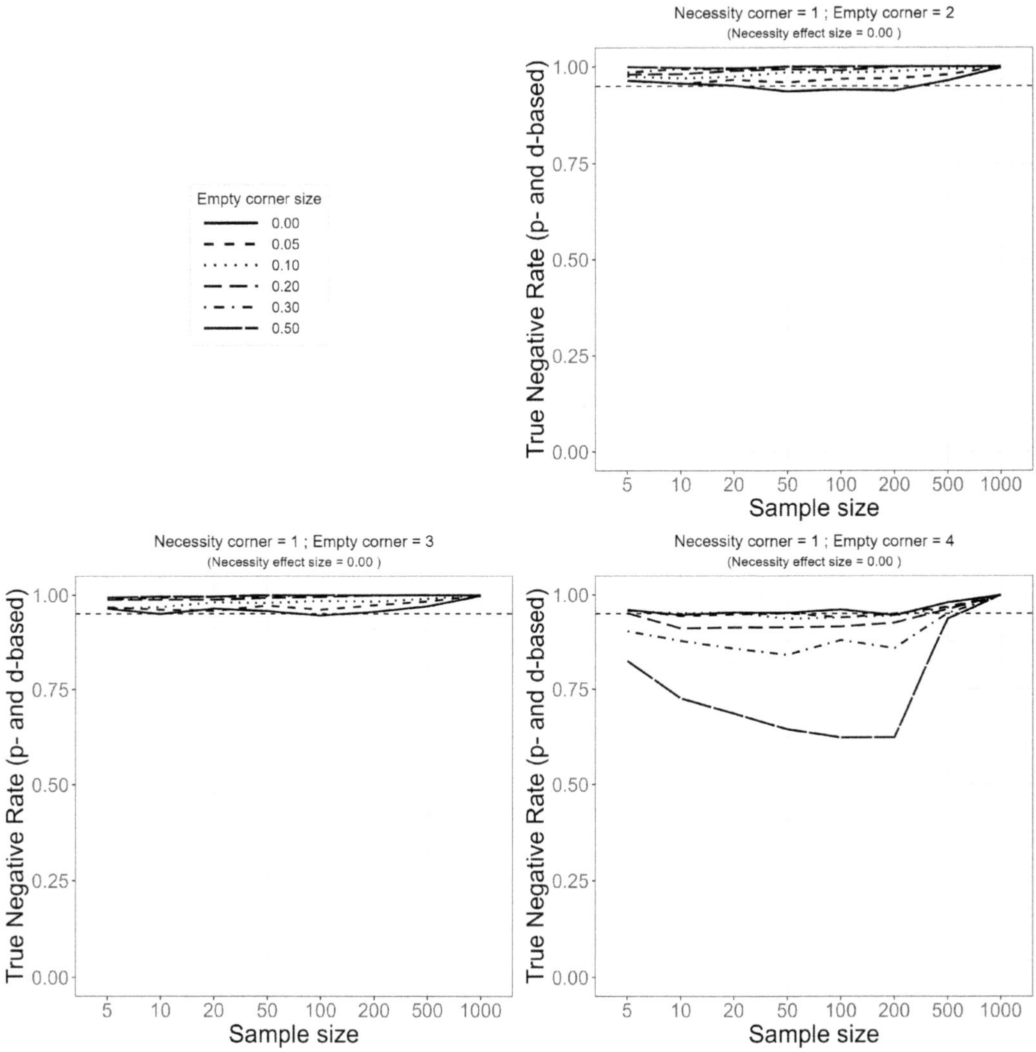

FIGURE E.4 True Negative Rate (TNR, specificity) of the NCA approach for identifying necessity in the population with the p and d-value criterion, with normal distribution of cases in the feasible space. True necessity effect size in corner $1 = 0$ for different sizes of other empty corners. Top-Right: other empty corner = 2. Bottom-Left: other empty corner = 3. Bottom-Right: other empty corner = 4.

F

Necessity and the traditional experiment

In a regression context, experiments are often considered the gold standard research design for making causal inference. The main reason is that confounders can bias the regression coefficients (omitted variable bias), which can lead to causal misinterpretations. In randomized experiments, the undesired effects of confounders are mitigated.

The traditional experiment is not suitable for testing a necessity relationship. Therefore, Section 8.2.4 presented the necessity experiment. Nevertheless, this Appendix shows that it is possible to learn something about necessity from traditional experiments.

F.1 Traditional experiment

The traditional experiment estimates the *average treatment effect*: the average effect of X on Y. The Randomized Controlled Trial (RCT) is considered the gold-standard for such an experiment. In an RCT, cases are randomly divided into a *treatment group* and a *control group*. In the treatment group, X (the treatment) is manipulated by the analyst and in the control group nothing is manipulated. After a certain period of time, the difference in outcomes Y between the two groups is observed, which is an indication of the causal effect of X on Y. To claim causality, the existence of a time difference between manipulation and measurement of the outcome is essential as it ensures that the cause came before the outcome. Although this time difference can be short (e.g., when there is an immediate effect of the manipulation) it will always exist in an experiment. During this time period, not only the treatment, but also other factors may change. The control group ensures that the 'time effect' of other factors is taken into account, and it is assumed that the time effect in the treatment group and in the control group are the same. Random allocation of cases to the two groups must ensure that there are no average time effect differences between the treatment group and the control group.

An RCT is used to evaluate the *average* effect of a treatment (a policy, an intervention, a medical treatment, etc.). The treatment group describes the real world of cases where the intervention has occurred, and the control group mimics the counterfactual world of *these* cases if the intervention had not occurred. The use of a control group and randomization is an attempt to compare the actual world of cases that got the intervention with the actual world of *similar* (but not the same) cases that did not get the intervention. "Similar" means that the two groups *on average* have the same characteristics with respect to factors that can also have an effect on the outcome, other than the treatment. Randomization must ensure a statistical similarity of the two groups: they have the same *average* level of potential confounders. So, randomization is an effort to control for potential confounders.

DOI: 10.1201/9781003430766-F

Figure F.1 represents the traditional experiment when X and Y are both dichotomous (absent $= 0$ or present $= 1$). The sample of 100 cases is divided into a treatment group of 50 cases (shown as $\times$) and a control group of 50 cases (shown as $\circ$). Figure F.1-left shows the situation before the treatment. All cases are in corner 3, and the treatment is absent. The manipulation for the treatment group consists of *adding* the treatment. This means that X is changed from absent ($X = 0$, no treatment) to present ($X = 1$, treatment). After the manipulation (Figure F.1-right), the effect Y of the treatment group is caused by a time effect and a manipulation effect. In the control group X is not changed (no manipulation, $X = 0$) but there is a time effect (see the control cases in corner 1, where $X = 0, Y = 1$). Because the time effects in both groups are considered the same (randomization) the difference between the outcome of the treatment group and the control group can be attributed to X, the manipulation. The results of the experiment after manipulation can be statistically evaluated by the *odds ratio* (OR), which is a population-level measure that summarizes how much more likely (in terms of odds) the outcome is in the treatment group compared to the control group[1].

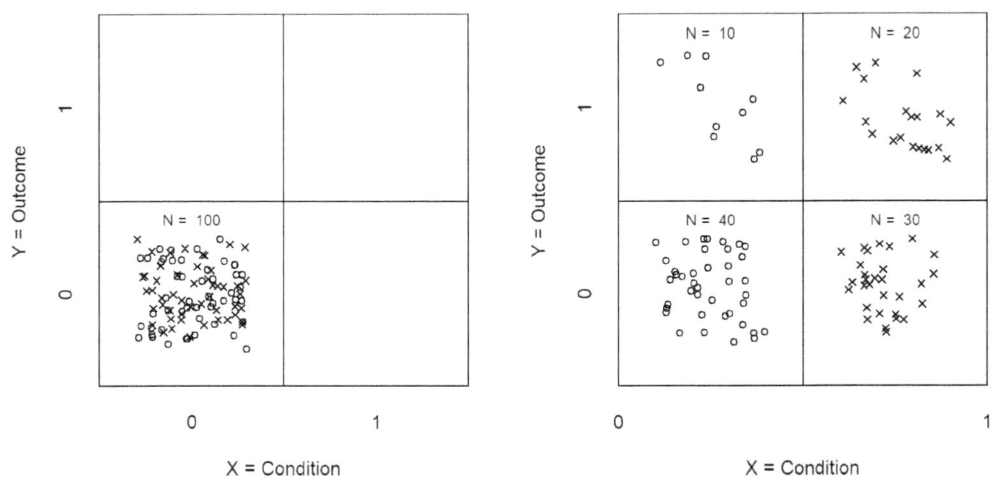

FIGURE F.1 XY-plot of the traditional experiment before (Left) and after (Right) manipulation. $X =$ Condition ($0 =$ no; $1 =$ yes); $Y =$ Outcome ($0 =$ no; $1 =$ yes). $\circ =$ control group; $\times =$ treatment group.

With a traditional RCT experiment, after the treatment usually all four corners have cases. In this example, 10 cases from the control group have moved from corner 3 to corner 2, suggesting a time effect. The cases from the treatment group have all moved to $X = 1$ (corner 2 or 4). Treatment and time had no effect on the 30 cases in corner 4. Treatment and time did have an effect on the 20 cases in corner 2. The data pattern in the four corners

[1]The mathematical formula for the odds ratio is $OR = \frac{c2/c4}{c1/c3} = \frac{c2 \cdot c3}{c1 \cdot c4}$ where $c1$, $c2$, $c3$, and $c4$ refer to the number of cases in corner 1, corner 2, corner 3, and corner 4, respectively. $c2$ is the number of cases in the treatment group with outcome success, $c4$ is the number of cases in the treatment group with outcome failure, $c1$ is the number of cases in the control group with outcome success (spontaneous success), and $c3$ is the number of cases in the control group with outcome failure. A value of $OR > 1$ indicates that the condition X is associated with a higher likelihood of Y, suggesting that X may be *probabilistically sufficient* to produce Y.

suggests that the treatment had an effect. There are more cases in corner 2 (resulting from a time effect and a treatment effect) than in corner 1. The odds ratio is an estimate of the *average treatment effect*: $(20/30)/(10/40) = 2.67$. Since the odds ratio is greater than 1, the conclusion is that X has a probabilistic sufficiency effect on Y.[2]

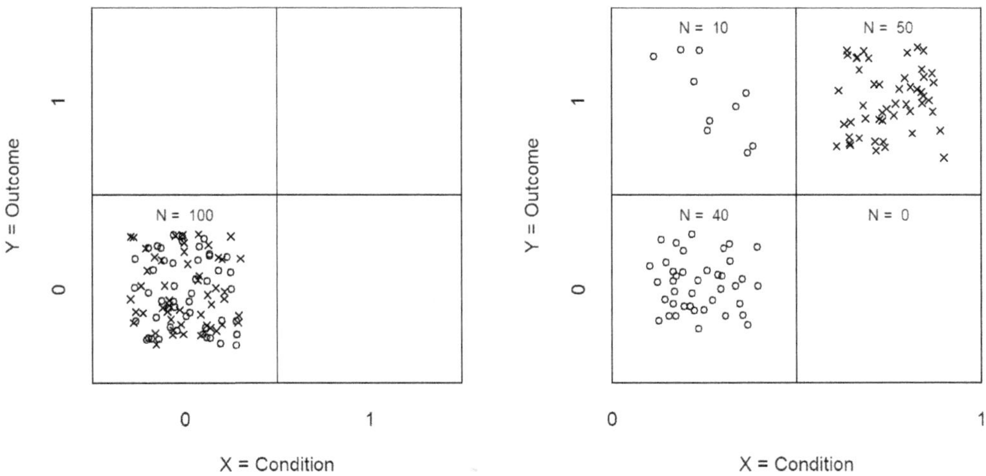

FIGURE F.2 *XY*-table of the sufficiency experiment before (Left) and after (Right) manipulation. X = Condition (0 = no; 1 = yes); Y = Outcome (0 = no; 1 = yes). ∘ = control group; × = treatment group.

F.2 Necessity lessons from the traditional experiment

Referring to Figures F.1-right and F.2-right, the (probabilistic) sufficiency experiment shows non-emptiness in corner 1 due to time effects. The non-emptiness reflects non-necessity. If X were necessary for Y, this corner would remain empty after the manipulation. If this corner has cases, it can be concluded that the condition $X = 1$ is not necessary for the effect $Y = 1$, because the effect is possible without the condition. If this corner remains empty, necessity is not rejected. This 'non-necessity test', as part of a (probabilistic) sufficiency experiment, can only be done with the control group. The treatment group cannot be used for testing necessity because it cannot test the emptiness of corner 1. The non-necessity test with the control group is not a necessity experiment since no manipulation is involved in necessity testing. It can be understood as an observational study embedded in an experiment (or even as a longitudinal study because x- and y-scores are observed for two time points: before and after the manipulation).

[2]The sufficiency experiment can be considered the deterministic version of the traditional experiment, namely when after the manipulation (Figure F.2), all cases of the treatment group are in corner 2. Then the odds ratio for estimating the average effect is infinite, regardless of the distribution of the cases in the control group: $(50/0)/(10/40) = \infty$. Note that this situation is named 'perfect separation' and is considered a problem for average effect statistical analysis. In this situation X is called a 'perfect predictor', which is considered a statistical problem.

G

Correlation by necessity

A correlation is a commonly used measure to express the linear relationship between two continuous variables. It offers a simple, standardized, and intuitive indication of how variables move together in the data. The *correlation coefficient* r captures both the direction and strength of a linear relationship in a single, unit-free number ranging from -1 to $+1$. Its popularity stems in part from its computational simplicity and its status as a widely taught first step in data analysis.

The Pearson correlation coefficient is closely related to the regression coefficient b from a simple linear regression equation: $y = a + bx + \epsilon$:

$$r = b \cdot \frac{sd_x}{sd_y} \tag{G.1}$$

where sd_x and sd_y are the standard deviations of x and y, respectively.

Like the regression coefficient, the observed correlation coefficient is often *interpreted* as evidence of a probabilistic sufficiency relationship, especially when testing hypotheses of that nature. However, this section shows that an observed non-zero correlation can also be produced by any relationship between X and Y, including a necessity relationship.

This implies that, without knowing the true underlying relationship in the population, an observed correlation coefficient could be interpreted both as an indication for a probabilistic sufficiency, or as a necessity relationship, or as something else. In other words, causality cannot be found from data, but data can be analyzed with different causal perspectives in mind (Section 2.7).

Figure G.1 Left shows an XY-plot in which the solid line represents a true linear probabilistic sufficiency relationship in the population. The population data were generated with a fixed X with values between 0 and 1, and normally distributed random variables Y and ϵ with zero averages and standard deviations of 1. A random sample of 100 cases was drawn from this population, and the relationship was estimated using ordinary least squares (OLS) regression (dashed line in the center). This shows that the estimated and the true average relationship are very close, as expected. The sample correlation coefficient is 0.25. The estimated ceiling line (dashed line at the top) illustrates that NCA does not capture the underlying probabilistic sufficiency relationship in the population. The empty space is only an artifact of finite sampling: the points could be there according to the normal distributions of X and Y, but were not observed.

Figure G.1 Right shows a similar XY-plot in which the solid line now represents a true linear *necessity* relationship in the population. The population data were generated with bounded variables X and Y with values between 0 and 1 and a uniform distribution below the ceiling (see Section 5.4). A sample of 100 cases was randomly drawn from this population and the relationship was estimated with the NCA's C-LP ceiling line (solid line at the

DOI: 10.1201/9781003430766-G

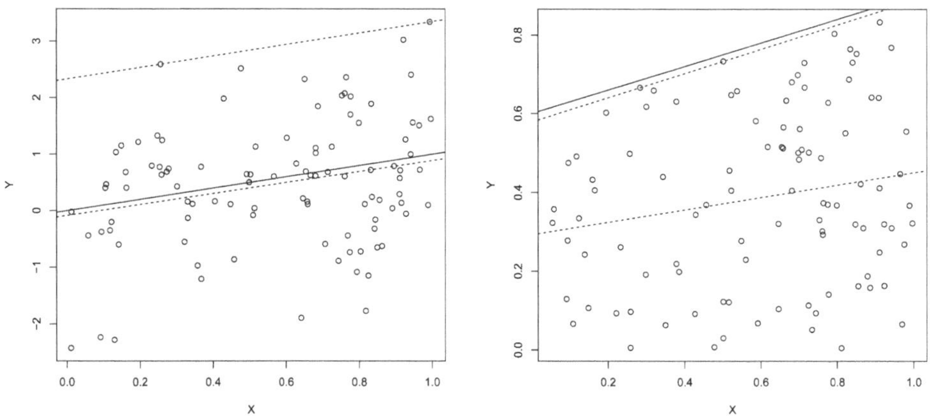

FIGURE G.1 Correlation by probabilistic sufficiency (Left) and by necessity (Right). Random samples (n = 100) from a population with probabilistic sufficiency (solid line, Left) and necessity (solid line, Right). The dashed lines are estimated lines by OLS regression (center lines) and by NCA (ceiling lines).

top). This shows that the estimated and the true necessity relationship are very close. The sample correlation coefficient is 0.19. The estimated OLS regression line in the center of the data (dotted line) illustrates that regression analysis does not capture the underlying necessity relationship in the population. The correlation is only an artifact of the necessity relationship due to the selected distribution of data below the ceiling.

The examples show that both types of relationships can produce a correlation coefficient. Since in these examples the data generation process is known, namely induced probabilistic sufficiency and necessity, respectively, it is easy to conclude in which situation the regression line and in which situation the ceiling line best represents the causal relationship in the population whereas the observed correlation is not informative for this. Unfortunately, in empirical studies the true causal relationship in the population is unknown, and an observed correlation does not tell what the underlying relationship was: probabilistic sufficiency, necessity or something else. Usually the analyst decides (e.g., based on theory) whether the underlying causality is probabilistic sufficiency, necessity (or both), and selects the appropriate estimation technique(s) accordingly. It is incorrect to assume a probabilistic relationship and apply NCA, just as it is inappropriate to assume a necessity relationship and apply regression analysis. Both represent a mismatch between theory and method, leading to a theory-method misfit (see Section 11.2). The often implicit assumption that the correlation coefficient indicates the presence of a probabilistic sufficiency relationship is incorrect if the underlying causality is necessity, but there is no way to evaluate this. The use of an (implicit) automatism to select probabilistic sufficiency as the underlying causality may have two reasons. First, this causal perspective is currently the main paradigm of causality. Second, the correlation coefficient r and the regression coefficient b are closely related and can be expressed by a simple mathematical equation (see Equation (G.1)).

A similar simple mathematical equation for the relationship between the regression coefficient and the correlation coefficient as in Equation (G.1) is not available for the relationship between necessity effect size and correlation coefficient. However, with certain assumptions, such a relationship can be derived. The general equation for the correlation coefficient can be written as:

$$\rho_{XY} = \frac{E(XY) - E(X)E(Y)}{\sqrt{E(X^2) - E(X)^2}\sqrt{E(Y^2) - E(Y)^2}}$$

where ρ_{XY} is the Pearson correlation coefficient, and E is the expected value.

For a linear ceiling line $f(x) = a + bx$ and a uniform distribution under the ceiling and within the bounding box (unit box), the correlation coefficient can be derived analytically, as shown below.[1]

Situation

Consider a linear case with a tight bounding box (unit box): $0 < a < 1 < a + b$.

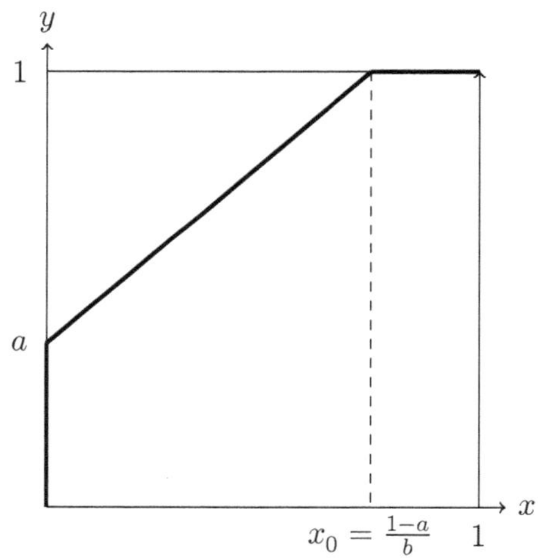

FIGURE G.2 Parameters: $0 < a < 1 < a + b$.

Define $L(x) = \min\{1, a + bx\}$.

$$F_{a,b} \equiv \int_0^1 L(x)\,\mathrm{d}x = \text{surface area below thick line} \tag{G.2}$$

$$= 1 - \text{surface area above thick line} \tag{G.3}$$

$$= 1 - \tfrac{1}{2}(1-a)x_0 = 1 - \frac{(1-a)^2}{2b} \tag{G.4}$$

[1] The derivation was done by Roelof Kuik.

Uniform distribution (X, Y) below the ceiling line:

$$X \propto \frac{1}{F_{a,b}} L(x)\,\mathrm{d}x \tag{G.5}$$

$$Y|X \propto \frac{I_{[0,L(x)]}(y)}{L(x)}\,\mathrm{d}y \tag{G.6}$$

where X is a random variable, $Y|X$ is a random variable Y conditioned on X and $\propto$ means 'distributed with a density function given as'. Further, $I_{[0,L(x)]}$ is the indicator function for the interval $[0, L(x)]$. Note that

$$(X,Y) \propto \frac{1}{F_{a,b}} I_{[0,L(x)]}(y)\,\mathrm{d}x\,\mathrm{d}y = \frac{1}{F_{a,b}} I_{y \le a+bx}(x,y)\,\mathrm{d}x\,\mathrm{d}y$$

is indeed the (normalized) uniform distribution below the line [2].

The text below uses four values, a, b, x_0, and $F_{a,b}$ to formulate expressions. Two are all that are needed. Indeed, for example,

$$\frac{1 - F_{a,b}}{x_0} = \frac{\frac{(1-a)^2}{2b}}{(1-a)/b} = \tfrac{1}{2}(1-a)$$

and so

$$a = 1 - 2\frac{1 - F_{a,b}}{x_0} \quad \text{or} \quad x_0 = 2\frac{1 - F_{a,b}}{1-a} \quad \text{or} \quad F_{a,b} = 1 - x_0(1-a)/2.$$

Likewise

$$b = (1-a)/x_0 = \left(\frac{1 - 2\frac{1-F_{a,b}}{x_0}}{x_0}\right).$$

Symmetry transformation

Consider the 'double flip + transposition' transformation:

$$(x, y) \mapsto (\alpha, \beta) \equiv (1-y, 1-x).$$

In $\alpha\beta$ coordinates Figure G.2 presents itself as Figure G.3.

The ceiling function in $\alpha\beta$ coordinates is $\hat{L}(\alpha) = \min(1, (1-x_0) + (1/b)\beta$. Now

$$\mathbb{E}Y = \mathbb{E}(1-\alpha) = 1 - \mathbb{E}(\alpha)$$

where, by symmetry,

$$\mathbb{E}(\alpha) = \mathbb{E}(X) \quad [\text{where } x_0 \leftrightarrow \hat{a}].$$

Note that under the symmetry $x_0 \leftrightarrow \hat{a}$ the value of $F_{a,b} = 1 - \frac{1}{2}x_0\hat{a}$ is evidently invariant.

---

[2] Also note that, in equation form, $X = G^{-1}(U_1)$ and $Y = L(X)U_2$ where U_1 and U_2 are independent uniform on the interval $[0, 1]$. The function G^{-1} is the inverse of

$$G(\theta) = \frac{1}{F_{a,b}} \int_0^{\theta} L(x)\,\mathrm{d}x.$$

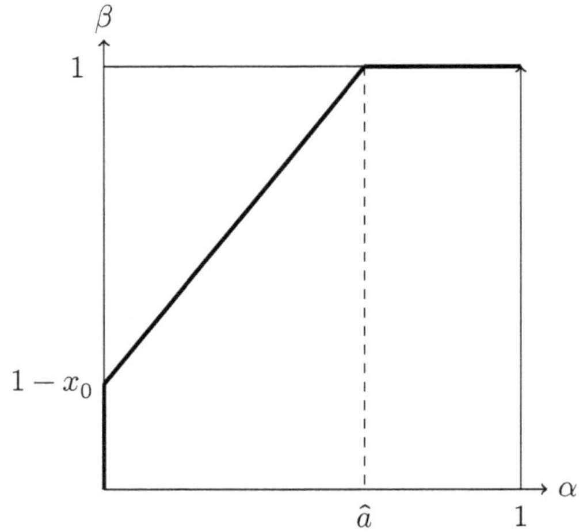

FIGURE G.3 Parameters: $\hat{a} = 1 - a$ with $0 < \hat{a} < 1$ and $0 < x_0 < 1$.

Correlation

$$\rho_{XY} = \frac{\mathbb{E}(XY) - \mathbb{E}(X)\mathbb{E}(Y)}{\sigma_X \sigma_Y}.$$

Five integrals:

1. Expectation of X.

$$\begin{aligned}
\mathbb{E}(X) &= \frac{1}{F_{a,b}} \int_0^1 x L(x)\,\mathrm{d}x == \frac{1}{F_{a,b}} \int_0^1 x \min\{1, a + bx\}\,\mathrm{d}x \\
&= \frac{1}{F_{a,b}} \left(\int_0^{(1-a)/b} (ax + bx^2)\,\mathrm{d}x + \int_{(1-a)/b}^1 x\,\mathrm{d}x \right) \\
&= \frac{1}{F_{a,b}} \left(\left(\tfrac{1}{2}ax^2 + \tfrac{1}{3}bx^3\right)\Big|_0^{(1-a)/b} + \tfrac{1}{2}x^2\Big|_{(1-a)/b}^1 \right) \\
&= \frac{1}{F_{a,b}} \left(\tfrac{1}{2}a \left(\frac{1-a}{b}\right)^2 + \tfrac{1}{3}b \left(\frac{1-a}{b}\right)^3 + \tfrac{1}{2} - \tfrac{1}{2}\left(\frac{1-a}{b}\right)^2 \right) \\
&= \frac{1}{2F_{a,b}} \left(1 + \left(a + \tfrac{2}{3}(1-a) + 1\right) \left(\frac{1-a}{b}\right)^2 \right) \\
&= \frac{1}{2F_{a,b}} \left(1 - \tfrac{1}{3}(1-a) \left(\frac{1-a}{b}\right)^2 \right) \\
&= \frac{1}{2F_{a,b}} \left(\tfrac{2}{3}\frac{(1-a)}{b} - \tfrac{2}{3}\frac{(1-a)}{b}\frac{(1-a)^2}{2b} + 1 - \tfrac{2}{3}\frac{(1-a)}{b} \right) \\
&= \frac{1}{2F_{a,b}} \left(\tfrac{2}{3}\frac{(1-a)}{b} F_{a,b} + 1 - \tfrac{2}{3}\frac{(1-a)}{b} \right)
\end{aligned}$$

$$= \tfrac{1}{2}\left(\tfrac{2}{3}\frac{(1-a)}{b} + \left(1 - \tfrac{2}{3}\frac{(1-a)}{b}\right)\frac{1}{F_{a,b}}\right)$$
$$= \tfrac{1}{2}\left(\tfrac{2}{3}x_0 + (1 - \tfrac{2}{3}x_0)\frac{1}{F_{a,b}}\right)$$
$$= \frac{1}{2F_{a,b}}\left(1 - \tfrac{2}{3}x_0 + \tfrac{2}{3}x_0 F_{a,b}\right).$$

Note. As $\frac{1}{F_{a,b}} \geq 1$ we have

$$\mathbb{E}(X) = \tfrac{1}{2}\left(\tfrac{2}{3}x_0 + (1 - \tfrac{2}{3}x_0)\frac{1}{F_{a,b}}\right) \geq \tfrac{1}{2}\left(\tfrac{2}{3}x_0 + 1 - \tfrac{2}{3}x_0\right) = \tfrac{1}{2}$$

as is evident from Figure G.2. Reworking $\mathbb{E}(X)$ (into $\hat{a}$ and x_0) we obtain

$$\begin{aligned}\mathbb{E}(X) &= \frac{1}{2F_{a,b}}\left(1 - \tfrac{2}{3}x_0 + \tfrac{2}{3}x_0 F_{a,b}\right)\\ &= \frac{1}{2F_{a,b}}\left(1 - \tfrac{2}{3}x_0 + \tfrac{2}{3}x_0(1 - \tfrac{1}{2}\hat{a}x_0)\right) = \frac{1}{2F_{a,b}}\left(1 - \tfrac{1}{3}\hat{a}\,x_0^2\right).\end{aligned}$$

2. Expectation of Y. Note that $\mathbb{E}(Y) = \mathbb{E}(\mathbb{E}(Y|X))$.
   Step 1:

$$\mathbb{E}(Y|X = x) = \int_0^{L(x)} y\frac{I_{[0,L(x)]}(y)}{L(x)}\,\mathrm{d}y = \tfrac{1}{2}\frac{y^2}{L(x)}\bigg|_0^{L(x)} = \tfrac{1}{2}L(x).$$

   Step 2:

$$\begin{aligned}\mathbb{E}(Y) &= \int_0^1 \tfrac{1}{2}L(x)\frac{L(x)}{F_{a,b}}\,\mathrm{d}x\\ &= \frac{1}{2F_{a,b}}\left(\int_0^{(1-a)/b}(a+bx)^2\,\mathrm{d}x + \int_{(1-a)/b}^1 \mathrm{d}x\right)\\ &= \frac{1}{2F_{a,b}}\left(\int_0^{(1-a)/b}(a^2 + 2abx + b^2x^2)\,\mathrm{d}x + \int_{(1-a)/b}^1 \mathrm{d}x\right)\\ &= \frac{1}{2F_{a,b}}\left(\left(a^2x + abx^2 + \tfrac{1}{3}b^2x^3\right)\Big|_0^{(1-a)/b} + x\Big|_{(1-a)/b}^1\right)\\ &= \frac{1}{2F_{a,b}}\left(a^2\frac{1-a}{b} + ab\left(\frac{1-a}{b}\right)^2 + \tfrac{1}{3}b^2\left(\frac{1-a}{b}\right)^3 + 1 - \frac{1-a}{b}\right)\\ &= \frac{1}{2F_{a,b}}\left(a^2x_0 + 2a\frac{(1-a)^2}{2b} + \tfrac{2}{3}(1-a)\frac{(1-a)^2}{2b} + 1 - x_0\right)\\ &= \frac{1}{2F_{a,b}}\left(a^2x_0 + (2a + \tfrac{2}{3}(1-a))(1 - F_{a,b}) + 1 - x_0\right)\\ &= \frac{1}{2F_{a,b}}\left(1 - x_0 + a^2x_0 + \tfrac{2}{3}(2a+1)(1 - F_{a,b})\right)\end{aligned}$$

$$\begin{aligned}
&= \frac{1}{2F_{a,b}} \left(1 + (a-1)(a+1)x_0 + \tfrac{2}{3}(2a+1)(1-F_{a,b})\right) \\
&= \frac{1}{2F_{a,b}} \left(1 - 2(a+1)(1-F_{a,b}) + \tfrac{2}{3}(2a+1)(1-F_{a,b})\right) \\
&= \frac{1}{2F_{a,b}} \left(1 - \tfrac{2}{3}(a+2)(1-F_{a,b})\right) \\
&= \frac{1}{2F_{a,b}} \left(1 - \tfrac{2}{3}(a+2) + \tfrac{2}{3}(a+2)F_{a,b}\right).
\end{aligned}$$

Set $\theta(a) = \frac{1}{3}(a+2) \leq \frac{1}{3}(1+2) = 1$. Also, evidently $\theta(a) \geq \frac{2}{3}$. Now

$$\begin{aligned}
\mathbb{E}(Y) &= \frac{1}{2F_{a,b}}(1 - 2\theta(a) + 2\theta(a)F_{a,b}) \\
&= \frac{1}{2F_{a,b}} \left(1 - 2\theta(a)(1-F_{a,b})\right) \\
&\leq \frac{1}{2F_{a,b}} \left(1 - \tfrac{4}{3}(1-F_{a,b})\right) \\
&= \frac{1}{2F_{a,b}} \left(-\tfrac{1}{3} + \tfrac{4}{3}F_{a,b})\right) \\
&= \frac{1}{2}\left(-\tfrac{1}{3}\frac{1}{F_{a,b}} + \tfrac{1}{3} + 1)\right) \leq \frac{1}{2}.
\end{aligned}$$

Note that the final inequality is strict for $F_{a,b} < 1$. The inequality $\mathbb{E}(Y) \leq \frac{1}{2}$ is evident from Figure G.2.

Alternative calculation.

$$\begin{aligned}
\mathbb{E}(Y) &= \int_0^1 \tfrac{1}{2}L(x)\frac{L(x)}{F_{a,b}}\,\mathrm{d}x \\
&= \frac{1}{2F_{a,b}} \left(\int_0^{(1-a)/b} (a+bx)^2\,\mathrm{d}x + \int_{(1-a)/b}^1 \mathrm{d}x\right) \\
&= \frac{1}{2F_{a,b}} \left(\frac{1}{b}\int_a^1 z^2\,\mathrm{d}z + x\Big|_{(1-a)/b}^1\right) \\
&= \frac{1}{2F_{a,b}} \left(\frac{1}{3b}\,z^3\Big|_a^1 + x\Big|_{(1-a)/b}^1\right) \\
&= \frac{1}{2F_{a,b}} \left(\frac{1}{3b}(1-a^3) + 1 - \frac{1-a}{b}\right)
\end{aligned}$$

Note that

$$\begin{aligned}
a^2(1-a) + a(1-a)^2 + \tfrac{1}{3}(1-a)^3 &= a^2 - a^3 + a(1-2a+a^2) + \tfrac{1}{3}(1-a)(1-2a+a^2) \\
&= a - a^2 + \tfrac{1}{3}(1 - 2a + a^2 - a + 2a^2 - a^3) \\
&= a - a^2 + \tfrac{1}{3}(1-a^3) - a + a^2 = \tfrac{1}{3}(1-a^3).
\end{aligned}$$

Also note

$$(1-a^3) = (1-a)(3a+(1-a)^2) = (1-a)(3a+1-2a+a^2) = (1-a)(1+a+a^2).$$

Reworking $\mathbb{E}(Y)$ (into $\hat{a}$ and x_0) using the symmetry

$$\begin{aligned}\mathbb{E}(Y) &= \frac{1}{2F_{a,b}}\left(1-\tfrac{2}{3}(a+2)+\tfrac{2}{3}(a+2)F_{a,b}\right)\\ &= \frac{1}{2F_{a,b}}\left(1-\tfrac{2}{3}(1-\hat{a}+2)+\tfrac{2}{3}(1-\hat{a}+2)(1-\tfrac{1}{2}x_0\hat{a})\right)\\ &= \frac{1}{2F_{a,b}}\left(1-\tfrac{1}{3}(3-\hat{a})x_0\hat{a}\right)\\ &= 1-\frac{1}{2F_{a,b}}\left(-1+2F_{a,b}+\tfrac{1}{3}(3-\hat{a})x_0\hat{a}\right)\\ &= 1-\frac{1}{2F_{a,b}}\left(-1+2-x_0\hat{a}+\tfrac{1}{3}(3-\hat{a})x_0\hat{a}\right)\\ &= 1-\frac{1}{2F_{a,b}}\left(1-\tfrac{1}{3}x_0\hat{a}^2\right) = 1-\mathbb{E}(X)[\text{where } x_0 \leftrightarrow \hat{a}].\end{aligned}$$

3. Second moment of X.

$$\begin{aligned}\mathbb{E}(X^2) &= \frac{1}{F_{a,b}}\int_0^1 x^2 L(x)\,\mathrm{d}x == \frac{1}{F_{a,b}}\int_0^1 x^2\min\{1,a+bx\}\,\mathrm{d}x\\ &= \frac{1}{F_{a,b}}\left(\int_0^{(1-a)/b}(ax^2+bx^3)\,\mathrm{d}x+\int_{(1-a)/b}^1 x^2\,\mathrm{d}x\right)\\ &= \frac{1}{F_{a,b}}\left(\left(\tfrac{1}{3}ax^3+\tfrac{1}{4}bx^4\right)\Big|_0^{(1-a)/b}+\tfrac{1}{3}x^3\Big|_{(1-a)/b}^1\right)\\ &= \frac{1}{F_{a,b}}\left(\tfrac{1}{3}a\left(\frac{1-a}{b}\right)^3+\tfrac{1}{4}b\left(\frac{1-a}{b}\right)^4+\tfrac{1}{3}-\tfrac{1}{3}\left(\frac{1-a}{b}\right)^3\right)\\ &= \frac{1}{F_{a,b}}\left(\tfrac{1}{3}-\tfrac{8}{3}(1-a)\left(\frac{1-a}{2b}\right)^3+2(1-a)\left(\frac{1-a}{2b}\right)^3\right)\\ &= \frac{1}{F_{a,b}}\left(\tfrac{1}{3}-\tfrac{8}{3}\left(\frac{(1-a)}{2b}\right)^2\left(\frac{(1-a)^2}{2b}\right)+2\left(\frac{(1-a)}{2b}\right)^2\left(\frac{(1-a)^2}{2b}\right)\right)\\ &= \frac{1}{F_{a,b}}\left(\tfrac{1}{3}-\tfrac{2}{3}x_0^2(1-F_{a,b})+\tfrac{1}{2}x_0^2\left(1-F_{a,b}\right)\right)\\ &= \frac{1}{F_{a,b}}\left(\tfrac{1}{3}-\tfrac{1}{6}x_0^2(1-F_{a,b})\right)\\ &= \frac{1}{3F_{a,b}}\left((1-\tfrac{1}{2}x_0^2)+\tfrac{1}{2}x_0^2F_{a,b}\right).\end{aligned}$$

4. Second moment of Y. Note $\mathbb{E}(Y^2)=\mathbb{E}(\mathbb{E}(Y^2|X))$.

Step 1:

$$\mathbb{E}(Y^2|X=x) = \int_0^{L(x)} y^2\frac{I_{[0,L(x)]}(y)}{L(x)}\,\mathrm{d}y = \tfrac{1}{3}\frac{y^3}{L(x)}\Bigg|_0^{L(x)} = \tfrac{1}{3}(L(x))^2.$$

Step 2:

$$\begin{aligned}
\mathbb{E}(Y^2) &= \int_0^1 \tfrac{1}{3}(L(x))^2 \frac{L(x)}{F_{a,b}}\,\mathrm{d}x \\
&= \frac{1}{3F_{a,b}} \left(\int_0^{(1-a)/b} (a+bx)^3\,\mathrm{d}x + \int_{(1-a)/b}^1 \mathrm{d}x \right) \\
&= \frac{1}{3F_{a,b}} \left(\int_a^1 z^3 \frac{1}{b}\,\mathrm{d}z + x\big|_{(1-a)/b}^1 \right) \\
&= \frac{1}{3F_{a,b}} \left(\frac{1}{b}\, \tfrac{1}{4} z^4\big|_a^1 + 1 - \frac{1-a}{b} \right) \\
&= \frac{1}{3F_{a,b}} \left(\frac{1}{4b}(1-a^4) + 1 - \frac{1-a}{b} \right)
\end{aligned}$$

Note that $1 - a^4 = (1-a)(1+a+a^2+a^3) = (1-a^3) + (1-a)a^3$. Also, $1 - a^4 = (1-a^2)(1+a^2) = (1-a)(1+a)(1+a^2)$. We continue:

$$\mathbb{E}(Y^2) = \frac{1}{3F_{a,b}} \left(\frac{1}{4b}(1-a)(1+a)(1+a^2)) + 1 - \frac{1-a}{b} \right)$$

Alternative:

$$\begin{aligned}
\mathbb{E}(Y^2) &= \int_0^1 \tfrac{1}{3}(L(x))^2 \frac{L(x)}{F_{a,b}}\,\mathrm{d}x \\
&= \frac{1}{3F_{a,b}} \left(\int_0^{(1-a)/b} (a+bx)^3\,\mathrm{d}x + \int_{(1-a)/b}^1 \mathrm{d}x \right) \\
&= \frac{1}{3F_{a,b}} \left(\int_0^{(1-a)/b} (a^3 + 3a^2bx + 3ab^2x^2 + b^3x^3)\,\mathrm{d}x + x\big|_{(1-a)/b}^1 \right) \\
&= \frac{1}{3F_{a,b}} \left(a^3x + \tfrac{1}{2}3a^2bx^2 + ab^2x^3 + \tfrac{1}{4}b^3x^4\big|_0^{(1-a)/b} + 1 - \frac{1-a}{b} \right) \\
&= \frac{1}{3F_{a,b}} \left(a^3\frac{1-a}{b} + \tfrac{3}{2}a^2b\left(\frac{1-a}{b}\right)^2 + ab^2\left(\frac{1-a}{b}\right)^3 + \tfrac{1}{4}b^3\left(\frac{1-a}{b}\right)^4 \right. \\
&\quad \left. + 1 - \frac{1-a}{b} \right) \\
&= \frac{1}{3F_{a,b}} (a^3x_0 + 3a^2(1-F_{a,b}) + 2a(1-a)(1-F_{a,b}) \\
&\quad + \tfrac{1}{2}(1-a)^2(1-F_{a,b}) + 1 - x_0) \\
&= \frac{1}{3F_{a,b}} \left(1 - (1-a^3)x_0 + \left(3a^2 + 2a(1-a) + \tfrac{1}{2}(1-a)^2\right)(1-F_{a,b})\right) \\
&= \frac{1}{3F_{a,b}} \left(1 - (1-a^3)x_0 + \left(3a^2 + 2a - 2a^2 + \tfrac{1}{2} - a \right.\right. \\
&\quad \left.\left. + \tfrac{1}{2}a^2\right)(1-F_{a,b})\right)
\end{aligned}$$

Continuing:

$$\begin{aligned}
\mathbb{E}(Y^2) &= \frac{1}{3F_{a,b}}\left(1-(1-a^3)x_0+\left(\tfrac{1}{2}+a+\tfrac{3}{2}a^2\right)(1-F_{a,b})\right)\\
&= \frac{1}{3F_{a,b}}\left(1-(1-a)(1+a+a^2)\frac{1-a}{b}+\left(\tfrac{1}{2}+a+\tfrac{3}{2}a^2\right)(1-F_{a,b})\right)\\
&= \frac{1}{3F_{a,b}}\left(1-2(1+a+a^2)(1-F_{a,b})+\left(\tfrac{1}{2}+a+\tfrac{3}{2}a^2\right)(1-F_{a,b})\right)\\
&= \frac{1}{3F_{a,b}}\left(1-\tfrac{1}{2}(3+2a+a^2)(1-F_{a,b})\right)\\
&= \frac{1}{3F_{a,b}}\left(-\tfrac{1}{2}-a-\tfrac{1}{2}a^2+\tfrac{1}{2}(3+2a+a^2)F_{a,b})\right)\\
&= \frac{1}{3F_{a,b}}\left(-\tfrac{1}{2}(1+a)^2+(1+\tfrac{1}{2}(1+a)^2)F_{a,b})\right)\\
&= \frac{1}{3F_{a,b}}\left(-\tfrac{1}{2}(2-\hat{a})^2+(1+\tfrac{1}{2}(2-\hat{a})^2)F_{a,b})\right)
\end{aligned}$$

Verification via symmetry:

$$\begin{aligned}
\mathbb{E}(Y^2) &= \mathbb{E}((1-\alpha)^2) = 1-2\mathbb{E}(\alpha)+\mathbb{E}(\alpha^2)\\
&= 1-2\mathbb{E}(X)[\text{where } x_0 \leftrightarrow \hat{a}]+\mathbb{E}(X^2)[\text{where } x_0 \leftrightarrow \hat{a}]\\
&= 1-\frac{1}{F_{a,b}}(1-\tfrac{1}{3}x_0\hat{a}^2)+\frac{1}{3F_{a,b}}(1-\tfrac{1}{2}\hat{a}^2+\frac{1}{2}\hat{a}^2F_{a,b})\\
&= \frac{1}{3F_{a,b}}\left(3F-3(1-\tfrac{1}{3}x_0\hat{a}^2)+1-\tfrac{1}{2}\hat{a}^2+\frac{1}{2}\hat{a}^2F_{a,b}\right)\\
&= \frac{1}{3F_{a,b}}\left(-2-\tfrac{1}{2}\hat{a}^2+x_0\hat{a}^2+(3+\tfrac{1}{2}\hat{a}^2)F_{a,b}\right)\\
&= \frac{1}{3F_{a,b}}\left(-2-\tfrac{1}{2}\hat{a}^2+2\hat{a}(1-F_{a,b})+(3+\tfrac{1}{2}\hat{a}^2)F_{a,b}\right)\\
&= \frac{1}{3F_{a,b}}\left(-2+2\hat{a}-\tfrac{1}{2}\hat{a}^2+(1+2-2\hat{a}+\tfrac{1}{2}\hat{a}^2)F_{a,b}\right)\\
&= \frac{1}{3F_{a,b}}\left(-\tfrac{1}{2}(2-\hat{a})^2+(1+\tfrac{1}{2}(2-\hat{a})^2)F_{a,b}\right)
\end{aligned}$$

5. Inner product $\mathbb{E}(XY)$. We have $\mathbb{E}(XY)=\mathbb{E}(\mathbb{E}(XY|X=x))$.
Step 1:

$$\mathbb{E}(XY|X=x) = x\int_0^{L(x)} y\frac{I_{[0,L(x)]}}{L(x)}(y)\,\mathrm{d}y = \frac{x}{L(x)}\;\tfrac{1}{2}y^2\Big|_0^{L(x)} = \tfrac{1}{2}xL(x)$$

Step 2:

$$
\begin{aligned}
\mathbb{E}(XY) &= \frac{1}{F_{a,b}} \int_0^1 \tfrac{1}{2} x (L(x))^2 \,\mathrm{d}x \\
&= \frac{1}{2F_{a,b}} \left(\int_0^{(1-a)/b} x(a+bx)^2 \,\mathrm{d}x + \int_{(1-a)/b}^1 x \,\mathrm{d}x \right) \\
&= \frac{1}{2F_{a,b}} \left(\int_0^{(1-a)/b} (a^2 x + 2abx^2 + b^2 x^3) \,\mathrm{d}x + \tfrac{1}{2}\, x^2 \Big|_{(1-a)/b}^1 \right) \\
&= \frac{1}{2F_{a,b}} \left((\tfrac{1}{2} a^2 x^2 + \tfrac{2}{3} abx^3 + \tfrac{1}{4} b^2 x^4) \Big|_0^{(1-a)/b} + \tfrac{1}{2} \left(1 - \left(\frac{1-a}{b} \right)^2 \right) \right) \\
&= \frac{1}{4F_{a,b}} \left(1 - \left(\frac{1-a}{b} \right)^2 + a^2 \left(\frac{1-a}{b} \right)^2 + \tfrac{4}{3} ab \left(\frac{1-a}{b} \right)^3 + \tfrac{1}{2} b^2 \left(\frac{1-a}{b} \right)^4 \right) \\
&= \frac{1}{4F_{a,b}} \left(1 - (1+a) \frac{(1-a)^3}{b^2} + \tfrac{4}{3} a(1-a)/b \frac{(1-a)^2}{b} + \tfrac{1}{2} (1-a)^2/b \frac{(1-a)^2}{b} \right) \\
&= \frac{1}{4F_{a,b}} \left(1 + \left(-2(1+a)x_0 + \tfrac{8}{3} ax_0 + (1-a)x_0 \right) (1 - F_{a,b}) \right) \\
&= \frac{1}{4F_{a,b}} \left(1 + \left(-x_0 - \tfrac{1}{3} ax_0 \right) (1 - F_{a,b}) \right) \\
&= \frac{1}{4F_{a,b}} \left(1 - x_0 - \tfrac{1}{3} ax_0 + (x_0 + \tfrac{1}{3} ax_0) F_{a,b} \right)
\end{aligned}
$$

Note that

$$
x_0 + \tfrac{1}{3} ax_0 = x_0 + \tfrac{1}{3}(1 - \hat{a}) x_0 = \tfrac{2}{3}(2x_0 - 1 + F_{a,b})
$$

and so

$$
\begin{aligned}
\mathbb{E}(XY) &= \frac{1}{4F_{a,b}} \left(1 + \tfrac{2}{3}(1 - 2x_0 - F_{a,b}) - \tfrac{2}{3}(1 - 2x_0 - F_{a,b}) F_{a,b} \right) \\
&= \frac{1}{4F_{a,b}} \left(1 - \tfrac{2}{3}(2x_0 + F_{a,b}) + \tfrac{2}{3}(2x_0 + F_{a,b}) F_{a,b} \right) \\
&= \frac{1}{4F_{a,b}} \left(1 - \tfrac{2}{3}(2x_0 + F_{a,b})(1 - F_{a,b}) \right)
\end{aligned}
$$

Therefore the correlation is:

$$
\rho_{XY} = \frac{E(XY) - E(X)E(Y)}{\sqrt{E(X^2) - E(X)^2}\sqrt{E(Y^2) - E(Y)^2}} =
$$

$$\frac{\frac{1}{4F_{a,b}}\left(1-x_0-\frac{1}{3}ax_0+(x_0+\frac{1}{3}ax_0)F_{a,b}\right)-\frac{1}{2F_{a,b}}\left(1-\frac{2}{3}x_0+\frac{2}{3}x_0F_{a,b}\right)\frac{1}{2F_{a,b}}\left(1-\frac{2}{3}(a+2)+\frac{2}{3}(a+2)F_{a,b}\right)}{\sqrt{\frac{1}{3F_{a,b}}\left((1-\frac{1}{2}x_0^2)+\frac{1}{2}x_0^2F_{a,b}\right)-\left(\frac{1}{2F_{a,b}}\left(1-\frac{2}{3}x_0+\frac{2}{3}x_0F_{a,b}\right)\right)^2}\sqrt{\frac{1}{3F_{a,b}}\left(-\frac{1}{2}(1+a)^2+(1+\frac{1}{2}(1+a)^2)F_{a,b}\right)-\left(\frac{1}{2F_{a,b}}\left(1-\frac{2}{3}(a+2)+\frac{2}{3}(a+2)F_{a,b}\right)\right)^2}}$$

H

Demonstration NCA with PLS-SEM

This section demonstrates the steps for combining NCA with SEM (Section 11.4) using an example. This example has been used in four published studies where NCA is combined with PLS-SEM (Hauff et al., 2024; Richter, Schubring, et al., 2020; Richter, Hauff, Ringle, et al., 2023; Sarstedt et al., 2024). The goal of the demonstration is fourfold:

- Show how the combined analysis can be done following the flowchart presented in this book.
- Show how the proposed BIPMA plot for prioritizing evidence-based action in practice can be produced and how it differs from IPMA and cIPMA.
- Show how robustness checks for NCA can be conducted in the context of SEM.
- Show how to conduct a combined NCA-SEM analysis with `R`.

This demonstration is a replication of the four above studies by using a different software package than the SmartPLS package. The NCA package for `R` includes all NCA functions and is free. The SmartPLS software may be more user-friendly for users who are not familiar with `R`, but it is not free and does not include all NCA functions.

The selected example evaluates an extended version of the Technology Acceptance Model (TAM) (Richter, Schubring, et al., 2020). Figure H.1 shows the model consisting of six variables (rectangles) and 9 relationships (arrows).

For the demonstration, the flowchart of Figure 11.4 is followed. The `R` packages `SEMinR` for conducting PLS-SEM, `NCA` for conducting NCA, and the general purpose `R` packages `ggplot2` and `ggrepel` for graphics need to be installed. Additional `R` code for conducting the analysis can be found in Section B.2.

H.1 Step 1: Theorize necessity and probabilistic relationships

It is assumed that the relationships shown in Figure H.1 are theoretically justified from both the perspective of probabilistic sufficiency as well as from the perspective of necessity (Richter, Schubring, et al., 2020). For enhancing the necessity theorizing see Section 7.5.3.

DOI: 10.1201/9781003430766-H

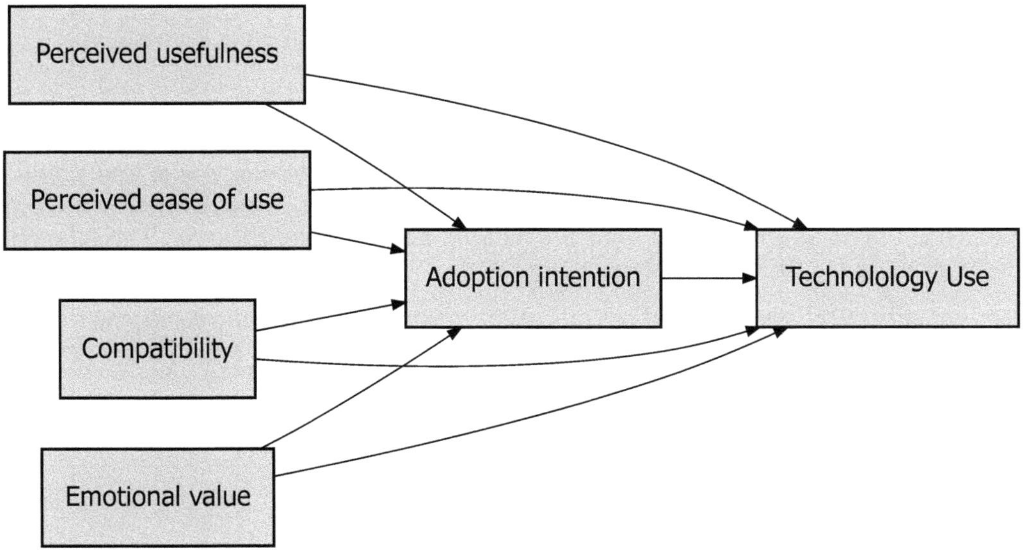

FIGURE H.1 The extended Technology Acceptance Model (TAM) according to Richter, Schubring, et al. (2020).

H.2 Step 2: Collect data

The dataset is provided by (Schubring & Richter, 2023) on Mendeley Data[1], and explained in Richter, Hauff, Kolev, et al. (2023). The dataset needs to be downloaded in the working directory. The helper function `get_indicator_data_TAM` imports the dataset, selects relevant columns, and changes the column names for use in `SEMinR`. The helper function can be downloaded from the book-page: https://jandul.github.io/NCA.

```
# Get indicator data
source("get_indicator_data_TAM.R")
df <- get_indicator_data_TAM()
head(df, 3)
```

```
##    PU1 PU2 PU3 CO1 CO2 CO3 EOU1 EOU2 EOU3 EMV1 EMV2 EMV3 AD1 AD2 AD3 USE
## 1    4   3   3   3   3   3    5    5    4    4    3    3   2   2   2   2
## 2    3   1   4   3   3   4    4    4    2    4    4    3   5   4   4   3
## 3    4   4   4   3   3   4    4    4    4    4    4    4   4   4   4   3
```

The original data consists of 174 rows (cases = individuals using e-readers) and 20 columns. The first 16 columns represent the raw indicators of the 6 variables of the TAM model. The last four columns are not considered in this demonstration and are removed from the dataset. After running the helper function, the resulting dataset `df` is used as input for the PLS-SEM model. The raw indicator scores are measured with 5-point Likert scales, except for the score for Technology use, which is measured with a 7-point Likert scale. The variable names are PU = Perceived usefulness, CO = Compatibility, EOU = Perceived ease of use, EMV = Emotional value, AD = Adoption intention, and USE = Technology use.

[1]https://data.mendeley.com/datasets/pd5dp3phx2/4

H.3 Step 3: Conduct SEM

Following the four original studies, the PLS-SEM approach is selected to estimate the SEM model using the SEMinR package. This package standardizes the raw scores and produces latent construct scores (mean = 0, sd = 1), path coefficients and other output. The helper function estimate_sem_model_TAM conducts the PLS-SEM model estimation. This R function can be downloaded from the book-page: https://jandul.github.io/NCA.

The summary of the results show the standardized path coefficients and some metrics about model fit:

```
# Estimate SEM model
# install.packages("seminr") # install the package if needed
source("estimate_sem_model_TAM.R")
TAM_pls <- estimate_sem_model_TAM(df)
summary(TAM_pls)
```

```
##
## Results from  package seminr (2.4.2)
##
## Path Coefficients:
##                         Adoption intention Technology use
## R^2                                  0.539          0.420
## AdjR^2                               0.528          0.403
## Perceived usefulness                 0.227          0.050
## Compatibility                        0.045          0.107
## Perceived ease of use                0.088          0.010
## Emotional value                      0.515          0.137
## Adoption intention                       .          0.437
##
## Reliability:
##                       alpha  rhoA  rhoC   AVE
## Perceived usefulness  0.723 0.753 0.842 0.642
## Compatibility         0.858 0.859 0.914 0.779
## Perceived ease of use 0.783 0.783 0.873 0.697
## Emotional value       0.914 0.917 0.946 0.853
## Adoption intention    0.938 0.939 0.960 0.889
## Technology use        1.000 1.000 1.000 1.000
##
## Alpha, rhoA, and rhoC should exceed 0.7 while AVE should exceed 0.5
```

The results correspond to those reported in the four original publications using the Smart-PLS software.

In addition, the 95% confidence interval of the path coefficients can be obtained using bootstrapping. A path is significant if its 95% bootstrap confidence interval does not include zero. The helper function get_significance_sem.R can be used to find significant direct effects, indirect effects, and total effects as follows. This R function can be downloaded from the book-page: https://jandul.github.io/NCA.

```
# Estimate significance SEM effects
source("get_significance_sem.R")

# Limit computation time
nboot <- 5000 # set lower for less computation time;
              # for final analysis select 5000
set.seed(123) # for reproducible results
TAM_sig <- get_significance_sem (sem = TAM_pls,
                                 nboot = nboot)
print(TAM_sig)
```

```
##                                             Path Direct Indirect Total
## 8  Perceived usefulness  -> Adoption intention  0.227    0.000 0.227
## 9      Perceived usefulness  -> Technology use  0.050    0.099 0.149
## 2         Compatibility  -> Adoption intention  0.045    0.000 0.045
## 3             Compatibility  -> Technology use  0.107    0.020 0.127
## 6 Perceived ease of use  -> Adoption intention  0.088    0.000 0.088
## 7     Perceived ease of use  -> Technology use  0.010    0.038 0.049
## 4       Emotional value  -> Adoption intention  0.515    0.000 0.515
## 5           Emotional value  -> Technology use  0.137    0.225 0.362
## 1        Adoption intention  -> Technology use  0.437    0.000 0.437
##   Sig_Direct Sig_Indirect Sig_Total pval_direct pval_total          Predicted
## 8       TRUE        FALSE      TRUE       0.030      0.030 Adoption intention
## 9      FALSE        FALSE     FALSE       0.602      0.180     Technology use
## 2      FALSE        FALSE     FALSE       0.690      0.690 Adoption intention
## 3      FALSE        FALSE     FALSE       0.383      0.294     Technology use
## 6      FALSE        FALSE     FALSE       0.268      0.268 Adoption intention
## 7      FALSE        FALSE     FALSE       0.886      0.501     Technology use
## 4       TRUE        FALSE      TRUE       0.000      0.000 Adoption intention
## 5      FALSE         TRUE      TRUE       0.121      0.000     Technology use
## 1       TRUE        FALSE      TRUE       0.000      0.000     Technology use
```

For the *direct* paths, three out of nine path coefficients are significant:

- Perceived usefulness → Adoption intention (direct effect: 0.227),
- Emotional value → Adoption intention (direct effect: 0.515)
- and Adoption intention → Technology use (direct effect: 0.437)

For the *indirect* paths, one out of four paths is significant:

- Emotional value → Adoption intention → Technology use (indirect effect: 0.225)

From the four combined direct and indirect paths (*total effect*) only one is significant: Emotional value → Adoption intention → Technology (total effect = 0.362).

These results suggest that for increasing Technology use on average, only Emotional value and Adoption intention are effective. Perceived usefulness, Compatibility and Perceived ease of use are not candidates for intervention based on the SEM results as they do not have a significant total effect on Technology use.

H.4 Step 4: Extract latent variable scores

By default, PLS-SEM model estimation produces standardized latent variable scores, and this is the only option in `SEMinR` version 2.4.2 that was used for the analysis. However, unstandardized scores are preferred as input for NCA, and normalized unstandardized scores are needed for IPMA, cIPMA and BIPMA. The helper function `unstandardize` calculates the unstandardized latent variable scores from the standardized latent variable scores using procedures described in (Ringle & Sarstedt, 2016), such that the scale of the latent variable scores matches the scale of the original indicator scores. This `R` function can be downloaded from the book-page: https://jandul.github.io/NCA.

```
# Unstandardize data
source("unstandardize.R")
dataset <- unstandardize(TAM_pls)
head(dataset, 3)

##   Perceived usefulness Compatibility Perceived ease of use Emotional value
## 1             3.276582      3.000000              4.676176        3.322971
## 2             2.668901      3.336436              3.352351        3.663039
## 3             4.000000      3.336436              4.000000        4.000000
##   Adoption intention Technology use
## 1           2.000000              2
## 2           4.329378              3
## 3           4.000000              3

data.frame (Mean = colMeans(dataset),
            Min = apply(dataset, 2, min),
            Max = apply(dataset, 2, max)
            )

##                           Mean      Min Max
## Perceived usefulness  3.569914 1.000000   5
## Compatibility         3.462276 1.000000   5
## Perceived ease of use 4.025616 1.674851   5
## Emotional value       3.806854 1.000000   5
## Adoption intention    3.881627 1.000000   5
## Technology use        3.982759 1.000000   7
```

The resulting dataset with unstandardized latent variable scores is called `dataset`. The variables are now interpretable according to the Likert scales that were used to measure its raw indicators. The results show that minimum and maximum latent variable *scores* correspond to the minimum and maximum values of the Likert *scale values*. There is one exception: the minimum observed latent variable score for Perceived ease of use is 1.67 whereas the minimum value of the scale is 1. Apparently, none of the 174 cases (individuals) scored a value 1 on all three indicators of this latent variable.

The unstandardized latent variable scores can be min-max normalized to obtain normalized unstandardized latent variable scores. Min-max normalization is essential if the predictor variables do not have the same scale values (e.g., one predictor variable is measured with a 1–5-Likert scale while another predictor variable is measured with a 1–7-Likert scale). If the same scale is used for all predictor variables, it is preferred not to normalize the scales and to keep the unstandardized latent variables scores, such that the results are interpretable according to the Likert scale values.

Data can be normalized using the helper function `nca_util_normalize` of the `NCA` package in `R`. The input arguments for this function are the data and the range (theoretical minimum and maximum) of the scale (the observed minimum or maximum values might be different). The range of the scale (e.g., 0–1 or 0–100). For example, (c)IPMA commonly uses a 0–100 normalization of the unstandardized latent variable scores:

```
# Min-max normalize data
data <- dataset
scale = c(0, 100) #new percentages scale
library(NCA) # use normalize utility of NCA
min_max <- c(rep(c(1,5), 5), c(1,7)) # min and max values of original scale
dataset1 <-  nca_util_normalize(data = data, scale = scale, min_max = min_max)
head(dataset1, 3)
```

```
##   Perceived usefulness Compatibility Perceived ease of use Emotional value
## 1             56.91456      50.00000              91.90439        58.07427
## 2             41.72252      58.41091              58.80878        66.57599
## 3             75.00000      58.41091              75.00000        75.00000
##   Adoption intention Technology use
## 1           25.00000       16.66667
## 2           83.23446       33.33333
## 3           75.00000       33.33333
```

The resulting dataset with 0–100 normalized unstandardized latent variable scores is called `dataset1`. The variable scores can be interpreted as percentages of the range of the scale. For example, the midpoint 3 of a non-normalized unstandardized 1–5-Likert scale becomes 50 after 0–100 normalization. Note that for the illustrative example normalization is not essential for conducting IPMA, cIPMA or BIPMA because the five predictor variables have the same scale values (1–5-Likert scales).

Since the results of NCA are insensitive to linear transformations (Appendix D), the results are the same for standardized, unstandardized, normalized or non-normalized scores. Unstandardized (normalized or non-normalized) latent variable scores are the preferred input for NCA because of interpretability. The reason is that, in contrast to SEM where a relationship is described in terms of a change of X producing a change of Y (e.g., more X produces more Y), NCA describes a relationship in terms of the necessity of a *level* of X for a *level* of Y (level x of X is necessary for level y of Y). By using interpretable levels linked to the original (unstandardized) scales, NCA results become more insightful.

H.5 Step 5: Conduct NCA

To estimate the necessity of the 9 relationships in the TAM model, 9 necessity analyses are done to obtain necessity effect sizes and p-values. These analyses are grouped into two multiple NCA's: one for the outcome Adoption intention (with four conditions: Perceived usefulness, Compatibility, Perceived ease of use, Emotional value), and one for the outcome Technology use (with five conditions: Perceived usefulness, Compatibility, Perceived ease of use, Emotional value, Adoption intention). The remainder of this demonstration focuses on the outcome Technology use, thus only five necessity relationships are analyzed.

In the original four studies that use the example, the analysts make following choices for the NCA analysis:

- Ceiling line = CE-FDH.
- Effect size threshold level = 0.10.
- *p*-value threshold level = 0.05.
- Scope: empirical scope.
- Outlier removal: none.
- Target outcome = 85% (only in Hauff et al., 2024; Sarstedt et al., 2024).

For this demonstration, the same choices for the primary NCA analysis are made (and different choices for the robustness checks).

The original studies use standardized scores (Richter, Schubring, et al., 2020), non-normalized unstandardized scores (Richter, Hauff, Ringle, et al., 2023), or 0–100 normalized unstandardized scores (Hauff et al., 2024; Sarstedt et al., 2024) as input for NCA. For this demonstration, the preferred scores from an NCA perspective are used as input data: unstandardized latent variable scores without normalization (because all indicator scales of the conditions use the same Likert scales). NCA can be conducted as follows with the `NCA` software in `R`.

```
library(NCA)
# Limit computation time
test.rep <- 10000 # set lower for less computation time;
                  # for final analysis select 10000
# Conduct NCA - Technology use
set.seed(123) # for reproducible results
model1.technology <- nca_analysis(dataset, 1:5, 6,
                                 ceilings = "ce_fdh",
                                 test.rep = test.rep)
print(model1.technology)

##
## --------------------------------------------------------------------------

## Effect size(s):

##                        ce_fdh p
## Perceived usefulness   0.24   0.001
## Compatibility          0.21   0.000
## Perceived ease of use  0.24   0.014
## Emotional value        0.33   0.000
## Adoption intention     0.29   0.000
## --------------------------------------------------------------------------

# Conduct NCA - Adoption intention
set.seed(123) # for reproducible results
model1.adoption <- nca_analysis(dataset, 1:4, 5,
                               ceilings = "ce_fdh",
```

```
                                    test.rep = test.rep)
print(model1.adoption)

##
## --------------------------------------------------------------------------
##                         ce_fdh p
## Perceived usefulness    0.12   0.003
## Compatibility           0.08   0.010
## Perceived ease of use   0.15   0.006
## Emotional value         0.21   0.000
## --------------------------------------------------------------------------
```

The results show NCA's effect size and p-value of the five conditions for Technology use. The results correspond to those reported in the four original studies using the SmartPLS software. It can be concluded that the five conditions are necessary conditions for Technology use because the three criteria for identifying necessity-in-kind are met:

1. A necessity hypothesis is formulated and theoretically justified (for the purpose of this demonstration this requirement is assumed to be met).
2. The effect size is relatively large ($d \geq 0.10$).
3. The p-value is relatively small ($p < 0.05$).

This means that the analysis can continue with an analysis of necessity-in-degree with a bottleneck table that includes all five necessary conditions. This analysis gives information about the minimum level of a condition that is necessary for the target level of the outcome. It also informs if in individual cases a selected target outcome level is achievable, given the observed levels of the conditions. If a case has a value of any condition below the minimum required level for that condition, this case is a 'bottleneck case' for that condition. If the case has a condition value above the minimum required level, the case is not a bottleneck case for the condition, but could be a bottleneck case for another conditions. Only when all condition values exceed their respective threshold levels, the case is not a bottleneck case, making the target outcome level achievable.

The use of non-normalized unstandardized latent variable scores and of 'actual values' in the bottleneck table ensures a direct link between the values in the bottleneck table and the values of the Likert scales that were used to measure the indicator scores. This means that in NCA's `nca_analysis` function, the values of conditions in the bottleneck table must be specified as 'actual' using the argument `bottleneck.x = 'actual'`. Since the outcome variable Technology use has 7 distinct levels, the steps in the bottleneck table preferably correspond to these levels of the outcome. This can be done with the argument `steps`.

```
# Bottleneck table Technology use: actual-actual
library(NCA)
bottleneck.y = "actual"
bottleneck.x = "actual"
steps = c(1,2,3,4,5,6,7)
model2.technology <- nca_analysis(dataset,# unstandardized (non-normalized)
                                  1:5, # five conditions
                                  6, # outcome
                                  ceilings = "ce_fdh",
```

```
                                          bottleneck.x = bottleneck.x,
                                          bottleneck.y = bottleneck.y,
                                          steps = steps)
nca_output (model2.technology, summaries = FALSE, plots = FALSE,
              bottlenecks = TRUE)
```

```
##
## --------------------------------------------------------------------------
## --------------------------------------------------------------------------
## Y      1     2     3     4     5
## 1     NN    NN    NN    NN    NN
## 2     NN    NN    2.015 NN    NN
## 3     NN    NN    2.015 NN    NN
## 4     1.628 2.021 2.339 2.986 2.353
## 5     1.628 2.348 2.339 2.986 2.353
## 6     2.925 2.348 2.355 2.986 2.353
## 7     3.648 2.348 3.676 2.986 4.000
```

The outcome variable Technology use is the perceived frequency of utilizing an e-book. The 7 anchor points are 1 = never, 2 = seldom; 3 = several times a month; 4 = once a week; 5 = several times a week; 6 = daily; 7 = several times daily. The condition variables are measured on a 5 point disagree-agree Likert scale where the anchor points range from 1 = strongly disagree to 5 = agree fully. In the original studies (and here) it is assumed that these Likert scales are interval scales, meaning the distances between anchor points are treated as being equal. This assumption facilitates quantitative analysis of the data, although it is often, like here, a strong assumption.

The bottleneck table can be evaluated row-wise. It shows that for a given target outcome level of Technology use (first column), the conditions must meet a certain minimum level (next columns). For example, for a level of 3 of Technology use ('several times a month'), it is only necessary to have a level 2.015 of Perceived ease of use; the other predictors cannot be a bottleneck (NN = Not Necessary). However, for a target level of 4 of Technology use ('once a week'), it is necessary to have a level 1.628 of Perceived usefulness, level 2.021 of Compatibility, level 2.339 of Perceived ease of use, level 2.986 of Emotional value, and level 2.353 of Adoption intention.

The bottleneck tables that are used by Hauff et al. (2024) and Sarstedt et al. (2024) have a large number of steps as they used 0–100 normalized unstandardized scores of the latent variable with bottleneck steps of 5% for the outcome (0% = original level 1, 100% = original level 7). For the purpose of cIPMA, 'actual' values for the outcome are selected (because the input data are already percentage of range) for the outcome and 'percentile' values for the conditions. Their analysis is replicated as follows:

```
# Bottleneck table Technology use: actual - percentile
library(NCA)
bottleneck.y = "actual"
bottleneck.x = "percentile"
steps=seq(0, 100, 5)
model3.technology <- nca_analysis(dataset1, 1:5, 6,
                                          ceilings = "ce_fdh",
                                          bottleneck.x = bottleneck.x,
                                          bottleneck.y = bottleneck.y,
```

```
                                        steps = steps)
nca_output (model3.technology, summaries = FALSE, plots = FALSE,
            bottlenecks = TRUE)

##
## ---------------------------------------------------------------------------

## ---------------------------------------------------------------------------
## Y       1         2         3         4         5
## 0       NN (0)    NN (0)    NN (0)    NN (0)    NN (0)
## 5       NN (0)    NN (0)    0.6 (1)   NN (0)    NN (0)
## 10      NN (0)    NN (0)    0.6 (1)   NN (0)    NN (0)
## 15      NN (0)    NN (0)    0.6 (1)   NN (0)    NN (0)
## 20      NN (0)    NN (0)    0.6 (1)   NN (0)    NN (0)
## 25      NN (0)    NN (0)    0.6 (1)   NN (0)    NN (0)
## 30      NN (0)    NN (0)    0.6 (1)   NN (0)    NN (0)
## 35      1.7 (3)   5.7 (10) 1.1 (2)    5.7 (10) 4.6 (8)
## 40      1.7 (3)   5.7 (10) 1.1 (2)    5.7 (10) 4.6 (8)
## 45      1.7 (3)   5.7 (10) 1.1 (2)    5.7 (10) 4.6 (8)
## 50      1.7 (3)   5.7 (10) 1.1 (2)    5.7 (10) 4.6 (8)
## 55      1.7 (3)   8.6 (15) 1.1 (2)    5.7 (10) 4.6 (8)
## 60      1.7 (3)   8.6 (15) 1.1 (2)    5.7 (10) 4.6 (8)
## 65      1.7 (3)   8.6 (15) 1.1 (2)    5.7 (10) 4.6 (8)
## 70      17.2 (30) 8.6 (15) 2.9 (5)    5.7 (10) 4.6 (8)
## 75      17.2 (30) 8.6 (15) 2.9 (5)    5.7 (10) 4.6 (8)
## 80      17.2 (30) 8.6 (15) 2.9 (5)    5.7 (10) 4.6 (8)
## 85      47.1 (82) 8.6 (15) 28.7 (50) 5.7 (10) 39.1 (68)
## 90      47.1 (82) 8.6 (15) 28.7 (50) 5.7 (10) 39.1 (68)
## 95      47.1 (82) 8.6 (15) 28.7 (50) 5.7 (10) 39.1 (68)
## 100     47.1 (82) 8.6 (15) 28.7 (50) 5.7 (10) 39.1 (68)
```

In this version of the bottleneck table, the outcome is expressed as 'actual values', which are the values used as input to NCA (here the 0–100% normalized values). The conditions are expressed as percentiles. The percentile value corresponds to the *percentage* of cases that are unable to achieve the level of the outcome (value in the first column of the row). The number between brackets refers to the *number* of cases that were unable to meet the required level of the condition for the given target level of the outcome. For example, for the target outcome level of Technology use of 85%, 47.1% of cases (82 cases) were unable to achieve the required level of Perceived usefulness. This means that these cases will not have a level of 85% Technology use. A case that is not able to achieve a target outcome because the required level of the necessary condition is not met is a *bottleneck case*.

The level of 85 on a 0–100 normalized scale corresponds to level 6.1 on the original 1–7-Likert scale. A value of 6 on the Likert scale (daily use of the e-reader) corresponds to 83.33 on the 0–100 normalized scale. It shows that although 0–100 normalization is common in the context of IPMA, the link with the original scales may be obscured.

The percentage of bottleneck cases for a given condition and target outcome can be extracted from the bottleneck table if percentiles were used as values for the conditions by using the helper function `get_bottleneck_cases`. This `R` function can be downloaded from the https://jandul.github.io/NCA

```
# Bottleneck cases - Technology use
source("get_bottleneck_cases.R")
data <-  dataset1
predicted <- "Technology use"
predictors <-  c("Perceived usefulness",
                 "Compatibility",
                 "Perceived ease of use",
                 "Emotional value",
                 "Adoption intention")
corner = 1 # default; expected empty space for all conditions
target_outcome <- 85
ceiling <- "ce_fdh"
bottlenecks_technology <- get_bottleneck_cases(data = data,
                                              conditions = predictors,
                                              outcome = predicted,
                                              corner = corner,
                                              target_outcome = target_outcome,
                                              ceiling = ceiling)
bottlenecks_technology

## $bottleneck_cases_per
##                        Bottlenecks (Y = 85)
## Perceived usefulness              47.126437
## Compatibility                      8.620690
## Perceived ease of use             28.735632
## Emotional value                    5.747126
## Adoption intention                39.080460
##
## $bottleneck_cases_num
##                        Bottlenecks (Y = 85)
## Perceived usefulness                     82
## Compatibility                            15
## Perceived ease of use                    50
## Emotional value                          10
## Adoption intention                       68
##
## $threshold
##  Perceived usefulness         Compatibility Perceived ease of use
##              66.21236              33.69644              66.90439
##       Emotional value    Adoption intention
##              49.65026              75.00000
```

These numbers correspond the number in row Y = 85 of the bottleneck table where the values of the conditions are expressed in percentiles. The percentage of bottleneck cases is used as input to the cIPMA.

H.6 Step 6: Produce NERT

The NCA Extended Regression Tables (NERTs) for SEM are shown in Tables H.1 and H.2 for the outcome Adoption intention and Technology use, respectively. The regression results are expressed as direct and total path coefficients with their p-values.

```
#NERT - Adoption intention
library(kableExtra)
library(knitr)

# variable names
vars <- c("Perceived usefulness",
          "Compatibility",
          "Perceived ease of use",
          "Emotional value")
n_vars <- length(vars)

# regression coefficients, p-values, model fit (character vectors)
reg   <- c("0.227", "0.045", "0.088", "0.515")
reg_p <- c("0.030", "0.690", "0.268", "< 0.001")
r2    <- 0.539
r2_adj <- 0.528   # NULL if not available

# necessity effect sizes and p-values
nec   <- c("0.12", "0.08", "0.15","0.21")
nec_p <- c("0.002", "0.010", "0.007", "< 0.001")

# Caption and footnote
caption = "Summary of the SEM and NCA results for outcome Adoption intention."
note_text <- "n = 174. NCA ceiling line: CE-FDH."

# build rows conditionally
row_labels <- c(vars, sprintf("$R^2$ = %.2f", r2))
if (!is.null(r2_adj) && !is.na(r2_adj)) {
  row_labels <- c(row_labels, sprintf("$R^2$(adj.) = %.2f", r2_adj))
}

# build NERT
nert <- data.frame(
  Row = row_labels,
  `Path coefficient` = c(reg,  rep("", length(row_labels) - length(reg))),
  reg_p_value              = c(reg_p, rep("", length(row_labels) - length(reg_p))),
  `Necessity effect size`  = c(nec,  rep("", length(row_labels) - length(nec))),
  nec_p_value              = c(nec_p, rep("", length(row_labels) - length(nec_p))),
  check.names = FALSE
)

kbl(
  nert,
  format = "latex",
  escape = FALSE,
```

```
    booktabs = TRUE,
    linesep= "",
    longtable = FALSE,
    caption = caption,
    col.names = c("",
                  "Path coefficient",
                  "\\emph{p}-value",
                  linebreak("Necessity\neffect size", align = "c"),
                  "\\emph{p}-value"),
    align = c("l", "c","c", "c", "c")

  ) %>%
    kable_styling(full_width = FALSE,
                  latex_options = c("HOLD_position", "scale_down")) %>%
    column_spec(1, width = "8em") %>%
    column_spec(2, width = "8em") %>%
    column_spec(3, width = "8em") %>%
    column_spec(4, width = "8em", border_left = TRUE) %>%
    column_spec(5, width = "8em") %>%
    pack_rows("Variables/Conditions", 1,n_vars) %>%
    pack_rows("Model fit (regression)", n_vars + 1, nrow(nert),
              hline_before = TRUE) %>%
    footnote(
      general = note_text
    )
```

TABLE H.1 Summary of the SEM and NCA results for outcome Adoption intention.

	Path coefficient	*p*-value	Necessity effect size	*p*-value
Variables/Conditions				
Perceived usefulness	0.227	0.030	0.12	0.003
Compatibility	0.045	0.690	0.08	0.010
Perceived ease of use	0.088	0.268	0.15	0.006
Emotional value	0.515	< 0.001	0.21	<0.001
Model fit (regression)				
$R^2 = 0.54$				
$R^2(\text{adj.}) = 0.53$				

Note:
n = 174. NCA ceiling line: CE-FDH.

Table H.1 for Adoption intention indicates that Perceived usefulness and Emotional value both have a probabilistic sufficiency effect on Adoption intention (path coefficients are significant) and are also necessary conditions for Adoption intention (effect sizes are large enough and significant). Compatibility has no average effect nor a necessity effect on Adoption intention (although its necessity effect size is statistically significant, it is too small $d < 0.10$). Perceived ease of use has also no average effect on Adoption intention, but it has a necessity effect.

```
#NERT - Technology use
library(kableExtra)
library(knitr)

# variable names
vars <- c("Perceived usefulness",
          "Compatibility",
          "Perceived ease of use",
          "Emotional value",
          "Adoption intention")
n_vars <- length(vars)

# regression coefficients, p-values, model fit (character vectors)
direct_coef <- c("0.050", "0.107", "0.010", "0.137", "0.437")
direct_p    <- c("0.602", "0.383", "0.886", "0.121", "< 0.001")

total_coef  <- c("0.149", "0.127", "0.049", "0.362", "0.437")
total_p     <- c("0.180", "0.294", "0.501", "< 0.001", "< 0.001")

nec_es <- c("0.24", "0.21", "0.24", "0.33", "0.29")
nec_p  <- c("0.001", "< 0.001", "0.015", "< 0.001", "< 0.001")

r2     <- 0.539
r2_adj <- 0.528

# Caption and footnote
caption <- "Summary of the SEM and NCA results for outcome Technology use."
note_text <- "n = 174. NCA ceiling line: CE-FDH."

# build rows conditionally
row_labels <- c(vars, sprintf("$R^2$ = %.2f", r2))
if (!is.null(r2_adj) && !is.na(r2_adj)) {
  row_labels <- c(row_labels, sprintf("$R^2$(adj.) = %.2f", r2_adj))
}

pad_to <- function(x, n) c(x, rep("", max(0, n - length(x))))
n_rows <- length(row_labels)

# build NERT
nert <- data.frame(
  Row = row_labels,
  direct_coef = pad_to(direct_coef, n_rows),
  direct_p    = pad_to(direct_p, n_rows),
  total_coef  = pad_to(total_coef, n_rows),
  total_p     = pad_to(total_p, n_rows),
  nec_es      = pad_to(nec_es, n_rows),
  nec_p       = pad_to(nec_p, n_rows),
  check.names = FALSE
)
```

```
kbl(
  nert,
  format = "latex",
  escape = FALSE,
  booktabs = TRUE,
  linesep= "",
  longtable = FALSE,
  caption = caption,
  col.names = c(" ",
                "Direct path coefficient",
                "\\emph{p}-value",
                "Total path coefficient",
                "\\emph{p}-value",
                linebreak("Necessity\neffect size", align = "c"),
                "\\emph{p}-value"),
  align = c("l", "c", "c", "c", "c", "c", "c"),

) %>%
  kable_styling(full_width = FALSE,
                latex_options = c("HOLD_position", "scale_down")) %>%
  column_spec(1, width = "8em") %>%
  column_spec(2, width = "5em") %>%
  column_spec(3, width = "4em") %>%
  column_spec(4, width = "5em") %>%
  column_spec(5, width = "4em") %>%
  column_spec(6, width = "4em", border_left = TRUE) %>%
  column_spec(7, width = "4em") %>%
  pack_rows("Variables/Conditions", 1, n_vars) %>%
  pack_rows("Model fit (SEM)", n_vars + 1, nrow(nert), hline_before = TRUE) %>%
  footnote(general = note_text, escape = FALSE)
```

Table H.2 for Technology use shows that none of the predictors except Adoption intention has a direct average effect on Technology use. The results of the total average effects show that only Emotional value and adoption intention have a probabilistic sufficiency effect on Technology use. However, all predictors are necessary conditions for Technology use.

H.7 Step 7: Produce BIPMA

Before producing BIPMA (Section 11.4.1.7), this demonstration first produces IPMA and cIPMA for comparison. To produce classic IPMA, Importance and Performance scores are extracted from the SEM results. The Importance score of each predictor variable corresponds to the total effect (path coefficient) on the predicted variable (Step 3). The Performance score of each predictor variable is the mean of the 0–100 normalized unstandardized variable score (Step 4). The helper function `get_ipma_df` can be used for extracting these values from the SEM results. This `R` function can be downloaded from the book-page: https://jandul.github.io/NCA.

TABLE H.2 Summary of the SEM and NCA results for outcome Technology use.

	Direct path coefficient	*p*-value	Total path coefficient	*p*-value	Necessity effect size	*p*-value
Variables/Conditions						
Perceived usefulness	0.050	0.602	0.149	0.180	0.24	0.001
Compatibility	0.107	0.383	0.127	0.294	0.21	< 0.001
Perceived ease of use	0.010	0.886	0.049	0.501	0.24	0.014
Emotional value	0.137	0.121	0.362	< 0.001	0.33	< 0.001
Adoption intention	0.437	< 0.001	0.437	< 0.001	0.29	< 0.001
Model fit (SEM)						
$R^2 = 0.54$						
$R^2(\text{adj.}) = 0.53$						

Note:
n = 174. NCA ceiling line: CE-FDH.

```
# IPMA  - dataset
source("get_ipma_df.R")
data <- dataset1 # 0-100 normalized unstandardized
sem <- TAM_pls
predicted <- "Technology use"
predictors <-  c("Perceived usefulness",
                 "Compatibility",
                 "Perceived ease of use",
                 "Emotional value",
                 "Adoption intention")
IPMA_df_technology <- get_ipma_df(data = data,
                                  sem = sem,
                                  predictors = predictors,
                                  predicted = predicted)
IPMA_df_technology
```

```
##               predictor Importance Performance
## 1  Perceived usefulness 0.14921041    64.24785
## 2         Compatibility 0.12699788    61.55690
## 3 Perceived ease of use 0.04863022    75.64040
## 4       Emotional value 0.36169649    70.17135
## 5    Adoption intention 0.43705233    72.04067
```

The IPMA is produced by mapping the predictor variables on a 2×2 plot with the horizontal axis representing Importance and the vertical axis Performance. This can be done with the helper function `get_ipma_plot`. This `R` function can be downloaded from the book-page: https://jandul.github.io/NCA.

```
# IPMA  - plot
# install.packages("ggplot2") # install the package if needed
# install.packages("ggrepel") # install the package if needed
source("get_ipma_plot.R")
ipma_df <- IPMA_df_technology
x_range <- c(0,0.6) # range of the Importance axis
y_range <- c(0,100) # range of the Performance axis
IPMA_plot_technology <- get_ipma_plot(ipma_df = ipma_df,
                                      x_range = x_range,
                                      y_range = y_range)
```

The results for Technology use are shown in Figure H.2. Although the SEM results show that Compatibility and Perceived ease of use have no significant effect on Technology use, the conventional IPMA plot includes the points for all predictors because this choice was made in the publications that previously used this example. The dashed lines represent mean values of the Importance and Performance scores of the predictors.

According to IPMA, priority of action should be given to a predictor with high Importance and low Performance scores. From the two predictors with the highest Importance (Emotional value and Adoption intention), Emotional value has a slightly lower Performance score and could be selected to prioritize action. This action (assuming other predictors constant) should result in an increase of its current Performance score (mean score of the latent variable), which is somewhat below 75%, to a higher score.

cIPMA adds the bottleneck dimension to IPMA. In selecting the priority predictor, cIPMA not only considers the Importance and Performance scores of the predictor, but also the percentage of bottleneck cases of the predictor: the percentage of cases that are unable to achieve a particular target outcome level. As shown above, the percentage of bottleneck cases per predictor can be obtained with the helper function `get_bottleneck_cases`. To produce cIPMA, the IPMA dataset with Importance and Performance scores is extended with the percentage of bottleneck cases. The function `get_cipma_df` is used for this. Note that the output includes the column 'Necessity' that indicates whether the predictor is necessary given the selected threshold values for the effect size d and the p-value. These predictors should not show up in the bottleneck table and therefore should not be part of cIPMA's bottleneck dimension. Therefore, in the `get_cipma_df` helper function a standard point size as in IPMA is given to such predictors. This `R` function can be downloaded from the book-page: https://jandul.github.io/NCA.

```
# cIPMA  - dataset
source("get_cipma_df.R")
d_threshold <- 0.10
p_threshold <- 0.05
model <- model1.technology

# Extract parameters
effect_size <- sapply(predictors, nca_extract, model = model, ceiling = "ce_fdh"
                      param   = "Effect size")
p_value <- sapply(predictors, nca_extract, model = model, ceiling = "ce_fdh",
                  param   = "p-value")

necessity <- ifelse(effect_size >= d_threshold & p_value <= p_threshold,
```

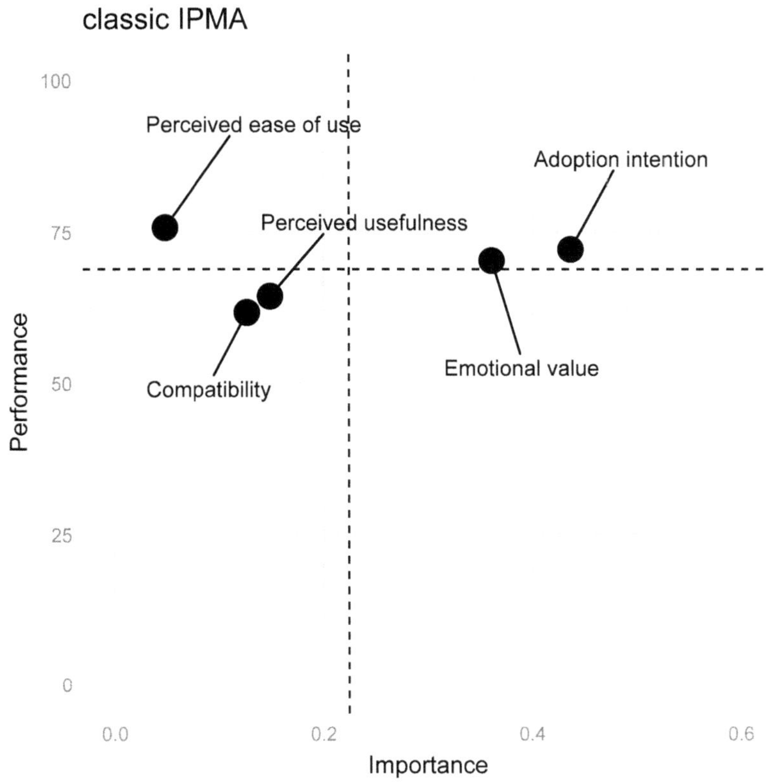

FIGURE H.2 Classic Importance Performance Map Analysis (IPMA) for the outcome Technology use.

```
                        "yes", "no")
ipma_df <- IPMA_df_technology
bottlenecks <- bottlenecks_technology
CIPMA_df_technology <- get_cipma_df(ipma_df = ipma_df,
                                    bottlenecks = bottlenecks,
                                    necessity = necessity)
CIPMA_df_technology

##                predictor Importance Performance Bottleneck_cases_per
## 1   Perceived usefulness 0.14921041    64.24785            47.126437
## 2          Compatibility 0.12699788    61.55690             8.620690
## 3  Perceived ease of use 0.04863022    75.64040            28.735632
## 4        Emotional value 0.36169649    70.17135             5.747126
## 5     Adoption intention 0.43705233    72.04067            39.080460
##   Bottleneck_cases_num Necessity            Predictor_with_cases
## 1                   82       yes  Perceived usefulness (47%, n=82)
## 2                   15       yes         Compatibility (9%, n=15)
## 3                   50       yes Perceived ease of use (29%, n=50)
## 4                   10       yes        Emotional value (6%, n=10)
## 5                   68       yes   Adoption intention (39%, n=68)
```

The plot can then be produced with the helper function `get_cipma_plot`. This `R` function can be downloaded from the book-page: https://jandul.github.io/NCA.

```
# cIPMA  - plot
source("get_cipma_plot.R")
cipma_df <- CIPMA_df_technology
x_range <- c(0,0.6)
y_range <- c(0,100)
size_limits = c(0,100)
size_range <- c(0.5,50) # the size of the bulbs
name_plot <- "Original"
cIPMA_plot_technology <- get_cipma_plot(cipma_df = cipma_df,
                                        x_range = x_range,
                                        y_range = y_range,
                                        size_limits = size_limits,
                                        size_range = size_range)
```

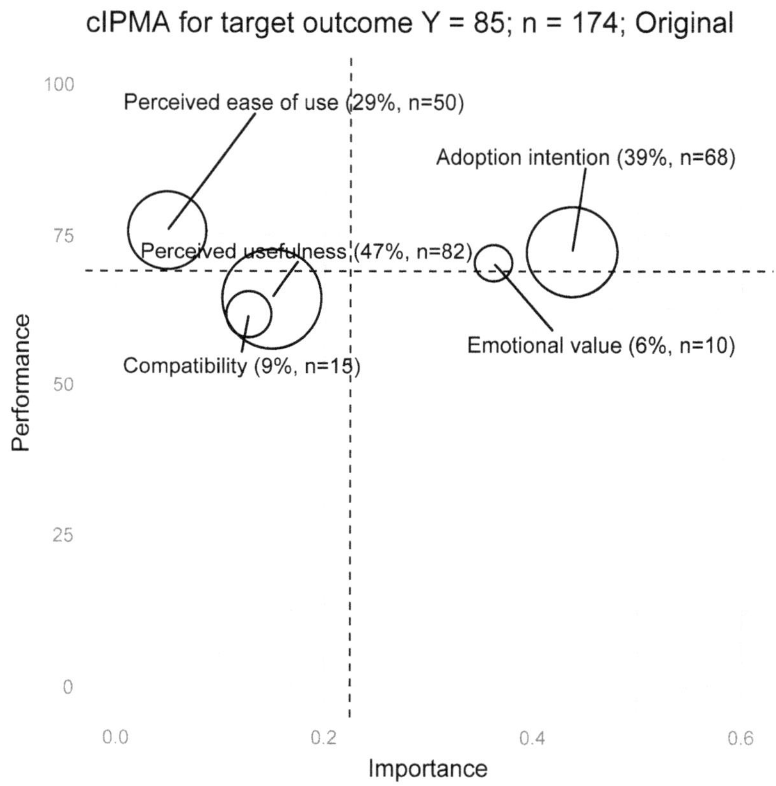

FIGURE H.3 Combined Importance Performance Map Analysis (cIPMA) for the outcome Technology use. The size of the dot is an indication of the number of cases that cannot achieve the target outcome of 85% of maximum outcome.

The results are shown in Figure H.3, which corresponds to the plots in Hauff et al. (2024) and Sarstedt et al. (2024). The IPMA plot is extended with point sizes that depend on the number or percentage of bottleneck cases. Since all predictors are necessary, each point's size is relevant. The percentage and number of bottleneck cases are included here between brackets. The NCA results show large differences in percentage and number of bottleneck cases between the five conditions. The order from high to low percentages is: Perceived usefulness, Adoption intention, Perceived ease of use, Compatibility and Emotional value. Perceived usefulness has 47% bottleneck cases, whereas Emotional value has only 6%. This means that from the perspective of NCA, Perceived usefulness is a more relevant than emotional value for a single predictor action. Successful interventions on a single condition (with the goal of bringing all bottleneck cases to at least the threshold level such that the target outcome is possible) will reduce more bottleneck cases when the intervention focuses on Perceived usefulness, rather than on emotional value. Note that such an intervention to eliminate bottlenecks does not guarantee that the target outcome level (85%) will be achieved: a necessary condition for an outcome is not a sufficient condition for it. Additionally, cases may be bottlenecks in multiple conditions. Note also the relevance of a necessary condition in terms of *effect size* may differ from the relevance of a necessary condition in terms of *number of bottleneck cases*. The first refers to the magnitude of the constraining role of the condition for the outcome in general, whereas the second refers to the size of the group of cases that cannot achieve a certain target level of the outcome. For emotional value the effect size is largest (0.33), but the percentage of bottleneck cases is lowest (6%).

The BIPMA extends the cIPMA by considering only single bottleneck cases per predictor as explained in Section 11.4.1.7. The number of single bottleneck cases can be extracted with the helper function `get_single_bottleneck_cases`. This `R` function can be downloaded from the book-page: https://jandul.github.io/NCA.

```
# Single bottleneck cases - Target outcome 85
source("get_single_bottleneck_cases.R")
data <- dataset1
ceilings = "ce_fdh"
target_outcome <- 85
predicted <- "Technology use"
predictors <-  c("Perceived usefulness",
                 "Compatibility",
                 "Perceived ease of use",
                 "Emotional value",
                 "Adoption intention")
corner = 1 # default expected empty space for all conditions
single_bottlenecks_technology <- get_single_bottleneck_cases(
                                   data,
                                   conditions = predictors,
                                   outcome = predicted,
                                   corner = corner,
                                   target_outcome=target_outcome,
                                   ceiling = ceilings)
single_bottlenecks_technology
```

```
## $single_bottleneck_cases_per
##                      Bottlenecks (Y = 85)
## Perceived usefulness           14.9425287
## Compatibility                   0.5747126
## Perceived ease of use           4.0229885
## Emotional value                 0.0000000
## Adoption intention              8.0459770
##
```

```
## $single_bottleneck_cases_num
##                       Bottlenecks (Y = 85)
## Perceived usefulness                    26
## Compatibility                            1
## Perceived ease of use                    7
## Emotional value                          0
## Adoption intention                      14
##
## $threshold
##  Perceived usefulness         Compatibility Perceived ease of use
##              66.21236              33.69644              66.90439
##       Emotional value    Adoption intention
##              49.65026              75.00000
##
## $single_bottleneck_case_names
## $single_bottleneck_case_names$`Perceived usefulness`
##  [1] "15"  "26"  "32"  "44"  "49"  "51"  "62"  "68"  "71"  "73"  "78"  "91"
## [13] "92"  "94"  "101" "102" "113" "115" "118" "126" "138" "152" "162" "163"
## [25] "164" "166"
##
## $single_bottleneck_case_names$Compatibility
## [1] "109"
##
## $single_bottleneck_case_names$`Perceived ease of use`
## [1] "7"   "24"  "40"  "45"  "65"  "79"  "105"
##
## $single_bottleneck_case_names$`Emotional value`
## character(0)
##
## $single_bottleneck_case_names$`Adoption intention`
##  [1] "11"  "23"  "25"  "27"  "38"  "39"  "74"  "85"  "89"  "106" "110" "136"
## [13] "140" "157"
##
##
## $any_bottleneck_case_names
##   [1] "1"   "2"   "6"   "7"   "8"   "9"   "11"  "14"  "15"  "19"  "20"  "22"
##  [13] "23"  "24"  "25"  "26"  "27"  "28"  "29"  "31"  "32"  "36"  "38"  "39"
##  [25] "40"  "42"  "43"  "44"  "45"  "47"  "48"  "49"  "50"  "51"  "52"  "54"
##  [37] "56"  "58"  "61"  "62"  "64"  "65"  "68"  "70"  "71"  "73"  "74"  "78"
##  [49] "79"  "83"  "85"  "86"  "88"  "89"  "90"  "91"  "92"  "94"  "95"  "96"
##  [61] "97"  "101" "102" "103" "105" "106" "108" "109" "110" "111" "113" "114"
##  [73] "115" "116" "117" "118" "119" "120" "121" "122" "123" "124" "125" "126"
##  [85] "132" "134" "135" "136" "138" "140" "141" "142" "143" "146" "148" "150"
##  [97] "151" "152" "153" "157" "158" "160" "162" "163" "164" "166" "168" "169"
## [109] "171" "172" "173" "174"
```

Next, the significant predictor variables for the predicted variable must be identified:

```
# Significant total effects (SEM)
sem_sig <- TAM_sig
for_select_predicted <- sem_sig[sem_sig$Predicted == predicted, ]
```

```
sufficiency <- ifelse(for_select_predicted$Sig_Total, "yes", "no")
names(sufficiency) <- predictors
```

The BIPMA dataset with Bottlenecks (single bottlenecks), Importance and Performance scores can be obtained with the helper function `get_bipma_df`. This `R` function can be downloaded from the book-page: https://jandul.github.io/NCA.

```
# BIPMA  - dataset
source("get_bipma_df.R")
ipma_df <- IPMA_df_technology
single_bottlenecks <- single_bottlenecks_technology
BIPMA_df_technology <- get_bipma_df(ipma_df = ipma_df,
                                    single_bottlenecks=single_bottlenecks,
                                    necessity = necessity,
                                    sufficiency = sufficiency)
BIPMA_df_technology
```

```
##              predictor Importance Performance Single_bottleneck_cases_per
## 1  Perceived usefulness 0.14921041    64.24785                  14.9425287
## 2         Compatibility 0.12699788    61.55690                   0.5747126
## 3 Perceived ease of use 0.04863022    75.64040                   4.0229885
## 4       Emotional value 0.36169649    70.17135                   0.0000000
## 5    Adoption intention 0.43705233    72.04067                   8.0459770
##   Single_bottleneck_cases_num Necessity Corner Target_outcome Sufficiency
## 1                          26       yes      1             85          no
## 2                           1       yes      1             85          no
## 3                           7       yes      1             85          no
## 4                           0       yes      1             85         yes
## 5                          14       yes      1             85         yes
##        Predictor_with_single_cases
## 1 Perceived usefulness (15%, n=26)
## 2           Compatibility (1%, n=1)
## 3  Perceived ease of use (4%, n=7)
## 4         Emotional value (0%, n=0)
## 5    Adoption intention (8%, n=14)
```

The BIPMA can be created with the helper function `get_bipma_plot`. This `R` function can be downloaded from the book-page: https://jandul.github.io/NCA.

```
# Bipma - plot
source("get_bipma_plot.R")
bipma_df <- BIPMA_df_technology
x_range <- c(0,0.6)
y_range <- c(0,100)
size_limits = c(0,100)
size_range <- c(0.5,50) # the size of the bulbs
name_plot <- "Original"
BIPMA_plot_technology <- get_bipma_plot(bipma_df = bipma_df,
                                        x_range = x_range,
                                        y_range = y_range,
                                        size_limits = size_limits,
```

```
                    size_range = size_range)
```

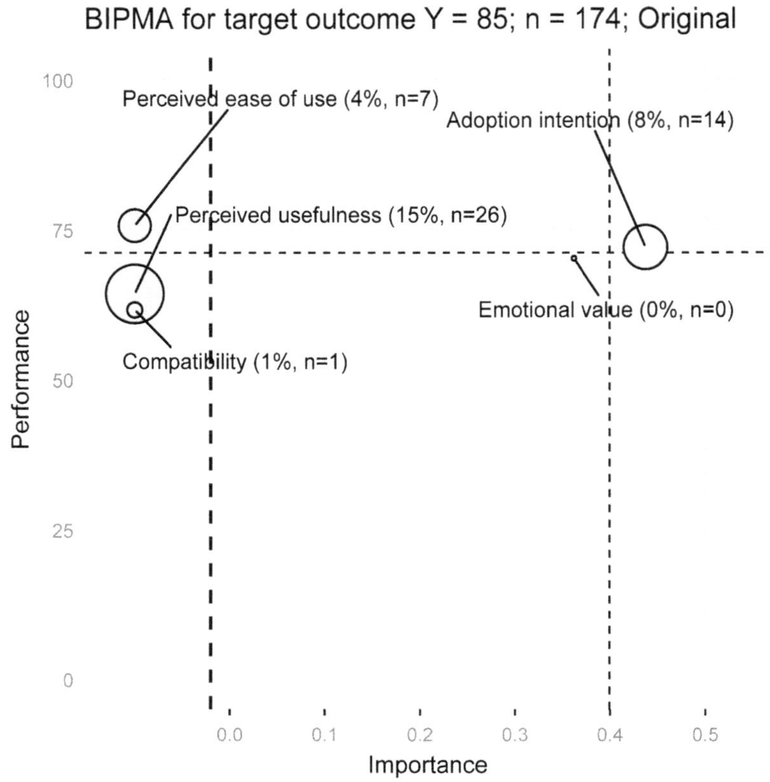

FIGURE H.4 Bottleneck Importance Performance Map Analysis (BIPMA) for the outcome Technology use. The size of the dot is an indication of the number of *single* bottleneck cases that cannot achieve the target outcome of 85% of maximum outcome. White dot: Significant average effect and significant necessity effect. Gray dot: Non-significant average effect and significant necessity effect. The horizontal and vertical thin dashed lines correspond to mean Performance and the mean Importance of the predictors with a significant total average effect, respectively.

The results are shown in Figure H.4. For each condition the percentage and number of *single* bottleneck cases is given between brackets. The results again show large differences in the percentage and number of single bottleneck cases between the five conditions. In this case, the *order* from high to low percentages is the same as for cIPMA as shown in Figure H.3. However, the *number and percentage* of bottleneck cases are in BIPMA considerably lower than in cIPMA. Perceived usefulness has most single bottleneck cases (15%, n=26), whereas Emotional value has none. This means that, in contrast to the conclusion from cIPMA, emotional value is not a relevant predictor for an action based on NCA if only one condition can be changed. Successful group interventions on a single predictor (with the goal to bring all bottleneck cases above the threshold level) could focus on Perceived usefulness, which has almost twice as many single bottlenecks as Adoption intention. Perceived usefulness, Compatibility and Perceived ease of use have non-significant average effects on Technology use according to SEM. Predictors that are necessary from the perspective of NCA, but are non-significant predictors from the perspective of SEM, are placed on the Importance axis a left of Importance = 0 and the corresponding points are shown in gray. Predictors that are significant from the perspective of SEM but non-significant from the perspective of NCA are shown as small black points, but none of the conditions in this analysis were non-significant from the perspective of NCA. Note, that it is possible that a condition without bottleneck cases is also displayed this way (Emotional value), but such a condition is irrelevant for an NCA-based intervention. Predictors that are non-significant from the perspective of both SEM and NCA are separately mentioned in the BIPMA plot in a text box (Figure 11.5) to indicate that actions on such predictors are not effective. However, in the TAM example non-significant predictors from the necessity perspective were not identified.

For selecting a proper group intervention strategy based on the BIPMA results, see Section 11.4.1.7. For a selecting a case-specific interventions strategy based only on the NCA results, see Chapter 12.

H.8 Step 8: Interpret results

The results of the SEM study suggest that for increasing the Technology use on average, only Emotional value and Adoption intention are effective. At the same time, the NCA study shows that a high level of Technology use is only achievable when all predictors meet a certain minimum level. To evaluate the robustness of these NCA results several checks can be done (Section 9.12). Robustness checks for SEM are not discussed here. For this demonstration eight checks are selected (Table H.3). In each check, one choice is changed and the new results are compared with the original results, and it is evaluated if the conclusion regarding necessity is changed.

The type of changes in several analyst's choices regarding necessity is shown in Table H.3.

The results of these robustness checks for each individual predictor variable are shown in the following set of tables:

TABLE H.3 Selected robustness checks. The black elements are changed compared to the original

Robustness check	Original	Ceiling change	*d*-threshold change	*p*-threshold change	Scope change	Outlier change-1	Outlier change-2	Target lower	Target higher
Ceiling line	CE-FDH	CR-FDH	CE-FDH	CE-FDH	CE-FDH	CE-FDH	CE-FDH	CE-FDH	CE-FDH
Threshold effect size	0.10	0.10	0.20	0.10	0.10	0.10	0.10	0.10	0.10
Threshold p-value	0.05	0.05	0.05	0.01	0.05	0.05	0.05	0.05	0.05
Scope	empirical	empirical	empirical	empirical	theoretical	empirical	empirical	empirical	empirical
Outliers	no	no	no	no	no	yes, single	yes, multiple	no	no
Target outcome	85	85	85	85	85	85	85	80	90

Predictor: Perceived usefulness

Robustness check	Effect size	p-value	Necessity	Single bottle-necks (%)	Single bottle-necks (n)	Priority
Original	0.24	< 0.001	yes	14.9	26	1/5
Ceiling line change	0.19	0.001	yes	5.2	9	2/5
Effect size thresh. change	0.24	< 0.001	yes	14.9	26	1/5
Significance level change	0.24	< 0.001	yes	14.9	26	1/5
Scope change	0.24	< 0.001	yes	14.9	26	1/5
Outlier change 1	0.30	< 0.001	yes	6.5	11	1/5
Outlier change 2	0.29	< 0.001	yes	6.6	11	1/5
Target outcome lower	0.24	< 0.001	yes	10.3	18	1/5
Target outcome higher	0.24	< 0.001	yes	14.9	26	1/5

Predictor: Compatibility

Robustness check	Effect size	p-value	Necessity	Single bottle-necks (%)	Single bottle-necks (n)	Priority
Original	0.21	< 0.001	yes	0.6	1	4/5
Ceiling line change	0.15	< 0.001	yes	2.3	4	3/5
Effect size thresh. change	0.21	< 0.001	yes	0.6	1	4/5
Significance level change	0.21	< 0.001	yes	0.6	1	4/5
Scope change	0.21	< 0.001	yes	0.6	1	4/5
Outlier change 1	0.28	< 0.001	yes	1.2	2	5/5
Outlier change 2	0.29	< 0.001	yes	1.2	2	5/5
Target outcome lower	0.21	< 0.001	yes	1.1	2	3/5
Target outcome higher	0.21	< 0.001	yes	0.6	1	4/5

Predictor: Perceived ease of use

Robustness check	Effect size	p-value	Necessity	Single bottle-necks (%)	Single bottle-necks (n)	Priority
Original	0.24	0.015	yes	4.0	7	3/5
Ceiling line change	0.20	0.013	yes	1.7	3	4/5
Effect size thresh. change	0.24	0.015	yes	4.0	7	3/5
Significance level change	0.24	0.015	no	4.0	7	3/5
Scope change	0.36	0.015	yes	4.0	7	3/5
Outlier change 1	0.18	0.047	yes	3.0	5	3/5
Outlier change 2	0.13	0.051	no	3.0	5	3/5
Target outcome lower	0.24	0.015	yes	0.6	1	5/5

(continued)

Robustness check	Effect size	*p*-value	Necessity	Single bottle-necks (%)	Single bottle-necks (n)	Priority
Target outcome higher	0.24	0.015	yes	4.0	7	3/5

Predictor: Emotional value

Robustness check	Effect size	*p*-value	Necessity	Single bottle-necks (%)	Single bottle-necks (n)	Priority
Original	0.33	< 0.001	yes	0.0	0	5/5
Ceiling line change	0.17	0.020	yes	0.0	0	5/5
Effect size thresh. change	0.33	< 0.001	yes	0.0	0	5/5
Significance level change	0.33	< 0.001	yes	0.0	0	5/5
Scope change	0.33	< 0.001	yes	0.0	0	5/5
Outlier change 1	0.44	< 0.001	yes	2.4	4	4/5
Outlier change 2	0.44	< 0.001	yes	2.4	4	4/5
Target outcome lower	0.33	< 0.001	yes	1.1	2	3/5
Target outcome higher	0.33	< 0.001	yes	0.0	0	5/5

Predictor: Adoption intention

Robustness check	Effect size	*p*-value	Necessity	Single bottle-necks (%)	Single bottle-necks (n)	Priority
Original	0.29	< 0.001	yes	8.0	14	2/5
Ceiling line change	0.20	0.001	yes	5.7	10	1/5
Effect size thresh. change	0.29	< 0.001	yes	8.0	14	2/5
Significance level change	0.29	< 0.001	yes	8.0	14	2/5
Scope change	0.29	< 0.001	yes	8.0	14	2/5
Outlier change 1	0.38	< 0.001	yes	5.3	9	2/5
Outlier change 2	0.42	< 0.001	yes	5.4	9	2/5
Target outcome lower	0.29	< 0.001	yes	1.7	3	2/5
Target outcome higher	0.29	< 0.001	yes	8.0	14	2/5

Without discussing details, the general results of the NCA robustness tables are as follows. The results for the necessity of Perceived usefulness are robust with respect to the conclusion about necessity-in-kind. For necessity-in-degree only a change of ceiling line from CE-FDH to CR-FDH resulted in a different priority level (from first to second). Most single bottleneck cases exist for this predictor. Generally, the original results for this predictor can be considered as robust. The same conclusion applies to Compatibility. The original conclusion about necessity-in-kind is stable and for all checks the priority level for action remains low. However, the results of Perceived ease of use are more variable. For example, the statistical significance of the original finding and therefore the conclusion of necessity is somewhat fragile. Also the number of bottleneck cases changes from 7 to 1 and is sensitive to small changes of the selected target level. The results for Emotional value and Adoption intention are relatively robust.

Note that the NCA robustness table is a tool for better understanding and communicating the credibility of the original results. The findings from the checks largely depend on the selected parameters for the checks. No hard rules exist to decide about the robustness or fragility of the results, as this is a judgement by the analyst and others.

I

Bottleneck distance table

In an NCA-based intervention, the focus is on removing cases from, or moving cases into, the bottleneck area. For an effective and efficient intervention, effort should correspond to the distance of the case from the threshold level x_c of the condition that is needed for the target outcome y_c. The bottleneck distance is $x - x_c$ such that a positive distance indicates that the case is in the non-bottleneck ($x > x_c$) and a negative distance indicates that the case is in the bottleneck area ($x \leq x_c$). When the outcome is desired the goal of the intervention is to remove a case from the bottleneck area by increasing its x-value. For an undesired outcome, the goal is to move the case from the non-bottleneck area to the bottleneck area by decreasing its x-value.

Table I.1 shows the bottleneck distances for the illustrative example discussed in 12.7.2. Here the outcome is desired, so the table only shows the cases that are in the bottleneck area (the black dots in Figure 12.6). A case can be a bottleneck case for one or more conditions. The 38 cases have 64 bottlenecks (i.e., the number of times cases fall in a condition's bottleneck area).

To obtain an ideal intervention, all bottlenecks must be resolved. This means that they must receive just enough individualized intervention effort. That is, the case's x-value must increase by an amount corresponding to the bottleneck distance, such that the cases can leave the bottleneck area and are no longer constrained by this condition to achieve the target outcome. With such an intervention, the case becomes an effective case after intervention. If all bottleneck cases are treated that way for all conditions, the effectiveness and efficiency of the intervention are both 100%.

The table is organized by the total bottleneck distance from low to high. The case's total bottleneck distance is the sum of the case's bottleneck distances. This assumes that the levels of the conditions and the intervention efforts are comparable, which may not always be correct. For example, it is possible that an intervention effort on one condition is more difficult or more costly than the same intervention effort on another condition.

TABLE I.1 Bottleneck distances

Case	Perceived usefulness	Perceived ease of use	Compatibility	Emotional value	Number of bottlenecks	Total bottleneck distance
10			0		1	0.00
42		−0.02			1	−0.02
115	−0.02				1	−0.02
109			−0.03		1	−0.03
62	−0.18				1	−0.18

DOI: 10.1201/9781003430766-I

8	−0.2		−0.02		2	−0.22
111	−0.2	−0.02			2	−0.22
2	−0.26				1	−0.26
68	−0.28				1	−0.28
119	−0.28				1	−0.28
143	−0.28		0		2	−0.28
169	−0.28				1	−0.28
125	−0.3				1	−0.30
61	−0.3		−0.03		2	−0.33
151				−0.34	1	−0.34
173	−0.02		−0.33		2	−0.35
163	−0.45				1	−0.45
138	−0.57				1	−0.57
31	−0.65	−0.02			2	−0.66
141	−0.67				1	−0.67
50			−0.68		1	−0.68
78	−0.92				1	−0.92
96	−0.95		−0.03		2	−0.98
90	−1.02				1	−1.02
52	−1.3		0		2	−1.30
122			−0.35	−0.99	2	−1.33
146	−0.02		−0.35	−0.99	3	−1.35
103	−1.2	−0.34			2	−1.54
158	−0.57		0	−0.99	3	−1.56
168			−1.35	−0.33	2	−1.67
88				−1.99	1	−1.99
14	−0.65		−1.35		2	−2.00
121	−0.37			−1.99	2	−2.36
95	−1.1		−1.35		2	−2.44
9	−1.55		−1.35		2	−2.90
47	−0.02		−1.35	−1.99	3	−3.35
29	−1.92		−1.35	−1.02	3	−4.29
19	−1.92	−0.68	−1.35	−1.99	4	−5.94

Bibliography

Abbass, A., Khan, J., & Tao, M. (in press). How digital doppelgänger ignites unstoppable brand love, loyalty & word-of-mouth in the metaverse. *International Journal of Human–Computer Interaction*. https://doi.org/10.1080/10447318.2025.2564271

Abdalla, M. J., Mwesiumo, D., Öztüren, A., & Kılıç, H. (2022). Perceived threat of informal players: Enhancing the operational performance of inbound tour operators through coopetition. *International Journal of Tourism Research*, *24*(6), 775–785. https://doi.org/10.1002/jtr.2544

Abner, G., Alam, A., & Cho, H. (2023). What is necessary for high employee engagement and satisfaction? A look at the best places to work in the federal government index. *Public Administration Review*, *83*(4), 974–981. https://doi.org/10.1111/puar.13576

Acquah, I. S. K. (2023). Modelling the importance of collaborative culture and its dimensions for supply chain collaboration: A necessary condition analysis. *RAUSP Management Journal*, *58*, 125–142. https://doi.org/10.1108/RAUSP-05-2022-0153

Acquah, I. S. K. (2024). Unravelling the asymmetric effects of procurement practices on firm performance: A complexity theory approach to complementing fsQCA with NCA. *Heliyon*, *10*(3), e25230. https://doi.org/10.1016/j.heliyon.2024.e25230

Acquah, I. S. K., Issau, K., Dei Mensah, R., & Vanderpuye, F. (2023). When stakeholder orientations matter: Modelling employee orientation, shareholder orientation and supply chain orientation as necessary and sufficient conditions for firm performance. *Heliyon*, *9*, e20359. https://doi.org/10.1016/j.heliyon.2023.e20359

Acquah, I. S. K., Naude, M. J., & Oguji, N. (2025). Harnessing the dimensions of supply chain collaboration to promote innovation and flexibility in buyer-supplier relationships: Insights from fsQCA and NCA. *Technological Forecasting and Social Change*, *221*, 124363. https://doi.org/10.1016/j.techfore.2025.124363

Acquah, I. S. K., Quaicoe, J., & Arhin, M. (2023). How to invest in total quality management practices for enhanced operational performance: Findings from PLS-SEM and fsQCA. *The TQM Journal*, *35*(7), 1830–1859. https://doi.org/10.1108/TQM-05-2022-0161

Adam, I. O., & Dzang Alhassan, M. (2022). The mediating role of ICT regulation on the effects of ICT access and ICT use on e-participation: Evidence from structural equation modelling and necessary condition analysis. *African Journal of Science, Technology, Innovation and Development*, *14*(5), 1161–1172. https://doi.org/10.1080/20421338.2021.1937815

Afrin, R., & Prybutok, G. (2022). Insights into the antecedents of cyberchondria: A perspective from the USA. *Health Promotion International*, *37*(4), daac108. https://doi.org/10.1093/heapro/daac108

Agmapisarn, C., Seetha, P., & Fakfare, P. (2024). Optimizing surplus food redistribution: Leveraging digital platforms to advance sustainable consumption. *Sustainable Development*, *33*, 3831–3846. https://doi.org/10.1002/sd.3322

Aguinis, H., Ramani, R. S., & Cascio, W. F. (2020). Methodological practices in international business research: An after-action review of challenges and solutions. *Journal of International Business Studies*, *51*(9), 1593–1608. https://doi.org/10.1057/s41267-020-00353-7

Agustí-Pérez, M. A., Vázquez-Sánchez, A., & Orta-Perez, M. (2025). Identifying the necessary elements in audit quality for customer retention. *International Journal of Quality & Reliability Management*, *43*, 212–233. https://doi.org/10.1108/ijqrm-03-2025-0104

Ahsan, M. J., & Tahir, M. Z. (2025). Building resilience at work: The role of employee development practices, employee competence and employee mindfulness. *Measuring Business Excellence*, *29*, 693–714. https://doi.org/10.1108/mbe-09-2024-0140

Ahsan, M., Gyensare, M. A., Adomako, S., & Zheng, C. (2024). Chronic health issues and entrepreneurship: Exploring the role of grit, bricolage and illness uncertainty on venture performance. *International Small Business Journal: Researching Entrepreneurship*, *43*, 246–273. https://doi.org/10.1177/02662426241297642

Ajzen, I. (1991). The theory of planned behavior. *Organizational Behavior and Human Decision Processes*, *50*, 179–211. https://doi.org/10.1016/0749-5978(91)90020-T

Al Mamun, A., Reza, M. N. H., Yang, Q., & Gao, J. (2025). Modeling the significance of green management practices on food waste reduction intention among small restaurants. *Mitigation and Adaptation Strategies for Global Change*, *30*, 74. https://doi.org/10.1007/s11027-025-10267-0

Alajhar, N. A., Bajaba, S., & Basahal, A. (2025). Leading with laughter: The impact of ebullient supervision on commitment to organizational career and employee time theft via psychological distress. *The International Journal of Human Resource Management*, *36*, 2743–2784. https://doi.org/10.1080/09585192.2025.2565655

Aldhamiri, A., Carlson, J., Vilches-Montero, S., Rahman, S. M., & Gudergan, S. P. (2024). What drives higher active customer engagement in luxury brands' social media? Measurement and contingencies. *Journal of Retailing and Consumer Services*, *79*, 103804. https://doi.org/10.1016/j.jretconser.2024.103804

Allard-Poesi, F., & Massu, J. (2023). Research note: Is urban nature necessary for well-being? For whom? A necessary condition analysis. *Landscape and Urban Planning*, *234*, 104728. https://doi.org/10.1016/j.landurbplan.2023.104728

An, M., Lin, J., & Benitez, J. (2026). Effects of artificial intelligence usage and knowledge-based dynamic capabilities on organizational innovation: A configurational approach. *Decision Support Systems*, *200*, 114573. https://doi.org/10.1016/j.dss.2025.114573

Analytics, R., & Weston, S. (2022a). *Foreach: Provides foreach looping construct.* https://github.com/RevolutionAnalytics/foreach

Analytics, R., & Weston, S. (2022b). *Iterators: Provides iterator construct.* https://github.com/RevolutionAnalytics/iterators

Anand, K., Arya, V., Suresh, S., & Sharma, A. (2023). Quality dimensions of augmented reality-based mobile apps for smart-tourism and its impact on customer satisfaction & reuse intention. *Tourism Planning & Development*, *20*(2), 236–259. https://doi.org/10.1080/21568316.2022.2137577

Andrés-Sánchez, J. de, Almahameed, A. A., Arias-Oliva, M., & Pelegrín-Borondo, J. (in press). Patients' acceptance of surgical robots: Findings with machine learning and necessary condition analysis. *International Journal of Human–Computer Interaction.* https://doi.org/10.1080/10447318.2025.2575896

Andrés-Sánchez, J. de, Arias-Oliva, M., Souto-Romero, M., & Llorens-Marín, M. (2025). A combined importance-performance map and necessary condition analysis of the acceptance of blockchain use in marketing: The case of loyalty programmes. *Sage Open*, *15*, 21582440251384122. https://doi.org/10.1177/21582440251384122

Andrevski, G., & Miller, D. (2022). Forbearance: Strategic nonresponse to competitive attacks. *Academy of Management Review*, *47*(1), 59–74. https://doi.org/10.5465/amr.2018.0248

Angrist, J. D., Imbens, G. W., & Rubin, D. B. (1996). Identification of causal effects using instrumental variables. *Journal of the American Statistical Association*, *91*(434), 444–455. https://doi.org/10.1080/01621459.1996.10476902

Arcidiacono, F., Podrecca, M., & Schupp, F. (2025). Unraveling the path to smart manufacturing advancement: insights from necessary condition and fuzzy-set qualitative comparative analyses. *Industrial Management & Data Systems*, *125*, 1790–1821. https://doi.org/10.1108/imds-09-2024-0954

Arenius, P., Engel, Y., & Klyver, K. (2017). No particular action needed? A necessary condition analysis of gestation activities and firm emergence. *Journal of Business Venturing Insights*, *8*, 87–92. https://doi.org/10.1016/j.jbvi.2017.07.004

Arpaci, I., Al-Sharafi, M. A., & Mahmoud, M. A. (2026). Modeling behavioral intentions for sustainable adoption of autonomous vehicles: Theory, measurement, and predictive insights. *Transportation Research Part F: Traffic Psychology and Behaviour*, *116*, 103378. https://doi.org/10.1016/j.trf.2025.103378

Asheq, A. A., Hasan, R., & Coughlan, J. (in press). Social networking site users' behavioral responses toward travel destinations promoted by virtual influencers: A multiple-model comparison approach. *Journal of Travel Research*. https://doi.org/10.1177/00472875251337718

Aw, E. C.-X., Thomas, S., Patel, R., Bhatt, V., & Cham, T.-H. (2024). Click to contribute: Understanding donation behaviour and well-being in donation-based crowdfunding mobile apps. *International Journal of Bank Marketing*, *42*(7), 1835–1866. https://doi.org/10.1108/IJBM-11-2023-0635

Aziz, F., Li, C., & Khan, A. U. (2025). Immersive learning in virtual worlds: A two-step analysis SEM and NCA for assessing the impact of Metaverse education on knowledge retention and student collaboration. *Technology in Society*, *81*, 102871. https://doi.org/10.1016/j.techsoc.2025.102871

Baah, C., Agyabeng-Mensah, Y., Afum, E., Acquah, I. S. K., & Agyeman, D. O. (2026). Digital co-transformation, interfirm knowledge creation, and collaborative innovation: Symmetrical and asymmetrical drivers of B2B SME performance under threats. *Industrial Marketing Management*, *132*, 117–134. https://doi.org/10.1016/j.indmarman.2025.12.003

Baah, C., Saleem, M. A., Greenland, S., Tenakwah, E. S., & Chakrabarty, D. (2024). Do the theories of planned behaviour and protection motivation provide probabilistic sufficient and necessary conditions for residential water conservation? Combined use of PLS_SEM and NCA. *Journal of Environmental Management*, *372*, 123354. https://doi.org/10.1016/j.jenvman.2024.123354

Baah, C., Saleem, M. A., Greenland, S., Tenakwah, E. S., & Chakrabarty, D. (2025). Place attachment and residential water conservation: Application of an extended Theory of Planned Behaviour in Australia's Northern Territory. *Environmental Development*, *55*, 101203. https://doi.org/10.1016/j.envdev.2025.101203

Baber, H., Fanea-Ivanovici, M., Pana, M.-C., & Manole, A. M. (2024). Do price consciousness, anti-socializing behavior and content add fuel to consuming online video streaming? A multigroup analysis and moderating role of gender and marital status. *Online Information Review*, *48*(7), 1389–1411. https://doi.org/10.1108/OIR-09-2023-0444

Bachmann, N., Rose, R., Maul, V., & Hölzle, K. (2024). What makes for future entrepreneurs? The role of digital competencies for entrepreneurial intention. *Journal of Business Research*, *174*, 114481. https://doi.org/10.1016/j.jbusres.2023.114481

Bajaba, S., & Basaad, S. (2025). Safeguarding well-being under exploitative leaders: The buffering effects of follower strategies on perceived injustice. *Stress and Health*, *41*, e70127. https://doi.org/10.1002/smi.70127

Bao, Q., Xie, G., & Yao, Y. (2026). Exploring psychological dependency on romantic companion robots in the single economy: Implications for digital intimacy and consumer services. *Journal of Retailing and Consumer Services*, *90*, 104665. https://doi.org/10.1016/j.jretconser.2025.104665

Batey, M., Hughes, D. J., Crick, L., & Toader, A. (2021). Designing creative spaces: An experimental examination of the effect of a nature poster on divergent thinking. *Ergonomics*, *64*(1), 139–146. https://doi.org/10.1080/00140139.2020.1811398

Battistoni, E., Gitto, S., Murgia, G., & Campisi, D. (2023). Adoption paths of digital transformation in manufacturing SME. *International Journal of Production Economics*, *255*, 108675. https://doi.org/10.1016/j.ijpe.2022.108675

Baumgartner, M. (2009). Inferring causal complexity. *Sociological Methods & Research*, *38*(1), 71–101. https://doi-org.eur.idm.oclc.org/10.1177/0049124109339369

Bawa, R., Jain, K., & Goel, P. (in press). Generative AI adoption in universities: How TAM–TTF and neuroticism influence sustained usage. *Interactive Technology and Smart Education*. https://doi.org/10.1108/itse-02-2025-0030

Bazan, C. (2025). Embedded necessity theorizing to understand the influence of the precursors of entrepreneurial intention of university students. *Sage Open*, *15*. https://doi.org/10.1177/21582440251392964

Becker, J.-M., Cheah, J.-H., Gholamzade, R., Ringle, C. M., & Sarstedt, M. (2023). PLS-SEM's most wanted guidance. *International Journal of Contemporary Hospitality Management*, *35*(1), 321–346. https://doi.org/10.1108/IJCHM-04-2022-0474

Becker, J.-M., Richter, N. F., Ringle, C. M., & Sarstedt, M. (2026). Must-have, or maybe not? A sensitivity-based extension to necessary condition analysis. *Journal of Business Research*, *206*, 115920. https://doi.org/10.1016/j.jbusres.2025.115920

Ben Jabeur, S. (2020). The relationship between renewable energy consumption and economic growth in France: A necessary condition analysis. *Environmental Modeling & Assessment*, *25*(3), 397–409. https://doi.org/10.1007/s10666-019-09678-6

Benjamin, D. J., Berger, J. O., Johannesson, M., Nosek, B. A., Wagenmakers, E.-J., Berk, R., Bollen, K. A., Brembs, B., Brown, L., Camerer, C., et al. (2018). Redefine statistical significance. *Nature Human Behaviour*, *2*(1), 6–10. https://doi.org/10.1038/s41562-017-0189-z

Bergh, D. D., Boyd, B. K., Byron, K., Gove, S., & Ketchen Jr, D. J. (2022). What constitutes a methodological contribution? *Journal of Management*, *48*(7), 1835–1848. https://doi.org/10.1177/01492063221088235

Berkelaar, M. et al. (2023). *lpSolve: Interface to lpsolve v. 5.5 to solve linear/integer programs*. https://github.com/gaborcsardi/lpSolve

Bhatt, V., Pandey, S., Patel, R., Ghodake, S., Jariwala, H., & Thomas, S. (2025). Predicting online donation intention in donation-based crowdfunding apps: A multi-stage SEM-ANN-NCA model integrating anthropomorphism, satisfaction, trust, and privacy concerns. *Journal of Nonprofit & Public Sector Marketing*, *37*(1), 80–104. https://doi.org/10.1080/10495142.2024.2351012

Bhattacharyya, J., Balaji, M. S., & Jiang, Y. (2023). Causal complexity of sustainable consumption: Unveiling the equifinal causes of purchase intentions of plant-based meat alternatives. *Journal of Business Research*, *156*, 113511. https://doi.org/10.1016/j.jbusres.2022.113511

Bi, H., Li, A., Hua, M., Zhu, H., & Ye, Z. (2022). Examining the varying influences of built environment on bike-sharing commuting: Empirical evidence from Shanghai. *Transport Policy*, *129*, 51–65. https://doi.org/10.1016/j.tranpol.2022.10.004

Blom, R., Voorn, B., & Borst, R. T. (2024). HRM autonomy, integration and performance in government agencies: Tests of necessity and sufficiency. *Public Management Review*, *26*(5), 1223–1241. https://doi.org/10.1080/14719037.2022.2145348

Bokhorst, J. A., Knol, W., Slomp, J., & Bortolotti, T. (2022). Assessing to what extent smart manufacturing builds on lean principles. *International Journal of Production Economics, 253*, 108599. https://doi.org/10.1016/j.ijpe.2022.108599

Bokrantz, J., & Dul, J. (2023). Building and testing necessity theories in supply chain management. *Journal of Supply Chain Management, 59*, 48–65. https://doi.org/10.1111/jscm.12287

Bokrantz, J., & Skoogh, A. (2023). Adoption patterns and performance implications of smart maintenance. *International Journal of Production Economics, 256*, 108746. https://doi.org/10.1016/j.ijpe.2022.108746

Bolívar, L. M., Roldán, J. L., Castro-Abancéns, I., & Casanueva, C. (2022). Speed of international expansion: The mediating role of network resources mobilisation. *Management International Review, 62*, 541–568. https://doi.org/10.1007/s11575-022-00478-x

Boon, C., Den Hartog, D. N., & Lepak, D. P. (2019). A systematic review of human resource management systems and their measurement. *Journal of Management, 45*(6), 2498–2537. https://doi.org/10.1177/0149206318818718

Bouncken, R. B., Fredrich, V., & Kraus, S. (2020). Configurations of firm-level value capture in coopetition. *Long Range Planning, 53*(1), 101869. https://doi.org/10.1016/j.lrp.2019.02.002

Bouncken, R. B., Fredrich, V., Ritala, P., & Kraus, S. (2020). Value-creation-capture-equilibrium in new product development alliances: A matter of coopetition, expert power, and alliance importance. *Industrial Marketing Management, 90*, 648–662. https://doi.org/10.1016/j.indmarman.2020.03.019

Bouncken, R. B., Fredrich, V., Sinkovics, N., & Sinkovics, R. R. (2023). Digitalization of cross-border R&D alliances: Configurational insights and cognitive digitalization biases. *Global Strategy Journal, 13*(2), 281–314. https://doi.org/10.1002/gsj.1469

Bouncken, R. B., Ratzmann, M., Barwinski, R., & Kraus, S. (2020). Coworking spaces: Empowerment for entrepreneurship and innovation in the digital and sharing economy. *Journal of Business Research, 114*, 102–110. https://doi.org/10.1016/j.jbusres.2020.03.033

Bowden, J. L.-H., Eshaghi, M. S., Afshardoost, M., & Rahman, S. M. (2025). Metaverse customer experience: Conditional mediation of avatar self-congruity, user similarity and well-becoming. *Journal of Consumer Marketing, 42*, 940–962. https://doi.org/10.1108/jcm-10-2024-7298

Braumoeller, B. F. (2015). Guarding against false positives in qualitative comparative analysis. *Political Analysis, 23*(4), 471–487. https://doi.org/10.1093/pan/mpv017

Brazo, P., Velicia-Martín, F., Palos-Sanchez, P., & Roldán, J. L. (2024). Thrive or dive: Navigating the waves of mandatory digital transformation. *European Business Review, 37*, 486–516. https://doi.org/10.1108/EBR-03-2024-0097

Bu, Y., Li, S., & Huang, Y. (2023). Research on the influencing factors of Chinese college students' entrepreneurial intention from the perspective of resource endowment. *The International Journal of Management Education, 21*(3), 100832. https://doi.org/10.1016/j.ijme.2023.100832

Calafat-Marzal, C., Piñeiro, V., Cervera, F. J., & García-Álvarez-Coque, J. M. (2025). Unravelling the functions of knowledge communities in Spain's agri-food system: An examination of Operational Groups. *New Medit, 5*. https://doi.org/10.30682/nm2505c

Calispa-Aguilar, E. (2024). The determinants of entrepreneurship in urban and non-urban regions: A fuzzy-set QCA approach. *Entrepreneurial Business and Economics Review, 12*(2), 143–156. https://doi.org/10.15678/EBER.2024.120209

Campbell, J. T., & Aguilera, R. V. (2022). Why I Rejected Your Paper: Common Pitfalls in Writing Theory Papers and How to Avoid Them. *Academy of Management Review, 47*(4), 521–527. https://doi.org/10.5465/amr.2022.0331

Campbell, J. T., & Fiss, P. C. (2026). Tackling the complexity challenge: When and how to engage in configurational and hybrid theorizing. *Academy of Management Review, ja*, amr–2024. https://doi.org/10.5465/amr.2024.0187

Cangialosi, N., & Odoardi, C. (2025). Team personality composition: A fsQCA and NCA approach to team innovation. *Small Group Research, 56*, 697–725. https://doi.org/10.1177/10464964251314196

Cao, D., Hassan, H., & Sampene, A. K. (2025). Decoding sustainability success: The role of environmental management accounting and environmental management initiatives under institutional pressure. *Journal of Environmental Management, 392*, 126789. https://doi.org/10.1016/j.jenvman.2025.126789

Cao, H., & Zhang, Y. (2025). Let farmers speak for themselves: A configurational study on farmers' participation in digital village construction. *Sage Open, 15*. https://doi.org/10.1177/21582440251393116

Carmona, P., Stef, N., Jabeur, S. B., & Zaied, Y. B. (2023). Climate change and government policy: Fresh insights from complexity theory. *Journal of Environmental Management, 338*, 117831. https://doi.org/10.1016/j.jenvman.2023.117831

Cartwright, N. (1989). *Nature's capacities and their measurement.* Clarendon Press, Oxford.

Casadiego-Alzate, R., Arroyo-Cañada, F. J., Castañeda-Quirama, T., Sánchez-Torres, J. A., & Luzardo Briceño, M. (in press). Sufficiency and necessity of influencers for digital collective memory: A study through combined IPMA. *Journal of Modelling in Management.* https://doi.org/10.1108/jm2-03-2025-0141

Cassia, F., & Magno, F. (2024). The value of self-determination theory in marketing studies: Insights from the application of PLS-SEM and NCA to anti-food waste apps. *Journal of Business Research, 172*, 114454. https://doi.org/10.1016/j.jbusres.2023.114454

Cezar, A. (2024a). A province-level configurational analysis of fixed and mobile broadband adoption. *Journal of Organizational Computing and Electronic Commerce, 34*(1), 27–45. https://doi.org/10.1080/10919392.2024.2303942

Cezar, A. (2024b). Pathways to e-participation diffusion: A societal and governance perspective. *Telematics and Informatics, 94*, 102177. https://doi.org/10.1016/j.tele.2024.102177

Chai, X., & Wu, Q. (2025). What kind of research network configurations lead to high academic productivity for young management scholars? A fuzzy-set qualitative comparative analysis (fsQCA). *Scientometrics, 130*, 2705–2748. https://doi.org/10.1007/s11192-025-05329-8

Chaity, N. S., Islam, K. M. Z., Masud, A. A., Islam, M. F., Rahman, Md. K. H., & Kabir, S. B. (2025). Sustainable pathways to circular economy adoption: Insights from the RMG sector of Bangladesh. *Business Strategy and the Environment, 34*, 5857–5868. https://doi.org/10.1002/bse.4277

Chaparro-Banegas, N., Sánchez-Garcia, M., Calafat-Marzal, C., & Roig-Tierno, N. (2024). Transforming the agri-food sector through eco-innovation: A path to sustainability and technological progress. *Business Strategy and the Environment, 33*(8), 9075–9097. https://doi.org/10.1002/bse.3968

Charles, G., & Kanani, R. (2025). Antecedents of social commerce purchase intention: Evidence from Tanzanian social media users. *Cogent Business & Management, 12*(1), 2447409. https://doi.org/10.1080/23311975.2024.2447409

Chaurasia, S. S., Kaul, N., Yadav, B., & Shukla, D. (2020). Open innovation for sustainability through creating shared value-role of knowledge management system, openness and organizational structure. *Journal of Knowledge Management, 24*(10), 2491–2511. https://doi.org/10.1108/JKM-04-2020-0319

Chen, P.-K. A. (2026). Entrepreneurial resilience learning in higher education: The role of metaverse-constructed ecosystems. *The International Journal of Management Education, 24*, 101300. https://doi.org/10.1016/j.ijme.2025.101300

Chen, X., Wu, X., & Wang, X. (2025). Two distinct innovation narrative strategies and firm growth: A configurational approach. *International Journal of Innovation Management, 29*, 2550014. https://doi.org/10.1142/s1363919625500148

Chen, Y., & Chen, Z. (2024). Can e-government online services offer enhanced governance support? A national-level analysis based on fsQCA and NCA. *Journal of Innovation & Knowledge, 9*(3), 100526. https://doi.org/10.1016/j.jik.2024.100526

Chen, Y., & Huang, J. (in press). Breakthrough path of strategic technology in developing countries: A case study of China's lithography technology. *Technology Analysis & Strategic Management.* https://doi.org/10.1080/09537325.2025.2547823

Chen, Y., Zhao, Y., Wang, L., & Shi, J. (in press). How does digital technology innovation improve land green use efficiency? Evidence from NCA and dynamic fsQCA method. *Environment, Development and Sustainability.* https://doi.org/10.1007/s10668-025-06205-4

Cheng, W., & Jiang, Y. (2023). Falling in the same predicament again? A network embeddedness perspective of organizational failure recidivism in special treatment firms. *Management Decision, 61*(3), 815–837. https://doi.org/10.1108/MD-11-2021-1426

Cheung, M. L., Leung, W. K., Aw, E. C.-X., & Koay, K. Y. (2022). "I follow what you post!": The role of social media influencers' content characteristics in consumers' online brand-related activities (COBRAs). *Journal of Retailing and Consumer Services, 66*, 102940. https://doi.org/10.1016/j.jretconser.2022.102940

Choi, K. (2026). Capturing causal configurations of tourism destination competitiveness conditions for tourism performance: Asymmetric and symmetric application. *Journal of Destination Marketing & Management, 39*, 101042. https://doi.org/10.1016/j.jdmm.2025.101042

Choi, S.-Y., Yang, J.-J., Ahmad, M. S., & Lee, Y.-K. (2026). Identifying necessary and sufficient conditions for enhancing loyalty in hybrid electronic vehicles: A combined PLS-SEM and NCA approach. *Travel Behaviour and Society, 43*, 101192. https://doi.org/10.1016/j.tbs.2025.101192

Choksi, M., & Bhatt, V. (2024). Unveiling the future: Decrypting mobile investment resistance with multi-stage SEM-ANN-NCA approach. *International Journal of Human–Computer Interaction, 41*, 6823–6840. https://doi.org/10.1080/10447318.2024.2384990

Chong, S.-E., Lim, X.-J., Cheah, J.-H., & Tee, K. K. (2024). From screen to scene: Exploring factors influencing flow state and actual destination visitation tendencies in short travel videos. *Current Issues in Tourism, 28*, 3563–3582. https://doi.org/10.1080/13683500.2024.2400183

Chong, S., & Ham, S. (2023). Evolutionary conservation of amino acids contributing to the protein folding transition state. *Journal of Computational Chemistry, 44*(9), 1002–1009. https://doi.org/10.1002/jcc.27060

Choudhary, S., & Ahuja, Y. (in press). A nuanced picture of young consumers' motivation for using intelligent voice assistants through a curvilinear analysis lens. *Journal of Global Marketing.* https://doi.org/10.1080/08911762.2025.2511090

Christogiannis, C., Nikolakopoulos, S., Pandis, N., & Mavridis, D. (2022). The self-fulfilling prophecy of post-hoc power calculations. *American Journal of Orthodontics and Dentofacial Orthopedics, 161*(2), 315–317. https://doi.org/10.1016/j.ajodo.2021.10.008

Chu, Y., Ai, Y., Wu, Z., & Zhang, H. (2025). Study on travel choice behavior of medium and long distance intercity passengers in the context of air-rail competition. *Transport Policy, 170*, 163–177. https://doi.org/10.1016/j.tranpol.2025.05.011

Ciacci, A., Balzano, M., & Marzi, G. (2024). Optimising business models through digital alignment and strategic flexibility: Evidence from the manufacturing industry. *Journal of Management & Organization, 31*, 1477–1497. https://doi.org/10.1017/jmo.2024.1

Cinelli, C., Forney, A., & Pearl, J. (2024). A crash course in good and bad controls. *Sociological Methods & Research*, *53*(3), 1071–1104. https://doi.org/10.1177/00491241221099552

Cohen, J. (1994). The earth is round (p<. 05). *American Psychologist*, *49*(12), 997. https://doi.org/10.1037/0003-066X.49.12.997

Colino, N., Maiche, A., & Antino, M. (2025). Which cognitive factors are necessary for the manifestation of mathematical talent? A necessary condition analysis approach. *High Ability Studies*, *36*, 129–149. https://doi.org/10.1080/13598139.2025.2570654

Colpizzi, I., Pössel, P., & Marchetti, I. (2025). Unveiling the necessary conditions for depressive symptoms in adolescent girls and boys: A Necessary Condition Analysis study. *Journal of Youth and Adolescence*, *54*, 2494–2507. https://doi.org/10.1007/s10964-025-02220-w

Conde, R. (2025). Uncovering sales agents' recruitment, hiring and training practices as necessary conditions to sales agents' exceeding quota and turnover intentions. *Journal of Business & Industrial Marketing*, *41*, 160–175. https://doi.org/10.1108/jbim-06-2024-0416

Contreras Pinochet, L. H., Pardim, V. I., Mangini, H., & Souza, C. A. de. (2025). Understanding the role of shared electric scooters in São Paulo's urban micromobility landscape. *Case Studies on Transport Policy*, *20*, 101419. https://doi.org/10.1016/j.cstp.2025.101419

Corbett, J., Nishant, R., Cherki El Idrissi, S., & Chung, A. (2025). Taking things out of context: A configuration analysis of the persuasiveness of digital technologies for sustainability across work-home boundaries. *Communications of the Association for Information Systems*, *57*, 476–512. https://doi.org/10.17705/1cais.05719

Corchuelo Martínez-Azúa, B., Dias, A., & Sama-Berrocal, C. (2024). Exploring the importance of innovation ambidexterity on performance: Insights from NCA and IPMA analysis. *International Journal of Innovation Science*, *17*, 1278–1300. https://doi.org/10.1108/IJIS-03-2024-0075

Corporation, M., & Weston, S. (2022). *doParallel: Foreach parallel adaptor for the parallel package*. https://github.com/RevolutionAnalytics/doparallel

Costa, J., Gomes, C., & Linardi, M. A. (in press). Is internal R&D a necessary condition to link open innovation and innovative performance? *Innovation*. https://doi.org/10.1080/14479338.2025.2517858

Costa, S., Daher, P., Neves, P., & Velez, M. J. (2022). The interplay between ethical leadership and supervisor organizational embodiment on organizational identification and extra-role performance. *European Journal of Work and Organizational Psychology*, *31*(2), 214–225. https://doi.org/10.1080/1359432X.2021.1952988

Credé, M., & Tynan, M. C. (2021). Should language acquisition researchers study "grit"? A cautionary note and some suggestions. *Journal for the Psychology of Language Learning*, *3*(2), 37–44. https://doi.org/10.52598/jpll/3/2/3

Cumming, G. (2013). *Understanding the new statistics: Effect sizes, confidence intervals, and meta-analysis*. Routledge. https://www.taylorfrancis.com/books/mono/10.4324/9780203807002/understanding-new-statistics-geoff-cumming

Czakon, W., Klimas, P., & Kawa, A. (2023). Re-thinking strategic myopia: A necessary condition analysis of heuristic and firm's performance. *Industrial Marketing Management*, *115*, 99–109. https://doi.org/10.1016/j.indmarman.2023.09.015

Dabić, M., Marzi, G., Vlačić, B., Daim, T. U., & Vanhaverbeke, W. (2021). 40 years of excellence: An overview of technovation and a roadmap for future research. *Technovation*, *106*, 102303. https://doi.org/10.1016/j.technovation.2021.102303

Damberg, S., Liu, Y., & Ringle, C. M. (2024). Does culture matter? Corporate reputation and sustainable satisfaction in the Chinese and German banking sector. *Journal of Marketing Analytics*, *12*(1), 6–24. https://doi.org/10.1057/s41270-023-00259-x

Dang, S., Quach, S., & Roberts, R. E. (in press). Human-generative AI interaction: Assessing futureproofing attributes from time perspectives and temporal distances. *International Journal of Human–Computer Interaction.* https://doi.org/10.1080/10447318.2025.2556233

Darwati, I., Yusron, M., Purnomo, J., Suryadi, R., Rusmin, D., Trisilawati, O., Yatno, E., Pitono, J., Syakir, M., Fidiyawati, E., & Hartati, R. S. (2026). Integrating SEM-PLS and NCA to reveal the mediating role of leaf nutrient status in linking soil nutrient availability and oil palm yield in peatlands. *Scientia Agropecuaria, 17*, 165–177. https://doi.org/10.17268/sci.agropecu.2026.012

Das, A., Kondasani, R. K. R., Deb, R., & Batabyal, D. (in press). Leveraging user-generated content (UGC) in pre-travel behavior: A PLS-SEM and necessary conditions analysis (NCA) approach. *Tourism and Hospitality Research.* https://doi.org/10.1177/14673584241292254

Dávila, G. A., Andreeva, T., & Sætre, A. S. (2024). Using social integration mechanisms to boost absorptive capacity: Walking a tightrope. *European Management Review, 22*, 619–639. https://doi.org/10.1111/emre.12667

De Andrés-Sánchez, J., & Gené-Albesa, J. (2024). Drivers and necessary conditions for chatbot acceptance in the insurance industry. Analysis of policyholders' and professionals' perspectives. *Journal of Organizational Computing and Electronic Commerce, 35*, 189–216. https://doi.org/10.1080/10919392.2024.2435118

De Sousa, T. M. F., Bizarrias, F. S., Raposo, L. L., & Cruz, D. L. da. (in press). When AI becomes afterlife: Psychological factors driving acceptance of thanabots services. *Journal of Services Marketing.* https://doi.org/10.1108/jsm-01-2025-0052

De Vries, J., De Koster, R., Rijsdijk, S., & Roy, D. (2017). Determinants of safe and productive truck driving: Empirical evidence from long-haul cargo transport. *Transportation Research Part E: Logistics and Transportation Review, 97*, 113–131. https://doi.org/10.1016/j.tre.2016.11.003

Dechaume, M., Mercier, M., Feybesse, C., Lubart, T., Chouvelon, G., Kermarrec, S., & Tordjman, S. (2024). Is intelligence necessary and sufficient for creativity? An analysis of convergent and divergent thinking. *Learning and Individual Differences, 116*, 102575. https://doi.org/10.1016/j.lindif.2024.102575

Deist, M. K., McDowell, W. C., & Bouncken, R. B. (2023). Digital units and digital innovation: Balancing fluidity and stability for the creation, conversion, and dissemination of sticky knowledge. *Journal of Business Research, 161*, 113827. https://doi.org/10.1016/j.jbusres.2023.113827

Del Sordo, E., & Zattoni, A. (2025). The role of employee ownership, financial participation, and decision-making in corporate governance: A multilevel review and research agenda. *Corporate Governance: An International Review, 33*(3), 529–549. https://doi.org/10.1111/corg.12614

Delgosha, M. S., Saheb, T., & Hajiheydari, N. (2021). Modelling the asymmetrical relationships between digitalisation and sustainable competitiveness: A cross-country configurational analysis. *Information Systems Frontiers, 23*, 1317–1337. https://doi.org/10.1007/s10796-020-10029-0

Della Corte, V., Aria, M., Del Gaudio, G., Barney, J. B., Cobanoglu, C., & Sepe, F. (2021). The relevance of relational capabilities in collaborative decisions. *International Journal of Contemporary Hospitality Management, 33*(12), 4391–4417. https://doi.org/10.1108/IJCHM-01-2021-0037

Delladio, S., Caputo, A., Magrini, A., & Pellegrini, M. M. (2023). Italian entrepreneurial decision-making under lockdown: The path to resilience. *Management Decision, 61*(13), 272–294. https://doi.org/10.1108/MD-12-2022-1746

Deng, Q., Zhuang, G., Li, S., & Yang, H. (2022). A configurational analysis of cross-channel integration. *Industrial Management & Data Systems*, *122*(7), 1686–1706. https://doi.org/10.1108/IMDS-06-2021-0388

Deprins, D., Simar, L., & Tulkens, H. (1984). *Measuring labor-efficiency in post offices. In M. Marchand, P. Pestieau and H. Tulkens (eds.), The Performance of Public Enterprises: Concepts and Measurement* (pp. 243–267). Amsterdam, North Holland. https://link.springer.com/chapter/10.1007/978-0-387-25534-7_16

Díez-de-Castro, E., Vázquez-Sánchez, A., & Agustí-Pérez, M. (in press). Legitimacy of mature microenterprises: essential dimensions for their stability and growth. *European Journal of Management and Business Economics.* https://doi.org/10.1108/ejmbe-01-2025-0031

Dijkstra, T. K., & Henseler, J. (2015). Consistent partial least squares path modeling. *MIS Quarterly*, *39*(2), 297–316. https://www.jstor.org/stable/10.2307/26628355

Ding, H. (2022). What kinds of countries have better innovation performance?–a country-level fsQCA and NCA study. *Journal of Innovation & Knowledge*, *7*(4), 100215. https://doi.org/10.1016/j.jik.2022.100215

Ding, H., & Kuvaas, B. (2023). Using necessary condition analysis in managerial psychology research: Introduction, empirical demonstration and methodological discussion. *Journal of Managerial Psychology*, *38*(4), 260–272. https://doi.org/10.1108/JMP-12-2022-0637

Ding, H., & Kuvaas, B. (2025). Exploring the necessary roles of basic psychological needs at work: A necessary condition analysis. *Journal of Occupational and Organizational Psychology*, *98*(1), e70012. https://doi.org/10.1111/joop.70012

Ding, X., Wang, L., & Tan, Y. (2026). Influence pathways of village livelihood capital on the idle homesteads reuse. *Land Use Policy*, *160*, 107834. https://doi.org/10.1016/j.landusepol.2025.107834

Dinh, E., Mühlbacher, H., & Torchia, M. (2024). Necessary perceptions of family values and luxury characteristics for brand luxuriousness: Evidence from luxury watch brands. *Journal of Brand Management*, *31*(1), 58–78. https://doi.org/10.1057/s41262-023-00336-7

Dionisio, E. A., Inacio Junior, E., Morini, C., & Carvalho, R. de Q. (2023). Identifying necessary conditions to deep-tech entrepreneurship. *RAUSP Management Journal*, *58*, 162–185. https://doi.org/10.1108/RAUSP-09-2022-0203

Do Prado, N. B., & De Moraes, G. H. S. M. (2024). An inquiry into the linkages between entrepreneurial perspectives and sustainable value creation in knowledge-intensive entrepreneurship. *The Journal of Technology Transfer*, *50*, 2076–2105. https://doi.org/10.1007/s10961-024-10167-6

Do, T. Q., Tho, N. D., & Le, N.-H. (2025). Learning generatively to facilitate marketing innovation of international joint ventures in emerging markets. *International Journal of Emerging Markets*, *20*, 3955–3976. https://doi.org/10.1108/IJOEM-02-2023-0171

Dong, Z., Jiang, P., Dai, Z., & Chi, R. (2024). What kind of urban brand ecology attracts talent best? Grey configuration analysis of 98 Chinese cities. *Grey Systems: Theory and Application*, *14*(4), 770–788. https://doi.org/10.1108/GS-03-2024-0035

Dos Santos Leite Gonella, J., Filho, M. G., Ganga, G. M. D., & Lizarelli, F. L. (2024). From awareness to action: Understanding the relationship between circular economy and favourable evaluation towards sustainable development. *Business Strategy and the Environment*, *33*(8), 8679–8694. https://doi.org/10.1002/bse.3934

Dou, Y., Hou, W., Yan, X., Jin, X., & Li, P. (2024). Optimizing policy mixes for promoting prefabricated construction in developing countries: An fsQCA approach. *Engineering, Construction and Architectural Management*, *33*, 810–831. https://doi.org/10.1108/ECAM-04-2024-0525

Du, Y., Kim, P. H., Fan, D., & Su, Y. (2025). Zooming in and out for strategic advantage: Slack configurations and R&D investments in high-growth firms. *Journal of Business Research*, *201*, 115729. https://doi.org/10.1016/j.jbusres.2025.115729

Duan, X., Si, H., & Xia, X. (2025). Understanding the non-users acceptability of autonomous vehicle hailing services using SEM-ANN-NCA approach. *Transportation Research Part F: Traffic Psychology and Behaviour*, *110*, 211–229. https://doi.org/10.1016/j.trf.2025.02.018

Duan, X., & Xiang, P. (2025). Exploring public acceptability of Mobility-as-a-Service (MaaS): A complementary dual-stage SEM-NCA perspective. *Transport Policy*, *174*, 103847. https://doi.org/10.1016/j.tranpol.2025.103847

Duarte, P., Meneses, R., Silva, S. C., & Tharakan, R. R. (2025). Cultural contrasts in vegan food choice: A multi-methods comparative analysis of consumption values in Portugal and India. *International Journal of Gastronomy and Food Science*, *42*, 101310. https://doi.org/10.1016/j.ijgfs.2025.101310

Dul, J. (2016a). Identifying single necessary conditions with NCA and fsQCA. *Journal of Business Research*, *69*(4), 1516–1523. https://doi.org/10.1016/j.jbusres.2015.10.134

Dul, J. (2016b). Necessary Condition Analysis (NCA): Logic and methodology of "necessary but not sufficient" causality. *Organizational Research Methods*, *19*(1), 10–52. https://doi.org/10.1177/1094428115584005

Dul, J. (2020). *Conducting Necessary Condition Analysis*. Sage. https://uk.sagepub.com/en-gb/eur/conducting-necessary-condition-analysis-for-business-and-management-students/book262898

Dul, J. (2021). *Advances in Necessary Condition Analysis*. online book. https://bookdown.org/ncabook/advanced_nca2/

Dul, J. (2022). Problematic applications of Necessary Condition Analysis (NCA) in tourism and hospitality research. *Tourism Management*, *93*, 104616. https://doi.org/10.1016/j.tourman.2022.104616

Dul, J. (2024a). A different causal perspective with Necessary Condition Analysis. *Journal of Business Research*, *177*, 114618. https://doi.org/10.1016/j.jbusres.2024.114618

Dul, J. (2024b). How to sample in Necessary Condition Analysis (NCA). *European Journal of International Management*, *23*(1), 1–12. https://doi.org/10.1504/EJIM.2024.138446

Dul, J. (2025). Identifying the perfect predictor of absence of disease: A shift toward necessary condition analysis in evidence-based medicine? In *Journal of the American Academy of Child and Adolescent Psychiatry* (pp. S0890–8567). https://doi.org/10.1016/j.jaac.2024.11.020

Dul, J., & Buijs, G. (2026). *NCA* (Version 5.0.2) [Computer software]. The Comprehensive R Archive Network. https://cran.r-project.org/web/packages/NCA/index.html

Dul, J., & Hak, T. (2008). *Case Study Methodology in Business Research*. Routledge. https://www.routledge.com/Case-Study-Methodology-in-Business-Research/Dul-Hak/p/book/9780750681964

Dul, J., Hak, T., Goertz, G., & Voss, C. (2010). Necessary condition hypotheses in operations management. *International Journal of Operations & Production Management*, *30*(11), 1170–1190. https://doi.org/10.1108/01443571011087378

Dul, J., Hauff, S., & Bouncken, R. B. (2023). Necessary C"ondition Analysis (NCA): Review of research topics and guidelines for good practice. *Review of Managerial Science*, *17*, 683–714. https://doi.org/10.1007/s11846-023-00628-x

Dul, J., Hauff, S., & Tóth, Z. (2021). Necessary Condition Analysis in marketing research. In R. Nunkoo, V. Teeroovengadum, & C. Ringle (Eds.), *Handbook of research methods for marketing management* (pp. 51–72). Edward Elgar Publishing. https://www.e-elgar.com/shop/gbp/handbook-of-research-methods-for-marketing-management-9781788976947.html

Dul, J., Karwowski, M., & Kaufman, J. C. (2020). Necessary Condition Analysis in creativity research. In V. Dörfler & M. Stierand (Eds.), *Handbook of research methods on creativity* (pp. 351–368). Edward Elgar Publishing. https://www.e-elgar.com/shop/gbp/handbook-of-research-methods-on-creativity-9781786439642.html

Dul, J., Laan, E. van der, Kuik, R., & Karwowski, M. (2019). Necessary condition analysis: Type i error, power, and over-interpretation of test results. A reply to a comment on NCA. Commentary: Predicting the significance of necessity. *Frontiers in Psychology, 10*, 1493. https://doi.org/10.3389/fpsyg.2019.01493

Dul, J., Raaij, E. van, & Caputo, A. (2024). Advancing scientific inquiry through data reuse: Necessary Condition Analysis with archival data. *Strategic Change, 33*(1), 35–40. https://doi.org/10.1002/jsc.2562

Dul, J., Van der Laan, E., & Kuik, R. (2020). A statistical significance test for Necessary Condition Analysis. *Organizational Research Methods, 23*(2), 385–395. https://doi.org/10.1177/1094428118795272

Dul, J., Vis, B., & Goertz, G. (2021). Necessary Condition Analysis (NCA) does exactly what it should do when applied properly: A reply to a comment on NCA. *Sociological Methods & Research, 50*(2), 926–936. https://doi.org/10.1177/0049124118799383

Dutta, D. S., Thomas, A., Shiva, A., Papa, A., & Cuomo, M. T. (2024). The hustle behind knowledge: Role of workplace ostracism and knowledge hiding towards quiet quitting in knowledge-intensive organisations. *Journal of Knowledge Management, 28*(10), 2911–2939. https://doi.org/10.1108/JKM-01-2024-0035

Eccarius, T., & Chen, C.-F. (2024). Examining trust as a critical factor for the adoption of electric vehicle sharing via necessary condition analysis. *Technological Forecasting and Social Change, 208*, 123681. https://doi.org/10.1016/j.techfore.2024.123681

Eggers, F., Risselada, H., Niemand, T., & Robledo, S. (2022). Referral campaigns for software startups: The impact of network characteristics on product adoption. *Journal of Business Research, 145*, 309–324. https://doi.org/10.1016/j.jbusres.2022.03.007

Einstein, A. (1934). On the method of theoretical physics. *Philosophy of Science, 1*(2), 163–169. https://doi.org/10.1086/286316

El Manzani, Y., & El Idrissi, M. (in press). Big data analytics capabilities and green innovation: A meta-analysis and necessary condition analysis. *Management Review Quarterly*. https://doi.org/10.1007/s11301-025-00530-8

Emmenegger, P. (2011). Job security regulations in Western democracies: A fuzzy set analysis. *European Journal of Political Research, 50*(3), 336–364. https://doi.org/10.1111/j.1475-6765.2010.01933.x

Erdmann, A., & Toro-Dupouy, L. (2025). The influence of the institutional environment on AI adoption in universities: Identifying value drivers and necessary conditions. *European Journal of Innovation Management, 28*, 4365–4398. https://doi.org/10.1108/ejim-04-2024-0407

Escadas, M., Jalali, M. S., Septianto, F., & Farhangmehr, M. (2024). Are emotions essential for consumer ethical decision-making: A necessary condition analysis. *Business Ethics, the Environment & Responsibility, 33*(3), 468–485. https://doi.org/10.1111/beer.12519

Faiella, A., Zielińska, A., Karwowski, M., & Corazza, G. E. (2025). Am I still creative? The effect of artificial intelligence on creative self-beliefs. *The Journal of Creative Behavior, 59*. https://doi.org/10.1002/jocb.70011

Fainshmidt, S., Witt, M. A., Aguilera, R. V., & Verbeke, A. (2020). The contributions of qualitative comparative analysis (QCA) to international business research. *Journal of International Business Studies, 51*(4), 455–466. https://doi.org/10.1057/s41267-020-00313-1

Fakfare, P., Manosuthi, N., Lee, J.-S., Han, H., Jarumaneerat, T., & Kim, J. J. (2024). Marine tourism and environmental preservation: Determinants of tourists' ecofriendly behavior and promotion through word-of-mouth. *Asia Pacific Journal of Tourism Research, 29*(7), 769–789. https://doi.org/10.1080/10941665.2024.2350411

Fakfare, P., Manosuthi, N., Lee, J.-S., Han, H., & Jin, M. (2025). Customer word-of-mouth for generative AI: Innovation and adoption in hospitality and tourism. *International Journal of Hospitality Management, 126*, 104070. https://doi.org/10.1016/j.ijhm.2024.104070

Fakfare, P., Manosuthi, N., Lee, J.-S., Han, H., & Jin, M. (in press). Exploring the drivers of hospitality and tourism customer loyalty for generative artificial intelligence (AI): A multi-analytic approach. *Current Issues in Tourism*. https://doi.org/10.1080/13683500.2025.2500726

Fakfare, P., Manosuthi, N., Lee, J.-S., Jin, M., Han, H., & Kim, J. J. (in press). Data vulnerability and privacy risk among hotel guests who share personal data. *Journal of Vacation Marketing*. https://doi.org/10.1177/13567667241276213

Fakfare, P., Manosuthi, N., Lee, J.-S., Lee, S.-M., & Han, H. (2024). Investigating the formation of ethical animal-related tourism behaviors: A self-interest and pro-social theoretic approach. *Journal of Hospitality & Tourism Research, 49*, 581–599. https://doi.org/10.1177/10963480241229237

Farmaki, A., Pappas, N., Stergiou, D., Apostolakis, A., & Camilleri, M. A. (in press). Green accommodation choices in the sharing economy. *Tourism Review*. https://doi.org/10.1108/tr-05-2025-0497

Fiss, P. C. (2011). Building better causal theories: A fuzzy set approach to typologies in organization research. *Academy of Management Journal, 54*(2), 393–420. https://doi.org/10.5465/amj.2011.60263120

Foroughi, B., Huy, T. Q., Iranmanesh, M., Ghobakhloo, M., Rejeb, A., & Nikbin, D. (2024). Why users continue e-commerce chatbots? Insights from PLS-fsQCA-NCA approach. *The Service Industries Journal, 45*, 935–965. https://doi.org/10.1080/02642069.2024.2371910

Franke, H., & Foerstl, K. (2020). Goals, conflict, politics, and performance of cross-functional sourcing teams—results from a social team experiment. *Journal of Business Logistics, 41*(1), 6–30. https://doi.org/10.1111/jbl.12225

Frazier, M. L., Fainshmidt, S., Klinger, R. L., Pezeshkan, A., & Vracheva, V. (2017). Psychological safety: A meta-analytic review and extension. *Personnel Psychology, 70*(1), 113–165. https://doi.org/10.1111/peps.12183

Fredrich, V., Bouncken, R. B., & Kraus, S. (2019). The race is on: Configurations of absorptive capacity, interdependence and slack resources for interorganizational learning in coopetition alliances. *Journal of Business Research, 101*, 862–868. https://doi.org/10.1016/j.jbusres.2018.11.038

Fredrich, V., Bouncken, R. B., & Tiberius, V. (2022). Dyadic business model convergence or divergence in alliances?–a configurational approach. *Journal of Business Research, 153*, 300–308. https://doi.org/10.1016/j.jbusres.2022.08.046

Frezatti, F., Mucci, D. M., & Bido, D. de S. (2022). Structure of planning and control artifacts and their accuracy in Brazilian family businesses. *BAR-Brazilian Administration Review, 19*(3). https://doi.org/10.1590/1807-7692bar2022210080

Frommeyer, B., Wagner, E., Hossiep, C. R., & Schewe, G. (2022). The utility of intention as a proxy for sustainable buying behavior–a necessary condition analysis. *Journal of Business Research, 143*, 201–213. https://doi.org/10.1016/j.jbusres.2022.01.041

Fu, S., Hu, X., Zhang, C., & Li, Z. (2025). A study on the influence of production and environmental information transparency on online consumers' purchase intention of green agricultural products. *British Food Journal, 127*, 1461–1479. https://doi.org/10.1108/bfj-09-2024-0889

Fujita, T., & Kusano, H. (2020). Denial or history? Yasukuni visits as signaling. *Journal of East Asian Studies*, *20*(2), 291–316. https://doi.org/10.1017/jea.2020.2

Gabay-Mariani, L., Bastian, B., Caputo, A., & Pappas, N. (2024). Hidden stories and the dark side of entrepreneurial commitment. *International Journal of Entrepreneurial Behavior & Research*, *30*(6), 1553–1575. https://doi.org/10.1108/IJEBR-03-2023-0248

Galindo-Martin, M.-A., Mendez-Picazo, M.-T., & Perez-Pujol, R.-. S. (2025). Open innovation and sustainable development: A micro and macroeconomic analysis using a mixed method research with PLS-SEM-NCA and Delphi. *International Journal of Information Management*, *82*, 102874. https://doi.org/10.1016/j.ijinfomgt.2025.102874

Galton, F. (1886). Regression towards mediocrity in hereditary stature. *The Journal of the Anthropological Institute of Great Britain and Ireland*, *15*, 246–263. https://www.jstor.org/stable/2841583

Gangakhedkar, R., & Khan, M. (2025). Antecedents of electricity saving intentions and behaviour of generation Y and Z of India – a combined use of structural equation modeling and necessary condition analysis. *Energy Efficiency*, *18*. https://doi.org/10.1007/s12053-025-10349-7

Gans, J., & Stern, S. (2003). *Assessing australia's innovative capacity on the 21st century* [Report]. Intellectual Property Research Institute of Australia (IPRIA). https://melbourneinstitute.unimelb.edu.au/outlook/assets/2003/JoshuaGans.pdf

Gantert, T. M., Fredrich, V., Bouncken, R. B., & Kraus, S. (2022). The moral foundations of makerspaces as unconventional sources of innovation: A study of narratives and performance. *Journal of Business Research*, *139*, 1564–1574. https://doi.org/10.1016/j.jbusres.2021.10.076

Gao, L., Luo, X., Wang, Y., Zhang, N., & Deng, X. (2024). Retention in challenging international construction assignments: Role of expatriate resilience. *Journal of Construction Engineering and Management*, *150*(2), 04023158. https://doi.org/10.1061/JCEMD4.COENG-13760

Gao, X., Wang, X., Sethanan, K., & Tseng, M.-L. (in press). Low-carbon transformation path in energy-intensive industries in China: A multilevel perspective and configuration effects. *Journal of Industrial and Production Engineering*. https://doi.org/10.1080/21681015.2025.2558709

Garg, A., Wadhwa, D., Gomes, N. M., & Sahasrabudhe, S. (in press). Effects of intervention implemented in MOOC discussion forums: contributing factors and participation analysis. *Education and Information Technologies*. https://doi.org/10.1007/s10639-025-13754-8

Garg, N. (2017). Workplace spirituality and employee well-being: An empirical exploration. *Journal of Human Values*, *23*(2), 129–147. https://doi.org/10.1177/0971685816689741

Garg, N. (2020a). Exploring role of gratitude in developing teacher leadership in Indian universities. *International Journal of Educational Management*, *34*(5), 881–901. https://doi.org/10.1108/IJEM-07-2019-0253

Garg, N. (2020b). Promoting organizational performance in Indian insurance industry: The roles of workplace spirituality and organizational citizenship behaviour. *Global Business Review*, *21*(3), 834–849. https://doi.org/10.1177/0972150918778983

Garg, N., & Sarkar, A. (2020). Vitality among university students: Exploring the role of gratitude and resilience. *Journal of Organizational Effectiveness: People and Performance*, *7*(3), 321–337. https://doi.org/10.1108/JOEPP-03-2020-0041

Garg, S., Ahmad, A., & Madsen, D. Ø. (2025). Understanding behavioral intentions in E-waste management: A combined PLS-SEM, NCA, and IPMA approach. *Cleaner and Responsible Consumption*, *19*, 100348. https://doi.org/10.1016/j.clrc.2025.100348

Gauss, C. (1809). *Theoria motus. English translation (1963): Theory of the motion of the heavenly bodies about the sun in the conic sections*. Dover, New York.

Gavín-Chocano, Ó., García-Martínez, I., & De La Rosa, A. L. (2024). Self-regulated learning as an inherent factor of academic self-concept in university students. *Current Psychology, 43*(34), 27469–27480. https://doi.org/10.1007/s12144-024-06352-3

Gavín-Chocano, Ó., Luque de la Rosa, A., & García-Martínez, I. (2025). Análisis de las condiciones necesarias de los estados de ánimo para el desarrollo de estrategias de aprendizaje en estudiantes universitarios – Analysis of the necessary conditions of mood states for the development of learning strategies in university students. *Psychology, Society & Education, 17*, 42–53. https://doi.org/10.21071/pse.v17i3.18117

Glück, J., & Scherpf, A. (2022). Intelligence and wisdom: Age-related differences and nonlinear relationships. *Psychology and Aging, 37*, 649–666. https://doi.org/10.1037/pag0000692

Glück, J., Weststrate, N. M., & Scherpf, A. (2022). Looking beyond linear: A closer examination of the relationship between wisdom and wellbeing. *Journal of Happiness Studies, 23*, 3285–3313. https://doi.org/10.1007/s10902-022-00540-3

Goertz, G., Hak, T., & Dul, J. (2013). Ceilings and floors: Where are there no observations? *Sociological Methods & Research, 42*(1), 3–40. https://doi.org/10.1177/0049124112460375

Goertz, G., & Mahoney, J. (2012). *A tale of two cultures: Qualitative and quantitative research in the social sciences.* Princeton University Press. https://press.princeton.edu/books/hardcover/9780691149707/a-tale-of-two-cultures

Goldstein, S. (2012). Typicality and notions of probability in physics. *Probability in Physics*, 59–71. https://link.springer.com/chapter/10.1007/978-3-642-21329-8_4

Golini, R., Deflorin, P., & Scherrer, M. (2016). Exploiting the potential of manufacturing network embeddedness: An OM perspective. *International Journal of Operations & Production Management, 36*(12), 1741–1768. https://doi.org/10.1108/IJOPM-11-2014-0559

Gollagari, R., Birega, T., & Mishra, S. S. (2024). Organizational justice, job satisfaction and academic rank: A moderating mediation study on employee commitment in Ethiopian public universities. *African Journal of Economic and Management Studies, 15*(2), 300–317. https://doi.org/10.1108/AJEMS-02-2023-0047

Gomes, G., Tontini, G., & Picolo, J. D. (2025). Unveiling the drivers of customer satisfaction and loyalty in cinema theatres: Integrating sufficiency and necessity perspectives. *International Journal of Services and Operations Management, 1*(1), 10060830. https://doi.org/10.1504/IJSOM.2023.10060830

Gómez-Gómez, C., Bustos-Gutierrez, M., Díaz Tautiva, J. A., & Quezada-Baier, C. (2025). Unseen barriers, unspoken limits: A necessary condition analysis of women's career advancement. *Gender Issues, 42*. https://doi.org/10.1007/s12147-025-09380-w

Gong, Y., Wu, Z., Tan, Y., He, X., & Lin, L. (in press). Antecedents of consumers' approach behavior in e-commerce information encountering: A configuration analysis. *Electronic Commerce Research.* https://doi.org/10.1007/s10660-025-10065-5

Gong, Z., Wang, Y., & Li, M. (2024). Determining the drivers of global innovation under COVID-19: An fsQCA approach. *PLOS ONE, 19*(2), e0295403. https://doi.org/10.1371/journal.pone.0295403

Gordon, J. A., Balta-Ozkan, N., Haq, A. U., & Nabavi, S. A. (2024a). Heterogeneous preferences for living in a hydrogen home: An advanced multigroup analysis. *Sustainable Energy & Fuels, 8*(12), 2601–2648. https://doi.org/10.1039/D4SE00392F

Gordon, J. A., Balta-Ozkan, N., Haq, A. U., & Nabavi, S. A. (2025). Fuelling hydrogen futures? A trust-based model of social acceptance. *Sustainable Energy & Fuels, 9*, 2510–2555. https://doi.org/10.1039/d4se01615g

Gordon, J. A., Balta-Ozkan, N., Haq, A., & Nabavi, S. A. (2024b). Necessary and sufficient conditions for deploying hydrogen homes: A consumer-oriented perspective.

International Journal of Hydrogen Energy, *69*, 982–1021. https://doi.org/10.1016/j.ijhydene.2024.04.352

Gorgievski, M. J., Bakker, A. B., Petrou, P., & Gawke, J. C. L. (2023). Antecedents of employee intrapreneurship in the public sector: A proactive motivation approach. *International Public Management Journal*, *26*(6), 852–873. https://doi.org/10.1080/1096749 4.2023.2255172

Greckhamer, T. (2016). CEO compensation in relation to worker compensation across countries: The configurational impact of country-level institutions. *Strategic Management Journal*, *37*(4), 793–815. https://doi.org/10.1002/smj.2370

Greco, A. M., Guilera, G., Maldonado-Murciano, L., Gómez-Benito, J., & Barrios, M. (2022). Proposing necessary but not sufficient conditions analysis as a complement of traditional effect size measures with an illustrative example. *International Journal of Environmental Research and Public Health*, *19*(15), 9402. https://doi.org/10.3390/ijerph19159402

Greene, W. H. (2012). *Econometric analysis. 7th. edition.* Pearson Education India.

Greenland, S., & Robins, J. M. (1986). Identifiability, exchangeability, and epidemiological confounding. *International Journal of Epidemiology*, *15*(3), 413–419. https://doi.org/10.1093/ije/15.3.413

Greenland, S., Senn, S. J., Rothman, K. J., Carlin, J. B., Poole, C., Goodman, S. N., & Altman, D. G. (2016). Statistical tests, p values, confidence intervals, and power: A guide to misinterpretations. *European Journal of Epidemiology*, *31*, 337–350. https://doi.org/10.1007/s10654-016-0149-3

Gu, J., Wu, C., Wu, X., He, R., Tao, J., Ye, W., Wu, P., Hao, M., & Qiu, W. (2022). Configurations for positive public behaviors in response to the COVID-19 pandemic: A fuzzy set qualitative comparative analysis. *BMC Public Health*, *22*, 1692. https://doi.org/10.1186/s12889-022-14097-6

Guenther, P., Guenther, M., Ringle, C. M., Zaefarian, G., & Cartwright, S. (2023). Improving PLS-SEM use for business marketing research. *Industrial Marketing Management*, *111*, 127–142. https://doi.org/10.1016/j.indmarman.2023.03.010

Guevara, M., Medina-Molina, C., Pérez-Macías, N., & Fabra Florit, M. E. (2025). In the face of difficulty: planning? Identifying the causes of success in project management from a multilevel perspective. *Quality & Quantity*, *59*, 2961–2989. https://doi.org/10.1007/s11135-025-02094-x

Guilford, J. P. (1967). *The nature of human intelligence.* McGraw-Hill. https://www.worldcat.org/title/nature-of-human-intelligence/oclc/204270

Guo, J., Fu, Y., & Sun, X. (2023). Green innovation efficiency and multiple paths of urban sustainable development in China: Multi-configuration analysis based on urban innovation ecosystem. *Scientific Reports*, *13*(1), 12975. https://doi.org/10.1038/s41598-023-40084-x

Guo, R., Wang, J., & Li, X. (2025). Intersectional analysis of discrimination on perceived job insecurity among young Chinese women—based on FsQCA method. *Scientific Reports*, *15*. https://doi.org/10.1038/s41598-025-09997-7

Gutiérrez-Broncano, S., Rubio-Andrés, M., Linuesa Langreo, J., & Sastre-Castillo, M. A. (2024). Hybrid strategy and its effect on the market performance of family-business SMEs: The mediating role of innovation. *International Journal of Organizational Analysis*, *33*, 848–878. https://doi.org/10.1108/IJOA-09-2023-4008

Gyensare, M. A., Miri, D., Zahoor, N., & Alajaty, M. (2024). Aspiring to go abroad: How and when international entrepreneurial aspiration fuel emerging markets entrepreneurial ventures' internationalisation speed. *Journal of International Management*, *30*(2), 101130. https://doi.org/10.1016/j.intman.2024.101130

Han, G., & Zhang, Z. (2022). Regional innovation paths selection in complex environments of China: A configurational perspective. *Managerial and Decision Economics*, *43*(6), 2596–2606. https://doi.org/10.1002/mde.3548

Harding, D. J., Fox, C., & Mehta, J. D. (2002). Studying rare events through qualitative case studies: Lessons from a study of rampage school shootings. *Sociological Methods & Research*, *31*(2), 174–217. https://doi.org/10.1177/0049124102031002003

Harun, M. D., Hogset, H., & Mwesiumo, D. (2023). Dynamic capabilities and sustainability performance: Exploring the moderating role of environmental dynamism in the Norwegian fishing industry. *Sustainable Development*, *31*(4), 2636–2655. https://doi.org/10.1002/sd.2536

Harun, M., Salema, G., & Matto, M. (2025). Achieving desired compliance performance in public procurement: Does embedding a culture of performance orientation matter? *Future Business Journal*, *11*. https://doi.org/10.1186/s43093-025-00444-7

Hassan, M. S., Mai, N. H., Wahab, N. S. A., Amin, M. B., Hassan, M. M., & Oláh, J. (2025). Decentralized fintech platforms adoption intention in cyber risk environment among GenZ: A dual-method approach using PLS-SEM and necessary condition analysis. *Computers in Human Behavior Reports*, *18*, 100687. https://doi.org/10.1016/j.chbr.2025.100687

Hasselwander, M. (2025). Who wants to use transport super apps? Insights from combined PLS-SEM and NCA methods. *Transportation Research Part F: Traffic Psychology and Behaviour*, *113*, 1–13. https://doi.org/10.1016/j.trf.2025.04.018

Hasselwander, M., Sunio, V., Lah, O., & Mogaji, E. (2026). Toward agentic AI: User acceptance of a deeply personalized AI super assistant (AISA). *Journal of Retailing and Consumer Services*, *89*, 104620. https://doi.org/10.1016/j.jretconser.2025.104620

Hauff, S. (2021). Analytical strategies in HRM systems research: A comparative analysis and some recommendations. *The International Journal of Human Resource Management*, *32*(9), 1923–1952. https://doi.org/10.1080/09585192.2018.1547779

Hauff, S., Guerci, M., Dul, J., & Rhee, H. van. (2021). Exploring necessary conditions in HRM research: Fundamental issues and methodological implications. *Human Resource Management Journal*, *31*(1), 18–36. https://doi.org/10.1111/1748-8583.12231

Hauff, S., Richter, N. F., Sarstedt, M., & Ringle, C. M. (2024). Importance and performance in PLS-SEM and NCA: Introducing the combined importance-performance map analysis (cIPMA). *Journal of Retailing and Consumer Services*, *78*, 103723. https://doi.org/10.1016/j.jretconser.2024.103723

Hausknecht, J. P., Hiller, N. J., & Vance, R. J. (2008). Work-unit absenteeism: Effects of satisfaction, commitment, labor market conditions, and time. *Academy of Management Journal*, *51*(6), 1223–1245. https://www.jstor.org/stable/40390270

Haverila, M., Twyford, J. C., McLaughlin, C., & Arora, M. (in press). Personnel capabilities and the quality of big data marketing analytics (BDMA). *Management Decision*. https://doi.org/10.1108/md-10-2024-2285

Hayat, N., Salameh, A. A., Malik, H. A., & Yaacob, M. R. (2022). Exploring the adoption of wearable healthcare devices among the Pakistani adults with dual analysis techniques. *Technology in Society*, *70*, 102015. https://doi.org/10.1016/j.techsoc.2022.102015

He, L., & Ismail, K. (2023). Do staff capacity and performance-based budgeting improve organisational performance? Empirical evidence from Chinese public universities. *Humanities and Social Sciences Communications*, *10*(29), 1–16. https://doi.org/10.1057/s41599-023-01523-2

He, Y., Cheng, C., & Wang, L. (2025). Unmasking the effects of E-leadership on virtual team effectiveness by an integrated fsQCA and NCA method. *PLOS ONE*, *20*, e0331500. https://doi.org/10.1371/journal.pone.0331500

He, Z., Chen, M., & Gu, D. (2025). How digital village construction affects to the effectiveness of rural governance? — research on the NCA and QCA methods. *Cities*, *156*, 105514. https://doi.org/10.1016/j.cities.2024.105514

Helinski, C., Westmattelmann, D., & Schewe, G. (2025). Necessary and sufficient values: A dual perspective on consumers' willingness to pay premiums and purchase intentions for circular fashion. *Journal of Cleaner Production*, *512*, 145220. https://doi.org/10.1016/j.jclepro.2025.145220

Hermans, M., Borda, A., Newburry, W., Cordova Chea, C. O., Finchelstein, D., Gonzalez-Perez, M. A., Montoya-Bayardo, M. A., Velasco, G., & Velez-Ocampo, J. (2024). Becoming a Multilatina: strategic capabilities as necessary conditions for the internationalization of Latin American firms. *European Business Review*, *36*(1), 50–70. https://doi.org/10.1108/EBR-02-2023-0040

Hernaus, T., & Černe, M. (2022). From the editors: The importance of data in research - best practices regarding data collection, processing and visualization. *Dynamic Relationships Management Journal*, *11*(2). http://sam-d.si/wp-content/uploads/2022/12/Editorial.pdf

Hernaus, T., Dragičević, N., & Hauff, S. (2024). The necessity of job design for employee creativity and innovation: Nothing happens without supervisor support. *European Journal of Work and Organizational Psychology*, *33*, 583–598. https://doi.org/10.1080/1359432x.2024.2348772

Hoeffding, W. (1952). The large-sample power of tests based on permutations of observations. *The Annals of Mathematical Statistics*, 169–192. https://www.jstor.org/stable/2236445

Höfer, T., Przyrembel, H., & Verleger, S. (2004). New evidence for the theory of the stork. *Paediatric and Perinatal Epidemiology*, *18*(1), 88–92. https://doi.org/10.1111/j.1365-3016.2003.00534.x

Hofstede, G. (1980). *Culture's consequences: International differences in work-related values*. Sage Publications. https://geerthofstede.com/culture-geert-hofstede-gert-jan-hofstede/6d-model-of-national-culture/

Hollenbeck, J. R., & Wright, P. M. (2017). Harking, sharking, and tharking: Making the case for post hoc analysis of scientific data. *Journal of Management*, *43*(1), 5–18. https://doi.org/10.1177/0149206316679487

Holtrop, J. S., Mullen, R., Curcija, K., Rubinson, C., Westfall, J. M., Nease Jr, D. E., & Zittleman, L. (2024). Increasing medication assisted treatment in rural primary care practice: A qualitative comparative analysis from IT MATTTRs Colorado. *Frontiers in Medicine*, *11*, 1450672. https://doi.org/10.3389/fmed.2024.1450672

Horvatinović, T., Mikic, M., & Dabić, M. (2024). The meeting points of team entrepreneurial passion, transactive memory systems and team performance: Examining mediation and necessity effects. *International Journal of Entrepreneurial Behavior & Research*, *30*(5), 1177–1196. https://doi.org/10.1108/IJEBR-04-2023-0383

Hu, S., Jie, Y., & Zhu, S. (in press). Digitalization and configurational effects on regional income inequality: Analysis of panel data from 134 economies. *Technological and Economic Development of Economy*. https://doi.org/10.3846/tede.2025.23950

Huang, L., Huang, Y., Rong, Z., & Xie, G. (2026). Knowledge and drivers of rural e-commerce entrepreneurial vibrancy: A fuzzy-set qualitative comparative analysis study. *Journal of Innovation & Knowledge*, *12*, 100911. https://doi.org/10.1016/j.jik.2025.100911

Huang, R., Wu, Y., Li, B., & Zhang, M. (in press). What configuration paths lead to a high utilization level of open government data (OGD) portals? *Aslib Journal of Information Management*. https://doi.org/10.1108/ajim-07-2024-0544

Huang, S., Ding, G., & Long, T. (2026). Exploring the synergistic effects of the effectiveness of carbon governance policy instruments. *Journal of Cleaner Production*, *538*, 147215. https://doi.org/10.1016/j.jclepro.2025.147215

Huang, X., Tung, C.-L., Wang, X., Xu, X., Lam, F.-I., & Zhang, T. (2022a). Configurations of the driving factors promoting China's commercial health insurance: A comparative qualitative analysis based on the technology–organization–environment framework. *Heliyon*, *8*, e11522. https://doi.org/10.1016/j.heliyon.2022.e11522

Huang, X., & Yang, X. (2025). Research on the configuration path of enterprise data asset value realization: An empirical analysis based on NCA and fsQCA. *Industrial Management & Data Systems*, *125*, 1979–2000. https://doi.org/10.1108/imds-05-2024-0453

Huang, Y., & Bu, Y. (2023). Institutional environment and college students' entrepreneurial willingness: A comparative study of Chinese provinces based on fsQCA. *Journal of Innovation & Knowledge*, *8*(1), 100307. https://doi.org/10.1016/j.jik.2023.100307

Huang, Y., Bu, Y., & Li, P. (in press). Pathways to prosperity: A configurational analysis of national entrepreneurship. *Small Business Economics.* https://doi.org/10.1007/s11187-025-01114-2

Huang, Y., Jiang, Y., & Hou, X. (2025). Multiple paths connecting poverty alleviation and rural revitalization from the perspective of functional configuration: An fsQCA analysis. *Sage Open*, *15*. https://doi.org/10.1177/21582440251385411

Huang, Y., Li, K., & Li, P. (2023). Innovation ecosystems and national talent competitiveness: A country-based comparison using fsQCA. *Technological Forecasting and Social Change*, *194*, 122733. https://doi.org/10.1016/j.techfore.2023.122733

Huang, Y., Li, P., Bu, Y., & Zhao, G. (2023). What entrepreneurial ecosystem elements promote sustainable entrepreneurship? *Journal of Cleaner Production*, *422*, 138459. https://doi.org/10.1016/j.jclepro.2023.138459

Huang, Y., Li, P., Chen, L., & Wang, J. (2023). Opportunity or necessity entrepreneurship? A study based on the national system of entrepreneurship. *Journal of Innovation & Knowledge*, *8*(4), 100448. https://doi.org/10.1016/j.jik.2023.100448

Huang, Y., Shuaib, M., Rahman, Md. M., Rahman, M., & Hossain, Md. E. (2024). Natural resources, digital financial inclusion, and good governance nexus with sustainable development: Fuzzy optimization to econometric modeling. *Natural Resources Forum*, *49*, 3763–3788. https://doi.org/10.1111/1477-8947.12549

Huang, Y., Sun, S., Yuan, Y., & Bu, Y. (2025). Pathways to high corporate environmental responsibility: A fuzzy-set and necessary condition analysis. *Corporate Social Responsibility and Environmental Management*, *33*, 534–546. https://doi.org/10.1002/csr.70192

Huang, Y., Zhang, J., Xu, Y., & Bu, Y. (2024). How can digital innovation capabilities of Chinese college students be cultivated? *The International Journal of Management Education*, *22*(3), 101085. https://doi.org/10.1016/j.ijme.2024.101085

Huang, Y., Zhang, J., Xu, Y., Yuan, Y., Zhang, M., & Huang, L. (2025). Digital entrepreneurial ecosystem and female entrepreneurial activity. *Humanities and Social Sciences Communications*, *12*. https://doi.org/10.1057/s41599-024-04345-y

Huang, Y., Zhang, M., Wang, J., Li, P., & Li, K. (2022b). Psychological cognition and women's entrepreneurship: A country-based comparison using fsQCA. *Journal of Innovation & Knowledge*, *7*(3), 100223. https://doi.org/10.1016/j.jik.2022.100223

Huang, Z., & Wu, Q. (2025). The critical factors in the innovation of Chinese urban AI technology: A necessary condition analysis. *Technology Analysis & Strategic Management*, *37*, 3732–3744. https://doi.org/10.1080/09537325.2024.2420618

Hui, G., Al Mamun, A., Masukujjaman, M., Makhbul, Z. K. M., & Ali, M. H. (2024). The relationship between mass customization and sustainable performance: The role of firm size and global e-commerce. *Heliyon*, *10*(6), e27726. https://doi.org/10.1016/j.heliyon.2024.e27726

Hulme, M. P. (2026). War and responsibility. *American Political Science Review*, *120*, 267–290. https://doi.org/10.1017/s0003055425000206

Hume, D. (1756). *Essays and treatises on several subjects. Volume 2, third edition.* London: A. Millar.

Hussain, K., Wong, P. P. W., Stephenson, M. L., & Chong, K. L. (in press). The impact of social media influencers on viewers' inspiration towards tourism destinations and revisit intention: A combined PLS-SEM and NCA approach. *Current Issues in Tourism.* https://doi.org/10.1080/13683500.2025.2571436

Ibrahimi, Q., Berisha, G., Krasniqi, B., & Ramosaj, B. (2024). Behaving with style: Investigating the influence of decision-making styles on the innovative work behavior of managers. *Journal of Organizational Effectiveness: People and Performance*, *12*, 939–960. https://doi.org/10.1108/JOEPP-01-2024-0031

Inacio Junior, E., Dionisio, E. A., & Gimenez, F. A. P. (2024). Entrepreneurial ecosystem patterns and its determinants and outputs: A necessary condition analysis to the Brazilian entrepreneurial cities. *Journal of Entrepreneurship in Emerging Economies*, *17*, 418–452. https://doi.org/10.1108/JEEE-07-2023-0316

İnce, F., & Orhan, G. (in press). Proposing a novel combined passenger satisfaction measurement model. *Kybernetes.* https://doi.org/10.1108/k-07-2024-1941

İnce, F., Şahin, İ. E., & Akbiyik, T. (2025). Uncovering non-substitutable value-for-money drivers in Skytrax top airlines: A MEREC-CoCoSo-NCA benchmarking framework. *DECISION*, *52*, 499–518. https://doi.org/10.1007/s40622-025-00449-9

İnce, F., Şahin, İ. E., & Durmaz, V. (in press). What today's airline passengers expect from airports in developed cities: A trilateral integrated approach. *Journal of Hospitality and Tourism Insights.* https://doi.org/10.1108/jhti-01-2025-0098

Indihar Štemberger, M., Bosilj Vuksic, V., Morelli, F., & Jaklič, J. (2024). Exploring the role of new and enhanced BPM capabilities in customer experience management: Does BPM matter? *Business Process Management Journal*, *30*(8), 120–143. https://doi.org/10.1108/BPMJ-10-2023-0838

Islam, M., Rahman, Md. M., Taher, Md. A., Quaosar, G. M. A. A., & Uddin, Md. A. (2024). Using artificial intelligence for hiring talents in a moderated mechanism. *Future Business Journal*, *10*(1), 13. https://doi.org/10.1186/s43093-024-00303-x

Ismael, O. A., Ibrahim, Z. K., Abdulmuhsin, A. A., Khan, S. T., Rehman, S. U., & Alkhwaldi, A. F. (in press). Robotic quality, lean leadership and sustainability healthcare: A multilevel investigative study using NCA perspective. *Journal of Health Organization and Management.* https://doi.org/10.1108/jhom-02-2025-0088

Jada, U. R., & Mukhopadhyay, S. (2019). Understanding the effects of empowering, transformational and ethical leadership on promotive and prohibitive voice: A moderated mediated examination. *Personnel Review*, *48*(3), 707–730. https://doi.org/10.1108/PR-11-2017-0365

Jain, N. K., Bhaskar, K., & Jain, S. (2022). What drives adoption intention of electric vehicles in India? An integrated UTAUT model with environmental concerns, perceived risk and government support. *Research in Transportation Business & Management*, 100730. https://doi.org/10.1016/j.rtbm.2021.100730

Jain, N. K., Chakraborty, K., & Choudhary, P. (2024). Building supply chain resilience through industry 4.0 base technologies: Role of supply chain visibility and environmental dynamism. *Journal of Business & Industrial Marketing*, *39*(8), 1750–1763. https://doi.org/10.1108/JBIM-09-2023-0550

Jain, N. K., Kaushik, K., & Sharma, A. (2023). What drives customers towards proximity payment services? An integrated theory of planned behaviour perspective. *International Journal of Consumer Studies*, *47*(3), 1095–1111. https://doi.org/10.1111/ijcs.12890

Jaiswal, M., & Zane, L. (2022a). National culture and attitudes' impact on diffusion of sustainable new technology-based products. *New England Journal of Entrepreneurship*, *25*(1), 5–25. https://doi.org/10.1108/NEJE-09-2021-0059

Jaiswal, M., & Zane, L. J. (2022b). Drivers of sustainable new technology diffusion in national markets: The case of electric vehicles. *Thunderbird International Business Review*, *64*(1), 25–38. https://doi.org/10.1002/tie.22243

Jankowska, D. M., & Karwowski, M. (2023). How vivid is your mental imagery? Validation of the vividness of visual imagery questionnaire in five Polish samples. *European Journal of Psychological Assessment*, *39*, 437–448. https://doi.org/10.1027/1015-5759/a000721

Ji, S., & Li, K. (2025). What factors promote Chinese English preservice teachers' development of digital literacy of teachers? A combination of SEM and NCA approaches. *Teaching and Teacher Education*, *168*, 105227. https://doi.org/10.1016/j.tate.2025.105227

Jiang, J., & Wang, C. (2024). Pathways to tourism eco-efficiency: Fresh insights from a configuration perspective. *Current Issues in Tourism*, *28*, 3865–3885. https://doi.org/10.1080/13683500.2024.2406413

Jiang, Y., & Huang, Y. (2023). How mumpreneurs achieve high entrepreneurial opportunity recognition and development: Evidence from China. *Women's Studies International Forum*, *100*, 102804. https://doi.org/10.1016/j.wsif.2023.102804

Jing, L., Qianqiang, L., Yantai, C., & Qi, A. (2025). How digital transformation facilitates ESG performance to heavy polluting enterprises: A panel fsQCA based on national big data comprehensive pilot zones. *Technological Forecasting and Social Change*, *221*, 124366. https://doi.org/10.1016/j.techfore.2025.124366

Jöreskog, K. G. (1970). A general method for estimating a linear structural equation system. *ETS Research Bulletin Series*, *1970*(2), i–41. https://doi.org/10.1002/j.2333-8504.1970.tb00783.x

Jovanovic, J., & Morschett, D. (2022). Under which conditions do manufacturing companies choose FDI for service provision in foreign markets? An investigation using fsQCA. *Industrial Marketing Management*, *104*, 38–50. https://doi.org/10.1016/j.indmarman.2022.03.018

Kakkar, S., Ghulyani, S., Kuril, S., & Chaudhuri, M. (2024). Understanding factors shaping engagement in remote internships: The role of institutional communication, faculty support and intern cynicism. *International Journal of Educational Management*, *38*(4), 1219–1240. https://doi.org/10.1108/IJEM-08-2023-0378

Kang, S., & Shang, Y. (2026). How does digital-real integration ignite urban entrepreneurship: Unpacking innovation's complex mediation role. *Technology in Society*, *84*, 103106. https://doi.org/10.1016/j.techsoc.2025.103106

Karwowski, M., Dul, J., Gralewski, J., Jauk, E., Jankowska, D. M., Gajda, A., Chruszczewski, M. H., & Benedek, M. (2016). Is creativity without intelligence possible? A necessary condition analysis. *Intelligence*, *57*, 105–117. https://doi.org/10.1016/j.intell.2016.04.006

Karwowski, M., Jankowska, D. M., Brzeski, A., Czerwonka, M., Gajda, A., Lebuda, I., & Beghetto, R. A. (2020). Delving into creativity and learning. *Creativity Research Journal*, *32*(1), 4–16. https://doi.org/10.1080/10400419.2020.1712165

Karwowski, M., Kaufman, J. C., Lebuda, I., Szumski, G., & Firkowska-Mankiewicz, A. (2017). Intelligence in childhood and creative achievements in middle-age: The necessary condition approach. *Intelligence*, *64*, 36–44. https://doi.org/10.1016/j.intell.2017.07.001

Kashive, N., & Raina, B. (in press). Digital leadership competencies and emotional intelligence for digital well-being: Examining through PLS-SEM and NCA. *International Journal of Organizational Analysis*. https://doi.org/10.1108/ijoa-03-2025-5330

Katz, J. (2001). Analytic induction. In N. J. Smelser & P. B. Baltes (Eds.), *International Encyclopedia of the Social & Behavioral sciences* (Vol. 1, pp. 480–484). Elsevier Oxford, UK. https://doi.org/10.1016/B0-08-043076-7/00774-9

Kaushik, K., Jain, N. K., & Choudhary, P. (2023). Public opinion about restrictive driving policy: Does political affiliation matter? A case study of odd-even restrictive driving policy in delhi, india. *Case Studies on Transport Policy*, *12*, 100999. https://doi.org/10.1016/j.cstp.2023.100999

Kayser, I., & Gradtke, M. (in press). Unlocking AI acceptance: An integration of NCA and PLS-SEM to analyse the acceptance of ChatGPT. *Journal of Decision Systems*. https://doi.org/10.1080/12460125.2024.2443231

Kazemzadeh, E., Fuinhas, J. A., Salehnia, N., Koengkan, M., & Silva, N. (2023). Exploring necessary and sufficient conditions for carbon emission intensity: A comparative analysis. *Environmental Science and Pollution Research*, *30*(43), 97319–97338. https://doi.org/10.1007/s11356-023-29260-8

Kazemzadeh, E., Fuinhas, J. A., Salehnia, N., Silva, N., & Radulescu, M. (2024). Revealing the essential and sufficient conditions for the energy consumption diversification. *Energy*, *313*, 133601. https://doi.org/10.1016/j.energy.2024.133601

Keimasi, M., Giwa, S., Chitsazan, H., & NasirpourOsgoei, S. N. (in press). Typologies of assessment journey experiences: Understanding what works for whom in block teaching through the lens of a 3D framework. *Assessment & Evaluation in Higher Education*. https://doi.org/10.1080/02602938.2025.2573090

Kellberg, S., Keller, M., Nordine, J., Moser, S., & Lewalter, D. (2024). Energy literacy for all? Exploring whether prior interest and energy knowledge mediate energy literacy development in a modern socio-scientific museum exhibition. *International Journal of Science Education, Part B*, *15*, 161–182. https://doi.org/10.1080/21548455.2024.2344129

Kennedy, F. E. (1995). Randomization tests in econometrics. *Journal of Business & Economic Statistics*, *13*(1), 85–94. https://www.jstor.org/stable/1392523

Kerwin, T., Hittle, B., Stredney, D., De Boeck, P., & Wiet, G. (2018). Cross-institutional evaluation of a mastoidectomy assessment instrument. *Journal of Surgical Education*, *75*(3), 678–687. https://doi.org/10.1016/j.jsurg.2017.10.006

Khan, T. H., Ali, S., Xiaobao, P., & Zhiying, L. (2024). Responsive to proactive market orientations: Unleashing the potential of effectuation–causation blending for business model innovation. *IEEE Transactions on Engineering Management*, *71*, 14307–14325. https://doi.org/10.1109/TEM.2024.3403571

Klimas, P., Czakon, W., & Fredrich, V. (2022). Strategy frames in coopetition: An examination of coopetition entry factors in high-tech firms. *European Management Journal*, *40*(2), 258–272. https://doi.org/10.1016/j.emj.2021.04.005

Klimas, P., Sachpazidu, K., & Stańczyk, S. (2023). The attributes of coopetitive relationships: What do we know and not know about them? *European Management Journal*. https://doi.org/10.1016/j.emj.2023.02.005

Knol, W. H., Slomp, J., Schouteten, R. L., & Lauche, K. (2018). Implementing lean practices in manufacturing SMEs: Testing "critical success factors" using necessary condition analysis. *International Journal of Production Research*, *56*(11), 3955–3973. https://doi.org/10.1080/00207543.2017.1419583

Knol, W. H., Slomp, J., Schouteten, R. L., & Lauche, K. (2019). The relative importance of improvement routines for implementing lean practices. *International Journal of Operations & Production Management*, *39*(2), 214–237. https://doi.org/10.1108/IJOPM-01-2018-0010

Koay, K. Y., Cheah, C. W., & Chang, Y. X. (2022). A model of online food delivery service quality, customer satisfaction and customer loyalty: A combination of PLS-SEM and

NCA approaches. *British Food Journal, 124*(12), 4516–4532. https://doi.org/10.1108/BFJ-10-2021-1169

Koay, K. Y., & Low, S. K. (2025). Understanding the influence of chain coffee shops' innovativeness on brand loyalty through brand satisfaction: A multi-analytical approach. *British Food Journal, 127*(8), 2868–2885. https://doi.org/10.1108/BFJ-01-2025-0088

Koay, K. Y., & Ngang, Y. M. (in press). Understanding consumers' purchase intentions for functional foods from the perspective of the theory of consumption values: A multi-analytical approach. *British Food Journal.* https://doi.org/10.1108/bfj-08-2025-1045

Kobarg, S., Stumpf-Wollersheim, J., Schlägel, C., & Welpe, I. M. (2020). Green together? The effects of companies' innovation collaboration with different partner types on ecological process and product innovation. *Industry and Innovation, 27*(9), 953–990. https://doi.org/10.1080/13662716.2020.1713733

Koenker, R. (2024). *Quantreg: Quantile regression.* https://www.r-project.org

Köhler, T., & Cortina, J. M. (2023). Constructive replication, reproducibility, and generalizability: Getting theory testing for JOMSR right. *Journal of Management Scientific Reports, 1*(2), 75–93. https://doi.org/10.1177/27550311231176016

Kopplin, C. S. (2023a). A configurational view on technology acceptance: The example of highly integrated collaboration platforms. *Australasian Journal of Information Systems, 27.* https://doi.org/10.3127/ajis.v27i0.4043

Kopplin, C. S. (2023b). Chatbots in the workplace: A technology acceptance study applying uses and gratifications in coworking spaces. *Journal of Organizational Computing and Electronic Commerce, 32*, 232–257. https://doi.org/10.1080/10919392.2023.2215666

Kopplin, C. S., Gantert, T. M., & Maier, J. V. (2022). Acceptance of matchmaking tools in coworking spaces: An extended perspective. *Review of Managerial Science, 16*, 1911–1943. https://doi.org/10.1007/s11846-021-00498-1

Kopplin, C. S., & Rausch, T. M. (2022). Above and beyond meat: The role of consumers' dietary behavior for the purchase of plant-based food substitutes. *Review of Managerial Science, 16*, 1335–1364. https://doi.org/10.1007/s11846-021-00480-x

Kopplin, C. S., & Rösch, S. F. (2021). Equifinal causes of sustainable clothing purchase behavior: An fsQCA analysis among generation Y. *Journal of Retailing and Consumer Services, 63*, 102692. https://doi.org/10.1016/j.jretconser.2021.102692

Korman, J. V., Van Quaquebeke, N., & Tröster, C. (2022). Managers are less burned-out at the top: The roles of sense of power and self-efficacy at different hierarchy levels. *Journal of Business and Psychology, 37*, 151–171. https://doi.org/10.1007/s10869-021-09733-8

Kuik, R., & Dul, J. (2026). *Meet the NCA band: Fitting NCA with a ribbon around the necessity ceiling.* Rotterdam School of Management, Erasmus University.

Kumar, D. (2021). Meteorological barriers to bike rental demands: A case of Washington DC using NCA approach. *Case Studies on Transport Policy, 9*(2), 830–841. https://doi.org/10.1016/j.cstp.2021.04.002

Kumar, S. P., Saha, S., & Anand, A. (2023). A green human resource management approach of participation in decision-making and behavioural outcomes – a moderated mediated model. *International Journal of Organizational Analysis, 31*(5), 1724–1747. https://doi.org/10.1108/IJOA-09-2021-2954

Kwon, S. P. (2021). Examination of design parameters affecting Nb3Sn CICC current sharing temperature using necessary condition analysis. *Superconductor Science and Technology, 34*, 085006. https://doi.org/10.1088/1361-6668/ac048f

Kwon, S. P. (2022). Critical bending wavelengths associated with current sharing temperature degradation in multistage twisted Nb3Sn cable CICC. *Cryogenics, 121*, 103375. https://doi.org/10.1016/j.cryogenics.2021.103375

Lai, S., Zhang, M., & Su, Z. (in press). Optimizing tourism ecological security: A study based on NCA and fsQCA methods. *Environment, Development and Sustainability*. https://doi.org/10.1007/s10668-024-05065-8

Lange, D., & Pfarrer, M. D. (2017). Editors' comments: Sense and structure—the core building blocks of an AMR article. *Academy of Management Review*, *42*(3), 407–416. https://journals.aom.org/doi/10.5465/amr.2016.0225

Lankoski, J., & Lankoski, L. (2023). Environmental sustainability in agriculture: Identification of bottlenecks. *Ecological Economics*, *204*, 107656. https://doi.org/10.1016/j.ecolecon.2022.107656

Ledesma-Chaves, P., Gil-Cordero, E., & Maldonado-López, B. (in press). Metaverse use intention as a function of psychographic segmentation: The VALS scale applied to tourist behavior. *Journal of Enterprise Information Management*. https://doi.org/10.1108/jeim-07-2024-0382

Lee, J., & Borgonovi, F. (2022). Relationships between family socioeconomic status and mathematics achievement in OECD and Non-OECD Countries. *Comparative Education Review*, *66*(2), 199–227. https://doi.org/10.1086/718930

Lee, W., & Jeong, C. (2020). Beyond the correlation between tourist eudaimonic and hedonic experiences: Necessary condition analysis. *Current Issues in Tourism*, *23*(17), 2182–2194. https://doi.org/10.1080/13683500.2019.1611747

Lee, W., & Jeong, C. (2021). Distinctive roles of tourist eudaimonic and hedonic experiences on satisfaction and place attachment: Combined use of SEM and necessary condition analysis. *Journal of Hospitality and Tourism Management*, *47*, 58–71. https://doi.org/10.1016/j.jhtm.2021.02.012

Lee, W., & Lu, L. (2023). Good theories predict: Unveiling the untapped potential of "necessity" theorizing. *Journal of Hospitality & Tourism Research*, *49*, 219–234. https://doi.org/10.1177/10963480231188666

Lee, W., Park, S., & Jeong, C. (2022). Repositioning risk perception as a necessary condition of travel decision: The case of North Korea tourism. *Journal of Hospitality and Tourism Management*, *52*, 252–263. https://doi.org/10.1016/j.jhtm.2022.07.001

Legendre, A. M. (1805). *Nouvelles méthodes la determination des orbites des comètes: Part 1-2: Mit einem supplement*. Didot. https://www.york.ac.uk/depts/maths/histstat/legendre2.pdf

Lehmann, E. L., Romano, J. P., & Casella, G. (2005). *Testing statistical hypotheses* (3rd ed.). Springer. https://www.springer.com/gp/book/9780387988641

Lenth, R. V. (2001). Some practical guidelines for effective sample size determination. *The American Statistician*, *55*(3), 187–193. https://doi.org/10.1198/000313001317098149

Li, C., & Larivière, B. (2023). The more extroverted the better? Unraveling the complex relationship between service robots' personality and the service robot experience. *Psychology & Marketing*, *40*(11), 2370–2386. https://doi.org/10.1002/mar.21896

Li, C., Sampene, A. K., Khan, A., & Wiredu, J. (2024a). Environmental orientations to innovation ecosystem: The role of green transition, perceived environmental uncertainty and green transformational leadership. *Operations Management Research*, *18*, 48–71. https://doi.org/10.1007/s12063-024-00527-8

Li, C., Wang, X., & Wang, L. (in pressa). Examining the crucial societal incentive approaches and their impact mechanisms to enhance the purchase intention of carbon-labeled products: A conjoined application of PLS-SEM and NCA methodologies. *Energy & Environment*, 0958305X241251405. https://doi.org/10.1177/0958305X241251405

Li, D., Yan, H., & Ma, S. (in press). ESG performance drivers and corporate growth: A life-cycle-based fsQCA–PSM study of China's construction and manufacturing enterprises. *Journal of Asian Architecture and Building Engineering*. https://doi.org/10.1080/13467581.2025.2517911

Li, G., Zhu, M. X., & Kong, S. (in press). The effectiveness of smart knowledge services in libraries in Chinese Mainland: A mixed methods study. *The Electronic Library.* https://doi.org/10.1108/el-03-2025-0104

Li, J., Du, Y., Sun, N., & Xie, Z. (2024). Ecosystems of doing business and living standards: A configurational analysis based on Chinese cities. *Chinese Management Studies, 18*(5), 1302–1323. https://doi.org/10.1108/CMS-04-2022-0139

Li, L., Che, W., Yao, Q., & Cui, Y. (2025). Research on high-quality development path of strategic emerging enterprises enabled by innovation. *PLOS ONE, 20*, e0328918. https://doi.org/10.1371/journal.pone.0328918

Li, L., Huang, M., & Yao, Q. (in pressb). How to crack the dilemma of enterprise green transformation?– Analysis of linkage effect based on fsQCA and NCA. *Environment, Development and Sustainability.* https://doi.org/10.1007/s10668-025-06438-3

Li, L., Ma, X., Zheng, Y., Ma, S., Zuo, J., & Goodsite, M. (2025). Facilitating urban green innovative efficiency from intergovernmental perspective in China. *Scientific Reports, 15.* https://doi.org/10.1038/s41598-025-08883-6

Li, M., Gong, D., & Song, Y. (2025). Research on multiple paths of transportation system resilience building in digital contexts. *Scientific Reports, 15.* https://doi.org/10.1038/s41598-025-14245-z

Li, R., Niu, Z., Liu, C., & Wu, B. (2024). The co-movement effect of managers' psychological factors on the BIM adoption decision in SMEs: A study based on fsQCA. *Engineering, Construction and Architectural Management, 31*(4), 1454–1472. https://doi.org/10.1108/ECAM-08-2022-0721

Li, S., Hu, S., Tong, N., Xie, W., Tong, L., & Hu, S. (2026). How urban resilience shapes residents' immediate emotions during heatwaves: A panel fuzzy-set qualitative comparative analysis (fsQCA)-based configurational analysis. *Sustainable Cities and Society, 136*, 107067. https://doi.org/10.1016/j.scs.2025.107067

Li, Y., Fan, D., Kumar, V., & Ananthram, S. (2024b). Configuring international growth of emerging market multinational enterprises: A compositional springboarding view in the context of India. *Journal of Management Studies, 61*(6), 2570–2603. https://doi.org/10.1111/joms.12981

Li, Y., Ma, M., Hao, Z., Xu, F., & Liu, J. (in pressc). What motivates the impulsive travel intentions of Generation Z? A study based on grounded theory and fsQCA. *Current Issues in Tourism.* https://doi.org/10.1080/13683500.2025.2469744

Li, Z., Yanrui, W., Fan, Z., Yuqing, Z., & Xu, S. (2025). Government attention, characteristics of officials-in-charge and performance of public pension funds in China: A configuration analysis. *China: An International Journal, 23*, 44–81. https://doi.org/10.56159/chn.2025.a976472

Liang, C., Wang, S.-J., Foley, M., & Ma, G.-H. (2023). The path selection on improving the quality of environmental information disclosure – configuration analysis based on fsQCA. *Applied Economics, 55*(19), 2207–2222. https://doi.org/10.1080/00036846.2022.2102134

Liang, S., Li, Y., & Yang, Y. (2025). Is student engagement without classroom social climate possible? A necessary condition analysis. *Humanities and Social Sciences Communications, 12.* https://doi.org/10.1057/s41599-025-05910-9

Liao, C.-P., Wu, C.-C., & Chen, C.-C. (in press). Necessary conditions in social media addiction: Identifying key antecedents. *Psychological Reports.* https://doi.org/10.1177/00332941241312314

Liao, Z., Li, W., Al Mamun, A., Tseng, M.-L., & Helmi Ali, M. (2025). SMEs' capabilities in relation to resilience, reconfiguration, and flexibility in the supply chain: A moderated mediation model. *International Journal of Logistics Research and Applications, 28*, 2098–2120. https://doi.org/10.1080/13675567.2025.2450365

Liébana-Cabanillas, F., Abbasi, G. A., Higueras-Castillo, E., & Wagner, R. (2025). Press to pay: The power of biometrics in financial transactions investigated by PLS-SEM, fsQCA, and NCA. *Technological Forecasting and Social Change*, *221*, 124365. https://doi.org/10.1016/j.techfore.2025.124365

Liehr, J., & Hauff, S. (2022). Must have or nice to have? Necessary leadership competencies to enable employees' innovative behaviour. *International Journal of Innovation Management*, *26*(10), 2250070. https://doi.org/10.1142/S1363919622500700

Lim, T.-Y., Leong, C.-M., Lim, L. T.-K., Lim, B. C.-Y., Lim, R. T.-H., & Heng, K.-S. (2023). Young adult tourists' intentions to visit rural community-based homestays. *Young Consumers*, *24*(5), 540–557. https://doi.org/10.1108/YC-11-2022-1637

Lim, W., Munikrishnan, U. T., Leong, C.-M., Hiew, L.-C., Leong, M.-W., & Yang, L. (2024). Do you want a secure e-wallet? Understanding the role of risk and security in e-wallet continuance intention. *Information & Computer Security*, *32*(3), 304–321. https://doi.org/10.1108/ICS-05-2023-0085

Lima, T. F., Bahli, B., Arbulu, A., Hamdi, A., & Saikouk, T. (2025). Differentiating "must–have" and "should–have" supply chain capabilities for enhanced performance: A necessary conditions analysis. *Journal of Business & Industrial Marketing*, *40*(2), 313–338. https://doi.org/10.1108/JBIM-06-2024-0453

Lin, Z., Yao, N., & Yu, C. (2024). Urban metro systems: Configurational analysis of antecedent conditions influencing high metro ridership. *Journal of Urban Planning and Development*, *150*(4), 04024030. https://doi.org/10.1061/JUPDDM.UPENG-5093

Linder, C., Moulick, A. G., & Lechner, C. (2023). Necessary conditions and theory-method compatibility in quantitative entrepreneurship research. *Entrepreneurship Theory and Practice*, *47*(5), 1971–1994. https://doi.org/10.1177/10422587221102103

Lipset, S. M. (1959). Some social requisites of democracy: Economic development and political legitimacy1. *American Political Science Review*, *53*(1), 69–105. https://doi.org/10.2307/1951731

Lisboa, A. C. C., Dias, Á. L., Abrantes, J. L., Ferreira, B., & Sousa, C. (in press). Using necessary condition analysis to understand digital storytelling's impact on consumer behavior. *Journal of Marketing Analytics*. https://doi.org/10.1057/s41270-025-00425-3

Liu, H., Wang, Y., He, G., Ma, R., & Fu, S. (2023). The impact of environmental information disclosure of origin using blockchain technology on online consumer behaviour: A combination of SEM and NCA approaches. *Journal of Cleaner Production*, *421*, 138449. https://doi.org/10.1016/j.jclepro.2023.138449

Liu, Q., Wen, X., Peng, H., & Cao, Q. (2023). Key technology breakthrough in new energy vehicles: Configuration path evolution from innovative ecosystem perspective. *Journal of Cleaner Production*, *423*, 138635. https://doi.org/10.1016/j.jclepro.2023.138635

Liu, X., Zhu, X., & Wu, J. (2025). Payment innovation diffusion among merchants: An integrated view using a tri-level hybrid method. *Electronic Commerce Research and Applications*, *74*, 101558. https://doi.org/10.1016/j.elerap.2025.101558

Liu, Y., & Meng, Q. (2024). How online reviews of sports tourism affect tourists' value co-creation behavior: Findings from PLS-SEM, NCA, and fsQCA. *Journal of Hospitality and Tourism Management*, *60*, 360–372. https://doi.org/10.1016/j.jhtm.2024.07.005

Liu, Y., Yan, B., Chen, B., Wu, L., & Zhang, X. (in press). The imperative of institutional trust and individualistic culture on early COVID-19 policy responses: A necessary condition analysis. *Policy Studies Journal*. https://doi.org/10.1111/psj.70037

Liu, Y., Yu, C., & Damberg, S. (2022). Exploring the drivers and consequences of the "awe" emotion in outdoor sports – a study using the latest partial least squares structural equation modeling technique and necessary condition analysis. *International Journal of Sports Marketing and Sponsorship*, *23*(2), 278–294. https://doi.org/10.1108/IJSMS-12-2020-0232

Liying, H., & Mengying, Z. (2024). How political influence and financial pressure contribute to performance-based budgeting and university performance: Evidence from SEM and NCA. *Sage Open*, *14*(2), 21582440241256637. https://doi.org/10.1177/21582440241256637

Loh, X.-M., Lee, V.-H., & Leong, L.-Y. (2022). Mobile-lizing continuance intention with the mobile expectation-confirmation model: An SEM-ANN-NCA approach. *Expert Systems with Applications*, *205*, 117659. https://doi.org/10.1016/j.eswa.2022.117659

Loh, X.-M., Shyu, W.-H., Lee, V.-H., Huang, H.-L., Tan, G. W.-H., & Ooi, K.-B. (in press). Super app: A multi-analytical cross-country approach on the "everything" app. *Journal of Computer Information Systems*. https://doi.org/10.1080/08874417.2025.2560379

Lourenco, C. E., Madeira, J. C., & Santos, V. M. (2025). Exploring the influence of flagship store experience and consumer interactions on purchase behaviour. *ReMark - Revista Brasileira de Marketing*, *24*, e27409. https://doi.org/10.5585/2025.27409

Low, M. P., Cham, T.-H., & Ramayah, T. (2025). Deciphering organisational resilience through a confluence analysis of PLS-SEM, necessary condition analysis, and combined importance-performance map. *Quality & Quantity*, *59*, 2961–2989. https://doi.org/10.1007/s11135-025-02112-y

Low, M. P., Rahim, F. A., & Wut, T. M. (2025). Leveraging artificial intelligence to foster pro-environmental and green behavior in organizations: Insights from PLS-SEM and necessary condition analysis. *Sustainable Futures*, *9*, 100786. https://doi.org/10.1016/j.sftr.2025.100786

Low, M. P., & Ramayah, T. (2023). It isn't enough to be easy and useful! Combined use of SEM and necessary condition analysis for a better understanding of consumers' acceptance of medical wearable devices. *Smart Health*, *27*, 100370. https://doi.org/10.1016/j.smhl.2022.100370

Low, M. P., Tang, K. L., & Yap, J. B. H. (in press). Switching gears: An in-depth understanding of electric vehicle adoption using the confluence approach of PLS-SEM and necessary condition analysis. *Management of Environmental Quality: An International Journal*. https://doi.org/10.1108/meq-10-2024-0458

Low, M. P., Wut, T. M., & Pok, W. F. (in press). Lifelong learning and artificial intelligence competency as drivers of career resilience in older workers: An integrative study using PLS-SEM and necessary condition analysis. *Educational Gerontology*. https://doi.org/10.1080/03601277.2025.2528673

Lu, Q., Yang, Y., & Yu, M. (2024). Can SMEs' quality management promote supply chain financing performance? An explanation based on signalling theory. *International Journal of Emerging Markets*, *19*(7), 1981–2002. https://doi.org/10.1108/IJOEM-03-2022-0456

Lu, X., & White, H. (2014). Robustness checks and robustness tests in applied economics. *Journal of Econometrics*, *178*, 194–206. https://doi.org/10.1016/j.jeconom.2013.08.016

Luo, G., Yang, Y., & Wang, L. (2023). Driving rural industry revitalization in the digital economy era: Exploring strategies and pathways in China. *PLOS ONE*, *18*(9), e0292241. https://doi.org/10.1371/journal.pone.0292241

Luo, L., Wang, Y., Liu, Y., Zhang, X., & Fang, X. (2022). Where is the pathway to sustainable urban development? Coupling coordination evaluation and configuration analysis between low-carbon development and eco-environment: A case study of the Yellow River Basin, China. *Ecological Indicators*, *144*, 109473. https://doi.org/10.1016/j.ecolind.2022.109473

Luther, L., Bonfils, K. A., Firmin, R. L., Buck, K. D., Choi, J., Dimaggio, G., Popolo, R., Minor, K. S., & Lysaker, P. H. (2017). Metacognition is necessary for the emergence of motivation in people with schizophrenia spectrum disorders: A necessary condition analysis. *The Journal of Nervous and Mental Disease*, *205*(12), 960. https://doi.org/10.1097/NMD.0000000000000753

Ly, B. (in press). Digital transformation, organizational culture, and public sector flexibility in emerging economies: Insights from NCA and PLS-SEM analysis. *Journal of the Knowledge Economy*. https://doi.org/10.1007/s13132-025-02801-x

Lyu, V. C., Chin, W., Zhang, H., & Liu, V. (2022). Value added or overload? A study of the countervailing effects of non-core features on mobile banking apps. *Journal of Consumer Behaviour*, *21*(3), 602–613. https://doi.org/10.1002/cb.2003

Ma, X., & Wang, J. (2001). A confirmatory examination of walberg's model of educational productivity in student career aspiration. *Educational Psychology*, *21*(4), 443–453. https://doi.org/10.1080/01443410120090821

Ma, Y., & Bennett, D. (2024). Analyzing predictors of perceived graduate employability from sufficiency and necessity perspectives. *Higher Education*, *88*(6), 2375–2397. https://doi.org/10.1007/s10734-024-01221-1

Machamer, P., Darden, L., & Craver, C. F. (2000). Thinking about mechanisms. *Philosophy of Science*, *67*(1), 1–25. https://www.jstor.org/stable/188611

Mackie, J. L. (1965). Causes and conditions. *American Philosophical Quarterly*, *2*(4), 245–264. https://www.jstor.org/stable/20009173

Magno, F., Cassia, F., et al. (2023). *Reviewing the SmartPLS 4 software: The latest features and enhancements*. Springer. https://doi.org/10.1057/s41270-023-00266-y

Magno, F., & Cassia, F. (2024). Predicting restaurants' surplus food platform continuance: Insights from the combined use of PLS-SEM and NCA and predictive model comparisons. *Journal of Retailing and Consumer Services*, *79*, 103820. https://doi.org/10.1016/j.jretconser.2024.103820

Mahoney, J. (2007). The elaboration model and necessary causes. In G. Goertz & J. S. Levy (Eds.), *Explaining war and peace: Case studies and necessary condition counterfactuals* (pp. 281–306). Routledge. https://doi.org/10.4324/9780203089101-18

Malibari, M., Bajaba, S., Bajaba, A., & Basahal, A. (2025). Cultivating innovative behaviors: How entrepreneurial leaders foster employees' antifragility within autonomous work settings. *Journal of Innovation & Knowledge*, *10*, 100701. https://doi.org/10.1016/j.jik.2025.100701

Manchiraju, S., Akbari, M., & Seydavi, M. (2023). Is entrepreneurial role stress a necessary condition for burnout? A necessary condition analysis. *Current Psychology*, *43*(5), 4766–4778. https://doi.org/10.1007/s12144-023-04704-z

Mandel, D. R., & Lehman, D. R. (1998). Integration of contingency information in judgments of cause, covariation, and probability. *Journal of Experimental Psychology: General*, *127*(3), 269. https://doi.org/10.1037/0096-3445.127.3.269

Manickavasagam, J., Ayyalusamy, K., Bhaiyat, F., Jha, S., Jain, T., & Poddar, S. (in press). Bitcoin prices: Configurational effects of technological drivers, macroeconomic fundamentals and economic agents' expectation. *International Journal of Finance & Economics*. https://doi.org/10.1002/ijfe.70032

Mansoor, S., Rahman, S. M., & Bowden, J. L. (2024). Purchase spillovers from the metaverse to the real world: The roles of social presence, trialability, and customer experience. *Journal of Consumer Behaviour*, *23*(5), 2501–2552. https://doi.org/10.1002/cb.2353

Manzi-Puertas, M. A., Agirre-Aramburu, I., & López-Pérez, S. (2025). Uncovering the hidden value of scarce resources: Frugality, resourcefulness and innovative behavior among student entrepreneurs. *Cogent Business & Management*, *12*. https://doi.org/10.1080/23311975.2025.2525500

Marchetti, I., Colpizzi, I., De Caro, E. F., Pedretti, L. M., Iannattone, S., Di Blas, L., & Bottesi, G. (in press). Is reduced self-esteem a necessary condition for eating disorder symptoms in adolescence? Preliminary evidence from a necessary condition analysis study. *Eating and Weight Disorders - Studies on Anorexia, Bulimia and Obesity*. https://doi.org/10.1007/s40519-025-01806-4

Marchetti, I., Koster, E. H. W., & Hankin, B. L. (2025). Which psychosocial risks are necessary for developing depression during adolescence? A novel approach applying necessary condition analysis. *Journal of the American Academy of Child & Adolescent Psychiatry*, *64*, 1201–1209. https://doi.org/10.1016/j.jaac.2024.11.001

Marchetti, I., Koster, E. H., & Dul, J. (2026). Necessity causality in mental health research: Applying necessary condition analysis in clinical psychology and psychiatry. *Clinical Psychology Review*, *123*, 102689. https://doi.org/10.1016/j.cpr.2025.102689

Marchetti, I., Pedretti, L. M., Iannattone, S., Colpizzi, I., Farina, A., Di Blas, L., Ghisi, M., & Bottesi, G. (2025). Is intolerance of uncertainty a necessary condition for anxiety symptoms in adolescents? A necessary condition analysis study. *Journal of Anxiety Disorders*, *112*, 102999. https://doi.org/10.1016/j.janxdis.2025.102999

Massu, J., Kuik, R., & Dul, J. (2020). *Simulations with NCA*. Rotterdam School of Management, Erasmus University.

Mayo, D. G. (1996). Hunting and snooping: Understanding the Neyman-Pearson predesignationist stance. In *Error and the growth of experimental knowledge* (pp. 294–318). University of Chicago Press.

Mazhar, W., & Zilahy, G. (2025). Pathways to green food purchases: Exploring the nexus of attitudes, habits and lifestyles using SEM and NCA. *British Food Journal*, *127*, 208–229. https://doi.org/10.1108/bfj-07-2024-0695

Mazumder, S., & Garg, S. (2021). Decoding digital transformational outsourcing: The role of service providers' capabilities. *International Journal of Information Management*, *58*, 102295. https://doi.org/10.1016/j.ijinfomgt.2020.102295

Meddeb, S., & St-Jean, É. (2025). Unraveling the mentor-mentee knot in entrepreneurship: Narcissism, emotional intelligence, and dyadic configurations. *Mentoring & Tutoring: Partnership in Learning*, *33*, 231–264. https://doi.org/10.1080/13611267.2025.2469732

Medina-Molina, C., Pérez-Macías, N., & Rey-Tienda, S. (2024). Strategies for urban cycling: An analysis through generalized analytic induction. *International Review on Public and Nonprofit Marketing*, *21*(4), 963–997. https://doi.org/10.1007/s12208-024-00411-y

Medina-Molina, C., & Rey-Moreno, M. (2024). The determinant conditions and causal mechanisms that explain the use of micromobility services. *Cuadernos de Gestión*, *24*(1), 7–20. https://doi.org/10.5295/cdg.231965cm

Meehl, P. E. (1962). Schizotaxia, schizotypy, schizophrenia. *American Psychologist*, *17*(12), 827–838. https://doi.org/10.1037/h0041029

Mello, P. A. (2021). *Qualitative comparative analysis: An introduction to research design and application*. Georgetown University Press. http://press.georgetown.edu/book/georgetown/qualitative-comparative-analysis

Mendel, J. M., & Korjani, M. M. (2010). *Charles Ragin's fuzzy set qualitative comparative analysis (fsQCA) applied to linguistic summarization* (Technical Report USC-SIPI Report No. 408). Signal; Image Processing Institute, Viterbi School of Engineering, University of Southern California. https://sipi.usc.edu/reports/pdfs/Originals/USC-SIPI-408.pdf

Mersmann, O., Trautmann, H., Steuer, D., & Bornkamp, B. (2023). *Truncnorm: Truncated normal distribution*. https://github.com/olafmersmann/truncnorm

Messner, W. (2020). Empirically assessing noisy necessary conditions with activation functions. *Computational Management Science*, *18*, 1–23. https://doi.org/10.1007/s10287-020-00377-2

Milovan, A.-M., Dobre, C., & Moisescu, O. (2025). Boosting brand behavioral intentions via integrated explicit product placements in podcasts. *Journal of Business Research*, *189*, 115129. https://doi.org/10.1016/j.jbusres.2024.115129

Min, Q., Zhu, R., & Peng, L. (2024). Pathways to improving carbon emission efficiency in provinces: A comparative qualitative analysis based on the technology-organization-environment framework. *Heliyon*, *10*(3), e25132. https://doi.org/10.1016/j.heliyon.2024.e25132

Mkedder, N., Das, M., Bakır, M., Lachachi, A., & Aldhabyani, Y. (in press). Buying local: How functional values, social identity and gender shape consumer preferences in developed and emerging markets. *International Journal of Emerging Markets*. https://doi.org/10.1108/ijoem-05-2024-0873

Mladenović, D., Monashev, M., Jirásek, M., & Bruni, R. (2025). When online review is not enough? Adoption of cryptocurrencies through the lenses of NCA and fsQCA. *Computers in Human Behavior*, *172*, 108764. https://doi.org/10.1016/j.chb.2025.108764

Mobeen, M., Kabir, K. H., Schneider, U. A., Ahmed, T., Ahmad, N., & Scheffran, J. (2026). Essential livelihood capitals for effective climate adaptation in Pakistan's irrigated agriculture: Insights from PLS-SEM and NCA. *Heliyon*, *12*, e44281. https://doi.org/10.1016/j.heliyon.2025.e44281

Mohammed, A., Al-Swidi, A. K., Ferraris, A., & Jabeen, F. (2025). Mobile payments in health crisis: A comprehensive approach to analyzing consumer behavior in emerging markets. *International Journal of Consumer Studies*, *49*. https://doi.org/10.1111/ijcs.70063

Mohsin, A., Rashed, Md., Gerschberger, M., Ahmed, S. F., Plasch, M., Rahman, A., & Noor, M. A. (2025). Can supply chains decarbonize? Exploring the potential of scope 3 emission reduction for sustainable transformation. *Sustainable Futures*, *10*, 101512. https://doi.org/10.1016/j.sftr.2025.101512

Mohsin, A., Rashed, Md., Gerschberger, M., Plasch, M., Ahmed, S. F., Dritan, O., & Islam, M. F. (2025). Smart supply chains: How do resilience and dynamic capabilities drive corporate performance? *The International Journal of Logistics Management*, *36*, 330–363. https://doi.org/10.1108/ijlm-12-2024-0783

Moon, J.-H., Lee, S.-T., & Lee, Y.-K. (2025). Impacts of general agents' customer relationship-building behaviors on trust, satisfaction, and renewal intention: A PLS-SEM and NCA approach. *Acta Psychologica*, *258*, 105162. https://doi.org/10.1016/j.actpsy.2025.105162

Morgado Oliveira, J., & Gomes, C. F. (2025). How EFQM Enablers unlock the full potential of TQM. *Total Quality Management & Business Excellence*, *36*, 1656–1670. https://doi.org/10.1080/14783363.2025.2589269

Muhlbacher, H., Einarsdóttir, B., & Castello, A. (2024). Necessary engagement in stakeholder relationships for successful membership in industrial clusters. *Industrial Marketing Management*, *118*, 1–11. https://doi.org/10.1016/j.indmarman.2024.02.001

Müller, K., Wickham, H., James, D. A., & Falcon, S. (2026). *RSQLite: SQLite interface for r*. https://rsqlite.r-dbi.org

Munir, M. A., Abbas, M. M., Qamar, A., Farooq, M., & Cohen, V. (2025). Managing volatile markets: A dynamic capability approach to analytics-driven performance in fast-moving supply chains for sustainable development. *Sustainable Futures*, *10*, 101338. https://doi.org/10.1016/j.sftr.2025.101338

Murtaza, S. H., Hussain, K., Salman, A., Mustafa, S. M., & Han, H. (in press). Exploring the nexus between green initiatives and hotel sustainable performance via circular economy practices. *European Journal of Management and Business Economics*. https://doi.org/10.1108/ejmbe-02-2025-0077

Mwesiumo, D. (in press). Identifying critical factors in higher education studies: Combining necessary condition and importance-performance map analyses. *Studies in Higher Education*. https://doi.org/10.1080/03075079.2025.2478948

Mwesiumo, D., Halpern, N., Bråthen, S., Budd, T., & Suau-Sanchez, P. (2023). Perceived benefits as a driver and necessary condition for the willingness of air passengers to provide personal data for non-mandatory digital services at airports. *Transportation*

Research Part A: Policy and Practice, *171*, 103659. https://doi.org/10.1016/j.tra.2023.103659

Mwesiumo, D., Harun, M., & Hogset, H. (2023). Unravelling the black box between coopetition and firms' sustainability performance. *Industrial Marketing Management*, *114*, 110–124. https://doi.org/10.1016/j.indmarman.2023.08.008

Mwesiumo, D., Similä, J. O., & Langseth, M. (2025). Performance orientation in public procurement: An empirical exploration. *Journal of Public Procurement*, *25*, 519–544. https://doi.org/10.1108/jopp-01-2024-0014

Newton, I. (1687). *Philosophiæ naturalis principia mathematica.* Royal Society. https://tile.loc.gov/storage-services/service/gdc/gdcwdl/wd/l_/17/84/2/wdl_17842/wdl_17842

Neyman, J., & Pearson, E. S. (1933). On the problem of the most efficient tests of statistical hypotheses. *Philosophical Transactions of the Royal Society of London. Series A, Containing Papers of a Mathematical or Physical Character*, *231*, 289–337. https://doi.org/10.1098/rsta.1933.0009

Nguyen, H. V., Nguyen, N., Le, B. N., Pervan, S., Vu, T. D., & Nguyen, T. M. N. (2026). Factors affecting consumers' energy curtailment behavior and purchases of energy-efficient appliances. *Journal of Retailing and Consumer Services*, *88*, 104504. https://doi.org/10.1016/j.jretconser.2025.104504

Nguyen, M. H., & Le, T. T. H. (2026). Playing business, achieving excellence: An NCA–MGA exploration of average and excellent business undergraduates in simulation-based learning. *The International Journal of Management Education*, *24*, 101347. https://doi.org/10.1016/j.ijme.2025.101347

Nguyen, N. S., Ngo, Q. D., Phan, T. M. C., Tao, V. A., & Bui, T. N. L. (2025). Understanding mHealth adoption among older adults: A novel NCA-fsQCA analysis. *Journal of Health Organization and Management*, *39*, 1565–1579. https://doi.org/10.1108/jhom-01-2025-0053

Nie, X., Qiu, L., Wu, T., Yang, X., Li, F., Feng, W., & Wang, H. (2025). Institutional pathways for enhancing villagers' participation in rural human settlement governance amid collective action dilemmas: Using NCA and fsQCA approaches. *Environmental Research Communications*, *7*, 075024. https://doi.org/10.1088/2515-7620/adf0ce

Nikookar, E., Ali, I., Stevenson, M., & Shokouhyar, S. (2025). Necessary antecedents of supply chain resilience: The nonnegotiable influence of supply chain responsiveness and collaboration. *Journal of Supply Chain Management*, *61*, 54–76. https://doi.org/10.1111/jscm.70000

Niranjan, S., Garg, V., Gligor, D. M., & Hawkins, T. G. (2025). Enhancing supply chain sustainability performance: The pivotal role of emerging technologies. *Journal of Business & Industrial Marketing*, *40*, 374–390. https://doi.org/10.1108/jbim-03-2024-0175

Niu, Z., Liu, W., Yao, J., Xiao, X., & Wu, J. (2025). A multidimensional analysis of potential suppliers' behavioral intentions in shared parking: Insights from SEM, NCA, and fsQCA. *Transport Policy*, *174*, 103861. https://doi.org/10.1016/j.tranpol.2025.103861

Norena-Chavez, D., Chavez-Limaylla, F., & Romani-Torres, R. (2025). Sustainable project success through entrepreneurial leadership: Insights from big data analytics, IPMA, and NCA integration. *Cogent Business & Management*, *12*, 2520913. https://doi.org/10.1080/23311975.2025.2520913

Novales, A., Mocker, M., Van Heck, E., & Dul, J. (2025). Realizing desired effects from digitized product affordances: A case study of key inhibiting factors. *Decision Support Systems*, *189*, 114365. https://doi.org/10.1016/j.dss.2024.114365

Ockham, W. of. (1323). *Summa logicae.*

Ofori, D. (2024). Necessary condition analysis of organisational capabilities for a resilient service operation in the hotel industry in Ghana. *Heliyon*, *10*(4), e26473. https://doi.org/10.1016/j.heliyon.2024.e26473

Ofori, D., Agyapong, G. K. Q., & Buer, S. B. (2025). Exploring the necessity of TQM practices for enhancing sustainability performance of manufacturing firms. *Journal of Manufacturing Technology Management*, *37*, 160–182. https://doi.org/10.1108/jmtm-01-2025-0008

Ogundipe, S. J., Peters, L. D., & Tóth, Z. (2022). Interfirm problem representation: Developing shared understanding within inter-organizational networks. *Industrial Marketing Management*, *100*, 76–87. https://doi.org/10.1016/j.indmarman.2021.11.004

Okeke, A., & Onyemere, I. (2025). Sustainable supply chain practices as catalyst for energy poverty alleviation in developing countries: A necessary condition analysis. *Discover Sustainability*, *6*. https://doi.org/10.1007/s43621-025-01003-4

Olafsen, A. H., & Marescaux, B. P. C. (in press). The bare necessities: A necessary condition analysis of the dual role of basic psychological need satisfaction and frustration on wellbeing at work. *Motivation and Emotion*. https://doi.org/10.1007/s11031-025-10113-5

Oliveira, E., Van der Heijden, B. I. J. M., Finsel, J. S., Axelrad, H., Derous, E., Gu, X., Marzec, I., Mykletun, R. J., Vignoli, M., Pajic, S., Wöhrmann, A. M., & Deller, J. (in press). Necessary HRM practices for extended working lives in tight and loose societies: A comparative perspective. *Human Resource Management*. https://doi.org/10.1002/hrm.70040

Oliveira, J. M., & Gomes, C. F. (2024). Critical factors for the success of sustainable development goal 12 in the digital transformation era: The effect of ISO 14001. *Business Strategy and the Environment*, *34*, 788–803. https://doi.org/10.1002/bse.4023

Oppong, D., & Bannor, R. K. (2024). Decoding influencer impact in food marketing: The must-haves, should-haves and the role of brand trust and price consciousness. *Journal of Food Products Marketing*, *30*(6–7), 210–235. https://doi.org/10.1080/10454446.2024.2406200

Osakwe, C. N., Říha, D., Elgammal, I. M. Y., & Ramayah, T. (2024). Understanding gen Z shoppers' interaction with customer-service robots: A cognitive-affective-normative perspective. *International Journal of Retail & Distribution Management*, *52*(13), 103–120. https://doi.org/10.1108/IJRDM-09-2023-0552

Osei, M. B., & Ofori, D. (2025). Cultural drivers of sustainability performance in emerging economies: Understanding the effect of supply chain learning. *Journal of Business Logistics*, *46*. https://doi.org/10.1111/jbl.70036

Ouazzani, Y., Calderón-García, H., & Tubillejas-Andrés, B. (2025). Unpacking the antecedents of word of mouth and electronic word of mouth in the opera sector: A multimethodological study based on PLS and NCA. *Empirical Studies of the Arts*, *43*(1), 77–104. https://doi.org/10.1177/02762374241245360

P., I., & Mekonnen, N. (2025). The role of ethical climate types in professional accountants' ethical decision-making: A necessary condition analysis (NCA). *Accounting Research Journal*, *38*, 581–608. https://doi.org/10.1108/arj-01-2025-0028

P., I., & Mekonnen, N. (in press). Moral intensity and ethical decision-making: A combined importance-performance map analysis for professional accountants. *International Journal of Ethics and Systems*. https://doi.org/10.1108/IJOES-05-2024-0120

Pandey, S., Thomas, S., Bhatt, V., Patel, R., & Malkar, V. (2024). An integrated SEM-ANN-NCA approach to predict the factors influencing CSR authenticity and CRM purchase intentions: An attribution theory perspective. *Journal of Marketing Theory and Practice*, *32*(4), 541–554. https://doi.org/10.1080/10696679.2023.2241158

Pangarso, A., Winarno, A., Aulia, P., & Ritonga, D. A. (2022). Exploring the predictor and the consequence of digital organisational culture: A quantitative investigation using sufficient and necessity approach. *Leadership & Organization Development Journal*, *43*(3), 370–385. https://doi.org/10.1108/LODJ-11-2021-0516

Pangarso, A., Yunani, A., Umbara, T., Hendriyanto, R., Sisilia, K., & Panutur, S. S. (2025). Uncovering the secrets of sustainable performance: Empirical studies on ethical behaviour, fortitude culture, ambidextrous governance and digital innovation using combined importance-performance map analysis. *Sustainable Futures*, *10*, 101181. https://doi.org/10.1016/j.sftr.2025.101181

Parker, S. K., & Knight, C. (2024). The SMART model of work design: A higher order structure to help see the wood from the trees. *Human Resource Management*, *63*(2), 265–291. https://doi.org/10.1002/hrm.22200

Pearl, J. (2009). *Causality*. Cambridge University Press. https://www.cambridge.org/core/books/causality/B0046844FAE10CBF274D4ACBDAEB5F5B

Peltokorpi, V., & Sekiguchi, T. (2023). The interaction of the five-factor personality traits and job embeddedness in explaining voluntary turnover: A necessary-condition perspective. *European Journal of Work and Organizational Psychology*, *32*(5), 688–701. https://doi.org/10.1080/1359432X.2023.2246216

Peng, B., Zhao, Y., Elahi, E., & Wan, A. (2022). Pathway and key factor identification of third-party market cooperation of China's overseas energy investment projects. *Technological Forecasting and Social Change*, *183*, 121931. https://doi.org/10.1016/j.techfore.2022.121931

Peng, F., & Hu, M. (2025). How Does Market Demand Incentivize Green Innovation in Resource-Based Industries? Evidence From China. *The American Journal of Economics and Sociology*, *85*, 113–127. https://doi.org/10.1111/ajes.70005

Pérez-Portabella, A., Arias-Oliva, M., Padilla-Castillo, G., & Andrés-Sánchez, J. de. (2026). Ethical conditions for university students' adoption of large language models in exam preparation contexts. *Computers and Education Open*, *10*, 100323. https://doi.org/10.1016/j.caeo.2025.100323

Pérez-Tapia, G., Cabrera-Sánchez, J.-P., & Hernández Correa, E. (in press). Innovative framework for the analysis of the effects of destination personality on behavioral results: The mediating role of self-congruity. *Journal of Vacation Marketing*. https://doi.org/10.1177/13567667251361151

Pillai, R., Sivathanu, B., & Rana, N. P. (2025). Unveiling virtual interactive marketplaces: Shopping motivations in the Metaverse through the lens of uses and gratifications theory. *Journal of Business Research*, *190*, 115219. https://doi.org/10.1016/j.jbusres.2025.115219

Pinochet, L. H. C., Souza, C. A. de, Viana, A. B. N., & Rodríguez-Abitia, G. (2024). Smart technologies in supermarket retail and their influence on citizens' quality of life during the COVID-19 pandemic. *Revista de Gestão*, *31*(1), 80–100. https://doi.org/10.1108/REGE-09-2021-0178

Pinthong, C., Inprasertkul, T., Rattanopas, C., Fakfare, P., Han, H., & Chua, B.-L. (2026). Spa social responsibility as a quality signal: Building customer loyalty through ethical and sustainable practices. *International Journal of Hospitality Management*, *133*, 104485. https://doi.org/10.1016/j.ijhm.2025.104485

Piriyakul, I., Kunathikornkit, S., & Piriyakul, R. (in press). Determining optimal marketing mix configurations: An integration of opinion-oriented language and necessary condition analysis. *Journal of Vacation Marketing*. https://doi.org/10.1177/13567667251351025

Poku, E., Nuertey, D., Agbemude, S., Owusu, F., & Buabeng, S. (in press). Building resilient supply chains through digitalization: The roles of supply chain traceability and desorptive capacity. *Benchmarking: An International Journal*. https://doi.org/10.1108/bij-01-2025-0057

Popper, K. (1959). *The logic of scientific discovery*. Hutchinson.

Popper, K. R. (1985). The problem of demarcation. In D. Miller (Ed.), *Popper selections* (pp. 118–130). Princeton. https://vdoc.pub/documents/popper-selections-29t8hfu8hp40

Porlán-Ariza, R., Villarejo-Ramos, Á. F., Ledesma-Chaves, P., & Gil-Cordero, E. (2026). Factors influencing the adoption of teaching innovations by university teachers: A comparative analysis using the FS/QCA-NCA methodology. *Thinking Skills and Creativity, 60*, 102059. https://doi.org/10.1016/j.tsc.2025.102059

Porter, M. E. (1985). *Competitive advantage of nations: Creating and sustaining superior performance*. The Free Press.

Pourazad, N., Stocchi, L., & Simmonds, L. (2025). Influencers and the choice of a travel destination: A customer journey and information processing perspective. *Information Technology & Tourism, 27*(4), 1185–1228. https://doi.org/10.1007/s40558-025-00330-6

Prakash, A., Jeyakumar Nathan, R., Kini, S., & Victor, V. (2022). Message framing and COVID-19 vaccine acceptance among millennials in South India. *PLOS ONE, 17*(7), e0269487. https://doi.org/10.1371/journal.pone.0269487

Qin, Y., Luo, C., Xu, Z., Wang, X., & Škare, M. (2025). Decoding tourist satisfaction for sustainable economic development: A multi-method configuration framework using online reviews. *Technological and Economic Development of Economy, 31*, 1801–1839. https://doi.org/10.3846/tede.2025.23251

Quaicoe, J., Acquah, I. S. K., & Gatsi, J. G. (2024). Unravelling the nice-to-have and must-have circular economy-oriented dynamic capabilities for sustainable product design and end-of-life management: Insights from PLS-SEM and NCA. *Journal of Cleaner Production, 479*, 144004. https://doi.org/10.1016/j.jclepro.2024.144004

Quan, Z., Wu, J., & Zhang, J. (in press). Sustainable learning in sight: A multi-analytical investigation into potential impact of extended reality (XR) on knowledge retention and skill development. *Asia Pacific Journal of Education*. https://doi.org/10.1080/02188791.2025.2596922

Quansah, T. K., Lang, M., & Mühlbacher, H. (2025). Deriving pleasure from other people's pain: The influence of vicarious pain on viewing intentions for violent sports. *Journal of Business Research, 199*, 115533. https://doi.org/10.1016/j.jbusres.2025.115533

Rafi, A., Roy, S. K., Rehman, M. A., & Hasni, M. J. S. (2025). Impact of multidimensional presence on user well-being in metaverse communities. *Decision Support Systems, 189*, 114382. https://doi.org/10.1016/j.dss.2024.114382

Ragin, C. C. (1987). *The comparative method: Moving beyond qualitative and quantitative strategies*. JSTOR. https://www.jstor.org/stable/10.1525/j.ctt1pnx57?turn_away=true

Ragin, C. C. (2000). *Fuzzy-set social science*. University of Chicago Press. https://press.uchicago.edu/ucp/books/book/chicago/F/bo3635786.html

Ragin, C. C. (2008). *Redesigning social inquiry: Fuzzy sets and beyond*. University of Chicago Press. https://press.uchicago.edu/ucp/books/book/chicago/R/bo5973952.html

Ragin, C. C. (2009). Qualitative comparative analysis using fuzzy-sets (fsQCA). In B. Rihoux & C. C. Ragin (Eds.), *Configurational comparative methods: Qualitative comparative analysis (QCA) and related techniques* (pp. 87–121). Sage. https://methods.sagepub.com/book/edvol/configurational-comparative-methods/toc#_

Rahimi, A. R., & Daneshvar Ghorbani, B. (2025). Toward the Language MOOC (LMOOC's) low dropout rate: The control-value theory of persistency in LMOOC (CVTPLMOOC). *Computers in Human Behavior Reports, 17*, 100593. https://doi.org/10.1016/j.chbr.2025.100593

Rahimi, A. R., & Sevilla-Pavón, A. (2024). The role of ChatGPT readiness in shaping language teachers' language teaching innovation and meeting accountability: A bisymmetric approach. *Computers and Education: Artificial Intelligence, 7*, 100258. https://doi.org/10.1016/j.caeai.2024.100258

Rahimi, A. R., & Sevilla-Pavón, A. (2025). The role of interactive, constructive, active, and passive learning activities (ICAPCALL) in shaping students' online engagement, and learning approaches to virtual exchange (SAVE): A bisymmetric approach. *Smart Learning Environments, 12*. https://doi.org/10.1186/s40561-024-00363-0

Rahman, Md. M., & Sadik, N. (2025). Nexus of digital financial service adoption, decent work and environmental sustainability: Results from SEM-NCA approaches. *Discover Sustainability, 6*. https://doi.org/10.1007/s43621-025-00972-w

Rahman, S. M. M., Tafsirun, U., Faisal-E-Alam, Md., Ferreira, P., Loures, L. s., & Castanho, R. A. (2025). Unveiling perceptions on academic leadership effectiveness: PLS-SEM, FSQCA, and NCA approaches. *PLOS ONE, 20*, e0320723. https://doi.org/10.1371/journal.pone.0320723

Rahman, S. M., Carlson, J., Chowdhury, N. H., Gudergan, S. P., Wetzels, M., Ringle, C. M., & Grewal, D. (2026). Omnichannel safe customer experience: How should it be measured? Does it affect customer well-being and retailers' performance? *Journal of Business Research, 202*, 115760. https://doi.org/10.1016/j.jbusres.2025.115760

Rashed, Md., Uddin, Md. K., Islam, M. F., Mohsin, A., Rahman, S., Kumar, S., & Rana, J. (2025). Building organizational resilience in emerging economies: Strategic insights from Bangladesh. *Sustainable Futures, 10*, 101327. https://doi.org/10.1016/j.sftr.2025.101327

Rath, S. P., & Jain, N. K. (in press). Validating the efficacy of the GreenSCOR model in creating sustainability performance for manufacturing companies: A mixed-method analysis. *Benchmarking: An International Journal*. https://doi.org/10.1108/bij-09-2024-0740

Ren, X., Hu, L., & Zhang, A. (2025). The impact of multilateral competitive and cooperative relationships in the platform ecosystem on the iterative innovation of platform-dependent enterprises: findings from NCA and fsQCA. *European Journal of Innovation Management, 28*, 5104–5127. https://doi.org/10.1108/ejim-12-2024-1425

Ren, Z., & Liu, Q. (2025). How digital transformation drives the value enhancement of SMTEs from the perspective of technology, business and social interaction. *Social Indicators Research, 178*, 1367–1390. https://doi.org/10.1007/s11205-025-03522-z

Richter, N. F., & Hauff, S. (2022). Necessary conditions in international business research–advancing the field with a new perspective on causality and data analysis. *Journal of World Business, 57*(5), 101310. https://doi.org/10.1016/j.jwb.2022.101310

Richter, N. F., Hauff, S., Kolev, A. E., & Schubring, S. (2023). Dataset on an extended technology acceptance model: A combined application of PLS-SEM and NCA. *Data in Brief, 48*, 109190. https://doi.org/10.1016/j.dib.2023.109190

Richter, N. F., Hauff, S., Ringle, C. M., & Gudergan, S. P. (2022). The use of partial least squares structural equation modeling and complementary methods in international management research. *Management International Review, 62*(4), 449–470. https://doi.org/10.1007/s11575-022-00475-0

Richter, N. F., Hauff, S., Ringle, C. M., Sarstedt, M., Kolev, A. E., & Schubring, S. (2023). How to apply necessary condition analysis in PLS-SEM. In *Partial least squares path modeling: Basic concepts, methodological issues and applications* (pp. 267–297). Springer. https://link.springer.com/chapter/10.1007/978-3-031-37772-3_10

Richter, N. F., Martin, J., Hansen, S. V., Taras, V., & Alon, I. (2021). Motivational configurations of cultural intelligence, social integration, and performance in global virtual teams. *Journal of Business Research, 129*, 351–367. https://doi.org/10.1016/j.jbusres.2021.03.012

Richter, N. F., Schlaegel, C., Bakel, M. V., & Engle, R. L. (2020). The expanded model of cultural intelligence and its explanatory power in the context of expatriation intention. *European Journal of International Management, 14*(2), 381–419. https://doi.org/10.1504/EJIM.2020.105545

Richter, N. F., Schubring, S., Hauff, S., Ringle, C. M., & Sarstedt, M. (2020). When predictors of outcomes are necessary: Guidelines for the combined use of PLS-SEM and NCA.

Industrial Management & Data Systems, *120*(12), 2243–2267. https://doi.org/10.1108/IMDS-11-2019-0638

Rico, J. P., Calvo-Mora, A., Medina-Molina, C., & Alves, H. (2024). Quality 4.0 social and strategic readiness factors: Sufficient and necessary condition analysis. *Total Quality Management & Business Excellence*, *35*(5–6), 559–583. https://doi.org/10.1080/14783363.2024.2323185

Riggs, R., Felipe, C. M., Roldán, J. L., & Real, J. C. (2025). Deepening big data sustainable value creation: Insights using IPMA, NCA, and cIPMA. *Journal of Marketing Analytics*, *13*, 1000–1019. https://doi.org/10.1057/s41270-024-00321-2

Ringle, C. M., & Sarstedt, M. (2016). Gain more insight from your PLS-SEM results: The importance-performance map analysis. *Industrial Management & Data Systems*, *116*(9), 1865–1886. https://doi-org.eur.idm.oclc.org/10.1108/IMDS-10-2015-0449

Rishi, B., Shiva, A., & Israney, R. S. (2024). Consumer willingness to buy and pay for dog-human companionship: A combination of SEM and NCA approaches. *Asia Pacific Journal of Marketing and Logistics*, *36*(12), 3533–3552. https://doi.org/10.1108/APJML-12-2023-1271

Robinson, S., Muratbekova-Touron, M., Linder, C., Bouncken, R. B., Findikoglu, M. N., Garbuio, M., Hartner-Tiefenthaler, M., Thanos, I. C., Aharonson, B. S., Strobl, A., et al. (2022). 40th anniversary editorial: Looking backwards to move forward in management research. *European Management Journal*, *40*(4), 459–466. https://doi.org/10.1016/j.emj.2022.07.002

Robinson, W. S. (1951). The logical structure of analytic induction. *American Sociological Review*, *16*(6), 812–818. https://doi.org/10.2307/2087508

Rocha, R. G., Ed-Dafali, S., & Sharma, N. (2025). Entrepreneurial orientation, agility and responsibility: A triad for sustainable competitive advantage. *Journal of Small Business and Enterprise Development*, *32*, 470–495. https://doi.org/10.1108/jsbed-04-2024-0224

Rogers, C. R. (1957). The necessary and sufficient conditions of therapeutic personality change. *Journal of Consulting Psychology*, *21*(2), 95. https://doi.org/10.1037/0033-3204.44.3.240

Rohlfing, I., & Schneider, C. Q. (2013). Improving research on necessary conditions: Formalized case selection for process tracing after QCA. *Political Research Quarterly*, *66*(1), 220–235. https://www.jstor.org/stable/23563606

Romero-Charneco, M., Casado-Molina, A.-M., Alarcón-Urbistondo, P., & Cabrera Sánchez, J. P. (2025). Customer intentions toward the adoption of WhatsApp chatbots for restaurant recommendations. *Journal of Hospitality and Tourism Technology*, *16*. https://doi.org/10.1108/JHTT-01-2024-0024

Rosli, N., Binti Mohamad, A. D., Johar, E. R., Hashim, S., & Lim, X.-J. (2024). Beyond borders: The transcendent effect of Korean celebrity credibility on brand perceptions among Malaysian youth - a necessary condition analysis. *Heliyon*, *11*, e41155. https://doi.org/10.1016/j.heliyon.2024.e41155

Rothenberg, T. J. (1971). Identification in parametric models. *Econometrica*, *39*(3), 577–591. https://doi.org/10.2307/1913267

Rothman, K. J. (1976). Causes. *American Journal of Epidemiology*, *104*(6), 587–592. https://doi.org/10.1093/oxfordjournals.aje.a112335

Rothman, K. J., & Greenland, S. (2005). Causation and causal inference in epidemiology. *American Journal of Public Health*, *95*(S1), S144–S150. https://doi.org/10.2105/AJPH.2004.059204

Roxas, B. (2024). E-governance and sustainable human development in Asia: A dynamic institutional path perspective. *Journal of Asian Business and Economic Studies*, *32*, 15–27. https://doi.org/10.1108/JABES-02-2024-0076

Rozenkowska, K. (2023). Theory of planned behavior in consumer behavior research: A systematic literature review. *International Journal of Consumer Studies*, *47*(6), 2670–2700. https://doi.org/10.1111/ijcs.12970

Rubinson, C. (2019). Presenting qualitative comparative analysis: Notation, tabular layout, and visualization. *Methodological Innovations*, *12*(2), 2059799119862110. https://doi.org/10.1177/2059799119862110

Rubinson, C. (2025). *Fiss*. https://grundrisse.org/qca.

Ruiz-Equihua, D., Romero, J., Loureiro, S. M. C., & Ali, M. (2023). Human–robot interactions in the restaurant setting: The role of social cognition, psychological ownership and anthropomorphism. *International Journal of Contemporary Hospitality Management*, *35*(6), 1966–1985. https://doi.org/10.1108/IJCHM-05-2022-0597

Rulandari, N., & Silalahi, A. D. K. (2025a). Achieving effectiveness of public service in AI-enabled service from public value theory: Does human–AI collaboration matters? *Transforming Government: People, Process and Policy*, *19*, 428–452. https://doi.org/10.1108/tg-01-2025-0004

Rulandari, N., & Silalahi, A. D. K. (2025b). Human-AI collaboration for efficiency and employee job satisfaction in public administration: Insights from a resource-based perspective. *Transforming Government: People, Process and Policy*, *19*, 264–287. https://doi.org/10.1108/tg-01-2025-0011

Safdar, N., Moazzam, M., Ahmed, W., Khan, A. S., Manzoor, W., & Raziq, M. M. (2024). The impact of adopting green procurement practices on the competitiveness of small and medium enterprises: A dual-mediation model. *Benchmarking: An International Journal*, *33*, 105–128. https://doi.org/10.1108/BIJ-04-2024-0297

Samagaio, A., Francisco, P. M., Felício, T., & Verga Matos, P. (in press). The relationship between time pressure, ethical culture and audit quality. *Management Decision*. https://doi.org/10.1108/md-04-2024-0870

Sánchez-Medina, A. J., Pellejero, M., C-Sánchez, E., & Cerviño-Cortínez, D. L. (in press). Attitude towards the profession, emotional exhaustion, cynicism, and intention to abandon tourism SMEs as an entrepreneur: A conditional mediation model. *Current Issues in Tourism*. https://doi.org/10.1080/13683500.2024.2421438

Santana, A. de S., Martens, C. D. P., Bizarrias, F. S., & Martens, M. L. (2025). The role of organizational factors in enabling, and of autonomy in potentializing intrapreneurial behavior: Mechanisms and preconditions. *BAR - Brazilian Administration Review*, *22*. https://doi.org/10.1590/1807-7692bar2025240065

Santos-Jaén, J. M., Pemartín, M., Gimeno-Arias, F., & León-Gómez, A. (2025). Necessary condition analysis for sustainable channel governance in fast-moving markets. *Journal of Business & Industrial Marketing*, *40*, 1863–1878. https://doi.org/10.1108/jbim-10-2024-0804

Sanyal, P., Jain, N. K., & Choudhury, D. (2024). What specific investments are critical in fostering buyer-supplier relationships: Evidence from MSMEs in emerging economies. *International Social Science Journal*, *74*(251), 219–239. https://doi.org/10.1111/issj.12456

Saraf, P. K., Saha, S., & Anand, A. (2022). Supportive culture and job involvement in public sector: The mediating role of participation in decision making and organizational learning. *International Journal of Public Sector Management*, *35*(5), 549–567. https://doi.org/10.1108/IJPSM-09-2021-0218

Sarstedt, M., & Liu, Y. (2023). Advanced marketing analytics using partial least squares structural equation modeling (PLS-SEM). In *Journal of Marketing Analytics* (pp. 1–5). Springer. https://doi.org/10.1057/s41270-023-00279-7

Sarstedt, M., Richter, N. F., Hauff, S., & Ringle, C. M. (2024). Combined importance–performance map analysis (cIPMA) in partial least squares structural equation modeling

(PLS–SEM): A SmartPLS 4 tutorial. *Journal of Marketing Analytics*, *12*, 746–760. https://doi.org/10.1057/s41270-024-00325-y

Sathatip, P., Agmapisarn, C., Suttipun, M., & Fakfare, P. (2025). Identifying probabilistic and necessary conditions for sustainable food purchasing behaviors: Insights from surplus food platforms. *British Food Journal*, *127*, 3073–3099. https://doi.org/10.1108/bfj-01-2025-0091

Sathatip, P., Senachai, P., Leruksa, C., & Fakfare, P. (2025). Cultivating ethical culinary practices and sustainability awareness in culinary education: Fostering responsible future chefs. *Journal of Hospitality, Leisure, Sport & Tourism Education*, *36*, 100531. https://doi.org/10.1016/j.jhlste.2024.100531

Schelvis, R. M. C., Veen, M. van, Tamminga, S. J., Oude Hengel, K. M., Nieuwenhuijsen, K., Boot, C. R. L., & Dul, J. (2025). Testing psychosocial work adversities as a necessary condition for work-related emotional exhaustion in young workers: a cross-sectional necessary condition analysis on a national general working population-based survey. *BMJ Open*, *15*, e094485. https://doi.org/10.1136/bmjopen-2024-094485

Schneider, C. Q., & Wagemann, C. (2012). *Set-theoretic methods for the social sciences: A guide to qualitative comparative analysis*. Cambridge University Press. https://www.cambridge.org/nl/academic/subjects/social-science-research-methods/qualitative-methods/set-theoretic-methods-social-sciences-guide-qualitative-comparative-analysis?format=PB

Schubring, S., & Richter, N. (2023). *Extended TAM* (Version V4). Mendeley Data. https://doi.org/10.17632/pd5dp3phx2.4

Sekar, D., Veerasamy, S., Michellin Kiruba, N. J., & Ranganathan, M. (in press). Beyond the threshold: Exploring intelligence and creative potential in Indian middle school students. *Journal of Human Behavior in the Social Environment*. https://doi.org/10.1080/10911359.2025.2550489

Sen, P., Roy, A., Laradi, S., Das, A., Ray, A., Parida, R., & Das, S. (2026). Young consumers' attitudes toward selecting green delivery option of ready-to-eat food products: A case of West Bengal, India using extended theory of planned behaviour model. *Journal of Cleaner Production*, *538*, 147291. https://doi.org/10.1016/j.jclepro.2025.147291

Sepe, F., Luongo, S., Di Gioia, L., & Della Corte, V. (2024). Cultural heritage experiences in the metaverse: Analyzing perceived value and behavioral intentions. *European Journal of Innovation Management*, *28*, 2054–2079. https://doi.org/10.1108/EJIM-05-2024-0557

Servajean-Hilst, R., & Suurmond, R. (2023). Purchasing team innovation climates as antecedents of purchasing involvement in open innovation. *International Journal of Innovation Management*, *27*(09n10), 2350048. https://doi.org/10.1142/S1363919623500482

Shahjehan, A., & Qureshi, J. A. (2019). Personality and impulsive buying behaviors. A necessary condition analysis. *Economic Research-Ekonomska Istraživanja*, *32*(1), 1060–1072. https://doi.org/10.1080/1331677X.2019.1585268

Shan, J., Deng, Z., Deng, Z., & Song, D. (2025). Emotional resonance and self-esteem: The role of humanlike competencies from mental health chatbots. *IEEE Transactions on Engineering Management*, *72*, 4281–4293. https://doi.org/10.1109/tem.2025.3625427

Shan, X., Wang, Y., & Luo, L. (2025). The effect of digital music education on non-cognitive abilities: Empirical evidence from China. *Acta Psychologica*, *260*, 105564. https://doi.org/10.1016/j.actpsy.2025.105564

Shao, Q., Jiang, C., Liou, J. J. H., Su, P., Yuan, Y., & Dan, Z. (2024). Exploring the influencing factors of digital transformation: Empirical results from SMEs in China. *Managerial and Decision Economics*, *46*, 2364–2380. https://doi.org/10.1002/mde.4463

Sharma, A., Dwivedi, R., Mariani, M. M., & Islam, T. (2022). Investigating the effect of advertising irritation on digital advertising effectiveness: A moderated mediation model.

Technological Forecasting and Social Change, *180*, 121731. https://doi.org/10.1016/j.techfore.2022.121731

Sharma, N., & Paço, A. (2025). Exploring the O zone effect: How own, others and outer influences shape green purchasing behaviour in e-commerce. *Journal of Information, Communication and Ethics in Society*, *23*, 234–261. https://doi.org/10.1108/jices-09-2024-0137

Sharma, P. N., Sarstedt, M., Ringle, C. M., Cheah, J.-H., Herfurth, A., & Hair, J. F. (2024). A framework for enhancing the replicability of behavioral MIS research using prediction oriented techniques. *International Journal of Information Management*, *78*, 102805. https://doi.org/10.1016/j.ijinfomgt.2024.102805

Sharma, S., Kar, A. K., & Gupta, M. P. (2024). Untangling the web between digital citizen empowerment, accountability and quality of participation experience for e-government: Lessons from India. *Government Information Quarterly*, *41*(3), 101964. https://doi.org/10.1016/j.giq.2024.101964

Shen, J., Zhang, H., Wong, C. U. I., Chen, J., & Ren, L. (2025). Relationship between childhood socio-economic status and wellness tourism intention: A combined PLS-SEM and NCA methods. *Humanities and Social Sciences Communications*, *12*. https://doi.org/10.1057/s41599-025-05177-0

Shen, J., Zhang, H., Zheng, J., & Ruan, M. (2025). How do self-identity and motivational foundation influence Chinese prosocial behavior intention? Findings from NCA and fsQCA. *SAGE Open*, *15*. https://doi.org/10.1177/21582440251326350

Shen, Y., Yang, H., & Zhu, Q. (2025). How does the institutional environment improve the entrepreneurial quality of returnees? A configuration analysis based on a complex system view. *PLOS ONE*, *20*, e0322863. https://doi.org/10.1371/journal.pone.0322863

Si, H., Li, N., Duan, X., Cheng, L., & Bao, Z. (2023). Understanding the public's willingness to participate in the carbon generalized system of preferences (CGSP): An innovative mechanism to drive low-carbon behavior in China. *Sustainable Production and Consumption*, *38*, 1–12. https://doi.org/10.1016/j.spc.2023.03.016

Siddik, A. B., Yong, L., Du, A. M., & Goodell, J. W. (2025). Exploring AI-driven digital banking platforms: Implications for business model innovation and sustainability in the financial sector. *IEEE Transactions on Engineering Management*, *72*, 3318–3334. https://doi.org/10.1109/tem.2025.3591773

Siemsen, E., Roth, A. V., & Balasubramanian, S. (2008). How motivation, opportunity, and ability drive knowledge sharing: The constraining-factor model. *Journal of Operations Management*, *26*(3), 426–445. https://doi.org/10.1016/j.jom.2007.09.001

Sievert, C., Parmer, C., Hocking, T., Chamberlain, S., Ram, K., Corvellec, M., & Despouy, P. (2024). *Plotly: Create interactive web graphics via plotly.js*. https://plotly-r.com

Silalahi, A. D. K. (in press). Frustration dissonance and the paradox of persistence in ChatGPT usage: Insights from Indonesia and Taiwan. *International Journal of Human–Computer Interaction*. https://doi.org/10.1080/10447318.2025.2567967

Similä, J. O., & Mwesiumo, D. (2024). Implementing public procurement of green innovations: Does structural alignment matter? *Journal of Cleaner Production*, *461*, 142562. https://doi.org/10.1016/j.jclepro.2024.142562

Simon, H. A. (1957). *Models of man: Social and rational*. Wiley.

Singh, A. K., Jain, N. K., Sharma, M. G., & Nigam, S. (2023). Reconceptualization of absorptive capacity as potential and realized absorptive capacity for project-based organizations. *International Journal of Project Management*, *41*(2), 102449. https://doi.org/10.1016/j.ijproman.2023.102449

Singh Yadav, S., & Kumar Kar, S. (2025). Influence of religious value and minimalism on repurchase intention of sustainable fashion: A combined PLS-SEM and NCA approach.

Journal of Cleaner Production, *486*, 144585. https://doi.org/10.1016/j.jclepro.2024.144585

Sirisety, M., Nekkanti, M. R., Datti, R. S., Bikkina, N., Patrakov, E. V., & Baturina, L. (2025). Necessary condition analysis on the relationship between fear of missing out, social networking addiction, and psychological well-being. *European Journal of Mental Health*, *20*, 1–16. https://doi.org/10.5708/ejmh.20.2025.0042

SmartPLS Development Team. (2023). *SmartPLS* (Version 4) [Computer software]. SmartPLS GmbH, Germany. https://www.smartpls.com/

Sohaib, O., Alshemeili, A., & Bhatti, T. (2025). Exploring AI-enabled green marketing and green intention: An integrated PLS-SEM and NCA approach. *Cleaner and Responsible Consumption*, *17*, 100269. https://doi.org/10.1016/j.clrc.2025.100269

Solaimani, S., & Swaak, L. (2023). Critical success factors in a multi-stage adoption of artificial intelligence: A necessary condition analysis. *Journal of Engineering and Technology Management*, *69*, 101760. https://doi.org/10.1016/j.jengtecman.2023.101760

Spinelli, D., Dul, J., & Buijs, G. (2023). *nca: Stata module to perform necessary condition analysis (NCA)* (Version 0.0.1) [Computer software]. Statistical Software Components S459269, Boston College Department of Economics. https://ideas.repec.org/c/boc/bocode/s459269.html

Stanford, K. (2023). Underdetermination of scientific theory. In E. N. Zalta & U. Nodelman (Eds.), *The stanford encyclopedia of philosophy (fall 2023 edition)*. Metaphysics Research Lab, Stanford University. https://plato.stanford.edu/archives/fall2023/entries/scientific-underdetermination/

Stek, K., & Schiele, H. (2021). How to train supply managers–necessary and sufficient purchasing skills leading to success. *Journal of Purchasing and Supply Management*, *27*(4), 100700. https://doi.org/10.1016/j.pursup.2021.100700

Su, D. N., Nguyen-Phuoc, D. Q., Tran, P. T. K., Van Nguyen, T., Luu, T. T., & Pham, H.-G. (2023). Identifying must-have factors and should-have factors affecting the adoption of electric motorcycles – a combined use of PLS-SEM and NCA approach. *Travel Behaviour and Society*, *33*, 100633. https://doi.org/10.1016/j.tbs.2023.100633

Su, Z., Lai, S., & Zhang, M. (2024). Post-pandemic outdoor music festival consumption behaviour intention: An integrative approach. *Leisure Studies*, *44*, 428–448. https://doi.org/10.1080/02614367.2024.2324824

Subramanian, B., Jain, N. K., & Bhattacharyya, S. S. (2024). Exploring the direct impact of environmental, social and governance (ESG) factors on organisational innovation output and institutional isomorphism: A study in the context of multinational life sciences organisations. *Benchmarking: An International Journal*, *33*, 129–166. https://doi.org/10.1108/BIJ-05-2024-0391

Suganthi, A., & Mohamed Jasim, K. (2025). Exploring the antecedents of users' intention on virtual tours and video walkthroughs in rental accommodation apps – an integrated PLS-SEM-ANN-NCA approach. *Journal of Internet Commerce*, *24*, 89–129. https://doi.org/10.1080/15332861.2025.2497970

Sukhov, A., Olsson, L. E., & Friman, M. (2022). Necessary and sufficient conditions for attractive public transport: Combined use of PLS-SEM and NCA. *Transportation Research Part A: Policy and Practice*, *158*, 239–250. https://doi.org/10.1016/j.tra.2022.03.012

Sun, C., Xia, E., Huang, J., & Tong, H. (2024). Coupled coordination and pathway analysis of food security and carbon emission efficiency under climate-smart agriculture orientation. *Science of the Total Environment*, *948*, 174706. https://doi.org/10.1016/j.scitotenv.2024.174706

Sun, H., Fu, L., & Yang, S. (2023). What contributes to the government-citizen knowledge sharing: Analysis of 293 cities in China. *Journal of Innovation & Knowledge*, *8*(2), 100362. https://doi.org/10.1016/j.jik.2023.100362

Sun, H., & Zhou, L. (in press). A configurational approach for local-government policy reinvention: A study on Chinese policy. *Review of Policy Research*. https://doi.org/10.1111/ropr.70046

Sun, K., Ooi, K.-B., Wei-Han Tan, G., & Lee, V.-H. (2024). Small and medium-sized enterprises' path to sustainable supply chains: Exploring the role of supply chain finance and risk management. *Supply Chain Management: An International Journal*, *30*, 1–18. https://doi.org/10.1108/SCM-08-2023-0437

Suryavanshi, A. K. S., Thomas, S., Trivedi, T., Patel, R., & Bhatt, V. (2024). Predicting user loyalty and repeat intention to donate towards fantasy sports gaming platforms: A large-sample study based on a model integrating philanthropic actions, well-being and flow experience. *Journal of Philanthropy and Marketing*, *29*(1), e1819. https://doi.org/10.1002/nvsm.1819

Talib, S., Papastathopoulo, A., & Ahmad, S. Z. (2024). Sufficiency and necessity of big data capabilities for decision performance in the public sector. *Digital Policy, Regulation and Governance*, *26*(1), 18–37. https://doi.org/10.1108/DPRG-05-2023-0057

Tan, J., & Wei, J. (2023). Configurational analysis of ESG performance, innovation intensity, and financial leverage: A study on total factor productivity in Chinese pharmaceutical manufacturing firms. *Journal of the Knowledge Economy*, *15*(3), 13803–13827. https://doi.org/10.1007/s13132-023-01678-y

Tan, K.-L., Leong, C.-M., & Richter, N. F. (2025). Navigating trust in mobile payments: Using necessary condition analysis to identify must-have factors for user acceptance. *International Journal of Human–Computer Interaction*, *41*, 3325–3339. https://doi.org/10.1080/10447318.2024.2338319

Tang, L., Zhao, H., Zhou, Z., Li, Q., Qian, Z., Hou, S., & Liu, B. (2025). Spatio-temporal distribution characteristics and driving factors of urban resilience in the Yangtze River Delta. *Environmental Geochemistry and Health*, *47*. https://doi.org/10.1007/s10653-025-02691-7

Tao, Y., Wang, H., He, J., Zhang, Z., & Liu, H. (2025). How does female representation affect long-term firm performance? Capturing causal complexity with a configurational approach. *Gender in Management: An International Journal*, *40*, 114–133. https://doi.org/10.1108/GM-07-2023-0258

Teece, D. J. (2007). Explicating dynamic capabilities: The nature and microfoundations of (sustainable) enterprise performance. *Strategic Management Journal*, *28*(13), 1319–1350. https://doi.org/10.1002/smj.640

Teece, D. J., Pisano, G., & Shuen, A. (1997). Dynamic capabilities and strategic management. *Strategic Management Journal*, *18*(7), 509–533. https://doi.org/10.1002/(SICI)1097-0266(199708)18:7<509::AID-SMJ882>3.0.CO;2-Z

Teston, S. de F., Zawadzki, P., Machado, H. P. V., & Cazella, C. F. (2025). What are the necessary psychological well-being conditions for women entrepreneurs to feel effective? *BAR - Brazilian Adminstration Review*, *22*, e240107. https://doi.org/10.1590/1807-7692bar2025240107

Tetteh, F. K., & Owusu Kwateng, K. (in press). Digital pathways to supply chain viability: Exploring nice and must-to have practices in dynamic environment – insight from SEM and NCA. *Journal of Enterprise Information Management*. https://doi.org/10.1108/jeim-05-2025-0367

Tetteh, F. K., Owusu Kwateng, K., Obiri-Yeboah, H., Amoako, D. K., & Nyamekye, B. (2025). Unpacking the nice-to-have and must-have antecedents and outcomes of supplier commitment. *Journal of Manufacturing Technology Management*, *36*, 1457–1491. https://doi.org/10.1108/jmtm-01-2025-0052

Tetteh, F. K., Owusu Kwateng, K., Owusu-Antwi, C., & Opoku-Mensah, S. (in press). Promoting sustainable end-of-life management and intelligence integration through supply chain digitalization: based on CB-SEM and NCA approaches. *Business Process Management Journal.* https://doi.org/10.1108/bpmj-05-2025-0643

Thien, L. M., Chuah, S. T., & Low, M. P. (2024). Understanding middle leaders' distributed leadership factors as necessary and sufficient conditions for fostering teacher commitment to change. *European Journal of Education, 59*(4), e12757. https://doi.org/10.1111/ejed.12757

Tho, N. D. (2018). Firm capabilities and performance: A necessary condition analysis *Journal of Management Development, 37*(4), 322–332. https://doi.org/10.1108/JMD-06-2017-0204

Tho, N. D. (2019a). Business students' hardiness and its role in quality of university life, quality of life, and learning performance. *Education + Training, 61*(3), 374–386. https://doi.org/10.1108/ET-03-2018-0068

Tho, N. D. (2019b). Strategic orientations and firm innovativeness: A necessary condition analysis. *Baltic Journal of Management, 14*(3), 427–442. https://doi.org/10.1108/BJM-07-2018-0280

Tho, N. D., Phong, N. D., Quan, T. H. M., & Trang, N. T. M. (2025). Corporate social responsibility and employees' quality of work life. *Baltic Journal of Management, 20*, 266–281. https://doi.org/10.1108/bjm-06-2024-0300

Tho, N. D., Trang, N. T. M., & Thu, N. N. Q. (2025). Ambidextrous leadership and innovation ambidexterity in a business function: The role of managers' psychological capital and proactive personality. *Journal of Knowledge Management, 29*, 1446–1464. https://doi.org/10.1108/jkm-09-2024-1109

Thu, N. N., Kadirov, D., Fam, K.-S., & Tho, N. D. (2023). Firms' sustainability marketing commitment: The roles of chief marketing officers' future focus and prosocial motivation. *Journal of Macromarketing, 43*(3), 306–321. https://doi.org/10.1177/02761467231174027

Tian, S., Wang, R., Li, H., Chen, L., Kuang, Y., & Mao, J. (2025). Predictive analysis of miners' group unsafe behavior based on group dynamics and institutional environment. *Scientific Reports, 15*, 9263. https://doi.org/10.1038/s41598-025-89860-x

Tingli, J., Aw, E. C.-X., Huam, H.-T., & Tan, G. W.-H. (in press). Artificial intelligence at the wheel: Exploring tourists' trust in autonomous vehicles through a time-lagged survey. *Journal of Hospitality and Tourism Technology.* https://doi.org/10.1108/jhtt-04-2025-0334

Tiwari, P., Kaurav, R. P. S., & Koay, K. Y. (2024). Understanding travel apps usage intention: Findings from PLS and NCA. *Journal of Marketing Analytics, 12*(1), 25–41. https://doi.org/10.1057/s41270-023-00258-y

Torres, P., & Godinho, P. (2022). Levels of necessity of entrepreneurial ecosystems elements. *Small Business Economics, 59*, 29–45. https://doi.org/10.1007/s11187-021-00515-3

Torres, P., & Godinho, P. (in press). National competitive advantage in the digital age: A necessary condition analysis. *Competitiveness Review: An International Business Journal.* https://doi.org/10.1108/cr-05-2025-0160

Tóth, Z., Dul, J., & Li, C. (2019). Necessary condition analysis in tourism research. *Annals of Tourism Research, 79*, 102821. https://doi.org/10.1016/j.annals.2019.102821

Tripathi, S. N., Husain, R., & Rana, N. P. (2025). Engineering emotional bonds: How anthropomorphic design shapes guest-robot relationships in hospitality. *Journal of Hospitality Marketing & Management, 35*, 45–72. https://doi.org/10.1080/19368623.2025.2552202

Troiville, J., Moisescu, O. I., & Radomir, L. (2025). Using necessary condition analysis to complement multigroup analysis in partial least squares structural equation modeling. *Journal of Retailing and Consumer Services, 82*, 104018. https://doi.org/10.1016/j.jretconser.2024.104018

Tuesta-Tapia, A. C., Vázquez-Sánchez, A., & Amador-Hidalgo, L. (2024). Excellence in the management of local action groups: An empirical investigation of critical success factors. *International Journal of Quality & Reliability Management*, *42*, 1249–1273. https://doi.org/10.1108/IJQRM-02-2024-0058

Tumaku, J., Ren, J., Boakye, K. G., Ofori, K. S., & Abubakari, A. (2023). Interplay between perceived value, trust and continuance intention: Evidence in the sharing economy. *International Journal of Quality and Service Sciences*, *15*(1), 74–96. https://doi.org/10.1108/IJQSS-05-2022-0048

Turner, J. R. (2009). *The Handbook of Project-based Management.* The McGraw-Hill Companies, Inc. https://www.accessengineeringlibrary.com/content/book/9780071549745

Tuuli, M. M., & Rhee, H. van. (2021). How ability, motivation, and opportunity drive individual performance behaviors in projects: Tests of competing theories. *Journal of Management in Engineering*, *37*(6), 04021070. https://doi.org/10.1061/(ASCE)ME.1943-5479.0000969

Twyford, J., Hosseini, E., Haverila, K., & Haverila, M. (in press). The interaction between Industry 4.0 practices, distributor sustainability development and performance: A necessary condition analysis. *Social Responsibility Journal.* https://doi.org/10.1108/srj-10-2024-0747

Tynan, M. C., Credé, M., & Harms, P. D. (2020). Are individual characteristics and behaviors necessary-but-not-sufficient conditions for academic success?: A demonstration of Dul's (2016) necessary condition analysis. *Learning and Individual Differences*, *77*, 101815. https://doi.org/10.1016/j.lindif.2019.101815

Ullibarriarana-Garate, A., Agirre-Aramburu, I., Miguel, M. M.-D. M., & Sánchez-Urien, N. (2024). Higher education apprenticeship. Analyzing the impact and essential conditions of workplace resources on apprentices' career satisfaction. *Journal of Infrastructure, Policy and Development*, *8*(13), 9040. https://doi.org/10.24294/jipd.v8i13.9040

Van der Valk, W., Sumo, R., Dul, J., & Schroeder, R. G. (2016). When are contracts and trust necessary for innovation in buyer-supplier relationships? A necessary condition analysis. *Journal of Purchasing and Supply Management*, *22*(4), 266–277. https://doi.org/10.1016/j.pursup.2016.06.005

Van Rhee, H., & Dul, J. (2017). The limiting-factor theory: AMO-factors individually necessary and jointly sufficient for behavior. *Academy of Management Proceedings*, *2017*, 16432. https://doi.org/10.5465/AMBPP.2017.16432abstract

Vargas-Zeledon, A. A., & Lee, S.-Y. (2024). Configurations of institutional enablers that foster inclusive entrepreneurship: A fuzzy-set qualitative comparative analysis. *Journal of Innovation & Knowledge*, *9*(4), 100549. https://doi.org/10.1016/j.jik.2024.100549

Varzi, A. (2019). Mereology. In E. N. Zalta (Ed.), *The stanford encyclopedia of philosophy. Edition spring 2019.* https://plato.stanford.edu/archives/spr2019/entries/mereology/

Vasist, P. N., & Krishnan, S. (2024). AI's impact on sustainability targets: A cross-country NCA and fsQCA study. *Information Systems Frontiers*, *27*, 1445–1470. https://doi.org/10.1007/s10796-024-10543-5

Vázquez-Sánchez, A., Tuesta-Tapia, A., & Amador-Hidalgo, L. (2025). The EFQM 2020 model and critical success factor for excellence in management for rural development. *Quality Management Journal*, *32*, 239–260. https://doi.org/10.1080/10686967.2025.2528598

Veiga, P. M., Ferreira, J. J., Zhang, J. Z., & Liu, Y. (2024). Exploring the connections: Ambidexterity, digital capabilities, resilience, and behavioral innovation. *Journal of Computer Information Systems.* https://doi.org/10.1080/08874417.2023.2297031

Velikova, N., Bouzdine-Chameeva, T., Ben Nasr, I., Sanz-Ibáñez, C., Anton Clavé, S., & Rabassa-Figueras, N. (in press). Drivers of green wine tourism among young adults:

Intentions, attitudes, and expansion to sustainable behaviour. *Current Issues in Tourism*. https://doi.org/10.1080/13683500.2025.2559313

Venciute, D., Lukosiute, N., Correia, R., Meneses, R., & Kuslys, M. (2025). Leveraging sustainability cues in fashion advertising: Comparative insights from Lithuania's slow and fast fashion markets. *European Business Review*, *37*, 898–932. https://doi.org/10.1108/ebr-01-2025-0013

Verma, S., & Kapoor, D. (2025). Sustainability on the plate: How influencers and social norms transform food choices. *British Food Journal*, *127*, 1942–1958. https://doi.org/10.1108/bfj-11-2024-1202

Vis, B., & Dul, J. (2018). Analyzing relationships of necessity not just in kind but also in degree: Complementing fsQCA with NCA. *Sociological Methods & Research*, *47*(4), 872–899. https://doi.org/10.1177/0049124115626179

Vu, U., & Tolstoy, D. (2025). Examining the complementary roles of market-driven and market-driving orientations in the geographical diversification strategies of e-commerce SMEs. *Journal of Business Research*, *194*, 115375. https://doi.org/10.1016/j.jbusres.2025.115375

Wagner, G. (2020). Typicality and minutis rectis laws: From physics to sociology. *Journal for General Philosophy of Science*, *51*, 447–458. https://doi.org/10.1007/s10838-020-09505-7

Wagner, G., Raymond, L., & Paré, G. (2023). Understanding prospective physicians' intention to use artificial intelligence in their future medical practice: Configurational analysis. *JMIR Medical Education*, *9*, e45631. https://doi.org/10.2196/45631

Walberg, H. J. (1984). Improving the productivity of america's schools. *Educational Leadership*, *41*(8), 19–27. https://eric.ed.gov/?id=EJ299536

Wan, S., Ke, S., & Liang, C. (2025). How the talent ecosystem of key state-owned forest areas in China empowers forestry scientific and technological talents aggregation. *Forest Policy and Economics*, *170*, 103400. https://doi.org/10.1016/j.forpol.2024.103400

Wand, M. (2023). *KernSmooth: Functions for kernel smoothing supporting wand & jones (1995)*. https://CRAN.R-project.org/package=KernSmooth

Wang, C., Rahman, Md. M., Siddik, A. B., Wen, Z. G., & Sobhani, F. A. (2024). Exploring the synergy of logistics, finance, and technology on innovation. *Scientific Reports*, *14*(1), 21918. https://doi.org/10.1038/s41598-024-72409-9

Wang, D., & Cheng, X. (2024). Study on the path of high-quality development of the construction industry and its applicability. *Scientific Reports*, *14*(1), 14727. https://doi.org/10.1038/s41598-024-64786-y

Wang, G., Li, X., Gao, Y., Zeng, C., Wang, B., Li, X., & Li, X. (2023). How does land consolidation drive rural industrial development? Qualitative and quantitative analysis of 32 land consolidation cases in China. *Land Use Policy*, *130*, 106664. https://doi.org/10.1016/j.landusepol.2023.106664

Wang, G., Zhang, H., Zeng, S., Meng, X., & Lin, H. (2023). Reporting on sustainable development: Configurational effects of top management team and corporate characteristics on environmental information disclosure. *Corporate Social Responsibility and Environmental Management*, *30*(1), 28–52. https://doi.org/10.1002/csr.2337

Wang, J. (2025). Exploring determinants and development paths of rural economy from sustainable common-pool resource governance perspective. *Economics of Governance*, *26*, 271–311. https://doi.org/10.1007/s10101-025-00330-6

Wang, J., & Wang, Y. (2024). Economic performance of rural collective-owned cooperatives: Determinants and influence mechanism. *Annals of Public and Cooperative Economics*, *95*(3), 629–653. https://doi.org/10.1111/apce.12454

Wang, L., Sun, Y., & Tiong, R. L. K. (2024). Institutional quality configuration for encouraging private capital participation in PPP projects: Evidence from 36 belt and road countries. *International Journal of Managing Projects in Business, 17*(6/7), 898–922. https://doi.org/10.1108/IJMPB-06-2024-0144

Wang, L., Tang, Y., & Zhao, X. (in press). From niche to green: Configurational pathways to enhance green innovation in Chinese chemical SRDI enterprises. *Environmental Progress & Sustainable Energy.* https://doi.org/10.1002/ep.70192

Wang, P., Wang, S., Foley, M., & Lin, F. (2025). Configurations for High-Quality Environmental Disclosure in China's Dual Carbon Era. *The American Journal of Economics and Sociology, 85*, 49–62. https://doi.org/10.1111/ajes.70001

Wang, T., Aw, E. C.-X., Tan, G. W.-H., Sthapit, E., & Li, X. (in press). AI colleagues: How AI influences hotel employees' service performance? *Current Issues in Tourism.* https://doi.org/10.1080/13683500.2025.2481486

Wang, T., Feng, M., Wang, H., & Shen, Y. (in press). Achieving low-carbon goals: Analysis of the configuration path for manufacturing enterprises to reduce carbon emissions intensity. *Polish Journal of Environmental Studies.* https://doi.org/10.15244/pjoes/196731

Wang, T., Xu, T., Wang, Z., Wang, H., Kang, J., Qiu, L., Xue, S., Fang, Z., & Zhang, Y. (2025a). Where do resilient cities grow? Exploring the pathways and mechanisms of resilience development. *Sustainable Cities and Society, 133*, 106856. https://doi.org/10.1016/j.scs.2025.106856

Wang, X., & Beh, L.-S. (2025). The Yin and Yang of social media: Employing PLS-SEM, fsQCA, and NCA to assess its effects on employee work performance. *Technological Forecasting and Social Change, 217*, 124164. https://doi.org/10.1016/j.techfore.2025.124164

Wang, X., & Beh, L.-S. (in press). Unveiling critical technostress bottlenecks: Social media's impact on work performance in social commerce. *Technology Analysis & Strategic Management.* https://doi.org/10.1080/09537325.2025.2497511

Wang, X., & Wang, C. (2025). How can scientific inquiry groups improve learning performance? Analysis based on the combination of NCA and fsQCA. *Learning and Instruction, 98*, 102135. https://doi.org/10.1016/j.learninstruc.2025.102135

Wang, X.-W., & Guo, Y. (2026). Beyond the screen: Understanding consumer engagement with live-stream shopping from the perspective of the information system success model. *Journal of Retailing and Consumer Services, 88*, 104515. https://doi.org/10.1016/j.jretconser.2025.104515

Wang, X., Zhang, T., Luo, S., & Abedin, M. Z. (2023). Pathways to improve energy efficiency under carbon emission constraints in iron and steel industry: Using EBM, NCA and QCA approaches. *Journal of Environmental Management, 348*, 119206. https://doi.org/10.1016/j.jenvman.2023.119206

Wang, Z., & Zhao, H. (2025). A configuration analysis of the driving path of corporate physical investments: Necessary condition analysis and qualitative comparative analysis based on fuzzy sets. *Managerial and Decision Economics, 46*(1), 681–697. https://doi.org/10.1002/mde.4397

Wang, Z., Zhao, J., Wang, H., Song, D., Feng, D., Yu, M., & Li, F. (2025b). The influencing factors of depression among Chinese male nursing students: Application of decision tree and FsQCA. *BMC Nursing, 24*, 877. https://doi.org/10.1186/s12912-025-03407-z

Warnes, G. R., Bolker, B., Bonebakker, L., Gentleman, R., Huber, W., Liaw, A., Lumley, T., Maechler, M., Magnusson, A., Moeller, S., Schwartz, M., & Venables, B. (2024). *Gplots: Various r programming tools for plotting data.* https://github.com/talgalili/gplots

Wasserstein, R. L., & Lazar, N. A. (2016). The ASA statement on p-values: Context, process, and purpose. *The American Statistician, 70*(2), 129–133. https://doi.org/10.1080/00031305.2016.1154108

Wattanacharoensil, W., Fakfare, P., Manosuthi, N., Lee, J.-S., Chi, X., & Han, H. (2024). Determinants of traveler intention toward animal ethics in tourism: Developing a causal recipe combining cognition, affect, and norm factors. *Tourism Management, 100*, 104823. https://doi.org/10.1016/j.tourman.2023.104823

Wattanacharoensil, W., Stettler, J., Zimmermann, L., & Wagenseil, U. (2024). Lucerne's crossroads: Residents' perspectives on tourism development and the imperative for regulatory oversight. *Tourism Recreation Research, 50*, 1352–1371. https://doi.org/10.1080/02508281.2024.2423116

Wei, W., & Yang, S. (2025). Study on the factors and paths of optimizing and upgrading the regional industrial structure: A hybrid analysis based on extended NCA and fsQCA. *Economic Analysis and Policy, 87*, 2083–2102. https://doi.org/10.1016/j.eap.2025.08.009

Wernerfelt, B. (1984). A resource-based view of the firm. *Strategic Management Journal, 5*(2), 171–180. https://doi.org/10.1002/smj.4250050207

Wickham, H., Chang, W., Henry, L., Pedersen, T. L., Takahashi, K., Wilke, C., Woo, K., Yutani, H., Dunnington, D., & van den Brand, T. (2024). *ggplot2: Create elegant data visualisations using the grammar of graphics.* https://ggplot2.tidyverse.org

Widge, A. S., Zhang, F., Gosai, A., Papadimitrou, G., Wilson-Braun, P., Tsintou, M., Palanivelu, S., Noecker, A. M., McIntyre, C. C., O'Donnell, L., McLaughlin, N. C. R., Greenberg, B. D., Makris, N., Dougherty, D. D., & Rathi, Y. (2022). Patient-specific connectomic models correlate with, but do not reliably predict, outcomes in deep brain stimulation for obsessive-compulsive disorder. *Neuropsychopharmacology, 47*(4), 965–972. https://doi.org/10.1038/s41386-021-01199-9

Wijesinghe, U., Mapitiyage, V., Wickramarathne, C., Wickramage, C., Wisenthige, K., & Aluthwala, C. (2025). Does management support drive sustained agile usage? a serial mediation model and cIPMA perspective. *PLOS ONE, 20*, e0316538. https://doi.org/10.1371/journal.pone.0316538

Wilhelm, I. (2022). Typical: A theory of typicality and typicality explanation. *The British Journal for the Philosophy of Science.* https://www.journals.uchicago.edu/doi/pdf/10.1093/bjps/axz016

Wiśniewska, A. (2025). Attitude towards sustainability as a predictor of consumer behavioral intention. A test of 3x3-item scale in the context of the food market. *Journal of Cleaner Production, 493*, 144947. https://doi.org/10.1016/j.jclepro.2025.144947

Wiśniewska, A., Liczmańska-Kopcewicz, K., & Żemigała, M. (2025). Cognitive and affective antecedents of consumer engagement in green innovation processes. *Sustainable Futures, 9*, 100750. https://doi.org/10.1016/j.sftr.2025.100750

Wold, H. (1985). Partial least squares. In S. Kotz & N. L. Johnson (Eds.), *Encyclopedia of statistical sciences* (Vol. 6, pp. 581–591). Wiley.

Wright, R. W. (1985). Causation in tort law. *Calif. L. Rev., 73*, 1735. https://scholarship.kentlaw.iit.edu/fac_schol/697

Wu, J.-W., Aw, E. C.-X., & Phau, I. (2025). Empowerment and psychological ownership in brand communities: A pathway to brand advocacy. *Journal of Internet Commerce, 24*, 153–183. https://doi.org/10.1080/15332861.2025.2525820

Wu, Z., Aw, E. C.-X., Tan, G. W.-H., & Liu, C.-C. (in press). Speaking to loyalty: A cIPMA approach to understanding AI voice assistant service outcomes. *Journal of Marketing Analytics.* https://doi.org/10.1057/s41270-025-00446-y

Wu, Z., Aw, E. C.-X., Tan, G. W.-H., & Ooi, K.-B. (in press). Speak and shop! Transforming retail experience with AI voice assistants. *Internet Research.* https://doi.org/10.1108/intr-02-2024-0149

Xia, Q., Cheng, J., Jiang, C., Anand, A., & He, P. (2025). A causal configurational analysis of knowledge hiding from a conversational constraint perspective: Evidence from China. *Journal of Knowledge Management, 29*, 1857–1879. https://doi.org/10.1108/jkm-04-2024-0412

Xiao, H., Dong, C., & Li, Y. (2025). Third driving force of green-oriented trajectory-transformed technology innovation realizing market orientation: Qualitative and quantitative empirical analyses. *Journal of Cleaner Production*, *509*, 145568. https://doi.org/10.1016/j.jclepro.2025.145568

Xie, G. (2026). The impact of generative AI shopping assistants on E-commerce consumer motivation and behavior: Consumer-AI interaction design. *International Journal of Information Management*, *86*, 102983. https://doi.org/10.1016/j.ijinfomgt.2025.102983

Xie, Y., Zainudin, R., & Mahdzan, N. S. (2025). Corporate governance and efficiency performance in China's life insurance industry: A configuration analysis based on DEA-fsQCA. *Future Business Journal*, *11*. https://doi.org/10.1186/s43093-025-00555-1

Xu, A., Siddik, A. B., Sobhani, F. A., & Rahman, Md. M. (2024). Driving economic success: Fintech, tourism, FDI, and digitalization in the top 10 tourist destinations. *Humanities and Social Sciences Communications*, *11*(1), 1549. https://doi.org/10.1057/s41599-024-03978-3

Xu, Z., Yang, T., Chen, Y., Chen, Y., & Nan, J. (in press). Determinants of Adopting Wearable Devices for Non-Immersive Virtual Reality Upper Limb Training Systems: An Interaction Experience Perspective. *International Journal of Human–Computer Interaction*. https://doi.org/10.1080/10447318.2025.2544003

Yadav, A., Parida, M., Choudhary, P., & Kumar, B. (2024). Investigating important and necessary conditions to analyse traffic noise levels at intersections in mid-sized cities. *Journal of Environmental Management*, *355*, 120515. https://doi.org/10.1016/j.jenvman.2024.120515

Yan, B., Liu, Y., Chen, B., Zhang, X., & Wu, L. (2023). What matters the most in curbing early COVID-19 mortality? A cross-country Necessary Condition Analysis. *Public Administration*, *101*(1), 71–89. https://doi.org/10.1111/padm.12873

Yan, C., & Wang, D. D. (2025). Digitalization and green innovation: A fuzzyset qualitative comparative analysis based on 30 provinces in China. *Chinese Management Studies*, *19*, 1852–1872. https://doi.org/10.1108/cms-05-2024-0302

Yan, J., Işık, C., Ongan, S., Pinzon, S., & Tillaguango, B. (2024). Analysis of green total factor productivity in China's forestry industry: Technological, organizational, and environmental framework for sustainable economic development. *Sustainable Development*, *32*(6), 7278–7291. https://doi.org/10.1002/sd.3080

Yang, J., & Hurmelinna-Laukkanen, P. (2022). Evolving appropriability–variation in the relevance of appropriability mechanisms across industries. *Technovation*, *118*, 102593. https://doi.org/10.1016/j.technovation.2022.102593

Yang, J., Lv, X., Xu, X., & Chen, X. (2024). Heterogeneous treatment effects of digital transformation on firms' innovation: Based on BART method. *Pacific-Basin Finance Journal*, *85*, 102372. https://doi.org/10.1016/j.pacfin.2024.102372

Yang, S., Bai, T., Li, Z., Feng, L., Zhang, J., & Jiang, W. (in press). Evaluation of indoor residential satisfaction of the older adults in old residential areas: A case study of Sichuan, China. *Journal of Asian Architecture and Building Engineering*. https://doi.org/10.1080/13467581.2025.2526196

Yang, Z. (in press). Research on the Driving Paths of Bilateral Renewable Energy Trade between Neighboring Countries and China from the Perspective of Behavioral Attribution. *Environmental Research Communications*. https://doi.org/10.1088/2515-7620/adac37

Ye, X., Ruan, Y., Xia, S., & Gu, L. (2025). Adoption of digital intangible cultural heritage: A configurational study integrating UTAUT2 and immersion theory. *Humanities and Social Sciences Communications*, *12*. https://doi.org/10.1057/s41599-024-04222-8

Yiling, L., He, Q., Wang, G., Deng, X., & Zhang, J. (2024). Exploring the missing link between top management teams' characteristics and environmental information disclosures of construction corporations. *Engineering, Construction and Architectural Management*, *32*, 6917–6950. https://doi.org/10.1108/ECAM-12-2023-1240

Yin, K. (2025). A Study on the Dissemination Pathways of Public Opinion in Urban Public Safety Emergencies Driven by Social Media. *Applied Spatial Analysis and Policy, 18*. https://doi.org/10.1007/s12061-025-09729-4

Yin, X., Ye, J., Yu, S., Li, H., & Liu, Q. (in press). The impact of AI literacy, perception of technology, and student engagement on academic performance in GenAI-supported learning: A study based on FsQCA and NCA methods. *Education and Information Technologies*. https://doi.org/10.1007/s10639-025-13831-y

Yoon, T., & Wang, S. (2025). When constraints become demotivators: Uncovering the psychological mechanisms behind parental travel avoidance and intention using PLS-SEM and NCA. *International Journal of Tourism Research*, *27*. https://doi.org/10.1002/jtr.70155

Yu, C., Cheah, J.-H., & Liu, Y. (2022). To stream or not to stream? Exploring factors influencing impulsive consumption through gastronomy livestreaming. *International Journal of Contemporary Hospitality Management*, *34*(9), 3394–3416. https://doi.org/10.1108/IJCHM-09-2021-1137

Yu, D., Fan, H., & Zhang, N. (2024). Driving factors and grouping paths of rural sports development in China - a qualitative comparative analysis based on fuzzy sets. *PLOS ONE*, *19*(3), e0300169. https://doi.org/10.1371/journal.pone.0300169

Yu, Y., Wang, J., Jiang, H., & Lu, H. (2025). How to manage a multifactor-driven crude oil market more effectively? A revisit based on the multiple criteria perspective. *Resources Policy*, *100*, 105446. https://doi.org/10.1016/j.resourpol.2024.105446

Yuan, B., Tao, F., Chen, M., Zhang, Y., Chen, Q., & Lai, S. (in press). Tailor-to-target: Configuring construction worker safety behavior on big five personality traits. *Engineering, Construction and Architectural Management*. https://doi.org/10.1108/ecam-07-2024-0864

Yuan, J., Shah, S. K., Tajeddini, K., Gamage, T. C., & Acevedo-Duque, Á. (2025). Unveiling psychosocial factors influencing metaverse-associated app adoption: Acumens from fsQCA approach. *International Journal of Human Computer-Interaction*, *41*, 14306–14320. https://doi.org/10.1080/10447318.2025.2482744

Yuan, M., Zhuang, T., & Liu, G. (in press). What types of urban regeneration projects catalyze urban development? A configurational perspective. *Engineering, Construction and Architectural Management*. https://doi.org/10.1108/ecam-05-2025-0800

Zaerpour, N., Gharehgozli, A., & Li, K. (in press). Impacts of additive manufacturing on manufacturing supply chain design. *Journal of Manufacturing Technology Management*. https://doi.org/10.1108/jmtm-06-2025-0611

Zahoor, N., Khan, Z., & Shenkar, O. (2023). International vertical alliances within the international business field: A systematic literature review and future research agenda. *Journal of World Business*, *58*(1), 101385. https://doi.org/10.1016/j.jwb.2022.101385

Zahrai, K., Veer, E., Ballantine, P. W., Vries, H. P. de, & Prayag, G. (2022). Either you control social media or social media controls you: Understanding the impact of self-control on excessive social media use from the dual-system perspective. *Journal of Consumer Affairs*, *56*, 806–848. https://doi.org/10.1111/joca.12449

Zang, X., & Guo, S. (2024). Influencing factors of spatial conflicts in the Chinese urban environment from the perspective of spatial politics: Empirical study based on NCA and fsQCA. *Journal of Chinese Governance*, *9*(1), 49–77. https://doi.org/10.1080/23812346.2023.2298012

Zeng, K., Murayama, T., Nishikizawa, S., & Suwanteep, K. (in press). Towards sustainable consumption: The influence of contextual factors on consumers' purchase intentions for biodegradable plastic products. *Sustainable Development.* https://doi.org/10.1002/sd.70257

Zhai, W., Zeng, Q., & Wang, L. (2025). Configurational pathways to port environmental efficiency: An fsQCA-NCA analysis of CO2 reduction drivers in port management. *Regional Studies in Marine Science, 92*, 104618. https://doi.org/10.1016/j.rsma.2025.104618

Zhang, H., Li, H., Li, C., & Lu, X. (2024). Disentangling user fatigue in WeChat use: The configurational interplay of fear of missing out and overload. *Internet Research, 34*(7), 160–186. https://doi.org/10.1108/INTR-07-2023-0565

Zhang, J. (2024a). A configurational analysis of innovation environment and industrial green total factor productivity. *Journal of Environmental Planning and Management, 68*, 3385–3410. https://doi.org/10.1080/09640568.2024.2351421

Zhang, J. (2024b). A multidimensional perspective on the relationship between tourism and green growth. *Journal of Sustainable Tourism, 32*(12), 2423–2441. https://doi.org/10.1080/09669582.2023.2276039

Zhang, J. (2024c). The effects of ecosystem of business environment on tourism: A configurational perspective. *Current Issues in Tourism, 27*(24), 4676–4692. https://doi.org/10.1080/13683500.2023.2291708

Zhang, J. (2025). Patterns of innovation-driven tourism competitiveness: Insights from 270 Chinese cities. *Tourism Management, 107*, 105063. https://doi.org/10.1016/j.tourman.2024.105063

Zhang, J. (in pressa). Determinants of tourism enterprises' low-carbon transition: Seen from the strategy tripod perspective. *Journal of Sustainable Tourism.* https://doi.org/10.1080/09669582.2025.2573702

Zhang, J. (in pressb). The necessary and sufficient relationships between tourism and regional gender equality: Empirical evidence from ASEAN countries. *Current Issues in Tourism.* https://doi.org/10.1080/13683500.2025.2495082

Zhang, J., Cui, J., & Chai, X. (2026). Job crafting in innovative digital contexts: A configurational analysis of behavioral pathways to performance. *Journal of Innovation & Knowledge, 11*, 100874. https://doi.org/10.1016/j.jik.2025.100874

Zhang, J., Yang, Z., Zhang, X., Sun, J., & He, B. (2024). Institutional configuration study of urban green economic efficiency – analysis based on fsQCA and NCA. *Polish Journal of Environmental Studies.* https://doi.org/10.15244/pjoes/187130

Zhang, K., & Cheng, X. (2025). Determinants of consumers intentions to use smart home devices from the perspective of perceived value: A mixed SEM, NCA, and fsQCA study. *Journal of Retailing and Consumer Services, 87*, 104399. https://doi.org/10.1016/j.jretconser.2025.104399

Zhang, L., Teirlinck, P., Su, T., & Yu, Y. (2024). How to configure the supply chain to stimulate firm innovation performance? *International Journal of Production Economics, 277*, 109384. https://doi.org/10.1016/j.ijpe.2024.109384

Zhang, X., & Hao, X. (2025). Linkage mechanism of antecedents for employees' continuous adoption of artificial intelligence virtual assistants. *Technological Forecasting and Social Change, 220*, 124317. https://doi.org/10.1016/j.techfore.2025.124317

Zhang, Y., Dong, Z., Wang, Y., Yao, Y., & Huang, Y. (in press). Multiple paths to enhancing total factor productivity in the construction industry from the configuration perspective: Evidence from China. *Engineering, Construction and Architectural Management.* https://doi.org/10.1108/ecam-12-2024-1758

Zhang, Y., Hedo, R., Rivera, A., Rull, R., Richardson, S., & Tu, X. M. (2019). Post hoc power analysis: Is it an informative and meaningful analysis? *General Psychiatry, 32*(4). https://doi.org/10.1136/gpsych-2019-100069

Zhang, Y., Lim, W., Cham, T.-H., Huang, H.-L., & Tapsir, S. H. (in press). Harnessing sustainability: The role of organizational resilience and big data in product innovation of Malaysian technology firms. *Asia Pacific Business Review*. https://doi.org/10.1080/13602381.2025.2507733

Zhang, Y., & Zhang, J. (2024). Examining the complex causal relationships between the digital economy and urban tourist destination competitiveness. *Economic Change and Restructuring*, *57*(4), 152. https://doi.org/10.1007/s10644-024-09739-1

Zhang, Z., Chen, H., & Wu, L. (2024). How does technological innovation drive green growth in China's complex environments? A configurational approach. *Journal of Environmental Planning and Management*, *67*(11), 2583–2604. https://doi.org/10.1080/09640568.2023.2195542

Zhao, H., & Furuoka, F. (2025). Linking social capital to green purchasing intentions in livestreaming sessions: The moderating role of perceived visible heterogeneity. *Humanities and Social Sciences Communications*, *12*. https://doi.org/10.1057/s41599-025-04810-2

Zhao, H., Ru, P., & Jia, H. (in press). TikTok users migration to Xiaohongshu (Rednote): Emotional dynamics, platform governance, and an NCA-SEM analysis in cross-cultural adaptation. *International Journal of Human–Computer Interaction*. https://doi.org/10.1080/10447318.2025.2530088

Zhao, H., Yang, Y., Chen, Y., Yu, H., Chen, Z., & Yang, Z. (2025). Configuration analysis of crop-pollination service management: A novel insight from the theory of planned behavior. *PLOS ONE*, *20*, e0326226. https://doi.org/10.1371/journal.pone.0326226

Zhao, Y., Ye, Q., & Cai, J. (2024). Psychic distance's impact on China's reverse M&A innovation. *Emerging Markets Finance and Trade*, *61*, 1249–1272. https://doi.org/10.1080/1540496X.2024.2412049

Zheng, J., Shi, L., & Jiang, T. (2022). What mechanism design helps to realize the innovation function of maker-spaces: A qualitative comparative analysis based on fuzzy sets. *PLOS ONE*, *17*(9), e0274307. https://doi.org/10.1371/journal.pone.0274307

Zheng, Y., Lin, C., Yan, J., & Guo, Y. (2025). How to enhance the innovation capacity of technology-based enterprises: A fuzzy set qualitative comparative analysis. *International Review of Economics & Finance*, *97*, 103817. https://doi.org/10.1016/j.iref.2024.103817

Zhong, W., & Luo, J. (in press). Befriend GenAI chatbots: Exploring the determinants of parasocial relationship in human-AI interaction through a hybrid PLS-ANN-NCA approach. *International Journal of Human–Computer Interaction*. https://doi.org/10.1080/10447318.2025.2537794

Zhou, X., Wu, W., Chen, K., & Li, J. (2025). How do national laboratories fulfil their mission? A synergistic driving analysis based on NCA and TJ-QCA. *Science, Technology and Society*, *30*, 690–709. https://doi.org/10.1177/09717218251376754

Zhou, X., Xing, S., Xu, J., Tian, J., Niu, A., & Lin, C. (2025). Impacts of climate change risk and economic policy uncertainty on carbon prices: Configuration analysis from a complex system perspective. *Journal of Environmental Management*, *373*, 123622. https://doi.org/10.1016/j.jenvman.2024.123622

Zhou, Y., Kumar, S., & Furuoka, F. (2024). Enhancing customer value co-creation and stickiness in social commerce: Integrating PLS-SEM and NCA for deeper insights into customer-to-customer dynamics. *Humanities and Social Sciences Communications*, *11*(1), 46. https://doi.org/10.1057/s41599-023-02586-x

Zhou, Z., Chen, Z., Jin, X.-L., & Yuan, Z. (2024). Unveiling motivational configurations in shaping meaningful engagement in green gamification. *European Journal of Information Systems*, *34*, 715–738. https://doi.org/10.1080/0960085X.2024.2404968

Zhu, P., Miao, C., Wang, Z., & Li, X. (2023). Informational cascade, regulatory focus and purchase intention in online flash shopping. *Electronic Commerce Research and Applications*, *62*, 101343. https://doi.org/10.1016/j.elerap.2023.101343

Zhu, Y., & Song, W. (2025). From cues to choices: The effect of eco-labels and quality excellence on sustainable online consumption in product–service systems. *Computers & Industrial Engineering*, *209*, 111480. https://doi.org/10.1016/j.cie.2025.111480

Zielińska, A., Lebuda, I., Gop, A., & Karwowski, M. (2024). Teachers as creative agents: How self-beliefs and self-regulation drive teachers' creative activity. *Contemporary Educational Psychology*, *77*, 102267. https://doi.org/10.1016/j.cedpsych.2024.102267

Zielińska, A., Lebuda, I., & Karwowski, M. (2023). Dispositional self-regulation strengthens the links between creative activity and creative achievement. *Personality and Individual Differences*, *200*, 111894. https://doi.org/10.1016/j.paid.2022.111894

Zitao, Y., Zixiao, Z., & Huaxia, H. (2025). Decoding multi-paths: Context, configuration and capability resonance of ESG performance in SRDI enterprises – configuration analysis based on fsQCA and NCA. *Sustainable Futures*, *10*, 100817. https://doi.org/10.1016/j.sftr.2025.100817

Znaniecki, F. (1934). *The Method of Sociology*. Farrar & Rinehart.

Index

Note: Page numbers in **bold** and *italics* refer to tables and figures, respectively.

For Product Safety Concerns and Information please contact our EU
representative GPSR@taylorandfrancis.com
Taylor & Francis Verlag GmbH, Kaufingerstraße 24, 80331 München, Germany

www.ingramcontent.com/pod-product-compliance
Ingram Content Group UK Ltd.
Pitfield, Milton Keynes, MK11 3LW, UK
UKHW052225190926
475702UK00007B/1215
* 9 7 8 1 0 3 2 5 5 4 4 7 1 *